Complex analysis

Written by a master of the subject, this textbook will be appreciated by students and experts. The author develops the classical theory of functions of a complex variable in a clear and straightfoward manner. In general, the approach taken here emphasises geometrical aspects of the theory in order to avoid some of the topological pitfalls associated with this subject. Thus, Cauchy's integral formula is first proved in a topologically simple case from which the author deduces the basic properties of holomorphic functions. Starting from the basics, students are led on to the study of conformal mappings, Riemann's mapping theorem, analytic functions on a Reimann surface, and ultimately the Riemann–Roch and Abel theorems.

Profusely illustrated and with plenty of examples and problems (solutions to many of which are included), this book should be a stimulating text for advanced courses in complex analysis.

KUNIHIKO KODAIRA (1915–1997) worked in many areas including harmonic integrals, algebraic geometry and the classification of compact complex analytic surfaces. He held faculty positions at many universities including Tokyo, Harvard, Stanford, and Johns Hopkins, and the Institute for Advanced Study in Princeton. He was awarded a Fields medal in 1954 and a Wolf Prize in 1984.

Complex analysis

KUNIHIKO KODAIRA

Translated by A. Sevenster

Edited by A. F. Beardon and T. K. Carne

CAMBRIDGE
UNIVERSITY PRESS

CAMBRIDGE
UNIVERSITY PRESS

Shaftesbury Road, Cambridge CB2 8EA, United Kingdom

One Liberty Plaza, 20th Floor, New York, NY 10006, USA

477 Williamstown Road, Port Melbourne, VIC 3207, Australia

314–321, 3rd Floor, Plot 3, Splendor Forum, Jasola District Centre, New Delhi – 110025, India

103 Penang Road, #05–06/07, Visioncrest Commercial, Singapore 238467

Cambridge University Press is part of Cambridge University Press & Assessment, a department of the University of Cambridge.

We share the University's mission to contribute to society through the pursuit of education, learning and research at the highest international levels of excellence.

www.cambridge.org
Information on this title: www.cambridge.org/9780521809375

Originally published in Japanese as *Complex Analysis,* Vols. I, II and III, by Iwanami Shoten, Publishers, Tokyo, 1977 and 1978

Volumes I and II published in English in 1984 as *Introduction to Complex Analysis.*
Combined three-volume edition first published in English 2007.

First published 1977
Fourth printing 2008

A catalogue record for this publication is available from the British Library

ISBN 978-0-521-80937-5 Hardback

Contents

Preface

This book aims to give a clear explanation of classical theory of analytic functions; that is, the theory of holomorphic functions of one complex variable. In modern treatments of function theory it is customary to call a function holomorphic if its derivative exists. However, we return to the old definition, calling a function holomorphic if its derivative exists *and* is continuous, since we believe this is a more natural approach.

The first difficulty one encounters in writing an introduction to function theory is the topology involved in Cauchy's Theorem and Cauchy's integral formula. In the first chapter of the book we prove the latter in a topologically simple case, and from that result we deduce the basic properties of holomorphic functions. In the second chapter we prove the general version of Cauchy's Theorem and integral formula. I have tried to replace the necessary topological considerations with elementary geometric considerations. This way turned out to be longer than I expected, so that in the original Japanese three-volume edition I had to end Volume 2 before Chapter 5 was completed. My original intention was to present classical many-valued analytic functions, in particular the Riemann surface of an algebraic function, and to introduce the general concept of a Riemann surface as its generalization. Now, with the appearance of the complete Japanese edition in a single volume, the link between the theory of Riemann surfaces and function theory is restored.

Similarly, for the theory of Riemann surfaces, with the assumption that the topology of curved surfaces is known, the plan was to introduce the content of Weyl's book: *The Concept of Riemann Surfaces. Part II: Functions on Riemann Surfaces*, but that would have been counter to the original policy of replacing the topological approach with elementary geometrical considerations. Thus, in Chapter 7, I have tried to illuminate topological characteristics of compact Riemann surfaces by using Riemann's mapping theorem. Consequently, Chapter 7 became longer than was planned, so Chapter 8 is limited to covering the Riemann–Roch theorem and Abel's theorem, which are the most basic theorems regarding analytic functions on compact Riemann surfaces.

1

Holomorphic functions

1.1 Holomorphic functions

a. The complex plane

An expression $z = x + iy$, where x and y are real numbers and $i = \sqrt{-1}$, is called a *complex number*. The sum, difference, and product of two complex numbers $z = x + iy$ and $w = u + iv$ are defined by

$$\begin{aligned} z + w &= (x+u) + i(y+v), \\ z - w &= (x-u) + i(y-v), \\ zw &= (xu - yv) + i(xv + yu) \end{aligned}$$

These expressions are obtained by first evaluating $z+w$, $z-w$, and zw as polynomials in the "variable" i and then replacing i^2 by -1. Therefore, addition, subtraction, and multiplication as defined above satisfy the associative, commutative, and distributive laws.

As usual, the real number line is represented by $\mathbb{R}$. The plane $\mathbb{R}^2$ is the product $\mathbb{R} \times \mathbb{R}$, that is, the collection of all pairs (x, y) of real numbers. If one identifies the point (x, y) of the plane $\mathbb{R}^2$ with the complex number $z = x + iy$, then $\mathbb{R}^2$ is called the *complex plane*. The complex plane is represented by $\mathbb{C}$.

The *absolute value* $|z|$ of the complex number $z = x + iy$ is defined by

$$|z| = \sqrt{x^2 + y^2}$$

For two complex numbers $z = x + iy$ and $w = u + iv$

$$|z - w| = \sqrt{(x-u)^2 + (y-v)^2}$$

is the distance between the points z and w in the plane $\mathbb{C}$. In particular, $|z|$ is the distance between the point z and the origin 0.

If one represents the complex number $z = x + iy$ by the vector $\mathbf{0z}$ from 0 to z, then (x, y) are the coordinates of z and $|z| = \sqrt{(x^2 + y^2)}$ is the length of $\mathbf{0z}$. Therefore, if z_1 and z_2 are complex numbers, and $w = z_1 + z_2$ is their sum, then the vector $\mathbf{0w}$ is equal to the sum of $\mathbf{0z}_1$ and $\mathbf{0z}_2$ (Fig. 1.1):

$$\mathbf{0w} = \mathbf{0z}_1 + \mathbf{0z}_2.$$

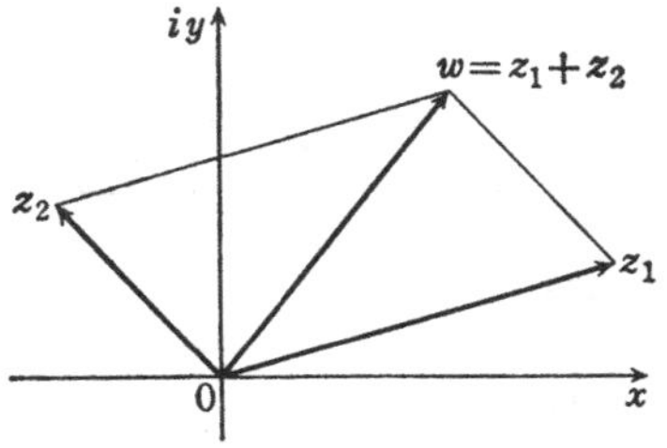

Fig. 1.1

For any complex number $z = x + iy$, one calls $x - iy$ the *conjugate* of z. The conjugate of z is represented by $\bar{z}$:

$$\bar{z} = x - iy.$$

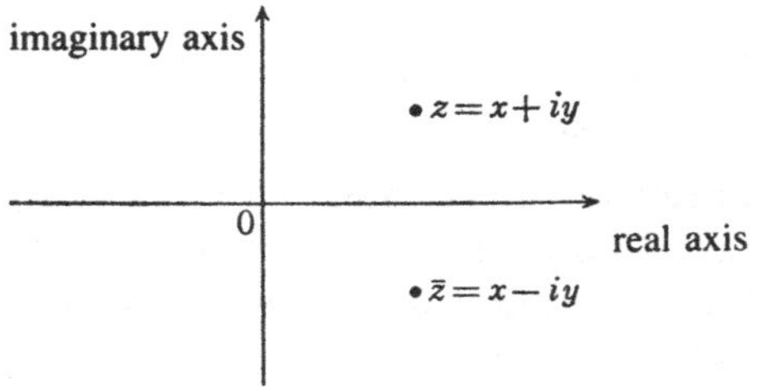

Fig. 1.2

Furthermore, x is called the *real part* of $z = x + iy$, and y is called the *imaginary part*. The real part of z is represented by Re z, the imaginary part by Im z:

$$\operatorname{Re} z = x = \frac{z + \bar{z}}{2}, \qquad \operatorname{Im} z = y = \frac{i(\bar{z} - z)}{2}$$

The line $\mathbb{R} \times \{0\}$ in the complex plane is called the *real axis* and the line $\{0\} \times \mathbb{R}$ is called the *imaginary axis*. The conjugate $\bar{z}$ of z and z are represented by points in the complex plane, that are symmetric with regard to the real axis. Obviously

$$\bar{\bar{z}} = z,$$

$$\overline{z + w} = \bar{z} + \bar{w}, \qquad \overline{z - w} = \bar{z} - \bar{w},$$

$$\overline{z \cdot w} = \bar{z} \cdot \bar{w}.$$

Moreover

$$|z|^2 = |\bar{z}|^2 = x^2 + y^2 = z \cdot \bar{z}.$$

Hence

$$|zw|^2 = zw\overline{zw} = zw\bar{z}\bar{w} = z\bar{z}w\bar{w} = |z|^2|w|^2,$$

and therefore

$$|zw| = |z|\,|w|. \tag{1.1}$$

If $z \neq 0$, then $|z| > 0$ and $z \cdot \bar{z}/|z|^2 = 1$. So if $z \neq 0$, then z has an inverse $1/z = \bar{z}/|z|^2$. Therefore, the collection of all complex numbers $\mathbb{C}$ is a field, called the field of complex numbers.

For $z \neq 0$ we have $\overline{(w/z)}\bar{z} = \overline{(w/z)z} = \bar{w}$, therefore

$$\overline{(w/z)} = \bar{w}/\bar{z}.$$

Since similar rules hold for addition, subtraction, and multiplication, as we saw above, it is now clear that if a complex number w is arrived at by a finite number of additions, subtractions, multiplications, and divisions applied to a finite number of complex numbers $z_1, z_2, \ldots, z_n$, then by applying the same operations in the same order to $\bar{z}_1, \bar{z}_2, \ldots, \bar{z}_n$ one arrives at $\bar{w}$. Therefore, the correspondence from $\mathbb{C}$ into $\mathbb{C}$ given by $z \to \bar{z}$ is an isomorphism.

For two arbitrary complex numbers, we have the following inequality

$$|z+w| \leqq |z| + |w|. \tag{1.2}$$

Proof: Using $\operatorname{Re} z \leqq |z|$ we have

$$\begin{aligned} |z+w|^2 &= (z+w)(\bar{z}+\bar{w}) = z\bar{z} + z\bar{w} + w\bar{z} + w\bar{w} \\ &= |z|^2 + 2\operatorname{Re}(z\bar{w}) + |w|^2 \leqq |z|^2 + 2|z\bar{w}| + |w|^2 \\ &= |z|^2 + 2|z|\,|w| + |w|^2 = (|z| + |w|)^2. \end{aligned}$$

From the inequality (1.2) the *triangle inequality*

$$|z_1 - z_3| \leqq |z_1 - z_2| + |z_2 - z_3|,$$

where z_1, z_2, z_3 are arbitrary points of the complex plane, follows at once.

From $|z| \leqq |z-w| + |w|$, we conclude $|z| - |w| < |z-w|$.

In the same way it is proved that $|w| - |z| \leqq |w-z|$. Hence

$$||z| - |w|| \leqq |z-w|. \tag{1.3}$$

Repeated application of (1.1) and (1.2) yields

$$|z_1 z_2 z_3 \cdots z_n| = |z_1|\,|z_2|\,|z_3| \cdots |z_n|,$$

$$|z_1 + z_2 + \cdots + z_n| \leqq |z_1| + |z_2| + \cdots + |z_n|.$$

Therefore

$$|a_0 + a_1 z + a_2 z^2 + \cdots + a_n z^n| \leqq |a_0| + |a_1|\,|z| + |a_2|\,|z|^2 + \cdots + |a_n|\,|z|^n.$$

Since the complex plane $\mathbb{C}$ can be identified with the real plane $\mathbb{R}^2$, definitions and theorems pertaining to subsets of $\mathbb{R}^2$ also apply to subsets of $\mathbb{C}$. For example, one says that the sequence $\{z_n\}$ of complex numbers *converges* to w, if the sequence $\{z_n\}$ of points converges to the point w, that is, if

$$\lim_{n\to\infty} |z_n - w| = 0.$$

Theorem 1.1 (Cauchy's criterion). The complex sequence $\{z_n\}$ converges if and only if for every real number $\varepsilon > 0$, there exists a natural number $n_0(\varepsilon)$ such that

$$|z_n - z_m| < \varepsilon \qquad \text{if} \quad n > n_0(\varepsilon) \text{ and } m > n_0(\varepsilon).$$

We have $||z_n| - |w|| \leqq |z_n - w|$ by (1.3), therefore from $\lim_{n\to\infty} z_n = w$ we can conclude $\lim_{n\to\infty} |z_n| = |w|$. Hence if the complex sequence $\{z_n\}$ converges, then the sequence $\{|z_n|\}$ converges too and we have

$$\lim_{n\to\infty} |z_n| = |\lim_{n\to\infty} z_n|.$$

The infinite series $\sum_{n=1}^{\infty} z_n = z_1 + z_2 + \cdots + z_n + \cdots$ is said to converge if the complex sequence $\{w_n\}$ of partial sums

$$w_n = z_1 + z_2 + \cdots z_n$$

converges. The complex number $w = \lim_{n\to\infty} w_n$ is called the *sum* of the series and we write

$$w = \sum_{n=1}^{\infty} z_n = z_1 + z_2 + \cdots + z_n + \cdots.$$

If the sequence $\{w_n\}$ does not converge, then the series $\sum_{n=1}^{\infty} z_n$ is called *divergent*.

Putting $\sigma_n = |z_1| + |z_2| + \cdots + |z_n|$, we have for $m < n$

$$|w_n - w_m| = \left| \sum_{k=m+1}^{n} z_k \right| \leqq \sum_{k=m+1} |z_k| = \sigma_n - \sigma_m.$$

Applying Cauchy's criterion we conclude that $\sum_{n=1}^{\infty} z_n$ converges if $\sum_{n=1}^{\infty} |z_n|$ converges. In this case, $\sum_{n=1}^{\infty} z_n$ is called *absolutely convergent*.

If $\sum_{n=1}^{\infty} z_n$ is absolutely convergent, then

$$\left|\sum_{n=1}^{\infty} z_n\right| = \left|\lim_{m\to\infty} w_m\right| = \lim_{m\to\infty} |w_m| \leqq \lim_{m\to\infty} \sum_{n=1}^{m} |z_n| = \sum_{n=1}^{\infty} |z_n|.$$

Since $\sum_{n=1}^{\infty} |z_n|$ either converges or diverges to $+\infty$, $\sum_{n=1}^{\infty} z_n$ is absolutely convergent if and only if $\sum_{n=1}^{\infty} |z_n| < +\infty$. If $w = \sum_{n=1}^{\infty} z_n$ and $\omega = \sum_{n=1}^{\infty} \zeta_n$ are both absolutely convergent, then

$$w \cdot \omega = z_1\zeta_1 + z_2\zeta_1 + z_1\zeta_2 + z_3\zeta_1 + z_2\zeta_2 + z_3\zeta_1 + z_2\zeta_2 + z_1\zeta_3 + \cdots. \tag{1.4}$$

Proof: Putting

$$\sigma_n = |z_n|\,|\zeta_1| + |z_{n-1}|\,|\zeta_2| + |z_{n-2}|\,|\zeta_3| + \cdots + |z_1|\,|\zeta_n|$$

we have $\sum_{n=1}^{\infty} \sigma_n = \sum_{n=1}^{\infty} |z_n| \sum_{n=1}^{\infty} |\zeta_n|$, so that the series of the right-hand side of (1.4) is absolutely convergent and

$$\left|\sum_{n=1}^{m} z_n \sum_{n=1}^{m} \zeta_n - \sum_{n=1}^{m} (z_n\zeta_1 + z_{n-1}\zeta_2 + \cdots + z_1\zeta_n)\right| \leqq \sum_{n=1}^{m} |z_n| \sum_{n=1}^{m} |\zeta_n| - \sum_{n=1}^{m} \sigma_n \to 0, \qquad m \to \infty.$$

b. Functions of a complex variable

Let D be a subset of $\mathbb{C}$, i.e., D is a point-set in the complex plane. A function f defined in D assigns to each element of D exactly one complex number. D is called the *domain* of f. For $\zeta \in D$ the complex number ω assigned to ζ by f is called the *value* of f at ζ. We write

$$\omega = f(\zeta).$$

If S is a subset of D, then the collection of all complex numbers $f(\zeta)$, where $\zeta \in S$, is written as $f(S)$

$$f(S) = \{f(\zeta) \colon \zeta \in S\}.$$

The set $f(D)$ of all values $\omega = f(\zeta)$ is called the *range* of f.

Writing $f(z)$ instead of f, one calls z a *variable* and $f(z)$ a *function of a complex variable*. Just as for functions of a real variable, z denotes an arbitrary element ζ of D, or, in other words, a symbol for which ζ has to be substituted. According to general custom, we will use the same letter z to denote points of D. Putting $w = f(z)$, we call w a function of z.

An open subset U of the complex plane is said to be *connected* if U is not the union of two nonempty, open subsets that have no points in common. An open subset U is connected if and only if each pair of points z, w of U can be connected by an arc lying in U.

A connected open subset of $\mathbb{C}$ is called a *region* (or a *domain*); the closure of a region is called a *closed region* (or a *closed domain*).

In this book we will mainly consider functions defined on regions or on closed regions, but we start by discussing limits, continuity, and other properties of functions, defined on arbitrary sets $D \subset \mathbb{C}$.

Definition 1.1. Let D be a point set in $\mathbb{C}$, c an accumulation point of D, and γ a complex number. We say that $f(z)$ *converges* to γ or that γ is the *limit* of $f(z)$ as z tends to c, if for every real $\varepsilon > 0$ there exists a real $\delta(\varepsilon) > 0$ satisfying

$$|f(z) - \gamma| < \varepsilon \qquad \text{if } 0 < |z - c| < \delta(\varepsilon). \tag{1.5}$$

This is written as

$$\lim_{z \to c} f(z) = \gamma$$

or

$$f(z) \to \gamma \quad \text{as } z \to c.$$

Since $f(z)$ is not defined if $z \notin D$, we have to assume that $z \in D$ in (1.5). The assumption that c is an accumulation point of D is necessary to exclude the possibility that there are no points z satisfying $z \in D$ and $0 < |z - c| < \delta(\varepsilon)$.

The proof of the following result is similar to the proof of the corresponding result for real functions.

The function $f(z)$ converges to γ as z tends to c if and only if for all complex sequences $\{z_n\}$, $z_n \in D$ and $z_n \neq c$, converging to c the complex sequence $\{f(z_n)\}$ converges to γ.

Combining this theorem with Cauchy's criterion for complex sequences, we arrive at Cauchy's criterion for functions.

Theorem 1.2 (Cauchy's criterion). Let $f(z)$ be a function of the complex variable z defined on $D \subset \mathbb{C}$ and let c be an accumulation point of D. Then $f(z)$ converges to some value if z tends to c if and only if for every real $\varepsilon > 0$ there exists a real $\delta(\varepsilon) > 0$ such that

$$|f(z) - f(w)| < \varepsilon \qquad \text{if} \quad 0 < |z - c| < \delta(\varepsilon) \text{ and } 0 < |w - c| < \delta(\varepsilon).$$

Let $f(z)$ be a complex function defined on $D \subset \mathbb{C}$, and assume that c belongs to D. If

$$\lim_{z \to c} f(z) = f(c), \tag{1.6}$$

then $f(z)$ is said to be *continuous* at c.

It follows at once from the definition that $f(z)$ is continuous at c if and only if for every real $\varepsilon > 0$ there exists a real $\delta(\varepsilon) > 0$ such that

$$|f(z)-f(c)| < \varepsilon \quad \text{if} \quad |z-c| < \delta(\varepsilon).$$

(If c is an isolated point of D, then for sufficiently small δ the only z satisfying $z \in D$ and $|z-c| < \delta(\varepsilon)$ is c, in which case $f(z)$ is certainly continuous at c.)

Putting $z = x+iy$ and $c = a+ib$, we can split $f(z)$ into a real and an imaginary part

$$f(z) = u(z)+iv(z), \quad u(z) = \operatorname{Re} f(z), \quad v(z) = \operatorname{Im} f(z).$$

The real part and the imaginary part can be considered as real functions of two real variables x and y where $z = x+iy$. From

$$|f(z)-f(c)| = \sqrt{|u(x, y)-u(a, b)|^2 + |v(x, y)-v(a, b)|^2}$$

we conclude that (1.6) is equivalent to

$$\lim_{(x, y)\to(a, b)} u(x, y) = u(a, b), \qquad \lim_{(x, y)\to(a, b)} v(x, y) = v(a, b).$$

Therefore, the function $f(z) = u(x, y)+iv(x, y)$ of the complex variable $z = x+iy$ is continuous at $c = a+ib$ if and only if its real part $u(x, y)$ and its imaginary part $v(x, y)$ are continuous at (a, b) as functions of the two real variables x and y.

If the complex function $f(z)$ is continuous at all points of its domain $D \subset \mathbb{C}$, then f is called a continuous function of z or simply a *continuous function*. The function $f(z) = u(x, y)+iv(x, y)$ of the complex variable $z = x+iy$ is continuous if and only if its real part $u(x, y)$ and its imaginary part $v(x, y)$ are continuous functions of the two real variables x and y.

Just as for functions of a real variable, limits of complex functions satisfy the following rules: let $f(z)$ and $g(z)$ be functions of a complex variable z defined on $D \subset \mathbb{C}$ and let c be an accumulation point of D. If both $f(z)$ and $g(z)$ converge to a limit as $z \to c$ then the linear combination $a_1 f(z)+a_2 f(z)$, where a_1 and a_2 are constants and the product $f(z)\cdot g(z)$ converge to a limit and these limits satisfy

$$\lim_{z\to c} (a_1 f(z)+a_2 g(z)) = a_1 \lim_{z\to c} f(z)+a_2 \lim_{z\to c} g(z),$$

$$\lim_{z\to c} f(z)g(z) = \lim_{z\to c} f(z) \cdot \lim_{z\to c} g(z).$$

If moreover $\lim_{z\to c} g(z) \neq 0$, then the quotient $f(z)/g(z)$ converges if $z \to c$ and the limit satisfies

$$\lim_{z \to c} \frac{f(z)}{g(z)} = \frac{\lim_{z \to c} f(z)}{\lim_{z \to c} g(z)}.$$

Hence if $f(z)$ and $g(z)$ are continuous functions of z, then the linear combination $a_1 f(z) + a_2 g(z)$ and the product $f(z)g(z)$ are continuous functions. If, moreover, $g(z) \neq 0$ for all $z \in D$, then the quotient $f(z)/g(z)$ is also a continuous function of z.

Continuity of the composite of two complex functions obeys the same rule as real functions do: If $f(z)$ is a continuous function of z defined on $D \subset \mathbb{C}$, if $g(w)$ is a continuous function of w defined on $E \subset \mathbb{C}$ and if $f(D) \subset E$, then the composite $g(f(z))$ is a continuous function of z on D. For, if c is an arbitrary point of D, then $\lim_{z \to c} f(z) = f(c)$ and $\lim_{w \to f(c)} g(w) = g(f(c))$; hence $\lim_{z \to c} g(f(z)) = g(f(c))$.

The functions z and $\bar{z}$ are obviously continuous functions of z defined on $\mathbb{C}$. According to the above, linear combinations of finite products of z and $\bar{z}$, that is polynomials in z and $\bar{z}$:

$$f(z) = \sum_{h=0}^{m} \sum_{k=0}^{n} a_{hk} z^g \bar{z}^k, \qquad a_{hk} \in \mathbb{C}$$

are continuous functions of z.

Definition 1.2. Let $f(z)$ be a continuous function of z defined on $D \subset \mathbb{C}$. If for every real $\varepsilon > 0$ there exists a real $\delta(\varepsilon) > 0$ such that

$$|f(z) - f(w)| < \varepsilon \qquad \text{if } |z - w| < \delta(\varepsilon) \text{ and } z \in D \text{ and } w \in D$$

then $f(z)$ is said to be *uniformly continuous* on D.

Theorem 1.3. A continuous function $f(z)$ defined on a bounded, closed set $D \subset \mathbb{C}$ is uniformly continuous on D.

Proof: Assume that $f(z)$ is not uniformly continuous on D. Then there exists an $\varepsilon > 0$, such that for each δ it is not true that $|f(z) - f(w)| > \varepsilon$ whenever $|z - w| < \delta$, $z \in D$, and $w \in D$. Hence there exist complex numbers z_n and w_n for each natural number n, satisfying

$$|z_n - w_n| < \frac{1}{n}, \qquad z_n \in D, \qquad w_n \in D, \qquad |f(z_n) - f(w_n)| \geqq \varepsilon. \quad (1.7)$$

Since D is bounded, there exists a subsequence $z_{n_1}, z_{n_2}, \ldots, z_{n_j}, \ldots, n_1 < n_2 < \cdots < n_j < \cdots$, of the complex sequence $\{z_n\}$, which converges.

Putting $\lim_{j\to\infty} z_{n_j} = c$ we conclude from $z_{n_j} \in D$ and the fact that D is closed that $c \in D$.

From

$$|z_{n_j} - w_{n_j}| < \frac{1}{n_j} \to 0 \qquad \text{as} \qquad j \to \infty$$

we conclude that $\lim_{j\to\infty} w_{n_j} = \lim_{j\to\infty} z_{n_j} = c$. Since f is continuous, we have

$$\lim_{j\to\infty} f(w_{n_j}) = \lim_{j\to\infty} f(z_{n_j}) = f(c)$$

and this contradicts (1.7). Hence $f(z)$ is uniformly continuous.

c. *Holomorphic functions*

Let $D \subset \mathbb{C}$ be a region and $f(z)$ a function of the complex variable z defined on D. The definition of the differential coefficient of a complex function is exactly the same form as the corresponding definition for real functions. Let z be a point belonging to D and let ρ be a positive real number such that the ρ-neighborhood of z:

$$U_\rho(z) = \{\zeta : |\zeta - z| < \rho\}$$

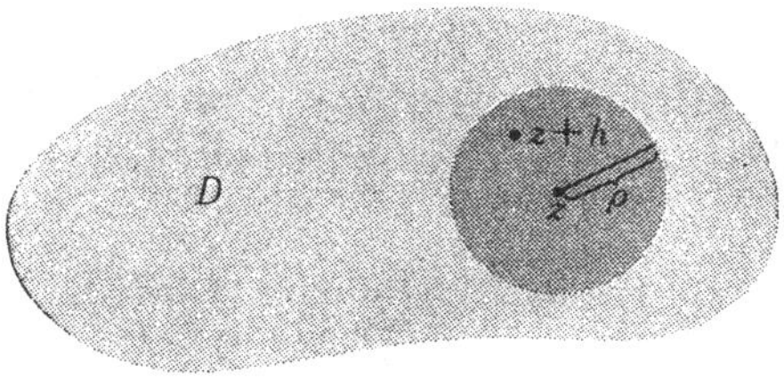

Fig. 1.3

is contained in D. Now, the expression $(f(z+h) - f(z))/h$ is a function of the complex variable h, $0 < |h| < \rho$. If

$$\lim_{h\to 0} \frac{f(z+h) - f(z)}{h}$$

exists, the function $f(z)$ is said to be *differentiable* at z. This limit is called the *differential coefficient* of $f(z)$ at z and denoted by $f'(z)$:

$$f'(z) = \lim_{h\to 0} \frac{f(z+h) - f(z)}{h}. \tag{1.8}$$

If $f(z)$ is differentiable at all points z of D, then $f(z)$ is called a *differentiable function* of z. In this case, $f'(z)$ is a function of z defined on D. This function $f'(z)$ is called the *derivative* of $f(z)$ and the process by which $f'(z)$ is obtained from $f(z)$ is called *differentiation*. Just as in the case of a function of a real variable the derivative of $w = f(z)$ is denoted by dw/dz or $df(z)/dz$. Similarly, a function $\alpha(h)$ such that $\lim_{h \to 0} \alpha(h) = \alpha(0) = 0$ is called an *infinitesimal*.

Representing the product $\varepsilon(h)\alpha(h)$ of the two infinitesimals $\varepsilon(h)$ and $\alpha(h)$ by $O(h)$, (1.8) can be rewritten as

$$f(z+h) - f(z) = f'(z)h + O(h). \tag{1.9}$$

Therefore, if $f(z)$ is differentiable, then $\lim_{h \to 0} f(z+h) = f(z)$. In other words, *differentiable functions are continuous*.

Putting

$$w = f(z) = f(x+iy) = u(x, y) + iv(x, y)$$

and

$$h = \Delta x + i\Delta y, \qquad \Delta w = f(z+h) - f(z) = \Delta u + i\Delta v$$

and $f'(z) = p + iq$, then (1.9) becomes

$$\Delta u + i\Delta v = (p+iq)\ (\Delta x + i\Delta y) + O(h).$$

Splitting this into its real and imaginary parts, we get

$$\Delta u = u(x+\Delta x, y+\Delta y) - u(x, y) = p\Delta x - q\Delta y + O(\sqrt{(\Delta x)^2 + (\Delta y)^2}),$$
$$\Delta v = v(x+\Delta x, y+\Delta y) - v(x, y) = q\Delta x + p\Delta y + O(\sqrt{(\Delta x)^2 + (\Delta y)^2}).$$

From this we conclude, if $f(z) = u(x, y) + iv(x, y)$ is a differentiable function of z, then $u(x, y)$ and $v(x, y)$ as functions of the real variables x and y are differentiable and

$$u_x(x, y) = v_y(x, y) = p, \qquad -u_y(x, y) = v_x(x, y) = q.$$

Hence,

$$\begin{aligned} u_x(x, y) &= v_y(x, y), \\ u_y(x, y) &= -v_x(x, y). \end{aligned} \tag{1.10}$$

The equations (1.10) are called the *Cauchy–Riemann equations*. As a direct consequence of the Cauchy–Riemann equations we note

$$f'(z) = u_x(x, y) + iv_x(x, y) = v_y(x, y) - iu_y(x, y). \tag{1.11}$$

Conversely, let $f(z) = u(x, y) + iv(x, y)$ be a function of $z = x + iy$ defined on the region $D \subset \mathbb{C}$, such that its real part $u(x, y)$ and its imaginary part $v(x, y)$ are differentiable functions of the real variables x and y satisfying the Cauchy–Riemann equations (1.10).

Putting

$$p = u_x(x, y) = v_y(x, y), \qquad q = -u_y(x, y) = v_x(x, y)$$

we have

$$\Delta u = p\Delta x - q\Delta y + O(\sqrt{(\Delta x)^2 + (\Delta y)^2}),$$
$$\Delta v = q\Delta x + p\Delta y + O(\sqrt{(\Delta x)^2 + (\Delta y)^2}).$$

Evaluating $f(z+h)-f(z)$ for $h = \Delta x + i\Delta y$ we get

$$f(z+h)-f(z) = \Delta u + i\Delta v = (p+iq)(\Delta x + i\Delta y) + O(|h|).$$

Hence

$$\lim_{h \to 0} \frac{f(z+h)-f(z)}{h} = p+iq,$$

i.e., $f(z)$ is differentiable and $f'(z) = p+iq$.

Summing up, the function $f(z) = u(x, y) + iv(x, y)$ of the complex variable $z = x+iy$ defined on the region $D \subset \mathbb{C}$ is a differentiable function of z if and only if $u(x, y)$ and $v(x, y)$ are differentiable functions of the real variables x and y satisfying the Cauchy–Riemann equations (1.10).

Just as for functions of a real variable, we have: If $f(z)$ is a function of the complex variable z defined on a region $D \subset \mathbb{C}$ such that $f(z)$ is differentiable at each point of D and such that $f'(z) = 0$ for each point z of D, then $f(z)$ is constant. For, putting $f(z) = u(x, y) + iv(x, y)$, we have $u_x(x, y) = u_y(x, y) = 0$ identically, by (1.11). Therefore, $u(x, y)$ is a constant function of x and y. Similarly, $v(x, y)$ is seen to be constant.

Definition 1.3. Let $f(z)$ be a function of the complex variable z, defined on the domain $D \subset \mathbb{C}$. If $f(z)$ is a differentiable function of z, such that the derivative $f'(z)$ is continuous, then $f(z)$ is called a *holomorphic function* of z or, briefly, *holomorphic*.

Remark: Usually, a function is called holomorphic if it is differentiable, but in this book we have adopted the above definition. This is a traditional definition, but since in modern theories of, for example, manifolds, continuously differentiable functions play a greater role than functions that are only differentiable, this older definition seems more natural. In Section 2.4 we will show that the derivative of a differentiable function of a complex variable is always continuous, so that our definition of a holomorphic function is equivalent with the usual one.

The following theorem follows immediately from (1.11).

Theorem 1.4. A function $f(z) = u(x, y) + iv(x, y)$ of the complex variable $z = x+iy$, defined on a domain $D \subset \mathbb{C}$, is holomorphic if and only if its real part $u(x, y)$ and its imaginary part $v(x, y)$ are continuously differentiable functions of the real variables x and y, satisfying the Cauchy–Riemann equations (1.10).

Example 1.1. Since $\lim_{h\to 0}[(z+h)-z]/h = \lim_{h\to 0} h/h = 1$, z is a holomorphic function of z. However, $\bar{z}$ is not a holomorphic function of z. For, putting $h = \Delta x + i\Delta y$, we have

$$\frac{\overline{z+h}-\bar{z}}{h} = \frac{\bar{h}}{h} = \frac{\Delta x + i\Delta y}{\Delta x - i\Delta y}.$$

If $\Delta y = 0$, then

$$\lim_{h\to 0}\frac{\overline{z+h}-\bar{z}}{h} = \lim_{h\to 0}\frac{\bar{h}}{h} = \lim_{\Delta x\to 0}\frac{\Delta x}{\Delta x} = 1,$$

and if $\Delta x = 0$, then

$$\lim_{h\to 0}\frac{\overline{z+h}-\bar{z}}{h} = \lim_{h\to 0}\frac{\bar{h}}{h} = \lim_{h\to 0}\frac{-i\Delta y}{i\Delta y} = -1.$$

Therefore, the limit $\lim_{h\to 0}[(\overline{z+h})-\bar{z}]/h$ does not exist.

Since Definition (1.8) of the differential coefficient is formally the same as that of the differential coefficient of a real function, the rules of differentiation of a linear combination, product, or quotient of differentiable functions of a complex variable are exactly the same as for functions of a real variable. Explicitly, if $f(z)$ and $g(z)$ are both differentiable functions of the complex variable z, defined on a certain domain, then the linear combination $a_1 f(z) + a_2 g(z)$, where a_1 and a_2 are constants, and the product $f(z)g(z)$ are differentiable and their derivatives are given by:

$$\frac{d}{dz}(a_1 f(z) + a_2 g(z)) = a_1 f'(z) + a_2 g'(z); \tag{1.12}$$

$$\frac{d}{dz}(f(z)g(z)) = f'(z)g(z) + f(z)g'(z). \tag{1.13}$$

If, moreover, $g(z) \neq 0$ for all points of that domain, then the quotient $f(z)/g(z)$ is differentiable and its derivative is given by

$$\frac{d}{dz}\left(\frac{f(z)}{g(z)}\right) = \frac{f'(z)g(z) - f(z)g'(z)}{g(z)^2}. \tag{1.14}$$

The proofs are also similar to the real case, but for completeness's sake we give the proofs for the derivatives of product and quotient. Since $f(z)$ and

$g(z)$ are differentiable, we have by (1.9)

$$f(z+h) = f(z) + f'(z)h + O(h),$$
$$g(z+h) = g(z) + g'(z)h + O(h).$$

Hence

$$f(z+h)g(z+h) - f(z)g(z) = (f'(z)g(z) + f(z)g'(z))h + O(h).$$

and therefore

$$\lim_{h\to 0} \frac{f(z+h)g(z+h) - f(z)g(z)}{h} = f'(z)g(z) + f(z)g'(z).$$

Thus, $f(z)g(z)$ is differentiable and (1.13) holds. If, moreover, $g(z) \neq 0$, then

$$\frac{1}{g(z+h)} - \frac{1}{g(z)} = \frac{-g'(z)h + O(h)}{(g(z) + g'(z)h + O(h))g(z)}.$$

Hence

$$\lim_{h\to 0} \frac{1}{h}\left(\frac{1}{g(z+h)} - \frac{1}{g(z)}\right) = -\frac{g'(z)}{g(z)^2}.$$

and therefore $1/g(z)$ is differentiable and its derivative is given by

$$\frac{d}{dz}\left(\frac{1}{g(z)}\right) = -\frac{g'(z)}{g(z)^2}.$$

Since $f(z)/g(z) = f(z) \cdot 1/g(z)$, we conclude that $f(z)/g(z)$ is differentiable and that (1.14) holds by using the product rule.

If $f(z)$ and $g(z)$ are holomorphic, then $f'(z)$ and $g'(z)$ are continuous, hence the right-hand sides of (1.12) and (1.13) are continuous too. If, moreover, $g(z) \neq 0$, then the right-hand side of (1.14) is continuous too. Therefore, we have the following theorem.

Theorem 1.5. If the complex functions $f(z)$ and $g(z)$, defined on a domain, are holomorphic, then their linear combination $a_1 f(z) + a_2 g(z)$, with a_1 and a_2 constants, and product $f(z)g(z)$ are also holomorphic. If, moreover, $g(z) \neq 0$ on that domain, then the quotient $f(z)/g(z)$ is holomorphic too.

As shown in Example 1.1, the function z, defined on the whole complex plane $\mathbb{C}$, is a holomorphic function of z. Therefore it follows from Theorem 1.5 that the polynomial $p(z)$ of z,

$$p(z) = a_0 + a_1 z + a_2 z^2 + \cdots + a_n z^n, \qquad a_0, a_1, \ldots, a_n \in \mathbb{C}$$

is holomorphic on $\mathbb{C}$. Moreover, the rational expression

$$\frac{p(z)}{q(z)}, \quad \text{when } p(z) \text{ and } q(z) \text{ are polynomials.}$$

is holomorphic on the domain, obtained by omitting from the complex plane $\mathbb{C}$ the roots of the algebraic equation $q(z) = 0$.

Let n be a natural number. Using induction on n and

$$\frac{d}{dz} z^n = \frac{d}{dz}(z \cdot z^{n-1}) = z^{n-1} + z\frac{d}{dz} z^{n-1}$$

it is easily proved that

$$\frac{d}{dz} z^n = nz^{n-1}.$$

Therefore, the derivative of the polynomial

$$p(z) = a_0 + a_1 z + a_2 z^2 + \cdots + a_n z^n$$

is given by

$$p'(z) = a_1 + 2a_2 z + \cdots + na_n z^{n-1}.$$

If D is a subset of the domain of a function $f(z)$, then the function obtained from $f(z)$ by restricting the domain of $f(z)$ to D is called the *restriction* of $f(z)$ to D and denoted by $f_D(z)$ or $(f \mid D)(z)$. Hence, $f_D(z)$ is a function defined on D by $f_D(z) = f(z)$ for each $z \in D$. If D is a domain, and if $f_D(z)$ is continuous, then $f(z)$ is called continuous on D; if $f_D(z)$ is holomorphic then $f(z)$ is called holomorphic on D. Generally, if $\mathscr{A}$ is some property of functions $f(z)$ and if $f_D(z)$ has $\mathscr{A}$, then $f(z)$ is said to have the property $\mathscr{A}$ on D.

If D is a closed domain, then the differential coefficient of $f_D(z)$ is not defined on the points of the boundary of D. Therefore, it is meaningless to say that $f(z)$ is holomorphic on D. What we mean if we say that $f(z)$ is holomorphic on the closed domain D *is* that $f(z)$ is holomorphic on some domain containing D.

Theorem 1.6. Let $f(z)$ be a holomorphic function of z defined on the domain $D \subset \mathbb{C}$ and let $g(w)$ be a holomorphic function of w, defined on the domain $E \subset \mathbb{C}$, such that $f(D) \subset E$. Then the composite function $g(f(z))$ is a holomorphic function of z defined on D and the derivative of $g(f(z))$ is given by

$$\frac{d}{dz} g(f(z)) = g'(f(z)) \cdot f'(z). \tag{1.15}$$

Proof: According to (1.9) we have

$$f(z+h)-f(z)=f'(z)h+O(h),$$
$$g(w+k)-g(w)=g'(w)k+O(k),$$

where $O(k)=\varepsilon(k)k$ with $\lim_{h\to 0}\varepsilon(k)=\varepsilon(0)=0$.

Substituting $w=f(z)$ and $k=f(z+h)-f(z)$, we get

$$g(f(z+h))-g(f(z))=(g'(f(z))+\varepsilon(f'(z)h+O(h)))(f'(z)h+O(h)).$$

Hence

$$g(f(z+h))-g(f(z))=g'(f(z))f'(z)h+O(h).$$

Therefore, $g(f(z))$ is differentiable and (1.15) holds. Since $g'(w)$ is a continuous function of w, $g'(f(z))$ is a continuous function of z. Since $f'(z)$ is a continuous function of z, the product $g'(f(z))f'(z)$ is a continuous function of z, so that $g(f(z))$ is a holomorphic function of z.

1.2 Power series

a. Series whose terms are functions

Let $\{f_n(z)\}$ be a sequence of functions, the terms $f_1(z)$, $f_2(z),\ldots,f_n(z),\ldots$, of which are functions of the complex variable z, all defined on the same domain, and let D be a subset of that domain. If this sequence converges for each point of D, then we say that $\{f_n(z)\}$ converges on D. In that case, a function $f(z)=\lim_{n\to\infty}f_n(z)$ is defined on D.

Definition 1.4. If for each $\varepsilon>0$, there exists a natural number $n_0(\varepsilon)$ such that

$$|f_n(z)-f(z)|<\varepsilon \qquad \text{if } n>n_0(\varepsilon) \text{ for all } z\in D,$$

then the sequence of functions $\{f_n(z)\}$ is said to *converge uniformly* to $f(z)$.

Similarly, if the series $\sum_{n=1}^{\infty}f_n(z)$, whose terms are functions of z, converges for each point $z\in D$, then the series $\sum_{n=1}^{\infty}f_n(z)$ is said to converge on D.

If the function sequence $\{s_m(z)\}$ of partial sums $s_m(z)=\sum_{n=1}^{m}f_n(z)$ is uniformly convergent on D, then the series $\sum_{n=1}^{\infty}f_n(z)$ is said to be uniformly convergent on D. Further, if the series $\sum_{n=1}^{\infty}|f_n(z)|$ is convergent on D, then the series $\sum_{n=1}^{\infty}f_n(z)$ is said to be absolutely convergent on D. In this case, obviously $\sum_{n=1}^{\infty}f_n(z)$ is convergent on D. If the series $\sum_{n=1}^{\infty}|f_n(z)|$ converges uniformly on D, then the series $\sum_{n=1}^{\infty}|f_n(z)|$ is said to be uniformly absolutely convergent. From

$$|s_m(z)-s(z)|=\left|\sum_{n=m+1}^{\infty}f_n(z)\right|\leq\sum_{n=m+1}^{\infty}|f_n(z)|$$

we conclude that if $\sum_{n=1}^{\infty} f_n'(z)$ is uniformly, absolutely convergent on D, then $\sum_{n=1}^{\infty} f_n(z)$ is uniformly convergent on D.

If there exists a convergent series $\sum_{n=1}^{\infty} A_n$ with positive terms A_n such that

$$|f_n(z)| \leqq A_n, \qquad n = 1, 2, 3, \ldots, \text{ for all } z \in D,$$

then $\sum_{n=1}^{\infty} f_n(z)$ converges uniformly absolutely on D.

Theorem 1.7. Let $f_1(z), f_2(z), \ldots, f_n(z), \ldots$ be continuous functions on D.

(1) If the sequence $\{f_n(z)\}$ converges uniformly on D, then its limit $f(z) = \lim_{n\to\infty} f_n(z)$ is continuous on D.

(2) If the series $\sum_{n=1}^{\infty} f_n(z)$ converges uniformly on D, then its sum $\sum_{n=1}^{\infty} f_n(z)$ is continuous on D.

Proof: Since (2) is a direct consequence of (1), it suffices to prove (1). Let c be an arbitrary point of D and let us prove that $f(z)$ is continuous at c. Choose a real, positive ε, then there exists a natural number n, such that

$$|f_n(z) - f(z)| < \varepsilon \qquad \text{for all points } z \in D.$$

Since $f_n(z)$ is continuous at c, there exists a real, positive $\delta(\varepsilon)$ such that

$$|f_n(z) - f_n(c)| < \varepsilon \qquad \text{if } |z - c| < \delta(\varepsilon).$$

From

$$|f(z) - f(c)| \leqq |f(z) - f_n(z)| + |f_n(z) - f_n(c)| + |f_n(c) - f(c)|$$

we conclude

$$|f(z) - f(c)| < 3\varepsilon \qquad \text{if } |z - c| < \delta(\varepsilon).$$

Therefore, $f(z)$ is continuous at c.

b. Power series

A series of the form

$$\sum_{n=0}^{\infty} a_n(z-c)^n = a_0 + a_1(z-c) + a_2(z-c)^2 + \cdots + a_n(z-c)^n + \cdots$$

is called a *power series with center c*. Power series play a fundamental role in complex analysis. In this section we consider power series,

$$\sum_{n=0}^{\infty} a_n z^n = a_0 + a_1 z + a_2 z^2 + \cdots + a_n z^n + \cdots,$$

with center at the origin 0.

By replacing z by $z - c$, all results obtained will be seen to be valid for power series with center c as well. We want to find a sufficient condition for the power series $\sum_{n=1}^{\infty} a_n z^n$ to converge. Suppose that $\sum_{n=0}^{\infty} a_n z^n$ converges for a certain complex number z. In that case, $\lim_{n\to\infty} a_n z^n = 0$, hence there exists a positive, real M such that

$$|a_n z^n| \leqq M \qquad \text{for all natural numbers } n.$$

For such an M we have

$$|z|\,|a_n|^{1/n} \leqq M^{1/n},$$

while $\lim_{n\to\infty} M^{1/n} = 1$. Therefore

$$|z| \limsup_{n\to\infty} |a_n|^{1/n} \leqq 1. \tag{1.16}$$

If $0 < \limsup_{n\to\infty} |a_n|^{1/n} < +\infty$, we put

$$r = \frac{1}{\limsup\limits_{n\to\infty} |a_n|^{1/n}}$$

and we obtain from (1.16) the inequality

$$|z| < r. \tag{1.17}$$

If $\limsup_{n\to\infty} |a_n|^{1/n} = +\infty$, then we put $r = 0$. If $|z| > 0$, then $\limsup_{n\to\infty} |a_n|^{1/n} < +\infty$ by (1.16), therefore if $\limsup_{n\to\infty} |a_n|^{1/n} = +\infty$, then $|z| = 0$, that is, (1.17) is valid. If $\limsup_{n\to\infty} |a_n|^{1/n} = 0$, then we put $r = +\infty$. In this case (1.17) is trivially satisfied.

The number r defined in this way has the following property.

Theorem 1.8. The power series $\sum_{n=0}^{\infty} a_n z^n$ converges absolutely if $|z| < r$ and diverges if $|z| > r$.

Proof: If $|z| < r$, we choose a real number κ such that $|z| < \kappa < r$. Then

$$\frac{1}{\kappa} > \frac{1}{r} = \limsup_{n\to\infty} |a_n|^{1/n}.$$

Therefore, there exists a natural number n_0 such that $|a_n|^{1/n} < 1/\kappa$ for $n > n_0$, i.e., $|a_n|\kappa^n < 1$ for $n > n_0$. Therefore, there exists a constant M, such that

$$|a_n| \leqq \frac{M}{\kappa^n}. \tag{1.18}$$

Since $|z| < \kappa$, we have

$$\sum_{n=0}^{\infty} |a_n z^n| \leqq M \sum_{n=0}^{\infty} \left(\frac{|z|}{\kappa}\right)^n = \frac{M\kappa}{\kappa - |z|} < +\infty.$$

Therefore, the series $\sum_{n=0}^{\infty} a_n z^n$ converges absolutely if $|z| < r$. The second part of the theorem follows from (1.17).

The number r is called the *radius of convergence* of the power series $\sum_{n=0}^{\infty} a_n z^n$. Putting $1/+\infty = 0$ and $1/0 = +\infty$, the radius of convergence is given in all cases by

$$r = \frac{1}{\limsup_{n\to\infty} |a_n|^{1/n}}. \tag{1.19}$$

(1.19) is called the Cauchy–Hadamard formula. For $0 < r < +\infty$ the circle C_r with center 0 and radius r is called the *circle of convergence* of the power series $\sum_{n=0}^{\infty} a_n z^n$. According to Theorem 1.8, the power series $\sum_{n=0}^{\infty} a_n z^n$ converges absolutely if z belongs to the interior of the circle of convergence and diverges if z belongs to the exterior of the circle of convergence.

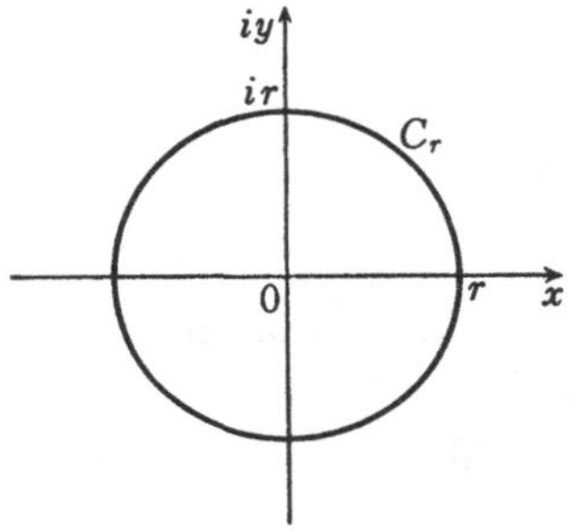

Fig. 1.4

If z is a point of C_r, then $\sum_{n=0}^{\infty} a_n z^n$ can be convergent or divergent. The interior of C_r is denoted by $U_r(0)$

$$U_r(0) = \{z: |z| < r\}. \tag{1.20}$$

If $r = +\infty$, the circle of convergence is not defined, but we put $U_{+\infty}(0) = \mathbb{C}$ and consider $\mathbb{C}$ as the interior of the circle of convergence. By doing so, the power series $\sum_{n=0}^{\infty} a_n z^n$ is absolutely convergent in the interior of the circle of convergence for all radii of convergence such that $0 < r \leqq +\infty$.

Theorem 1.9. Let $\sum_{n=0}^{\infty} a_n z^n$ be a power series with radius of convergence r, such that $0 < r \leqq +\infty$. If ρ is a real number such that $0 < \rho < r$, $\sum_{n=0}^{\infty} a_n z^n$ is uniformly absolutely convergent on the interior $U_\rho(0)$ of the circle with center 0 and radius ρ.

Proof: Choose a real κ, such that $\rho < \kappa < r$. According to (1.18), there exists a constant M such that

$$|a_n|\rho^n < M(\rho/\kappa)^n.$$

Since $\sum_{n=0}^{\infty} M(\rho/\kappa)^n = M\kappa/(\kappa-\rho) < +\infty$, the sum $\sum_{n=0}^{\infty} |a_n|\rho^n < +\infty$. From $|a_n z^n| \leqq |a_n|\rho^n$ for all z with $|z| < \rho$, we conclude that $\sum_{n=0}^{\infty} a_n z^n$ converges uniformly absolutely on $U_\rho(0) = \{z\colon |z| < \rho\}$.

Since all terms $a_n z^n$ are continuous functions of z on $U_\rho(0)$, by Theorem 1.7(2), the sum $f(z) = \sum_{n=0}^{\infty} a_n z^n$ is also continuous on $U_\rho(0)$. Since this is true for each ρ with $0 < \rho < r$, we conclude that $f(z) = \sum_{n=0}^{\infty} a_n z^n$ is a continuous function of z on the interior $U_r(0)$ of the circle of convergence.

Theorem 1.10. Let $\sum_{n=0}^{\infty} a_n z^n$ be a power series with radius of convergence r such that $0 < r \leqq +\infty$. Then $f(z) = \sum_{n=0}^{\infty} a_n z^n$ is a holomorphic function of z on the interior $U_r(0)$ of the circle of convergence and the derivative of $f(z)$ is given by "termwise differentiation of $\sum_{n=0}^{\infty} a_n z^n$":

$$f'(z) = \sum_{n=1}^{\infty} n a_n z^{n-1} = a_1 + 2a_2 z + 3a_3 z^2 + \cdots + n a_n z^{n-1} + \cdots. \tag{1.21}$$

The radius of convergence of the power series $\sum_{n=1}^{\infty} n a_n z^{n-1}$ is also r.

Proof: First we prove that the radius of convergence of $\sum_{n=1}^{\infty} n a_n z^{n-1}$ equals r.

Differentiating both sides of the equality

$$\sum_{n=0}^{m} t^n = \frac{1-t^{m+1}}{1-t}$$

yields

$$\sum_{n=1}^{m} n t^{n-1} = \frac{1-(m+1)t^m + m t^{m+1}}{(1-t)^2}.$$

Since $\lim_{m\to\infty} m t^m = 0$ for $0 < t < 1$, we find

$$\sum_{n=1}^{\infty} n t^{n-1} = \frac{1}{(1-t)^2}. \tag{1.22}$$

In order to prove that $\sum_{n=1}^{\infty} n a_n z^{n-1}$ converges absolutely if $|z| < r$, we fix a z with $|z| < \tau$. Next we choose a real κ such that $|z| < \kappa < r$. According to (1.18) there exists a constant M, such that

$$|a_n| \leqq M/\kappa^n.$$

Using (1.22) we get

$$\sum_{n=1}^{\infty} n|a_n z^{n-1}| \leqq \frac{M}{\kappa} \sum_{n=1}^{\infty} n\left(\frac{|z|}{\kappa}\right)^{n-1} = \frac{M\kappa}{(\kappa-|z|)^2} < +\infty.$$

Therefore, $\sum_{n=1}^{\infty} n a_n z^{n-1}$ converges absolutely for $|z| < r$.

From

$$\sum_{n=0}^{\infty} |a_n z^n| \leqq |a_0| + |z| \sum_{n=1}^{\infty} n|a_n z^{n-1}|$$

we conclude that $\sum_{n=0}^{\infty} a_n z^n$ converges absolutely if $\sum_{n=1}^{\infty} na_n z^{n-1}$ converges absolutely. Therefore, the radius of convergence of $\sum_{n=1}^{\infty} na_n z^{n-1}$ equals r. Hence, the sum $\sum_{n=1}^{\infty} na_n z^{n-1}$ is a continuous function of z on $U_r(0)$.

Next we prove that $f(z) = \sum_{n=0}^{\infty} a_n z^n$ is differentiable on $U_r(0)$ and that $f'(z) = \sum_{n=0}^{\infty} na_n z^{n-1}$. Fix z such that $|z| < r$ and choose real numbers ρ and κ such that $|z| < \rho < \kappa < r$. Consider, for $|w| < \rho$,

$$\frac{f(w) - f(z)}{w - z} = \sum_{n=1}^{\infty} a_n \frac{w^n - z^n}{w - z}$$

as a function of w. Putting

$$p_n(w) = a_n \frac{w^n - z^n}{w - z} = a_n(w^{n-1} + w^{n-2}z + w^{n-3}z^2 + \cdots + z^{n-1})$$

we see that $p_n(w)$ is a continuous function of w. Let M be a constant such that $|a_n| \leqq M/\kappa^n$, then

$$|p_n(w)| \leqq |a_n| (|w|^{n-1} + |w|^{n-2}|z| + \cdots + |z|^{n-1}) < Mn\rho^{n-1}/\kappa^n.$$

By (1.22) we have

$$\sum_{n=1}^{\infty} \frac{M}{\kappa^n} n\rho^{n-1} = \frac{M\kappa}{(\kappa - \rho)^2} < +\infty.$$

Therefore, the series on the right-hand side of

$$\frac{f(w) - f(z)}{w - z} = \sum_{n=1}^{\infty} p_n(w)$$

converges uniformly absolutely for $|w| < \rho$ and hence its sum is a continuous function of w for $|w| < \rho$, i.e.,

$$\lim_{w \to z} \frac{f(w) - f(z)}{w - z} = \sum_{n=1}^{\infty} p_n(z) = \sum_{n=1}^{\infty} na_n z^{n-1}.$$

Therefore $f(z)$ is a differentiable function of z on $U_r(0)$ and

$$f'(z) = \sum_{n=1}^{\infty} na_n z^{n-1}.$$

Since, as we proved above, $\sum_{n=1}^{\infty} na_n z^{n-1}$ is a continuous function of z on $U_r(0)$, we conclude that $f(z)$ is a holomorphic function of z on $U_r(0)$.

Since the circle of convergence of $\sum_{n=1}^{\infty} na_n z^{n-1}$ equals r, we can apply Theorem 1.10 to conclude that $f'(z)$ is a holomorphic function of z on $U_r(0)$

and

$$f''(z) = \sum_{n=2}^{\infty} n(n-1)a_n z^{n-2}.$$

Since the radius of convergence of this power series again equals r, $f''(z)$ is also holomorphic on $U_r(0)$ and

$$f'''(z) = \sum_{n=3}^{\infty} n(n-1)(n-2)\,a_n z^{n-3}.$$

Continuing in this way, we find that $f(z)$ is infinitely often differentiable on $U_r(z)$, that the higher-order derivatives $f''''(z), \ldots, f^{(m)}(z), \ldots$ are all holomorphic on $U_r(0)$ and that $f^{(m)}(z)$ is represented by

$$f^{(m)}(z) = \sum_{n=m}^{\infty} n(n-1)(n-2)\cdots(n-m+1)a_n z^{n-m}.$$

Replacing z by $z-c$ in $\sum_{n=0}^{\infty} a_n z^n$ and applying the above results to the power series $\sum_{n=0}^{\infty} a_n(z-c)^n$, we get the following results.

Let $\sum_{n=0}^{\infty} a_n(z-c)^n$ be a power series with radius of convergence r, such that $0 < r \leqq +\infty$. Then $\sum_{n=0}^{\infty} a_n(z-c)^n$ is absolutely convergent for $|z-c| < r$ and divergent for $|z-c| > r$. The radius of convergence is given by the Cauchy–Hadamard formula, (1.19). If $0 < r < +\infty$, then the circle with center c and radius r is called the circle of convergence of $\sum_{n=0}^{\infty} a_n(z-c)^n$. For $0 < r \leqq +\infty$, $U_r(c) = \{z: |z-c| < r\}$ is called the interior of the circle of convergence. Consequently,

Theorem 1.11. The sum of the power series

$$f(z) = \sum_{n=0}^{\infty} a_n(z-c)^n = a_0 + a_1(z-c) + a_2(z-c)^2 + \cdots$$

is a holomorphic function of z on the interior $U_r(c)$ of the circle of convergence. The function $f(z)$ is infinitely often differentiable on $U_r(c)$ and the derivatives $f'(z), f''(z), \ldots, f^{(m)}(z), \ldots$ are all holomorphic on $U_r(c)$. The mth derivative of $f(z)$ is obtained by differentiating $\sum_{n=0}^{\infty} a_n(z-c)^n$ termwise m times:

$$f^{(m)}(z) = \sum_{n=m}^{\infty} n(n-1)(n-2)\cdots(n-m+1)a_n(z-c)^{n-m}$$

and the radius of convergence of this power series is also r.

Substituting $z = c$ in the above expression, we find

$$f^{(m)}(c) = m!\,a_m. \tag{1.23}$$

Therefore, the original power series can be written as

$$f(z) = \sum_{n=0}^{\infty} a_n(z-c)^n = \sum_{n=0}^{\infty} \frac{f^{(n)}(c)}{n!}(z-c)^n.$$

The power series on the right-hand side is called the *Taylor series of* $f(z)$ *with center* c.

Now let $f(x)$ be a real analytic function of x, defined on some interval of the real line $\mathbb{R}$. Let c be a point from the interior of that interval and let $f(x)$ be represented by its Taylor series with center c:

$$f(x) = f(c) + \sum_{n=1}^{\infty} \frac{f^{(n)}(c)}{n!}(x-c)^n.$$

Assuming that its radius of convergence r satisfies $0 < r \leqq +\infty$, then the power series obtained from this Taylor series by replacing x by the complex variable z converges absolutely in the interior $U_r(c)$ of a circle with center c and radius r in the complex plane $\mathbb{C}$ and represents a holomorphic function of z on $U_r(c)$. Denoting this function also by f, we have

$$f(z) = f(c) + \sum_{n=1}^{\infty} \frac{f^{(n)}(c)}{n!}(z-c)^n.$$

The original real analytic function is obtained by replacing the complex variable z by the real variable x, i.e., by restricting $f(z)$ to the real axis. A real analytic function can always be extended to a holomorphic function.

Example 1.2. The *exponential function* e^x, or exp **x**, has a Taylor expansion $e^x = 1 + \sum_{n=1}^{\infty} x^n/n!$ with radius of convergence equal to $+\infty$. Therefore, the exponential function of the complex variable z,

$$e^z = 1 + \frac{z}{1!} + \frac{z^2}{2!} + \frac{z^3}{3!} + \cdots + \frac{z^n}{n!} + \cdots,$$

is holomorphic on the whole complex plane $\mathbb{C}$.

The complex *sine function*,

$$\sin z = z - \frac{z^3}{3!} + \frac{z^5}{5!} - \frac{z^7}{7!} + \cdots + (-1)^n \frac{z^{2n+1}}{(2n+1)!},$$

and the complex *cosine function*,

$$\cos z = 1 - \frac{z^2}{2!} + \frac{z^4}{4!} - \frac{z^6}{6!} + \cdots + (-1)^n \frac{z^{2n}}{(2n)!} + \cdots,$$

are also holomorphic on $\mathbb{C}$.

Since $\log(1+x)$ has the Taylor expansion $\log(1+x) = \sum_{n=1}^{\infty} (-1)^{n-1} x^n/n$

with radius of convergence 1, the complex extension,

$$\log(1+z) = z - \frac{z^2}{2} + \frac{z^3}{3} - \cdots + (-1)^{n-1}\frac{z^n}{n} + \cdots,$$

is a holomorphic function of the complex variable z on the interior of the unit circle $U_1(0)$.

We note as an obvious consequence of the above, the following formula

$$e^{iz} = \cos z + i \sin z.$$

In particular for real θ

$$e^{i\theta} = \cos\theta + i\sin\theta$$

so that $|e^{i\theta}| = 1$. Hence, the transformation given by

$$z \to z' = e^{i\theta} z$$

is a rotation of the complex plane $\mathbb{C}$ with center 0 and angle θ.

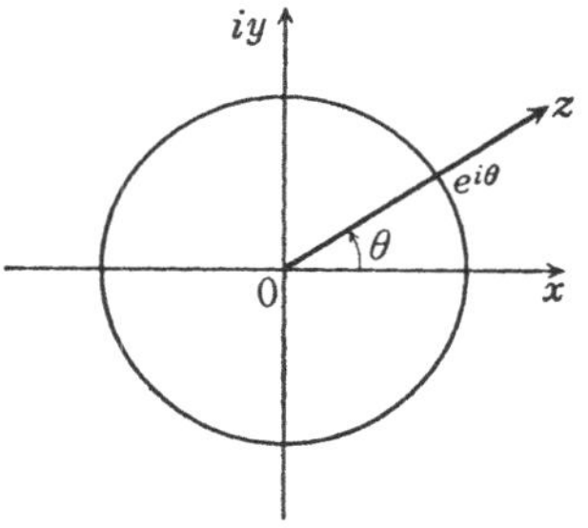

Fig. 1.5

Every complex number $z \neq 0$ can be represented as

$$z = |z| e^{i\theta},$$

where θ is the angle between the segment $0z$ and the positive real axis. One calls θ the *argument* (or *amplitude*) of z. If one represents the complex number z by the vector $\mathbf{0z}$, then the argument of z is the angle between the vector $\mathbf{0z}$ and the real axis.

Using (1.4) and the binomial theorem, we have

$$\begin{aligned}\sum_{n=0}^{\infty}\frac{z^n}{n!}\sum_{n=0}^{\infty}\frac{w^n}{n!} &= \sum_{n=0}^{\infty}\left(\frac{z^n}{n!} + \frac{z^{n-1}w}{(n-1)!\,1!} + \frac{z^{n-2}w^2}{(n-2)!\,2!} + \cdots + \frac{w^n}{n!}\right)\\ &= \sum_{n=0}^{\infty}\frac{1}{n!}\left[z^n + \binom{n}{1}z^{n-1}w + \binom{n}{2}z^{n-2}w^2 + \cdots + w^n\right]\\ &= \sum_{n=0}^{\infty}\frac{(z+w)^n}{n!}.\end{aligned}$$

So we have proved

$$e^z e^w = e^{z+w}. \tag{1.24}$$

Therefore the product of two complex numbers $z_1 = |z_1| \exp(i\theta_1)$ and $z_2 = |z_2| \exp(i\theta_2)$ is given by

$$z_1 z_2 = |z_1| |z_2| \exp[i(\theta_1 + \theta_2)].$$

From this we see that the argument of the product of two complex numbers is equal to the sum of the arguments of those numbers.

At this point, we insert some remarks concerning coordinate transformations of the complex plane. The complex number $x = x + iy$ is called the *coordinate* of the point z of the complex plane $\mathbb{C}$. Given a linear expression $aw + b$, a and b constants, $a \neq 0$, of the complex variable w, the value of w satisfying $z = aw + b$ can be thought of as a new coordinate of the point z. The point z has now two coordinates z and w, related with each other through the so-called coordinate transformation

$$z \to w = \frac{z}{a} - \frac{b}{a}.$$

The origin of the new system of coordinates is b and the angle α between the real axis of the new system of coordinates and the real axis of the old system of coordinates equals the argument of a.

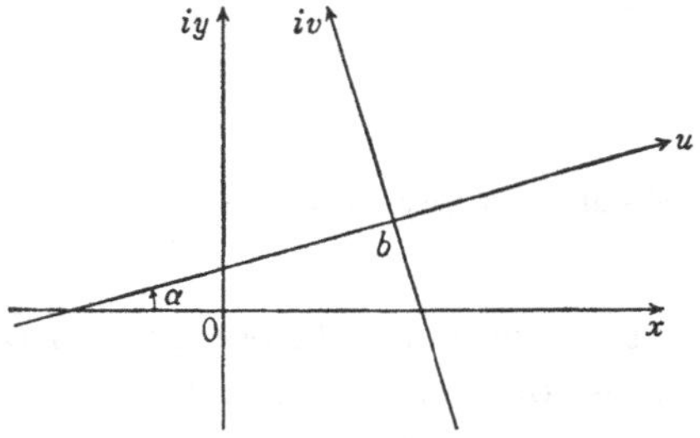

Fig. 1.6

1.3 Integrals

In the previous section we have seen that the sum $f(z) = \sum_{n=0}^{\infty} a_n(z-c)^n$ is a holomorphic function of z in the interior of the circle of convergence. Conversely, it is possible to represent a holomorphic function, defined on a certain region, in some neighborhood of an arbitrary point c of that region as the sum of a power series, i.e.,

$$f(z) = \sum_{n=0}^{\infty} a_n(z-c)^n \qquad \text{if } |z-c| < r \text{ for some } r > 0.$$

a. Curves

Consider a complex-valued function $\gamma(t)$, defined on some interval of the real line $\mathbb{R}$.

The following definitions are obvious.

If for each t in the interior of the interval

$$\lim_{\Delta t \to 0} |\gamma(t+\Delta t) - \gamma(t)| = 0$$

then $\gamma(t)$ is called a *continuous function* of t.

If for each t in the interior of the interval the limit

$$\gamma'(t) = \lim_{\Delta t \to 0} \frac{\gamma(t+\Delta t) - \gamma(t)}{\Delta t}$$

exists, then $\gamma(t)$ is called a *differentiable function* of t. (Here, Δt represents an increment of t.)

If $\gamma(t)$ is differentiable and its derivative $\gamma'(t)$ is continuous, then $\gamma(t)$ is called *continuously differentiable* or *a function of class* C^1.

The function $\gamma(t)$ is continuous, of class C^1 and so on if and only if its real part $\operatorname{Re}\gamma(t)$ and its imaginary part $\operatorname{Im}\gamma(t)$ are both continuous, continuously differentiable, and so on. Let $\gamma(t)$ be a continuous function defined on the closed interval $I = [a, b]$. If t moves from a to b over the real axis, the point $\gamma(t)$ moves in the complex plane $\mathbb{C}$, describing a "curve" $C = \gamma(I) = \{\gamma(t): a \leqq t \leqq b\}$.

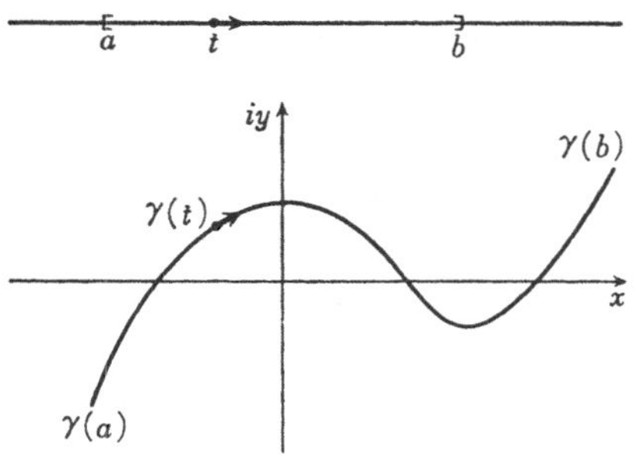

Fig. 1.7

Definition 1.5. A continuous map $\gamma: t \to \gamma(t)$ from the closed interval $I = [a, b]$ into the complex numbers is called an *arc* or *curve* in the complex plane. The point $\gamma(a)$ is called the *initial point* and the point $\gamma(b)$ the *final point* of the curve. If $\gamma(t)$ is continuously differentiable on I, γ is called a *continuously differentiable curve* or a *curve of class* C^1. If, moreover, $\gamma'(t) \neq 0$ for all $t \in I$, then γ is called a *smooth arc.* Points belonging to the subset $C = \gamma(I) = \{\gamma(t): a \leqq t \leqq b\}$ are said to be *points on* γ.

(Formally speaking, there is no difference between a "map" and a "function," but "map" seems to better convey the idea of a curve.)

If α and β are two points of $\mathbb{C}$, then a curve $\gamma: t \to \gamma(t)$, $a \leqq t \leqq b$, such that $\gamma(a) = \alpha$ and $\gamma(b) = \beta$ is said to *connect* α and β.

If D is a region in $\mathbb{C}$, $[D]$ its closure, $\alpha \in D$ and $\beta \notin [D]$, then all curves connecting α and β intersect the boundary $[D] - D$ of D, that is, for some s with $a < s < b$ we have $\gamma(s) \in [D] - D$. To see this, let S be the subset of I consisting of all points t such that $\gamma(t) \notin [D]$ and let s be the supremum of S. Suppose $\gamma(s) \notin [D]$, then by the continuity of γ there exists an $\varepsilon > 0$ such that $\gamma(t) \notin [D]$ if $t > s - \varepsilon$. This is a contradiction and therefore $\gamma(s) \in [D]$. Suppose $\gamma(s) \notin D$, then there exists an $\varepsilon > 0$ such that $\gamma(t) \notin D$ for $t < s + \varepsilon$. This is a contradiction and therefore $\gamma(s) \notin D$.

Hence, if it is possible to connect the points α and β of $\mathbb{C}$ with a curve γ not intersecting the boundary of D, then either α and β both belong to D or to the complement of $[D]$.

In complex analysis one frequently encounters curves $\gamma: \theta \to \gamma(\theta) = c + re^{i\theta}$, where $c \in \mathbb{C}$, $0 < r < +\infty$, and $\theta \in I = [0, 2\pi]$. In this case we write θ rather than t, since θ represents an angle.

In Definition 1.5 we have defined a curve as a map γ, rather than as a subset $\gamma(I)$ of $\mathbb{C}$, because the way the point $\gamma(t)$ moves when t moves from a to b is also characteristic of a certain curve. For example, if we represent the direction on the curve corresponding with increasing t by arrows $\to$ then the two curves γ_1 and γ_2 (see Fig. 1.8) are different, although the point sets $\gamma_1(I)$ and $\gamma_2(I)$ are the same.

Of course, if a curve γ is given, we also have in mind the point set $C = \gamma(I)$ and this is made concrete by saying that points belonging to $C = \gamma(I)$ are on γ. For example, if we say that the boundary of a certain region D is formed by the curve γ, we mean that the boundary of D is the point set $C = \gamma(I)$.

If $\gamma: t \to \gamma(t)$, $a \leqq t \leqq b$, is a curve such that for all s, t with $a \leqq s < t \leqq b$ we have $\gamma(s) \neq \gamma(t)$, that is, if γ is a one-to-one map from I into $\mathbb{C}$, then γ is called a *Jordan arc.*

If γ is such that $\gamma(a) = \gamma(b)$ then γ is called a *closed curve* and if γ is such

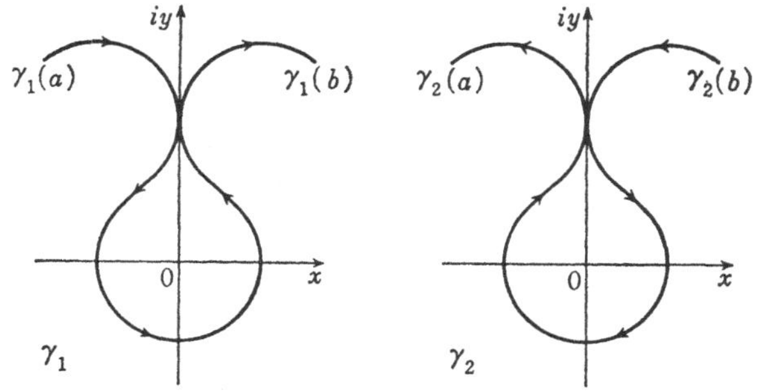

Fig. 1.8

that for all s, t with $a \leqq s < t < b$ we have $\gamma(s) \neq \gamma(t)$, while $\gamma(a) = \gamma(b)$, then γ is called a *Jordan curve.*

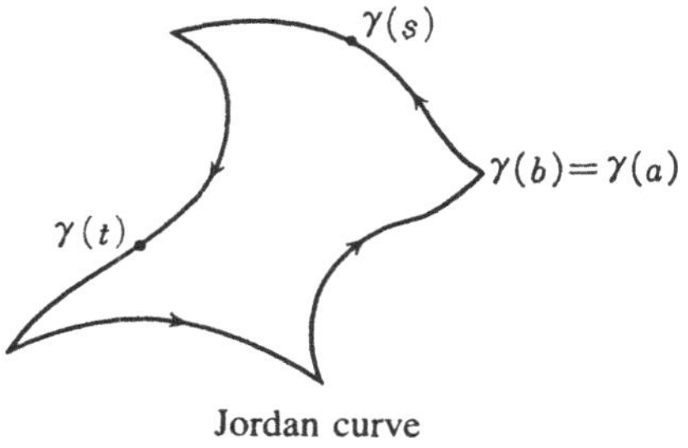

Fig. 1.9

There are examples of curves, which are not "curvelike" at all, such as a curve, γ, the corresponding point set, $C = \gamma(I)$, of which is a triangle, together with all of its interior points. In his case, the idea of a curve is only carried by the map and not by the point set. This curve is not a Jordan arc but there are also unintuitive Jordan arcs.

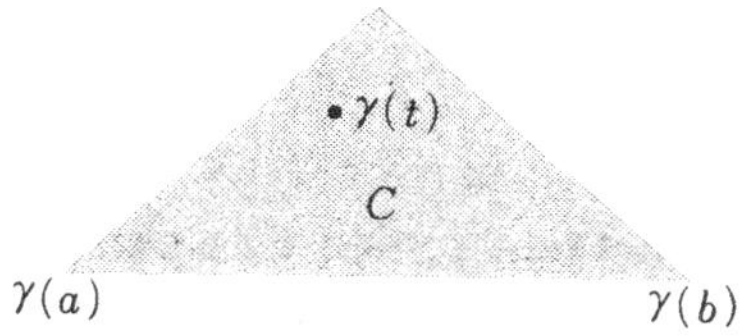

Fig. 1.10

Jordan arcs of class C^1, however, have properties that are worthy of curves. Let $\gamma: t \to \gamma(t)$, $a \leqq t \leqq b$, be a Jordan arc of class C^1 and let $t_0 \in [a, b]$ be such that $\gamma'(t_0) \neq 0$. Writing $\gamma(t)$ as

$$\gamma(t) = \gamma(t_0) + \gamma'(t_0)(t - t_0) + O(t - t_0)$$

the linear part of the right-hand side defines a line through $\gamma(t_0)$:

$$l: t \to l(t) = \gamma(t_0) + \gamma'(t_0)(t - t_0)$$

This line l is called the *tangent* to the curve γ at $\gamma(t_0)$. The vector $\mathbf{0\gamma'}(t_0)$ is called the *tangent vector* to the curve γ at $\gamma(t_0)$. If we agree to call the direction of the tangent corresponding with increasing t the positive direction, it is clear that the angle θ between the positive direction of the tangent and the positive direction of the real axis is just the argument of $\gamma'(t_0)$:

$$\gamma'(t_0) = |\gamma'(t_0)| e^{i\theta}.$$

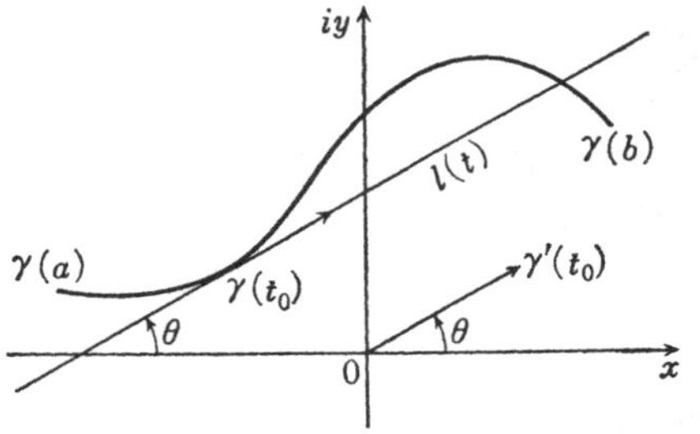

Fig. 1.11

If $\gamma'(t_0) = 0$, then the tangent at $\gamma(t_0)$ is not defined. If $\gamma'(t_0) = 0$, curves can exhibit a number of different behaviors in the vicinity of $\gamma(t_0)$.

Example 1.3. Consider $\gamma: t \to \gamma(t) = t^3$, $-1 \leqq t \leqq 1$. This is a Jordan arc of class C^1. In this case $\gamma'(0) = 0$, but $\gamma(0) = 0$ is not a singular point on $\gamma(I) = [-1, 1]$.

Example 1.4. Consider $\gamma: t \to \gamma(t) = t^2 + it^3$, $-1 \leqq t \leqq 1$. This is a Jordan arc of class C^1 and $\gamma'(0) = 0$. In this case the point $\gamma(0) = 0$ is a singular point (a so-called *cusp*) of $\gamma(I)$.

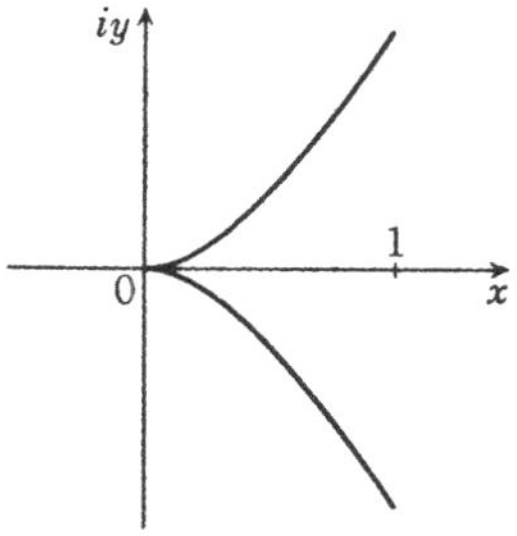

Fig. 1.12

Example 1.5. Consider γ defined on $[0,1]$ by $\gamma(t) = t^3 e^{2\pi i/t}$ if $0 < t \leqq 1$ and $\gamma(0) = 0$. Now, $\gamma'(0) = \lim_{t \to +0} \gamma(t)/t = 0$ and $\gamma'(t) = t(3t - 2\pi i)e^{2\pi i/t}$ for $0 < t < 1$. Since $\lim_{t \to +0} \gamma'(t) = 0$, $\gamma(t)$ is a continuously differentiable function of t on $I = [0,1]$. For s, t with $0 \leqq s < t \leqq 1$ we have $|\gamma(s)| = s^3 < t^3 = |\gamma(t)|$, hence $\gamma: t \to \gamma(t), 0 \leqq t \leqq 1$ is a Jordan arc of class C^1. If t tends to 0, $\gamma(t)$ "spirals infinitely often" around 0 and it is clear that $\gamma(0) = 0$ is a singular point of $\gamma(I)$.

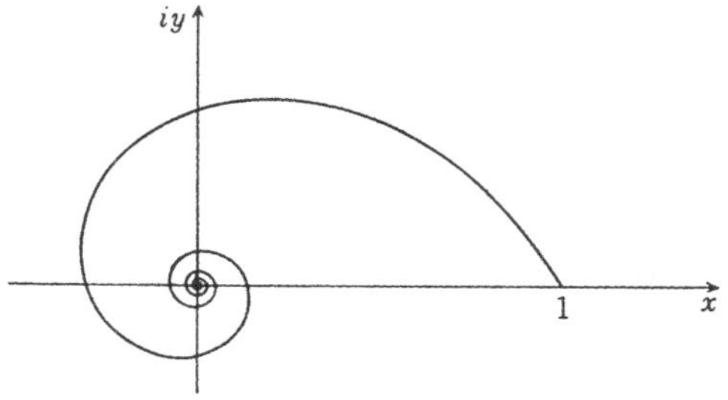

Fig. 1.13

If $\gamma: t \to \gamma(t)$, $a \leqq t \leqq b$, is a smooth Jordan arc, then by definition $\gamma'(t) \neq 0$ for all $t \in I$, so that a tangent is defined at each point. A Jordan curve $\gamma: t \to \gamma(t)$, $a \leqq t \leqq b$, satisfying γ is continuously differentiable, $\gamma'(t) \neq 0$ for all t and $\gamma'(a) = \gamma'(b)$ is called a *smooth Jordan curve.* (If $\gamma'(a) \neq \gamma'(b)$, then there would be two different tangents at the point $\gamma(a) = \gamma(b)$.)

If $\gamma: t \to \gamma(t)$ and $\lambda: s \to \lambda(s)$ are two smooth Jordan curves intersecting in $z_0 = \gamma(t_0) = \lambda(s_0)$, then the angle θ between the tangent to γ at $\gamma(t_0)$ and the tangent to λ at $\lambda(s_0)$, that is, the argument of $\lambda'(s_0)/\gamma'(t_0)$ is called the angle between λ and γ. If $\theta = m\pi$ for some integer m then the curves λ and γ are said to *touch* at z_0.

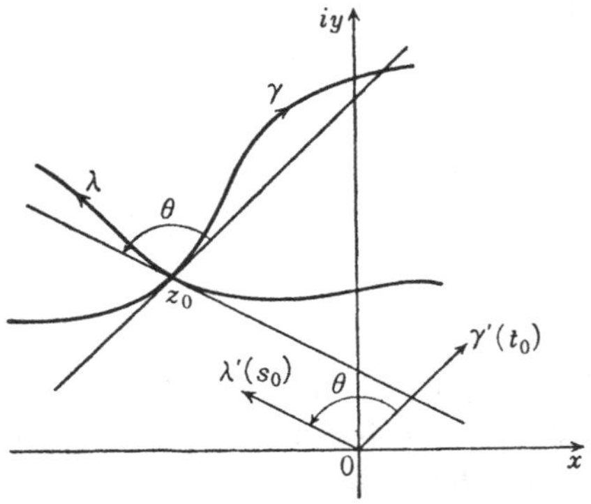

Fig. 1.14

b. Integrals

Generally, if $\phi(t) = u(t) + iv(t)$, $a \leqq t \leqq b$, is a continuous function of t, the definite integral $\int_a^b \phi(t)dt$ is defined by

$$\int_a^b \phi(t)dt = \int_a^b u(t)dt + i \int_a^b v(t)dt. \tag{1.25}$$

If $\phi(t)$ and $\psi(t)$ are both continuous functions, defined on $[a, b]$ and if c_1 and c_2 are constants, the following formula holds.

$$\int_a^b (c_1\phi(t) + c_2\psi(t))\,dt = c_1 \int_a^b \phi(t)dt + c_2 \int_a^b \psi(t)\,dt. \tag{1.26}$$

Proof: By (1.25) we have

$$\int_a^b (c_1\phi(t) + c_2\psi(t))dt = \int_a^b c_1\phi(t)dt + \int_a^b c_2\psi(t)dt.$$

Putting $c_1 = \rho + i\sigma$, where ρ and σ are real numbers, we have

$$c_1\phi(t) = \rho u(t) - \sigma v(t) + i(\sigma u(t) + \rho v(t)).$$

Hence

$$\begin{aligned}\int_a^b c_1\phi(t)dt &= \rho \int_a^b u(t)dt - \sigma \int_a^b v(t)dt + i\sigma \int_a^b u(t)dt + i\rho \int_a^b v(t)dt \\ &= (\rho + i\sigma)\left(\int_a^b u(t)dt + i \int_a^b v(t)dt \right),\end{aligned}$$

that is,

$$\int_a^b c_1\phi(t)dt = c_1\int_a^b \phi(t)dt.$$

Similarly for the other term. This proves (1.26).

We want to derive another useful formula

$$\left|\int_a^b \phi(t)dt\right| \leqq \int_a^b |\phi(t)|dt. \tag{1.27}$$

Let θ be the argument of the complex number $\int_a^b \phi(t)dt$, then

$$\left|\int_a^b \phi(t)\,dt\right| = e^{-i\theta}\int_a^b \phi(t)\,dt = \int_a^b e^{-i\theta}\phi(t)dt.$$

(This defines θ only if $\int_a^b \phi(t)dt \neq 0$; however (1.27) is trivially true if this integral is zero.) By (1.25) we have

$$\int_a^b e^{-i\theta}\phi(t)dt = \int_a^b \operatorname{Re}(e^{-i\theta}\phi(t))\,dt + i\int_a^b \operatorname{Im}(e^{-i\theta}\phi(t))\,dt$$

and since $\operatorname{Re}(e^{-i\theta}\phi(t)) \leqq |e^{-i\theta}\phi(t)| = |\phi(t)|$ we conclude

$$\left|\int_a^b \phi(t)dt\right| = \int_a^b \operatorname{Re}(e^{-i\theta}\phi(t))dt \leqq \int_a^b |\phi(t)|\,dt.$$

The chain rule is also true for the composition of a complex-valued continuously differentiable function $\gamma(t)$ of a real variable t and a holomorphic function $f(z)$ of a complex variable z:

Theorem 1.12. Let $f(z)$ be a holomorphic function defined on a region D and let $\gamma(t)$ be a continuously differentiable function defined on an interval I. If $\gamma(t)\in D$ for all $t\in I$, then the composite function $f(\gamma(t))$ is a continuously differentiable function of t and the chain rule is valid:

$$\frac{d}{dt}f(\gamma(t)) = f'(\gamma(t))\cdot\gamma'(t). \tag{1.28}$$

Proof: Let Δt be an increment of t, then

$$\gamma(t+\Delta t)-\gamma(t) = \gamma'(t)\Delta t + O(\Delta t).$$

By (1.9)

$$f(z+h)-f(z) = f'(z)h + O(h).$$

Substituting $z = \gamma(t)$ and $h = \gamma(t+\Delta t) - \gamma(t)$ in the last expression, we get

$$f(\gamma(t+\Delta t)) - f(\gamma(t)) = f'(\gamma(t))\gamma'(t)\Delta t + O(\Delta t).$$

This proves (1.28). Since $f'(\gamma(t))\gamma'(t)$ is obviously a continuous function of t, $f(\gamma(t))$ is a continuously differentiable function.

Let $\gamma\colon t \to \gamma(t)$, $a \leqq t \leqq b$, be a curve in $\mathbb{C}$ and let D be a region in $\mathbb{C}$. If $\gamma(t) \in D$ for all $t \in [a, b]$, we say that γ is in D and we write $\gamma \subset D$. Let $f(z)$ be a continuous function on a region D. A holomorphic function $F(z)$, defined on D and satisfying $F'(z) = f(z)$, is called a *primitive function* of $f(z)$ on D. If, for example, $f(z)$ is given by

$$f(z) = \sum_{n=0}^{\infty} a_n (z-c)^n,$$

where the power series on the right converges on $U_r(c)$ for some r with $0 < r \leqq +\infty$, then, by Theorem 1.11,

$$F(z) = \sum_{n=0}^{\infty} \frac{a_n}{n+1} (z-c)^{n+1}$$

is a primitive function of $f(z)$ on $U_r(c)$.

If $F(z)$ is a primitive function of $f(z)$ on D and if $\gamma\colon t \to \gamma(t)$, $a \leqq t \leqq b$, is a curve of class C^1 in D, we have by (1.28)

$$\frac{d}{dt} F(\gamma(t)) = f(\gamma(t))\gamma'(t).$$

Writing $F(\gamma(t)) = U(t) + iV(t)$, we get

$$\frac{d}{dt} U(t) + i \frac{d}{dt} V(t) = f(\gamma(t))\gamma'(t).$$

Integrating both sides from a to b, we get

$$F(\gamma(b)) - F(\gamma(a)) = \int_a^b f(\gamma(t))\gamma'(t)\,dt. \tag{1.29}$$

This formula suggests the following definition.

Definition 1.6. Let $\gamma\colon t \to \gamma(t)$, $a \leqq t \leqq b$, be a curve of class C^1 on D and let $f(z)$ be a continuous function of z on D. The integral $\int_a^b f(\gamma(t))\gamma'(t)dt$ is called the *integral of $f(z)$ along* γ and denoted by $\int_\gamma f(z)dz$. Thus

$$\int_\gamma f(z)dz = \int_a^b f(\gamma(t))\gamma'(t)\,dt.$$

The curve γ is called the *path of integration* of this integral.

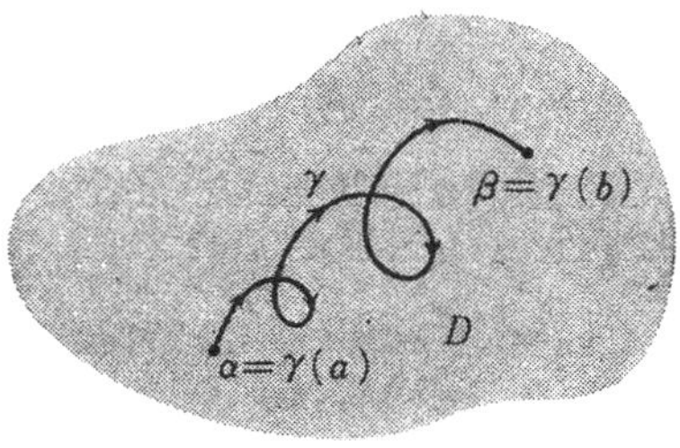

Fig. 1.15

If a primitive function $F(z)$ of $f(z)$ exists on D, then, by (1.29), the integral of $f(z)$ along γ equals the difference of the values of $F(z)$ at $\alpha = \gamma(a)$ and $\beta = \gamma(b)$:

$$\int_\gamma f(z)dz = F(\beta) - F(\alpha). \tag{1.30}$$

Therefore, in this case, the value of $\int_\gamma f(z)dz$ depends only on the points $\alpha = \gamma(a)$ and $\beta = \gamma(b)$ and is independent of the choice of curve connecting α and β on D. In particular, if γ is a closed curve, then $\int_\gamma f(z)dz = 0$.

If $f(z)$ is continuous on D, the following inequality:

$$\left|\int_\gamma f(z)dz\right| \leqq \int_a^b |f(\gamma(t))|\,|\gamma'(t)|dt \tag{1.31}$$

is an obvious consequence of (1.27).

Similarly, if $f(z)$ and $g(z)$ are both continuous on D and if c_1 and c_2 are constants, the following formula:

$$\int_\gamma (c_1 f(z) + c_2 g(z))dz = c_1 \int_\gamma f(z)dz + c_2 \int_\gamma g(z)dz. \tag{1.32}$$

is a direct consequence of (1.28).

An integral $\int_\gamma f(z)dz$ along a curve γ of class C^1 can be obtained as a limit of finite sums $\sum_{k=1}^m f(\zeta_k)(z_k - z_{k-1})$. The details of this procedure are as follows. We choose $m+1$ points $t_0, t_1, t_2, \ldots, t_m$ in $[a, b]$ such that $a = t_0 < t_1 < t_2 < \cdots < t_m = b$ and we put $\Delta = \{t_0, t_1, t_2, \ldots, t_m\}$. The interval $[a, b]$ is divided into m intervals $[t_{k-1}, t_k]$, $k = 1, 2, \ldots, m$, by these $m+1$ points. We call Δ a *partition* and we denote the maximal length of the intervals $[t_{k-1}, t_k]$ by $\delta[\Delta]$:

$$\delta[\Delta] = \max_k (t_k - t_{k-1}).$$

For each k we pick $\tau_k \in [t_{k-1}, t_k]$ and putting $\zeta_k = \gamma(\tau_k)$ and $z_k = \gamma(t_k)$ we find a finite sequence of points on γ

$$z_0 = \gamma(a), \qquad \zeta_1, z_1, \zeta_2, z_2, \ldots, z_{k-1}, \zeta_k, z_k, \ldots, \zeta_m, z_m = \gamma(b),$$

which are ordered according to increasing values of t. This sequence determines a finite sum:

$$\sigma_\Delta = \sum_{k=1}^{m} f(\zeta_k)(z_k - z_{k-1}).$$

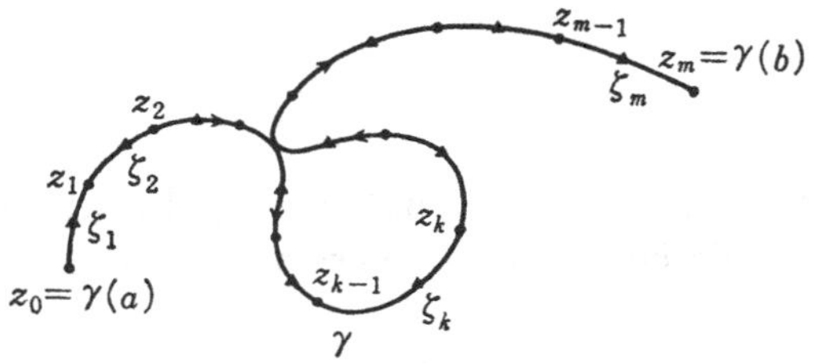

Fig. 1.16

Now, the integral $\int_\gamma f(z)\,dz$ equals the limit of all these sums σ_Δ, where we take all possible partitions Δ of $[a, b]$ and let $\delta[\Delta]$ tend to 0:

$$\int_\gamma f(z)\,dz = \lim_{\delta[\Delta]\to 0} \sum_{k=1}^{m} f(\zeta_k)(z_k - z_{k-1}). \tag{1.33}$$

For, since $f(\gamma(t))$ is defined on the closed interval $[a, b]$, it is a uniformly continuous function of t, that is, for each $\varepsilon > 0$, there exists a $\delta(\varepsilon) > 0$ such that

$$|f(\gamma(t)) - f(\gamma(s))| < \varepsilon \quad \text{if } |t - s| < \delta(\varepsilon) \quad \text{and } t, s \in [a, b].$$

Since

$$\int_\gamma f(z)\,dz = \int_a^b f(\gamma(t))\gamma'(t)\,dt = \sum_{k=1}^{m} \int_{t_{k-1}}^{t_k} f(\gamma(t))\gamma'(t)\,dt,$$

and since from

$$z_k - z_{k-1} = \gamma(t_k) - \gamma(t_{k-1}) = \int_{t_{k-1}}^{t_k} \gamma'(t)\,dt$$

we have

$$f(\zeta_k)(z_k - z_{k-1}) = \int_{t_{k-1}}^{t_k} f(\zeta_k)\gamma'(t)\,dt = \int_{t_{k-1}}^{t_k} f(\gamma(\tau_k))\gamma'(t)\,dt,$$

we deduce

$$\int_\gamma f(z)\,dz - \sum_{k=1}^{m} f(\zeta_k)(z_k - z_{k-1}) = \sum_{k=1}^{m} \int_{t_{k-1}}^{t_k} (f(\gamma(t)) - f(\gamma(\tau_k)))\gamma'(t)\,dt.$$

If Δ is a partition such that $\delta[\Delta] < \delta(\varepsilon)$, then for all $t \in [t_{k-1}, t_k]$ we have $|t - \tau_k| < \delta(\varepsilon)$. Hence $|f(\gamma(t)) - f(\gamma(\tau_k))| < \varepsilon$. Using (1.31) we have

$$\left|\int_\gamma f(z)dz - \sum_{k=1}^{m} f(\zeta_k)(z_k - z_{k-1})\right| \leqq \sum_{k=1}^{m} \left|\int_{t_{k-1}}^{t_k} (f(\gamma(t)) - f(\gamma(\tau_k)))\gamma'(t)\,dt\right|$$
$$\leqq \sum_{k=1}^{m} \int_{t_{k-1}}^{t_k} |f(\gamma(t)) - f(\gamma(\tau_k))|\,|\gamma'(t)|\,dt$$
$$\leqq \sum_{k=1}^{m} \int_{t_{k-1}}^{t_k} \varepsilon|\gamma'(t)|\,dt = \varepsilon \int_a^b |\gamma'(t)|\,dt.$$

This proves (1.33).

The parameter t of the curve $\gamma: t \to \gamma(t)$ seems to be absent from the sums $\sum_{k=1}^{m} f(\zeta_k)(z_k - z_{k-1})$ occurring on the right-hand side of (1.33). This is not true, since the order of the points $z_0, \zeta_1, z_1, \zeta_2, z_2, \ldots, z_{k-1}, \zeta_k, z_k, \ldots, z_{m-1}, \zeta_m, z_m$, which is determined by the parametrization, is essential.

c. *Cauchy's integral formula for circles*

Let c be a point of complex plane $\mathbb{C}$ and let r be such that $0 < r < +\infty$, *then the interior* $U_r(c) = \{z:|z-c| < r\}$ of the circle with center c and radius r is called the *disk with center c and radius r*. The closure $[U_r(c)] = \{z:|z-c| \leqq r\}$ is called the *closed disk with center c and radius r*.

Let $f(z)$ be a holomorphic function on the region D, c a point of D, and $U_r(c)$ a disk with center c and radius r, $0 < r < +\infty$, satisfying $[U_r(c)] \subset D$. Under these circumstances the following theorem holds.

Theorem 1.13. Let $\gamma: \theta \to \gamma(\theta) = c + re^{i\theta}$, $0 \leqq \theta \leqq 2\pi$, be the circle with center c and radius r. If w is a point of $U_r(c)$, then the value $f(w)$ is given by

$$f(w) = \frac{1}{2\pi i}\int_\gamma \frac{f(z)}{z-w}\,dz. \tag{1.34}$$

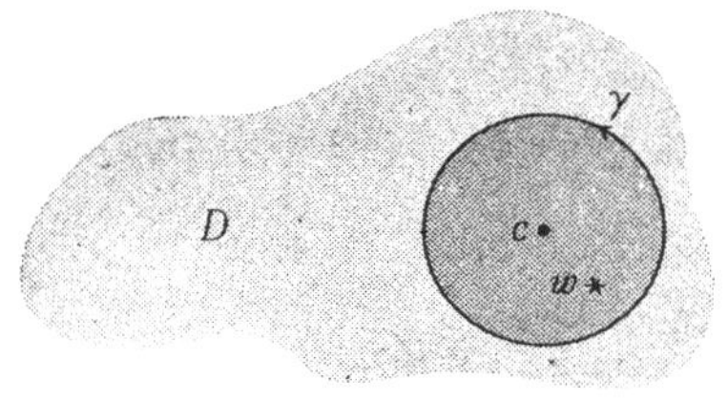

Fig. 1.17

Formula (1.34) is a special case of Cauchy's integral formula, which we will prove in Section 2.3.

Putting $w = c$ in (1.34) and using $\gamma(\theta) - c = re^{i\theta}$ and $\gamma'(\theta) = rie^{i\theta}$ we get

$$f(c) = \frac{1}{2\pi i}\int_\gamma \frac{f(z)}{z-c}dz = \frac{1}{2\pi i}\int_0^{2\pi}\frac{f(\gamma(\theta))}{\gamma(\theta)-c}\gamma'(\theta)d\theta = \frac{1}{2\pi}\int_0^{2\pi} f(\gamma(\theta))d\theta.$$

Hence

$$f(c) = \frac{1}{2\pi}\int_0^{2\pi} f(c + re^{i\theta})d\theta. \tag{1.35}$$

Formula (1.35) shows that the value $f(c)$ of a holomorphic function $f(z)$ at a point c is equal to the average of the values $f(c + re^{i\theta})$ taken on the points of a circle with center c. Therefore, (1.35) is called the Mean Value Theorem.

It is also easy to prove the Mean Value Theorem directly. Putting

$$\mu(r) = \frac{1}{2\pi}\int_0^{2\pi} f(c + re^{i\theta})d\theta$$

we have

$$\mu(r) - f(c) = \frac{1}{2\pi}\int_0^{2\pi} (f(c + re^{i\theta}) - f(c))d\theta.$$

If $r \to +0$, $|f(c + re^{i\theta}) - f(c)|$ tends to 0 uniformly with respect to θ and so

$$|\mu(r) - f(c)| \leqq \frac{1}{2\pi}\int_0^{2\pi} |f(c + re^{i\theta}) - f(c)|d\theta \to 0, \qquad r \to +0.$$

In other words,

$$\lim_{r \to +0} \mu(r) = f(c).$$

Therefore, in order to prove (1.35) it is sufficient to prove that $\mu(r)$ is a constant, independent of r, for sufficiently small values of r.

Applying the chain rule, (1.28), gives

$$\frac{\partial}{\partial r} f(c + re^{i\theta}) = f'(c + re^{i\theta})e^{i\theta},$$

$$\frac{\partial}{\partial \theta} f(c + re^{i\theta}) = f'(c + re^{i\theta})rie^{i\theta}.$$

Hence

$$\frac{\partial}{\partial r} f(c + re^{i\theta}) = \frac{1}{ir}\frac{\partial}{\partial \theta} f(c + re^{i\theta}), \qquad r > 0. \tag{1.36}$$

Since, by assumption, $f(z)$ is holomorphic, $f'(z)$ is continuous, hence

$\partial f(c+re^{i\theta})/\partial r = f'(c+re^{i\theta})e^{i\theta}$ is a continuous function of the two variables θ and r. Therefore, $\mu(r)$ is a differentiable function of r and its derivative can be found by differentiating "under the integral sign":

$$2\pi \frac{d}{dr}\mu(r) = \frac{d}{dr}\int_0^{2\pi} f(c+re^{i\theta})d\theta = \int_0^{2\pi} \frac{\partial}{\partial r} f(c+re^{i\theta})d\theta$$

$$= \frac{1}{ir}\int_0^{2\pi} \frac{\partial}{\partial \theta} f(c+re^{i\theta})d\theta = \frac{1}{ir}[f(c+re^{2\pi i}) - f(c+r)] = 0.$$

This proves that $\mu(r)$ is independent of r.

The vital part of this proof is (1.36), which gives a relation between the partial derivatives of $f(c+re^{i\theta})$ with respect to r and θ.

Now we return to the proof of the integral formula (1.34). Let $\gamma_\varepsilon : \theta \to \gamma_\varepsilon(\theta) = w + \varepsilon e^{i\theta}$, $0 \leqq \theta \leqq 2\pi$, represent a circle with center w and radius ε satisfying $0 < \varepsilon < r - |w-c|$. From

$$\frac{1}{2\pi i}\int_{\gamma_\varepsilon} \frac{f(z)}{z-w}dz = \frac{1}{2\pi i}\int_0^{2\pi} \frac{f(\gamma_\varepsilon(\theta))}{\gamma_\varepsilon(\theta)-w}\gamma_\varepsilon'(\theta)d\theta = \frac{1}{2\pi}\int_0^{2\pi} f(w+\varepsilon e^{i\theta})d\theta$$

we get

$$\lim_{\varepsilon\to 0}\frac{1}{2\pi i}\int_{\gamma_\varepsilon} \frac{f(z)}{z-w}dz = \lim_{\varepsilon\to 0}\frac{1}{2\pi}\int_0^{2\pi} f(w+\varepsilon e^{i\theta})d\theta = f(w).$$

(Of course, $1/2\pi \int_0^{2\pi} f(w+\varepsilon e^{i\theta})d\theta = f(w)$ by the Mean Value Theorem, but we do not need that result in the present proof.)

Therefore, in order to prove (1.34) it suffices to prove

$$\frac{1}{2\pi i}\int_{\gamma_\varepsilon} \frac{f(z)}{z-w}dz = \frac{1}{2\pi i}\int_{\gamma} \frac{f(z)}{z-w}dz. \tag{1.37}$$

We first need a lemma.

Lemma 1.1. Let $f(z)$ be a holomorphic function on a region D and let $\gamma_0 : t \to \gamma_0(t)$ and $\gamma_1 : t \to \gamma_1(t)$, $a \leqq t \leqq b$, be closed curves of class C^1 in D. If, for all $t \in [a,b]$, the segments L_t connecting $\gamma_0(t)$ and $\gamma_1(t)$, represented by

$$L_t = \{(1-s)\gamma_0(t) + s\gamma_1(t) : 0 \leqq s \leqq 1\}$$

are in D, then

$$\int_{\gamma_0} f(z)dz = \int_{\gamma_1} f(z)dz \tag{1.38}$$

holds.

Proof: Putting

$$\gamma_s(t) = (1-s)\gamma_0(t) + s\gamma_1(t), \qquad a \leqq t \leqq b$$

for each $s \in [0, 1]$, the $\gamma_s(t)$ are continuously differentiable, such that $\gamma_s(t) \in L_t \subset D$. Since γ_0 and γ_1 are closed curves, $\gamma_0(a) = \gamma_0(b)$ and $\gamma_1(a) = \gamma_1(b)$, hence $\gamma_s(a) = \gamma_s(b)$ for all s. (The curves $\gamma_s : t \to \gamma_s(t)$, $a \leqq t \leqq b$, are a family of closed curves in D. If s moves from 0 to 1 the curve γ_0 is gradually "transformed" via the curves γ_s into the curve γ_1.) It suffices to prove that $\int_{\gamma_s}(z)dz$ is independent of s, and to prove this it is sufficient to prove that

$$\frac{d}{ds}\int_{\gamma_s} f(z)dz = 0.$$

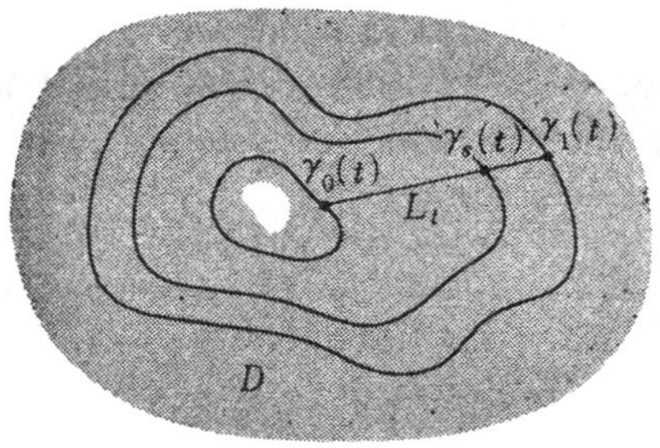

Fig. 1.18

Putting

$$\Gamma(t, s) = \gamma_s(t) = (1-s)\gamma_0(t) + s\gamma_1(t) \tag{1.39}$$

we see that $\Gamma(t, s)$ is a function of the two variables t and s defined on the rectangle $K = \{(t, s): a \leqq t \leqq b, 0 \leqq s \leqq 1\}$ with continuous partial derivatives on K:

$$\Gamma_t(t, s) = (1-s)\gamma_0'(t) + s\gamma_1'(t),$$
$$\Gamma_s(t, s) = \gamma_1(t) - \gamma_0(t),$$
$$\Gamma_{ts}(t, s) = \Gamma_{st}(t, s) = \gamma_1'(t) - \gamma_0'(t).$$

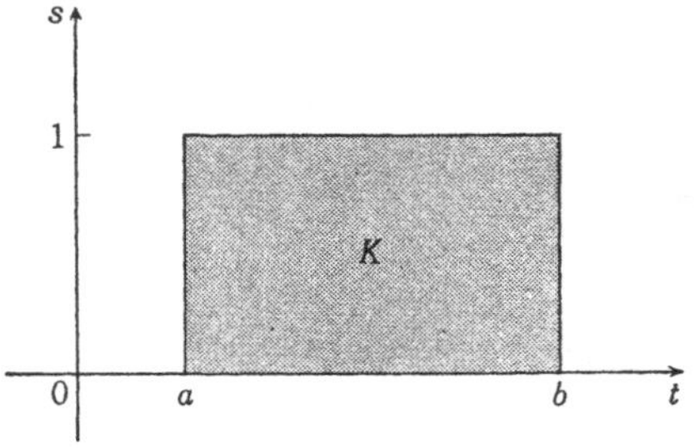

Fig. 1.19

For fixed s, $\Gamma(t, s)$ is a function of t defined on the closed interval $[a, b]$ and the derivative $\Gamma_t(t, s)$ of this function is also defined at the end points (a, s) and (b, s). Similarly, fixing t, the derivative $\Gamma_s(t, s)$ is also defined at the points $(t, 0)$ and $(t, 1)$. The same is true for $\Gamma_{ts}(t, s)$ and $\Gamma_{st}(t, s)$. From $\gamma_s(b) = \gamma_s(a)$ we have

$$\Gamma(b, s) = \Gamma(a, s). \tag{1.40}$$

Since

$$\int_{\gamma_s} f(z)dz = \int_a^b f(\gamma_s(t))\gamma_s'(t)dt = \int_a^b f(\Gamma(t, s))\Gamma_t(t, s)dt,$$

in order to prove that $\int_{\gamma_s} f(z)dz$ is independent of s, it suffices to prove

$$\frac{d}{ds}\int_a^b f(\Gamma(t, s))\Gamma_t(t, s)dt = 0.$$

To do this, we will establish a relation between the partial derivatives with regard to t and s, which will play a similar role as (1.36) in the proof of the Mean Value Theorem.

Since $\Gamma(t, s)$ is continuously differentiable with regard to s, $f(\Gamma(t, s))$ is also continuously differentiable with regard to s, by Theorem 1.12, and the derivative is given by

$$\frac{\partial}{\partial s} f(\Gamma(t, s)) = f'(\Gamma(t, s))\Gamma_s(t, s).$$

Therefore

$$\frac{\partial}{\partial s}(f(\Gamma)\Gamma_t) = f'(\Gamma)\Gamma_s\Gamma_t + f(\Gamma)\Gamma_{ts}, \tag{1.41}$$

where we have written Γ for $\Gamma(t, s)$. Writing

$$f(\Gamma(t, s))\Gamma_t(t, s) = U(t, s) + iV(t, s)$$

and observing that the right-hand side of (1.41) is a continuous function of t and s defined on K, we conclude that $U_s(t, s)$ and $V_s(t, s)$ are continuous on K.

"Differentiating under the integral sign" yields

$$\frac{d}{ds}\int_a^b f(\Gamma(t, s))\Gamma_t(t, s)dt = \int_a^b \frac{\partial}{\partial s}(f(\Gamma(t, s))\Gamma_t(t, s))dt.$$

On the other hand

$$\frac{\partial}{\partial t}(f(\Gamma)\Gamma_s) = f'(\Gamma)\Gamma_t\Gamma_s + f(\Gamma)\Gamma_{st}.$$

Hence

$$\frac{\partial}{\partial s}(f(\Gamma(t,s))\Gamma_t(t,s)) = \frac{\partial}{\partial t}(f(\Gamma(t,s))\Gamma_s(t,s)) \tag{1.42}$$

and therefore

$$\int_a^b \frac{\partial}{\partial s}(f(\Gamma(t,s))\Gamma_t(t,s))dt = \int_a^b \frac{\partial}{\partial t}(f(\Gamma(t,s))\Gamma_s(t,s))dt$$
$$= f(\Gamma(b,s))\Gamma_s(b,s) - f(\Gamma(a,s))\Gamma_s(a,s),$$

and finally

$$\frac{d}{ds}\int_{\gamma_s} f(z)dz = f(\Gamma(b,s))\Gamma_s(b,s) - f(\Gamma(a,s))\Gamma_s(a,s). \tag{1.43}$$

$\Gamma(b,s) = \Gamma(a,s)$ by (1.40), hence $\Gamma_s(b,s) = \Gamma_s(a,s)$. Thus, the right-hand side of (1.43) equals zero, which proves (1.38).

Proof of Theorem 1.13: It suffices to prove the equality (1.37):

$$\frac{1}{2\pi i}\int_{\gamma_\varepsilon} \frac{f(z)}{z-w}dz = \frac{1}{2\pi i}\int_{\gamma} \frac{f(z)}{z-w}dz.$$

By Theorem 1.5, $f(z)/(z-w)$ is a holomorphic function of z on the region $D-\{w\}$. The circles γ_ε and γ are both in $D-\{w\}$ and the line segment L_θ connecting $\gamma_\varepsilon(\theta)$ and $\gamma(\theta)$ is in $D-\{w\}$ for all θ. This is because the distance d between a point $(1-s)\gamma_\varepsilon(\theta)+s\gamma(\theta)$, $0 \leqq s \leqq 1$, on L_θ and w satisfies

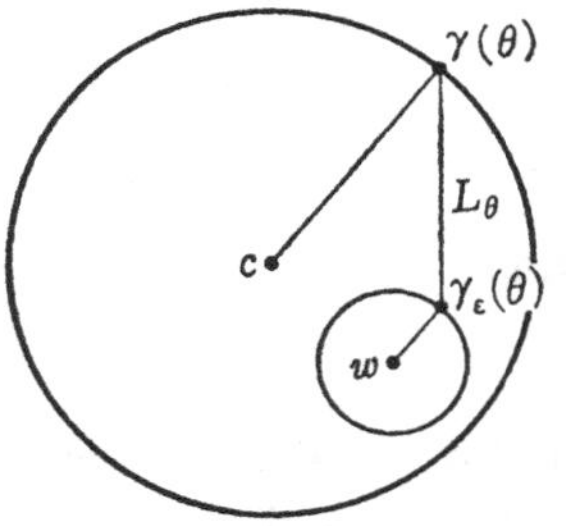

Fig. 1.20

$$d = |(1-s)\gamma_\varepsilon(\theta)+s\gamma(\theta)-w| = |(sr-s\varepsilon+\varepsilon)e^{i\theta}-s(w-c)|$$
$$\geqq sr - s\varepsilon + \varepsilon - s|w-c|$$
$$= s(r-|w-c|-\varepsilon)+\varepsilon \geqq \varepsilon.$$

Equality (1.37) is now a direct consequence of Lemma 1.1.

If the curve γ_0 of Lemma 1.1 is defined by $\gamma_0(t) = c$, where c is a constant, $a \leqq t \leqq b$, then $\gamma_0'(t) = 0$, hence $\int_{\gamma_0} f(z)dz = 0$. This gives the following corollary to Lemma 1.1.

Corollary. Let $f(z)$ be holomorphic on the region D, $\gamma: t \to \gamma(t)$, $a \leqq t \leqq b$, a closed curve of class C^1 in D, and c a point in D. If all segments connecting c and $\gamma(t)$, $a \leqq t \leqq b$, are in D then

$$\int_\gamma f(z)dz = 0.$$

In the proof of Lemma 1.1 we did not use the explicit formula (1.39) for $\Gamma(t, s)$. We used the existence and continuity of the partial derivatives $\Gamma_t(t, s)$, $\Gamma_s(t, s)$, and $\Gamma_{ts}(t, s) = \Gamma_{st}(t, s)$ and equality (1.40) asserting that $\Gamma(b, s) = \Gamma(a, s)$.

Therefore, let us assume that there exists a continuous map Γ: $(t, s) \to \Gamma(t, s)$ defined on the rectangle $K = \{(t, s): a \leqq t \leqq b, 0 \leqq s \leqq 1\}$, taking values in D such that the partial derivatives $\Gamma_t(t, s)$, $\Gamma_s(t, s)$ and $\Gamma_{ts}(t, s) = \Gamma_{st}(t, s)$ exist and are continuous and let us apply the same reasoning as above to this map $\Gamma(t, s)$ and the holomorphic function $f(z)$ defined on D.

Putting

$$\gamma_s(t) = \Gamma(t, s) \qquad \text{for} \qquad s \in [0, 1]$$

$\gamma_s: t \to \gamma_s(t)$, $a \leqq t \leqq b$ is a curve of class C^1 in D for each $t \in [0, 1]$. Since we did not assume that the equality $\Gamma(b, s) = \Gamma(a, s)$ holds, the curves γ_s are not necessarily closed.

Up to equality (1.43)

$$\frac{d}{ds}\int_{\gamma_s} f(z)dz = f(\Gamma(b, s))\Gamma_s(b, s) - f(\Gamma(a, s))\Gamma_s(a, s)$$

the above proof can be repeated without any changes.

Now defining

$${}_t\gamma(s) = \Gamma(t, s) \qquad \text{for all } t \in [a, b]$$

${}_t\gamma: s \to {}_t\gamma(s)$, $0 \leqq s \leqq 1$, is a curve of class C^1 in D for all $t \in [a, b]$ and

$$\int_{{}_t\gamma} f(z)dz = \int_0^1 f(\Gamma(t, s))\Gamma_s(t, s)ds.$$

Integrating both sides of the equality with regard to s from 0 to 1 and using the last equality, we get

$$\int_{\gamma_1} f(z)dz - \int_{\gamma_0} f(z)dz = \int_{{}_b\gamma} f(z)dz - \int_{{}_a\gamma} f(z)dz.$$

So, we have proved:

Theorem 1.14. Let $f(z)$ be a holomorphic function on a region D and let Γ: $(t, s) \to \Gamma(t, s)$ be a continuous map defined on the rectangle $K = \{(t, s)$: $a \leqq t \leqq b, 0 \leqq s \leqq 1\}$ taking values in D, such that the partial derivatives $\Gamma_t(t, s)$, $\Gamma_s(t, s)$, and $\Gamma_{ts}(t, s) = \Gamma_{st}(t, s)$ exist and are continuous on K. Putting γ_s: $t \to \gamma_s(t) = \Gamma(t, s)$, $a \leqq t \leqq b$, and ${}_t\gamma$: $s \to {}_t\gamma(s) = \Gamma(s, t)$, $0 \leqq s \leqq 1$, the equality

$$\int_{\gamma_1} f(z)dz - \int_{\gamma_0} f(z)dz = \int_{{}_b\gamma} f(z)dz - \int_{{}_a\gamma} f(z)dz \tag{1.44}$$

is valid.

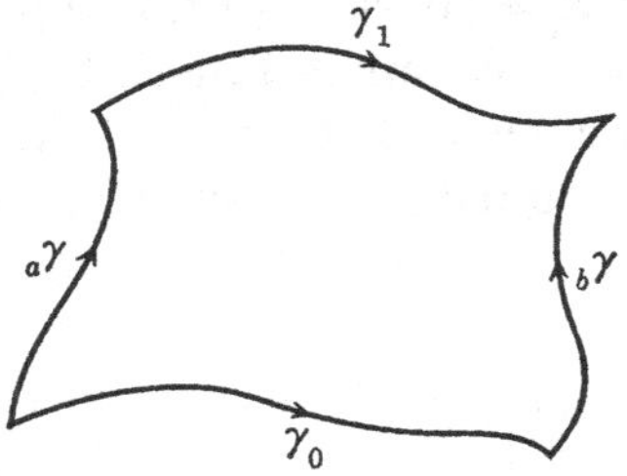

Fig. 1.21

Theorem 1.14 is a special case of Cauchy's Theorem, which will be proved in Section 2.3. The vital part of the proof of Theorem 1.14 is equality (1.42):

$$\frac{\partial}{\partial s}(f(\Gamma(t, s))\Gamma_t(t, s)) = \frac{\partial}{\partial t}(f(\Gamma(t, s))\Gamma_s(t, s)).$$

Actually, this equality is a generalization of the Cauchy–Riemann equations, (1.10). To see this, replace t and s by x and y and put $\Gamma(x, y) = x + iy$. Since $\Gamma_x = 1$ and $\Gamma_y = i$, (1.42) becomes

$$\frac{\partial}{\partial y} f(x+iy) = i\frac{\partial}{\partial x} f(x+iy).$$

Writing $f(x+iy) = u + iv$, we get

$$u_y + iv_y = i(u_x + iv_x),$$

from which the Cauchy–Riemann equations, $u_x = v_y$ and $u_y = -v_x$, follow directly.

By integrating both sides of the generalized Cauchy–Riemann equations over the rectangle K,

$$\iint_K \frac{\partial}{\partial s}(f(\Gamma(t,s))\Gamma_t(t,s))dt\,ds = \iint_K \frac{\partial}{\partial t}(f(\Gamma(t,s))\Gamma_s(t,s))dt\,ds,$$

we arrive at Cauchy's Theorem (1.44).

d. Power series expansions

Let $f(z), f_1(z), f_2(z), \ldots, f_n(z), \ldots$ be continuous functions of z defined on a region D and let $\gamma: t \to \gamma(t)$, $a \leqq t \leqq b$, be a curve in D. If the sequence of functions

$$|f_n(\gamma(t)) - f(\gamma(t))|$$

tends uniformly to 0 on $[a, b]$ if $n \to \infty$, we say that the sequence of functions $\{f_n(z)\}$ *converges uniformly* to $f(z)$ on γ. Similarly, if the sequence

$$\left|\sum_{n=1}^{m} f_n(\gamma(t)) - f(\gamma(t))\right|$$

tends uniformly to 0 on $[a, b]$ as $m \to \infty$, we say that the series $\sum_{n=1}^{\infty} f_n(z)$ *converges uniformly* to $f(z)$ on γ. If for all points $\gamma(t)$ of γ the series $\sum_{n=1}^{\infty} f_n(\gamma(t))$ converges to $f(\gamma(t))$ and if $\sum_{n=1}^{\infty} |f_n(\gamma(t))|$ converges uniformly on $[a, b]$, then $\sum_{n=1}^{\infty} f_n(z)$ *converges uniformly* to $f(z)$ on γ. This follows from

$$\left|\sum_{n=1}^{m} f_n(\gamma(t)) - f(\gamma(t))\right| = \left|\sum_{n=m+1}^{\infty} f_n(\gamma(t))\right| \leqq \sum_{n=m+1}^{\infty} |f_n(\gamma(t))|.$$

If $\sum_{n=1}^{\infty} |f_n(\gamma(t))|$ converges uniformly on $[a, b]$, then $\sum_{n=1}^{\infty} f_n(z)$ is said to *converge uniformly absolutely* on γ.

Theorem 1.15. Let γ be a curve of class C^1.

(1) If the sequence of functions $\{f_n(z)\}$ converges uniformly to $f(z)$ on γ,

$$\int_\gamma f(z)dz = \lim_{n \to \infty} \int_\gamma f_n(z)dz.$$

(2) If the series of functions $\sum_{n=1}^{\infty} f_n(z)$ converges uniformly to $f(z)$ on γ,

$$\int_\gamma f(z)dz = \sum_{n=1}^{\infty} \int_\gamma f_n(z)dz.$$

Proof: The second part follows directly from the first part, so it suffices to prove (1). By assumption, there exists a sequence $\{\varepsilon_n\}$ satisfying: $|f_n(\gamma(t)) - f(\gamma(t))| < \varepsilon_n$ for $a \leqq t \leqq b$ and $\lim_{n\to\infty} \varepsilon_n = 0$.

Using (1.31) we have

$$\left| \int_\gamma f_n(z)dz - \int_\gamma f(z)dz \right| = \left| \int_a^b (f_n(\gamma(t)) - f(\gamma(t)))\gamma'(t)dt \right|$$
$$\leqq \int_a^b |f_n(\gamma(t)) - f(\gamma(t))|\,|\gamma'(t)|dt$$
$$\leqq \varepsilon_n \int_a^b |\gamma'(t)|dt \to 0, \qquad n \to \infty.$$

Hence

$$\lim_{n\to\infty} \int_\gamma f_n(z)dz = \int_\gamma f(z)dz.$$

Let D be a region in $\mathbb{C}$ and let $c \in D$. We want to consider disks $U_r(c)$ with center c and radius r. First let us assume that $D \neq \mathbb{C}$. Then there exists a maximum value of r for which $U_r(c) \subset D$, that is, there exists an $r(c)$ satisfying

$$U_{r(c)}(c) \subset D, \qquad [U_{r(c)}(c)] \not\subset D.$$

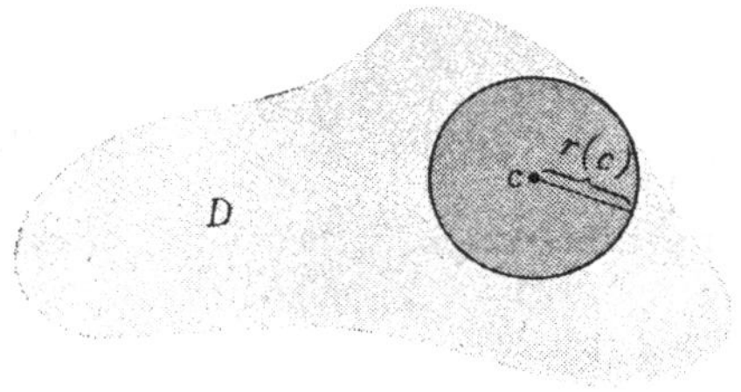

Fig. 1.22

(*Proof*: Let $r(c)$ denote the supremum of all values r for which $U_r(c) \subset D$. If $r < r(c)$, then $U_r(c) \subset D$, hence $U_{r(c)}(c) = \bigcup_{r<r(c)} U_r(c) \subset D$. Next we prove that $[U_{r(c)}(c)] \not\subset D$. If $r > r(c)$, then $U_r(c) \not\subset D$, hence there exist points $w_1, w_2, \ldots, w_n, \ldots$ such that $w_n \notin D$ and $|w_n - c| < r(c) + 1/n$. Select a convergent subsequence $\{w_{n_j}\}$, $n_1 < n_2 < \cdots < n_j < \cdots$, and denote the limit of this subsequence by w. Since D is open, $w = \lim_{j\to\infty} w_{n_j} \notin D$ and since $|w-c| = \lim_{j\to\infty}|w_{j_n}-c| \leqq r$, $w \in [U_{r(c)}(c)]$. If $D = \mathbb{C}$, we put $r(c) = +\infty$. In this case, $U_{r(c)}(c) = U_{+\infty}(c) = \mathbb{C} = D$.)

If $f(z)$ is holomorphic on D and if $r < r(c)$, then Cauchy's formula (1.34) holds with regard to the circle γ_r: $\theta \to \gamma_r(\theta) = c + re^{i\theta}$, $0 \leqq \theta \leqq 2\pi$. If we replace w by z and z by ζ in (1.34), we get

$$f(z) = \frac{1}{2\pi i}\int_{\gamma_r}\frac{f(\zeta)}{\zeta - z}\,d\zeta, \qquad z \in U_r(c). \tag{1.45}$$

Theorem 1.16. Let $f(z)$ be a holomorphic function on a region D, and let c be a point of D. Then $f(z)$ can be expanded in a power series

$$\begin{aligned} f(z) &= \sum_{n=0}^{\infty} a_n(z-c)^n \\ &= a_0 + a_1(z-c) + a_2(z-c)^2 + \cdots + a_n(z-c)^n + \cdots, \end{aligned} \tag{1.46}$$

which converges on $U_{r(c)}(c)$. The coefficients a_n are given by

$$a_n = \frac{1}{2\pi i}\int_{\gamma_r}\frac{f(\zeta)}{(\zeta - c)^{n+1}}\,d\zeta, \qquad 0 < r < r(c). \tag{1.47}$$

Proof: We may assume without loss of generality that $c = 0$. Let $z \in U_{r(0)}(0)$ and choose r such that: $|z| < r < r(0)$. By (1.45) we have

$$f(z) = \frac{1}{2\pi i}\int_{\gamma_r}\frac{f(\zeta)}{\zeta - z}\,d\zeta.$$

Since ζ is a point of the circle γ_r, we can write $\zeta = \gamma_r(\theta) = re^{i\theta}$, hence $|\zeta| = r > |z|$, and therefore $|z/\zeta| < 1$. So we can write

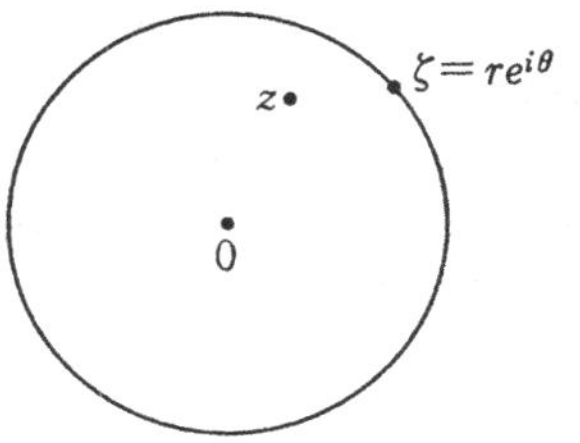

Fig. 1.23

$$\frac{1}{\zeta - z} = \frac{1}{\zeta}\cdot\frac{1}{1 - z/\zeta} = \frac{1}{\zeta}\sum_{n=0}^{\infty}\left(\frac{z}{\zeta}\right)^n = \sum_{n=0}^{\infty}\frac{z^n}{\zeta^{n+1}}.$$

Hence

$$\frac{f(\zeta)}{\zeta - z} = \sum_{n=0}^{\infty}\frac{f(\zeta)z^n}{\zeta^{n+1}}.$$

The continuous function $|f(\gamma(\theta))|$ of θ assumes a maximum on $[0, 2\pi]$. Denoting this maximum by M, we have

$$\left|\frac{f(\zeta)\, z^n}{\zeta^{n+1}}\right| \leqq M \frac{|z|^n}{r^{n+1}}, \qquad \sum_{n=1}^{\infty} \frac{|z|^n}{r^{n+1}} = \frac{1}{r-|z|} < +\infty.$$

Therefore the series $\sum_{n=0}^{\infty} (f(\zeta)/\zeta^{n+1}) z^n$ converges uniformly on the circle γ_r and thus, by Theorem 1.15(2), we have

$$\int_{\gamma_r} \frac{f(\zeta)}{\zeta - z} d\zeta = \sum_{n=0}^{\infty} \left(\int_{\gamma_r} \frac{f(\zeta)}{\zeta^{n+1}} d\zeta \right) z^n.$$

Putting $a_n = (1/2\pi i) \int_{\gamma_r} (f(\zeta)/\zeta^{n+1}) d\zeta$ we arrive at

$$f(z) = \frac{1}{2\pi i} \int_{\gamma_r} \frac{f(\zeta)}{\zeta - z} d\zeta = \sum_{n=0}^{\infty} a_n z^n. \tag{1.48}$$

The coefficients $a_n = (1/2\pi i) \int_{\gamma_r} (f(\zeta)/\zeta^{n+1}) d\zeta$ are independent of the choice of r, because we can select r and s such that $0 < r < r(0)$ and $0 < s < r(0)$. Since $f(\zeta)/\zeta^{n+1}$ is holomorphic on the region $U_{r(0)}(0) - \{0\}$ and the segments connecting $\gamma_r(\theta)$ and $\gamma_s(\theta)$ are clearly in this region, we have by Lemma 1.1

$$a_n = \frac{1}{2\pi i} \int_{\gamma_r} \frac{f(\zeta)}{\zeta^{n+1}} d\zeta = \frac{1}{2\pi i} \int_{\gamma_s} \frac{f(\zeta)}{\zeta^{n+1}} d\zeta.$$

Since z is an arbitrary point satisfying $|z| < r(0)$, the power series $\sum_{n=0}^{\infty} a_n z^n$ converges for $|z| < r(0)$. Hence, by Theorem 1. 8, the radius of convergence of $\sum_{n=0}^{\infty} a_n z^n$ is not less than $r(0)$. Therefore, $\sum_{n=0}^{\infty} a_n z^n$ converges absolutely on $U_{r(0)}(0)$.

Corollary. A function $f(z)$ defined on a region D is holomorphic if and only if for each $c \in D$ there exists a neighborhood $U_\varepsilon(c) \subset D$ on which $f(z)$ can be expanded in an absolutely convergent power series, $f(z) = \sum_{n=0}^{\infty} a_n (z-c)^n$ (ε depends on c). If $f(z)$ is holomorphic on D, $f(z)$ is arbitrarily often differentiable on D and the derivatives $f'(z), f''(z), \ldots, f^{(m)}(z), \ldots$ are all holomorphic on D.

Proof: The result follows directly from Theorems 1.16 and 1.11.

A complex function, $f(z)$, defined on a region D, that can be expanded in a power series $f(z) = \sum_{n=0}^{\infty} a_n (z-c)^n$ on a neighborhood of each point c of D is called an *analytic function* (or a *complex analytic function* if it is necessary to distinguish it from a real analytic function). By the above corollary, analytic functions are holomorphic and vice versa.

1.4 Properties of holomorphic functions

In this section we will study the fundamental properties of holomorphic functions.

a. *mth-order derivatives*

By the corollary to Theorem 1.16, a holomorphic function $f(z)$ is arbitrarily often differentiable and the derivatives $f'(z)$, $f''(z), \ldots, f^{(m)}(z), \ldots$ are all homomorphic. This was proved using the power series expansion of $f(z)$. In this section we want to deduce this fact directly from Cauchy's integral formula (1.34) and find integral formulas for the mth derivatives as well.

Lemma 1.2. Let $\gamma(t)$ and $\psi(t)$ be complex-valued continuous functions defined on the integral $I = [a, b]$.

For each natural number n

$$g(z) = \int_a^b \frac{\psi(t)}{(\gamma(t) - z)^n} dt \tag{1.49}$$

is a holomorphic function of z defined on $\mathbb{C} - \gamma(I)$ and its derivative is given by

$$g'(z) = n \int_a^b \frac{\psi(t)}{(\gamma(t) - z)^{n+1}} dt. \tag{1.50}$$

Proof: Differentiating the integrand on the right-hand side of (1.49) with regard to z yields (1.50), so we actually prove that the order of differentiation and integration can be interchanged.

If $z \neq 0$, we obviously have

$$\lim_{w \to z} \frac{1}{z - w}\left(\frac{1}{w^n} - \frac{1}{z^n}\right) = -\frac{d}{dz}\left(\frac{1}{z^n}\right) = \frac{n}{z^{n+1}}.$$

If ρ is a positive real number, the convergence of this limit is uniform for $|z| > \rho$ and $|w| > \rho$. Actually,

$$\frac{1}{z - w}\left(\frac{1}{w^n} - \frac{1}{z^n}\right) = \frac{1}{wz^n} + \frac{1}{w^2 z^{n-1}} + \cdots + \frac{1}{w^n z}.$$

Hence

$$\left|\frac{1}{w^n} - \frac{1}{z^n}\right| = |z - w| \left|\sum_{k=1}^{n} \frac{1}{w^k z^{n-k}}\right| \leqq |z - w| \frac{n}{\rho^{n+1}}.$$

Since

$$\frac{1}{z - w}\left(\frac{1}{w^n} - \frac{1}{z^n}\right) - \frac{n}{z^{n+1}} = \sum_{k=1}^{n} \left(\frac{1}{w^k} - \frac{1}{z^k}\right) \frac{1}{z^{n-k+1}},$$

we get

$$\left|\frac{1}{z-w}\left(\frac{1}{w^n}-\frac{1}{z^n}\right)-\frac{n}{z^{n+1}}\right| \leqq |z-w|\frac{n^2}{\rho^{n+2}}. \tag{1.51}$$

Fix $z \in \mathbb{C}-\gamma(I)$, and put $2\rho = \min_{a \leqq t \leqq b}|\gamma(t)-z|$, then $\rho > 0$ and $|\gamma(t) - z - h| > \rho$ for all h with $|h| < \rho$. Replacing z by $\gamma(t)-z$ and w by $\gamma(t)-z-h$ in (1.51) we get

$$\left|\frac{1}{h}\left(\frac{\psi(t)}{(\gamma(t)-z-h)^n}-\frac{\psi(t)}{(\gamma(t)-z)^n}\right)-\frac{n\psi(t)}{(\gamma(t)-z)^{n+1}}\right| \leqq \frac{|h|n^2}{\rho^{n+2}}|\psi(t)|$$

hence, using (1.27)

$$\left|\frac{g(z+h)-g(z)}{h}-n\int_a^b\frac{\psi(t)}{(\gamma(t)-z)^{n+1}}dt\right| \leqq \frac{|h|n^2}{\rho^{n+2}}\int_a^b|\psi(t)|dt.$$

Therefore

$$\lim_{h\to 0}\frac{g(z+h)-g(z)}{h} = n\int_a^b\frac{\psi(t)}{(\gamma(t)-z)^{n+1}}dt,$$

i.e., $g(z)$ is differentiable on $\mathbb{C}-\gamma(I)$ and $g'(z)$ is given by (1.50). If one replaces $\psi(t)$ by $n\psi(t)$ and n by $n+1$ in (1.49), one gets (1.50), so that $g'(z)$ is also differentiable on $\mathbb{C}-\gamma(I)$. Therefore, $g'(z)$ is certainly continuous and hence $g(z)$ is holomorphic on $\mathbb{C}-\gamma(I)$.

Theorem 1.17. Let $\gamma(t)$ and $\psi(t)$ be complex-valued continuous functions defined on the interval $[a, b]$.

The function

$$f(z) = \int_a^b\frac{\psi(t)}{\gamma(t)-z}dt$$

is a holomorphic function of z on the region $\mathbb{C}-\gamma(I)$. $f(z)$ is arbitrarily often differentiable on $\mathbb{C}-\gamma(I)$ and all its derivatives $f'(z), f''(z), \ldots,$ $f^{(m)}(z), \ldots$ are holomorphic on $\mathbb{C}-\gamma(I)$. The mth derivative is given by

$$f^{(m)}(z) = m!\int_a^b\frac{\psi(t)}{(\gamma(t)-z)^{m+1}}dt. \tag{1.52}$$

Proof: By Lemma 1.1, $f(z)$ is holomorphic on $\mathbb{C}-\gamma(I)$ and $f'(z) = \int_a^b[\psi(t)/(\gamma(t)-z)^2]dt$. Therefore, by the same lemma, $f'(z)$ is holomorphic and $f''(z) = 2!\int_a^b[\psi(t)/(\gamma(t)-z)^3]dt$. Again by the same lemma, $f'(z)$ is holomorphic and $f'''(z) = 3!\int_a^b[\psi(t)/(\gamma(t)-z)^4]dt$, and so on.

Let $f(z)$ be a holomorphic function on a region $D \subset \mathbb{C}$ and let $c \in D$. Let $U_r(c)$, $0 < r < +\infty$, be a disk such that $[U_r(c)] \subset D$ and let γ_r be defined by

$\gamma_r: \theta \to \gamma_r(\theta) = c + re^{i\theta}, 0 \leqq \theta \leqq 2\pi$. By Cauchy's integral formula (1.45) we have for $z \in U_r(c)$

$$f(z) = \frac{1}{2\pi i}\int_{\gamma_r} \frac{f(\zeta)}{\zeta - z}\,d\zeta = \frac{1}{2\pi i}\int_0^{2\pi} \frac{f(\gamma_r(\theta))\gamma_r'(\theta)}{\gamma_r(\theta) - z}\,d\theta.$$

Therefore, by Theorem 1.17, $f(z)$ is arbitrarily often differentiable on $U_r(c)$ and all its derivatives $f'(z), f''(z), \ldots, f^{(m)}(z), \ldots$ are holomorphic. Since $U_r(c)$ is an arbitrary disk in D, we have proved that $f(z)$ is arbitrarily often differentiable on D and that all its derivatives $f'(z)$, $f''(z), \ldots, f^{(m)}(z), \ldots$ are holomorphic.

By (1.52) we have

$$f^{(m)}(z) = \frac{m!}{2\pi i}\int_0^{2\pi} \frac{f(\gamma_r(\theta))\gamma_r'(\theta)}{(\gamma_r(\theta) - z)^{m+1}}\,d\theta, \qquad z \in U_r(c).$$

So we have proved.

Theorem 1.18. The mth derivative of $f(z)$ is given by

$$f^{(m)}(z) = \frac{m!}{2\pi i}\int_{\gamma_r} \frac{f(\zeta)}{(\zeta - z)^{m+1}}\,d\zeta \tag{1.53}$$

on $U_r(c)$.

By Theorem 1.17 the integral $\int_{\gamma_r}[f(\zeta)/(\zeta - z)^{m+1}]\,d\zeta$ on the right-hand side of (1.53) is also a holomorphic function of z on $D - [U_r(c)]$. This function is identically equal to 0 by the corollary to Lemma 1.1, since $f(\zeta)/(\zeta - z)^{m+1}$ is a holomorphic function of ζ on $D - \{z\}$ and all segments connecting c and the points of γ_r are in $D - \{z\}$.

b. *Limits of sequences of holomorphic functions*

Theorem 1.19. Let $\{f_n(z)\}$ be a sequence of holomorphic functions defined on a region D, converging uniformly on all compact subsets of D. Then $f(z) = \lim_{n\to\infty} f_n(z)$ is holomorphic on D and its mth derivative is given by

$$f^{(m)}(z) = \lim_{n\to\infty} f_n^{(m)}(z), \qquad m = 1, 2, 3, \ldots,$$

and the sequences $\{f_n^{(m)}(z)\}$ converge uniformly to $f^{(m)}(z)$ on all compact subsets of D.

Proof: Let $U_r(c)$, $0 < r < +\infty$ be an arbitrary disk such that $[U_r(c)] \subset D$. Since, by assumption, $\{f_n(z)\}$ converges uniformly on $[U_r(c)]$, $f(z) = \lim_{n\to\infty} f_n(z)$ is a continuous function of z on $[U_r(c)]$. Let γ_r be the circle with center c and radius r, then, by the integral formula (1.45), we have

$$f_n(z) = \frac{1}{2\pi i}\int_{\gamma_r} \frac{f_n(\zeta)}{\zeta - z}\, d\zeta, \qquad z \in U_r(c).$$

Since $\{f_n(\zeta)\}$ converges uniformly to $f(\zeta)$ on $[U_r(c)]$, $\{f_n(\zeta)/(\zeta - z)\}$ converges uniformly to $f(\zeta)/(\zeta - z)$ on γ_r for a fixed $z \in U_r(c)$. Therefore, by Theorem 1.15, (1)

$$\lim_{n\to\infty} \int_{\gamma_r} \frac{f_n(\zeta)}{\zeta - z}\, d\zeta = \int_{\gamma_r} \frac{f(\zeta)}{\zeta - z}\, d\zeta.$$

Hence

$$f(z) = \lim_{n\to\infty} f_n(z) = \frac{1}{2\pi i}\int_{\gamma_r} \frac{f(\zeta)}{\zeta - z}\, d\zeta, \qquad z \in U_r(0).$$

Therefore $f(z)$ is holomorphic on $U_r(c)$ by Theorem 1.17 and since $U_r(c)$ is an arbitrary disk in D, we see that $f(z)$ is holomorphic on D.

Next, by (1.53)

$$f_n^{(m)}(z) - f^{(m)}(z) = \frac{m!}{2\pi i}\int_{\gamma_r} \frac{f_n(\zeta) - f(\zeta)}{(\zeta - z)^{m+1}}\, d\zeta.$$

Since $\{f_n(\zeta)\}$ converges uniformly to $f(\zeta)$ on γ_r, there exists a sequence $\{\varepsilon_n\}$ such that $\lim_{n\to\infty} \varepsilon_n = 0$ and $|f_n(\zeta) - f(\zeta)| < \varepsilon_n$ for all $\zeta \in \gamma_r$.

Choosing z such that $|z - c| \leqq r/z$, we have $|\gamma_r(\theta) - z| \geqq r/2$. Using $\gamma_r'(\theta) = rie^{i\theta}$, we get, by (1.31),

$$|f_n^{(m)}(z) - f^{(m)}(z)| \leqq \frac{m!}{2\pi}\int_0^{2\pi} \frac{\varepsilon_n r\, d\theta}{|\gamma_r(\theta) - z|^{m+1}} \leqq \frac{m!2^{m+1}}{\gamma^m}\varepsilon_n,$$

i.e., $\{f_n^{(m)}(z)\}$ converges uniformly to $f^{(m)}(z)$ on $[U_{r/2}(c)]$. Let K be a compact set of D and choose for each $c \in K$ a closed disk $[U_r(c)] \subset D$. As shown above, $\{f_n^{(m)}(z)\}$ converges uniformly to $f^{(m)}(z)$ on all disks $[U_{r/2}(c)]$. Since K can be covered by a finite number of these disks, $\{f_n^{(m)}(z)\}$ converges uniformly to $f^{(m)}(z)$ on K.

Corollary. Let $\sum_{n=1}^{\infty} f_n(z)$ be a series of holomorphic functions defined on a region D, converging uniformly on all compact subsets of D. Then $f(z) = \sum_{n=1}^{\infty} f_n(z)$ is holomorphic on D and its mth derivative is given by

$$f^{(m)}(z) = \sum_{n=1}^{\infty} f_n^{(m)}(z), \qquad m = 1, 2, 3, \ldots$$

and the series $\sum_{n=1}^{\infty} f_n^{(m)}(z)$ converges uniformly on each compact subset of D.

Let us apply this corollary to the case of power series. Let $\sum_{n=1}^{\infty} a_n z^n$ be a power series with radius of convergence $r, 0 < r \leqq +\infty$. All terms $a_n z^n$

are holomorphic on $D = U_r(0)$ and $\sum_{n=1}^{\infty} a_n z^n$ converges uniformly on all disks $U_\rho(0)$ for $0 < \rho < r$ by Theorem 1.9. For each compact subset K of D it is possible to find a $\rho < r$ such that $K \subset U_\rho(0)$, hence $\sum_{n=1}^{\infty} a_n z^n$ converges uniformly on K. Therefore, by our corollary, the sum $\sum_{n=1}^{\infty} a_n z^n = f(z)$ is holomorphic on $D = U_r(0)$ and its mth derivative $f^{(m)}(z)$ is given by

$$f^{(m)}(z) = \sum_{n=m}^{\infty} n(n-1)(n-2)\ldots(n-m+1)a_n z^{n-m}.$$

Thus, we have given another proof of Theorem 1.11, using the corollary to Theorem 1.19.

If $\sum_{n=1}^{\infty} f_n(z)$ is a series of holomorphic functions defined on a region D and converging uniformly on all compact subsets of D, then $\sum_{n=1}^{\infty} f_n^{(m)}(z)$ is also uniformly convergent on all compact subsets of D by the above corollary. However, uniform convergence of $\sum_{n=1}^{\infty} f_n(z)$ on D does not necessarily imply uniform convergence of $\sum_{n=1}^{\infty} f_n^{(m)}(z)$ on D. For example, $\sum_{n=1}^{\infty} z^{n+1}/(n+1)n$ converges uniformly on $U_1(0) = \{z: |z| < 1\}$ (because $\sum_{n=1}^{\infty} 1/(n+1)n = \sum_{n=1}^{\infty} [1/n - 1/(n+1)] = 1$) but $\sum_{n=1}^{\infty} z^n$, obtained by differentiating the above series term by term twice, is not uniformly convergent on $U_1(0)$.

c. *The Mean Value Theorem and the maximum principle*

We have already seen in Section 1.3c that the Mean Value Theorem is a direct consequence of Cauchy's integral formula (1.34).

Theorem 1.20 (*Mean Value Theorem*). If $f(z)$ is a holomorphic function on a region D and $c \in D$ and r are such that $[U_r(c)] \subset D$, then

$$f(c) = \frac{1}{2\pi}\int_0^{2\pi} f(c + re^{i\theta})\, d\theta.$$

By taking absolute values we obtain the inequality

$$|f(c)| \leqq \frac{1}{2\pi}\int_0^{2\pi} |f(c + re^{i\theta})|\, d\theta. \tag{1.54}$$

Next, let $f(z)$ be a holomorphic function on a region D. We want to consider the real-valued function $|f(z)|$ of z, also defined on D.

Theorem 1.21 (maximum principle). Unless $f(z)$ is a constant function on D, the function $|f(z)|$ assumes no maximum on D.

Proof: Let us assume that $f(z)$ assumes a maximum value M, at, say, $z = c$. We have to show that $f(z)$ is a constant function on D. Choose r, such that

$[U_r(c)] \subset D$. Then, by (1.54),

$$\frac{1}{2\pi}\int_0^{2\pi}(M-|f(c+re^{i\theta})|)\,d\theta \leqq M-|f(c)|=0.$$

Since $M-|f(c+re^{i\theta})|$ is a continuous function of θ satisfying $M-|f(c+re^{i\theta})| \geqq 0$, we conclude that $M-|f(c+re^{i\theta})|=0$. Choose $\varepsilon>0$ such that $[U_\varepsilon(c)] \subset D$, then $|f(c+re^{i\theta})|=M$ for all $c+re^{i\theta}$ with $0<r<\varepsilon$, i.e., $|f(z)|=M$ on $U_\varepsilon(c)$. Put $V=\{z\in D: |f(z)|=M\}$. If $c\in V$, then there exists an $\varepsilon>0$ such that $U_\varepsilon(c)\subset V$, hence V is open. On the other hand, since $|f(z)|$ is a continuous function of z, V is also closed. Since D is a region, we conclude that $D=V$, i.e., $|f(z)|=M$ for all $z\in D$.

If $M=0$, we are finished, so we assume $M>0$. Writing $f(z)=f(x+iy)=u(x,y)+iv(x,y)$, we have $u^2+v^2=|f(z)|^2=M^2$ on D, hence

$$uu_x+vv_x=\tfrac{1}{2}\frac{\partial}{\partial x}(u^2+v^2)=0,$$

$$uu_y+vv_y=\tfrac{1}{2}\frac{\partial}{\partial y}(u^2+v^2)=0.$$

Combining these with the Cauchy–Riemann equations, $u_x=v_y$ and $u_y=-v_x$, we get a system of equations with "unknowns" v_y and u_y. Since $\left|\begin{smallmatrix} u & -v \\ v & u\end{smallmatrix}\right| = u^2+v^2=M^2>0$, we conclude $v_y=u_y=0$. Hence $u_x=v_x=0$ and $f(z)=u+iv$ has to be a constant.

Corollary. Let D be a bounded region and let $f(z)$ be defined and continuous on $[D]$ and holomorphic on D. Then $|f(z)|$ assumes a maximum at a point of the boundary of D.

Proof: We may assume that $f(z)$ is not constant. Since $[D]$ is compact and $|f(z)|$ is continuous on $[D]$, $|f(z)|$ assumes a maximum at a point $c\in[D]$. However, since $f(z)$ is a nonconstant holomorphic function, we have $c\notin D$, hence $c\in[D]-D$.

d. Isolated singularities

Let D be a region, $c\in D$ and $r(c)$ the number defined in Section 1.3d as the maximum real number satisfying $U_{r(c)}(c)\subset D$ if $D\neq\mathbb{C}$ and as $+\infty$ if $D=\mathbb{C}$. Let $\gamma_r\colon \theta\to\gamma_r(\theta)=c+re^{i\theta}$, $0\leqq\theta<2\pi$, as usual represent the circle with center c and radius r. In this section we consider functions that are holomorphic on $D-\{c\}$.

Theorem 1.22. A function $f(z)$ that is holomorphic on the region $D-\{c\}$ can be expanded in a power series

$$f(z)=\sum_{n=-\infty}^{\infty} a_n(z-c)^n, \tag{1.55}$$

which converges absolutely on $U_{r(c)}(c)-\{c\}$.

Its coefficients are given by

$$a_n=\frac{1}{2\pi i}\int_{\gamma_r}\frac{f(\zeta)}{(\zeta-c)^{n+1}}\,d\zeta, \qquad 0<r<r(c). \tag{1.56}$$

Remark: The power series $\sum_{n=-\infty}^{+\infty} a_n(z-c)^n$, i.e.,

$$\cdots+\frac{a_{-n}}{(z-c)^n}+\cdots+\frac{a_{-1}}{z-c}+a_0+a_1(z-c)+a_2(z-c)^2+\cdots,$$

is said *to converge absolutely* on $U_{r(c)}(c)-\{c\}$ if the series $\sum_{n=1}^{\infty} a_n(z-c)^n$ and $\sum_{n=1}^{\infty} a_{-n}/(z-c)^n$ both converge absolutely for $0<|z-c|<r(c)$. A power series of the form $\sum_{n=-\infty}^{+\infty} a_n(z-c)^n$ is called a *Laurent series.*

Proof: We may assume without loss of generality that $c=0$. Choose a point w such that $0<|w|<r(0)$ and real numbers ε and r such that $0<\varepsilon<|w|<r<r(0)$. Let γ_ε: $\theta\to\gamma_\varepsilon(\theta)=\varepsilon e^{i\theta}$ and γ_r: $\theta\to\gamma_r(\theta)=re^{i\theta}$, $0<\theta<2\pi$ represent the circles with center 0 and radii ε and, r respectively.

Putting

$$g(z)=\frac{f(z)-f(w)}{z-w},$$

$g(z)$ is a holomorphic function on $D-\{0, w\}$. By Theorem 1.16, $f(z)$ can be expanded in an absolutely converging power series in some neighborhood of w:

$$f(z)=f(w)+b_1(z-w)+b_2(z-w)^2+\cdots+b_n(z-w)^n+\cdots.$$

Therefore

$$g(z)=\frac{f(z)-f(w)}{z-w}=b_1+b_2(z-w)+\cdots+b_n(z-w)^{n-1}+\cdots.$$

Therefore, $g(z)$ can be defined at $z=w$ in such a way that $g(z)$ becomes a holomorphic function on some neighborhood of w and therefore on the whole of $D-\{0\}$.

Therefore, we can apply Lemma 1.1:

$$\int_{\gamma_\varepsilon} g(z)\,dz=\int_{\gamma_r} g(z)\,dz,$$

that is,

$$\int_{\gamma_\varepsilon} \frac{f(z)-f(w)}{z-w}\,dz = \int_{\gamma_r} \frac{f(z)-f(w)}{z-w}\,dz. \tag{1.57}$$

Applying the integral formula (1.34) to the function $f(z) = 1$, we get

$$1 = \frac{1}{2\pi i}\int_{\gamma_r} \frac{1}{z-w}\,dz,$$

while by the corollary to Lemma 1.1:

$$\int_{\gamma_\varepsilon} \frac{1}{z-w}\,dz = 0.$$

Combining these results with (1.57) we get

$$f(w) = \frac{1}{2\pi i}\int_{\gamma_r} \frac{f(z)}{z-w}\,dz - \frac{1}{2\pi i}\int_{\gamma_\varepsilon} \frac{f(z)}{z-w}\,dz.$$

Replacing z by ζ and w by z yields

$$f(z) = \frac{1}{2\pi i}\int_{\gamma_r} \frac{f(\zeta)}{\zeta - z}\,d\zeta - \frac{1}{2\pi i}\int_{\gamma_\varepsilon} \frac{f(\zeta)}{\zeta - z}\,d\zeta. \tag{1.58}$$

For the first term of the right-hand side we have by (1.48)

$$\frac{1}{2\pi i}\int_{\gamma_r} \frac{f(\zeta)}{\zeta - z}\,d\zeta = \sum_{n=0}^{\infty} a_n z^n, \qquad a_n = \frac{1}{2\pi i}\int_{\gamma_r} \frac{f(\zeta)}{\zeta^{n+1}}\,d\zeta.$$

Since the ζ occurring in the second term is a point of γ_ε, we have $|\zeta/z| = \varepsilon/|z| < 1$, hence

$$\frac{1}{\zeta - z} = -\frac{1}{z}\cdot\frac{1}{(1-\zeta/z)} = -\sum_{n=0}^{\infty} \frac{\zeta^n}{z^{n+1}},$$

and therefore

$$\frac{f(\zeta)}{\zeta - z} = -\sum_{n=0}^{\infty} \frac{\zeta^n f(\zeta)}{z^{n+1}}$$

with the power series on the right-hand side converging uniformly absolutely on γ_ε.

Hence, by Theorem 1.15 (2)

$$\frac{1}{2\pi i}\int_{\gamma_\varepsilon} \frac{f(\zeta)}{\zeta - z}\,d\zeta = -\sum_{n=0}^{\infty} \frac{a_{-n-1}}{z^{n+1}}, \qquad a_{-n-1} = \frac{1}{2\pi i}\int_{\gamma_\varepsilon} \zeta^n f(\zeta)\,d\zeta.$$

The series $\sum_{n=1}^{\infty} a_{-n-1}/z^{n+1}$ on the right-hand side converges for $|z| > \varepsilon$. Hence, by (1.58)

$$f(z) = \sum_{n=0}^{\infty} a_n z^n + \sum_{n=0}^{\infty} \frac{a_{-n-1}}{z^{n+1}} = \sum_{n=-\infty}^{\infty} a_n z^n.$$

Since $\zeta^n f(\zeta)$ is a holomorphic function of ζ on D-$\{0\}$ and since all segments connecting $\gamma_\varepsilon(\theta)$ and $\gamma_r(\theta)$ are in D-$\{0\}$, we conclude from Lemma 1.1

$$\int_{\gamma_\varepsilon} \zeta^n f(\zeta)d\zeta = \int_{\gamma_r} \zeta^n f(\zeta)d\zeta.$$

Replacing $-n-1$ by n in the expression for a_{-n-1}, we see that we can write for all indices

$$a_n = \frac{1}{2\pi i}\int_{\gamma_r} \frac{f(\zeta)}{\zeta^{n+1}}d\zeta.$$

It is clear from Lemma 1.1 that a_n is independent of the choice of the radius $r, 0 < r < r(0)$, of the path of integration γ_r. This proves that for all z with $0 < |z| < r(0)$ the series $\sum_{n=-\infty}^{\infty} a_n z^n$ converges and that

$$f(z) = \sum_{n=-\infty}^{\infty} a_n z^n.$$

The power series $\sum_{n=-\infty}^{\infty} a_n z^n$ is composed of two power series, $\sum_{n=0}^{\infty} a_n z^n$, which converges absolutely for $|z| < r(0)$ and $\sum_{n=1}^{\infty} a_{-n}(1/z)^n$, which converges absolutely for $|z| > 0$. Therefore, $\sum_{n=-\infty}^{\infty} a_n z^n$ converges absolutely for $0 < |z| < r(0)$.

The Laurent series on the right-hand side of (1.55) is called the *Laurent expansion* of $f(z)$ about c. The point c is called an *isolated singular point* of $f(z)$ and the part of the Laurent expansion with negative exponents:

$$\sum_{n=-\infty}^{-1} a_n(z-c)^n = \cdots + \frac{a_{-3}}{(z-c)^3} + \frac{a_{-2}}{(z-c)^2} + \frac{a_{-1}}{z-c}$$

is called the *principal part* of $f(z)$ at c.

We distinguish the following three cases:

Case (1): There is no principal part. In this case we have for $0 < |z-c| < r(c)$,

$$f(z) = a_0 + a_1(z-c) + a_2(z-c)^2 + \cdots.$$

By defining $f(c) = a_0$, we make $f(z)$ into a function which is also holomorphic at c. In this case, c is called a *removable singularity* of $f(z)$. Usually, one removes removable singularities from the start making $f(z)$ into a function which is holomorphic on the whole of D.

Let $f(z)$ be holomorphic on D. A point $c \in D$ such that $f(c) = 0$ is called a *zero* of $f(z)$. Consider the Taylor expansion of $f(z)$ around the zero c:

$$f(z) = a_0 + a_1(z-c) + a_2(z-c)^2 + \cdots + a_n(z-c)^n + \cdots.$$

Then $a_0 = f(c) = 0$, and it is possible that some of the other coefficients $a_1, a_2, \ldots$ are also equal to 0. If all coefficients $a_1, a_2, \ldots, a_n, \ldots$ are equal to 0, the function $f(z)$ is identically equal to 0 in a neighborhood of c (and, in fact, in D. See Theorem 3.1). Let us assume that not all coefficients are equal to 0 and that a_m is the first coefficient which is not equal to 0. Then

$$f(z) = (z-c)^m (a_m + a_{m+1}(z-c) + a_{m+2}(z-c)^2 + \cdots), \qquad a_m \neq 0.$$

In this case, m is called the *order* or the *multiplicity of the zero c*. Writing

$$f(z) = (z-c)^m g(z) \tag{1.59}$$

the function $g(z) = a_m + a_{m+1}(z-c) + a_{m+2}(z-c)^2 + \cdots$ is holomorphic on $U_{r(c)}(c)$ and $g(c) = a_m \neq 0$. Therefore, $g(z) \neq 0$ on a sufficiently small neighborhood $U_\varepsilon(c)$ of c, hence

$$f(z) = (z-c)^m g(z) \neq 0 \qquad \text{for } 0 < |z-c| < \varepsilon. \tag{1.60}$$

We have proved that there are no other zeros in a sufficiently small neighborhood of a zero of a holomorphic function. The zero c is "isolated."

If one counts the number of zeros of a function $f(z)$ in a certain region, a zero of order m is counted m times.

Case (2): The principal part has only a finite number of terms, that is, for $0 < |z-c| < r(c)$ we have

$$f(z) = \frac{a_{-m}}{(z-c)^m} + \cdots + \frac{a_{-1}}{(z-c)} + \sum_{n=0}^{+\infty} a_n (z-c)^n, \qquad a_{-m} \neq 0.$$

In this case c is called a *pole* of order m of the function $f(z)$. The function $g(z)$, defined by

$$g(z) = (z-c)^m f(z) = a_{-m} + a_{-m+1}(z-c) + a_{-m+2}(z-c)^2 + \cdots$$

is holomorphic on $U_{r(c)}(c)$ and $g(c) = a_{-m} \neq 0$. Therefore, $g(z) \neq 0$ on a sufficiently small neighborhood $U_\varepsilon(c)$ of c. From

$$f(z) = \frac{1}{(z-c)^m} g(z) \tag{1.61}$$

we conclude

$$\lim_{z \to c} |f(z)| = \frac{|g(z)|}{|z-c|^m} = +\infty.$$

In this case, we write $f(c) = \infty$. We will return to the meaning of ∞ later.

Putting $h(z) = 1/g(z)$ we have

$$\frac{1}{f(z)} = (z-c)^m h(z)$$

and $h(z)$ is holomorphic on $U_\varepsilon(c)$ and $h(z) \neq 0$ for $z \in U_\varepsilon(c)$. Therefore, the pole of the mth-order c of $f(z)$ is a zero of the mth order of $1/f(z)$. Conversely, it is easy to prove that a zero of the mth order of $1/f(z)$ is a pole of the mth order of $f(z)$.

Case (3): The principal part is an infinite series, that is, $a_{-n} \neq 0$ for infinitely many natural numbers n. In this case, c is called an *essential singularity* of $f(z)$. As the following theorem shows, the behavior of $f(z)$ in the neighborhood of an essential singularity is very complicated.

Theorem 1.23 (*Weierstrass's Theorem*). Let c be an essential singularity of $f(z)$ and let w be a complex number. It is possible to find a sequence $\{z_n\}$ of points converging to c such that $\lim_{n\to\infty} f(z_n) = w$. It is also possible to find a sequence $\{z_n\}$ converging to c such that $\lim_{n\to\infty} |f(z_n)| = +\infty$.

Proof: We first prove that if c is an isolated singularity of $f(z)$, such that $f(z)$ is bounded on some neighborhood of c, c is a removable singularity. By assumption there exist $\delta > 0$ and M such that $|f(z)| \leqq M$ for $0 < |z - c| < \delta$.

According to (1.56)

$$a_{-n} = \frac{1}{2\pi i}\int_{\gamma_r} (\zeta - c)^{n-1} f(\zeta)\,d\zeta$$

hence, for $0 < r < \delta$

$$|a_{-n}| \leqq \frac{1}{2\pi}\int_0^{2\pi} |\gamma_r(\theta) - c|^{n-1} M\,|\gamma_r'(\theta)|\,d\theta = Mr^n.$$

Since r can be made arbitrarily small, we conclude $a_{-n} = 0$, that is, c is a removable singularity. If there does not exist a sequence $\{z_n\}$ converging to c such that $|f(z_n)| \to +\infty$, $f(z)$ is bounded on some neighborhood of c. By the above, c would be a removable singularity contradicting the assumption.

If there does not exist a sequence $\{z_n\}$ converging to c such that $f(z_n) \to w$, there exist positive ε and δ such that

$$|f(z) - w| \geqq \varepsilon \qquad \text{for} \qquad 0 < |z - c| < \delta.$$

Putting $g(z) = 1/(f(z) - w)$, $g(z)$ is holomorphic for $0 < |z - c| < \delta$ and $|g(z)| \leqq 1/\varepsilon$, hence c is a removable singularity of $g(z)$, so that we may assume that $g(z)$ is holomorphic for $|z - c| < \delta$. If $g(c) \neq 0$, $f(z) = w + 1/g(z)$ is also holomorphic for $|z - c| < \delta$, contradicting the assumption. If c is an mth-order zero of $g(z)$, c is an mth-order pole of $f(z)$, also contradicting the assumption.

Example 1.6. If we define $f(z)$ by

$$f(z) = e^{1/z} = 1 + \frac{1}{z} + \frac{1}{2!z^2} + \frac{1}{3!z^3} + \cdots,$$

then $f(z)$ is holomorphic on $\mathbb{C} - \{0\}$ and 0 is an essential singularity of $f(z)$. Let $w \neq 0$ be a complex number. Putting $q = |w|e^{i\theta}$, where θ is a real number, we have $w = \exp(\log|w| + i\theta + 2n\pi i)$ for all natural numbers n. Putting $z_n = 1/(\log|w| + i\theta + 2n\pi i)$, we see that $\lim_{n\to\infty} z_n = 0$ and $\lim_{n\to\infty} f(z_n) = w$. (In this case, we actually have $f(z_n) = w$ for all z_n. We will return to this phenomenon later.)

e. Entire functions

A function holomorphic on the whole complex plane $\mathbb{C}$ is called an *entire function*. If $f(z)$ is an entire function, then, by Theorem 1.16, $f(z)$ can be expanded in a power series

$$f(z) = a_0 + a_1 z + a_2 z^2 + \cdots + a_n z^n + \cdots$$

which converges absolutely on the whole complex plance $\mathbb{C}$. We distinguish the following three cases:

(1) $f(z) = a_0$,

(2) $f(z) = a_0 + a_1 z + \cdots + a_m z^m, \quad a_m \neq 0,\ m \geqq 1,$

(3) $f(z) = \sum_{n=0}^{\infty} a_n z^n \qquad a_n \neq 0$ for infinitely many n.

The first case is trivial.

In the second case $f(z)$ is a polynomial of degree m. From

$$|f(z)| \geqq |z|^m \left(|a_m| - \frac{|a_{m-1}|}{|z|} - \frac{|a_{m-2}|}{|z^2|} - \cdots - \frac{|a_0|}{|z|^m} \right),$$

we conclude that there exists an $r_0 > 0$ such that

$$|f(z)| > \frac{|a_m||z|^m}{2} \qquad \text{if } |z| > r_0, \tag{1.62}$$

that is, $|f(z)| \to +\infty$ if $|z| \to +\infty$.

In the third case, $f(z)$ is called a *transcendental entire function*. If $f(z)$ is a transcendental entire function and w is a complex number, there exists a sequence $\{z_n\}$ such that $\lim_{n\to\infty}|z_n| = +\infty$ and $\lim_{n\to\infty} f(z_n) = w$. It is also possible to find a sequence $\{z_n\}$ such that $\lim_{n\to\infty}|z_n| = +\infty$ and $\lim_{n\to\infty}|f(z_n)| = +\infty$. To prove this, put $g(\zeta) = f(1/\zeta)$. Then $g(\zeta)$ is holomorphic on $\mathbb{C} - \{0\}$ and 0 is an essential singularity of $g(\zeta)$, so that we can apply Theorem 1.23 to $g(\zeta)$.

Theorem 1.24 (Liouville's Theorem). A bounded entire function is a constant function.

Proof: If $f(z)$ is a bounded entire function, cases (2) and (3) above cannot occur, that is, $f(z) = a_0$.

The fundamental theorem of algebra, saying that an equation of the form $a_m z^m + a_{m-1} z^{m-1} + \cdots + a_0 = 0$, $a_m \neq 0$, $m \geqq 1$, has at least one root, can be deduced from Liouville's theorem as follows. Suppose $f(z) = a_m z^m + a_{m-1} z^{m-1} + \cdots + a_0$ has no zeros, then $1/f(z)$ is an entire function. By (1.62) we have $|1/f(z)| < 2/|a_m| r_0^m$ for $|z| > r_0$, hence $1/f(z)$ is a bounded function. Therefore, $1/f(z)$ is a constant by Liouville's Theorem, contradicting the assumption that $f(z)$ is not a constant function.

It is also possible to deduce the Fundamental Theorem of Algebra from the Mean Value Theorem. Suppose $f(z) = a_m z^m + a_{m-1} z^{m-1} + \cdots + a_0 \neq 0$ for all $z \in \mathbb{C}$. Then $1/f(z)$ is an entire function and by the Mean Value Theorem (1.35) we have

$$\frac{1}{f(0)} = \frac{1}{2\pi i} \int_0^{2\pi} \frac{1}{f(re^{i\theta})}\, d\theta.$$

Using (1.62) and letting r tend to $+\infty$ we arrive at a contradiction

$$0 < \left|\frac{1}{f(0)}\right| \leqq \frac{1}{2\pi} \int_0^{2\pi} \left|\frac{1}{f(re^{i\theta})}\right| d\theta \leqq \frac{2}{|a_m| r^m} \to 0.$$

2

Cauchy's Theorem

2.1 Piecewise smooth curves

a. Smooth Jordan curves

According to Definition 1.5 of Section 1.3, a continuous map $\gamma: t \to \gamma(t)$ from the closed interval $[a, b]$ into the complex plane $\mathbb{C}$ is called a curve. In this section we will first investigate to what extent the map γ is determined by its image $\gamma(I)$ if γ is a smooth Jordan arc, that is, if $\gamma(t)$ is a one-to-one continuously differentiable function of t such that $\gamma'(t) \neq 0$.

Lemma 2.1. A one-to-one and continuous map $\phi(t)$ from $I = [a, b]$ onto $J = [\alpha, \beta]$ is monotone.

Proof: If $r < s < t$ and $\phi(r) < \phi(t)$, then $\phi(r) < \phi(s) < \phi(t)$. Because if $\phi(s) < \phi(r) < \phi(t)$, then there exists a u, such that $s < u < t$ and $\phi(u) = \phi(r)$ by the intermediate value theorem, contradicting the fact that ϕ is one-to-one. In the same way, the assumption $\phi(r) < \phi(t) < \phi(s)$ leads to a contradiction.

First consider the case $\phi(a) < \phi(b)$. If $a \leqq s < t \leqq b$, then $\phi(a) \leqq \phi(s) < \phi(b)$, hence $\phi(s) < \phi(t) \leqq \phi(b)$, i.e., ϕ is monotone.

The case $\phi(a) > \phi(b)$ is reduced to the previous case by putting $\psi(t) = -\phi(t)$.

Let $\lambda: \tau \to \lambda(\tau), \tau \in J = [\alpha, \beta]$, and $\gamma: t \to \gamma(t), t \in I = [a, b]$, be Jordan arcs, such that $\lambda(J) = \gamma(I) = C$. Since λ and γ are both one-to-one, a function ϕ can be defined by assigning to each $t \in I$ the value $\tau \in J$ with $\gamma(t) = \lambda(\tau)$. This function ϕ is monotone. This is because according to Lemma 2.1 it suffices to show that ϕ is continuous. Assume that ϕ is not continuous at some point s. Then there exists an $\varepsilon > 0$ such that for each $\delta > 0$ there exists at least one point t' such that $|t' - s| < \delta$ and $|\phi(t') - \phi(s)| \geqq \varepsilon$. Hence there exist $t_1, t_2, t_3, \ldots$ such that $|t_n - s| < 1/n$ and $|\phi(t_n) - \phi(s)| \geqq \varepsilon$. Putting $\tau_n = \phi(t_n)$, there exists a subsequence $\{\tau_{n_j}\}$, $n_1 < n_2 < \cdots < n_j < \cdots$, converging to a limit, say, τ. Putting $\sigma = \phi(s)$, we have $|\tau - \sigma| \geqq \varepsilon$. Since $\lim_{j\to\infty} t_{n_j} = s$ and $\gamma(t_{n_j}) = \lambda(\tau_{n_j})$ and $\gamma(s) = \lambda(\sigma)$, we have

$$\lambda(\sigma) = \gamma(s) = \lim_{j\to\infty} \gamma(t_{nj}) = \lim_{j\to\infty} \lambda(\tau_{nj}) = \lambda(\tau)$$

contradicting the fact that λ is one-to-one. Hence ϕ is a continuous function of t.

The Jordan arcs γ and λ are therefore connected by the formula $\gamma(t) = \lambda(\phi(t))$, where ϕ is a continuous and monotone function. One can say that γ is obtained from λ by a change of variable $\tau = \phi(t)$.

If λ and γ are both smooth Jordan arcs, $\phi(t)$ is a continuously differentiable function of t with $\phi'(t) \neq 0$.

To see this, let Δt be an increment of t and put $\Delta\tau = \phi(t + \Delta t) - \phi(t)$. We have

$$\gamma(t + \Delta t) - \gamma(t) = (\gamma'(t) + \varepsilon(\Delta t))\Delta t, \quad \lim_{\Delta t\to 0} \varepsilon(\Delta t) = 0,$$

$$\lambda(\tau + \Delta\tau) - \lambda(\tau) = (\lambda'(\tau) + \varepsilon_1(\Delta\tau))\Delta\tau, \quad \lim_{\Delta t\to 0} \varepsilon_1(\Delta\tau) = 0.$$

Since $\lambda(\tau) = \gamma(t)$ and $\lambda(\tau + \Delta\tau) = \gamma(t + \Delta t)$, $\Delta\tau \to 0$ if $\Delta t \to 0$ and $\lambda'(\tau) \neq 0$, we arrive at

$$\lim_{\Delta t\to 0} \frac{\Delta\tau}{\Delta t} = \lim_{\Delta t\to 0} \frac{\gamma'(t) + \varepsilon(\Delta t)}{\lambda'(\tau) + \varepsilon_1(\Delta\tau)} = \frac{\gamma'(t)}{\lambda'(\tau)}.$$

Hence $\phi(t)$ is a differentiable function of t and $\phi'(t)$ is given by

$$\phi'(t) = \frac{\gamma'(t)}{\lambda'(\tau)}. \tag{2.1}$$

Since $\tau = \phi(t)$ is a continuous function of t, while $\lambda'(\tau)$ and $\gamma'(t)$ are continuous functions of τ and t respectively, $\phi'(t)$ is a continuous function of t, while obviously $\phi'(t) \neq 0$.

We can say that the smooth Jordan arc $\gamma\colon t \to \gamma(t)$ is obtained from the smooth Jordan arc $\lambda\colon \tau \to \lambda(\tau)$ by a change of variable $\tau = \phi(t)$, where $\phi(t)$ is a continuously differentiable and monotone function of t, with $\phi'(t) \neq 0$.

Formula (2.1) is nothing but the chain rule for this change of variable

$$\gamma'(t) = \lambda'(\tau)\phi'(t), \qquad \tau = \phi(t). \tag{2.2}$$

If $\phi(t)$ is a monotone increasing function, we have $\phi(a) = \alpha$ and $\phi(b) = \beta$, if ϕ is monotone decreasing, we have $\phi(a) = \beta$ and $\phi(b) = \alpha$.

Let $f(z)$ be a continuous function defined on a region and let γ be a smooth Jordan curve such that $\gamma(I)$ is contained in that region. We want to investigate the influence of a change of variables on the value of $\int_\gamma f(z)dz$. We distinguish two cases:

Case 1. If $\phi(t)$ is monotone increasing, we have

$$\int_\alpha^\beta f(\lambda(\tau))\lambda'(\tau)d\tau = \int_a^b f(\lambda(\phi(t)))\lambda'(\phi(t))dt = \int_a^b f(\gamma(t))\gamma'(t)dt$$

and therefore

$$\int_\gamma f(z)dz = \int_\lambda f(z)dz. \tag{2.3}$$

Case 2. If $\phi(t)$ is monotone decreasing, we have

$$\int_\alpha^\beta f(\lambda(\tau))\lambda'(\tau)d\tau = \int_b^a f(\gamma(t))\gamma'(t)dt = -\int_a^b f(\gamma(t))\gamma'(t)dt$$

and therefore

$$\int_\gamma f(z)dz = -\int_\lambda f(z)dz. \tag{2.4}$$

Therefore, $\int_\gamma f(z)dz = \pm\int_\lambda f(z)dz$, where the $+$ sign holds if ϕ is monotone increasing and the $-$ sign if ϕ is monotone decreasing. If we call the direction of increasing t the *orientation* of the Jordan arc γ, we can say that the Jordan arcs γ and λ have the same orientation if $\phi(t)$ is monotone increasing and opposite orientation if $\phi(t)$ is monotone decreasing. If we could assign an orientation to the point set $C = \gamma(I)$, $\int_\gamma f(z)dz$ would be unambiguously determined by C together with its orientation.

We can assign an orientation to C in the following way. On the interval I there is defined an order $<$. This order is transferred to C by a one-to-one map γ and denoted by $\prec$, that is, if $t < s$, and $z = \gamma(t)$ and $w = \gamma(s)$, then $z \prec w$. The order $\prec$ determines a linear order on C. If we call this linear order on C the *orientation* of C, $\int_\gamma f(z)dz$ is determined unambiguously by C together with its orientation.

We summarize the above in the following new definition of a Jordan arc.

Definition 2.1. Let $\gamma: t \to \gamma(t)$ be a one-to-one and continuous map from the interval $I = [a, b]$ into the complex plane $\mathbb{C}$. On the point set $\gamma(I)$ a linear order $\prec$ is defined by: if $t < s$ then $\gamma(t) \prec \gamma(s)$. This linearly ordered set $C = \gamma(I)$ is called a *Jordan arc*, the map γ is called a *parameter representation of* C and t is called a *parameter*. If $\gamma(t)$ is continuously differentiable and $\gamma'(t) \neq 0$ for $t \in I$, then C is called a *smooth Jordan arc*.

By the above results, if $\lambda: \tau \to \lambda(\tau)$, $\alpha \leqq \tau \leqq \beta$, is a parameter representation of the Jordan arc C, then an arbitrary parameter representation $\gamma: t \to \gamma(t)$, $a \leqq t \leqq b$, of C has the form $\gamma(t) = \lambda(\phi)$, $a \leqq t \leqq b$, where ϕ is a continuous and monotone increasing function. If C is a smooth Jordan arc, then $\phi(t)$ is continuously differentiable and $\phi'(t) > 0$. In particular, taking $\phi(t) = \alpha + (\beta - \alpha)t$ we arrive at a parameter representation $\gamma: t \to \gamma(t)$ with domain [0, 1].

If $f(z)$ is a continuous function, defined on a region which contains the smooth Jordan arc C, then $\int_C f(z)dz$ is defined by:

$$\int_C f(z)dz = \int_\gamma f(z)dz = \int_a^b f(\gamma(t))\gamma'(t)dt \tag{2.5}$$

where $\gamma: t \to \gamma(t)$, $a \leqq t \leqq b$, is some parameter representation of C. We have seen already that the value of this integral is independent of the choice of the parameter representation. Using the order $\prec$ given on C it is possible to give a definition of $\int_C f(z)dz$ which does not use any parameter representation. Let us call $\alpha = \gamma(a)$ and $\beta = \gamma(b)$ the initial and the terminal points of C, respectively. For all $z \in C$ we have $\alpha \prec z \prec \beta$. Let $\Delta = \{z_0, z_1, z_2, \ldots, z_m\}$ be a finite set of points on C such that

$$\alpha = z_0 \prec z_1 \prec z_2 \prec \cdots \prec z_{m-1} \prec z_m = \beta$$

and let $\delta[\Delta] = \max_k |z_k - z_{k-1}|$. Choose for each k a point ζ_k on C such that $z_{k-1} \prec \zeta_k \prec z_k$. The integral $\int_C f(z)dz$ as defined in (2.5) can be obtained as

$$\int_C f(z)dz = \lim_{\delta[\Delta] \to 0} \sum_{k=1}^m f(\zeta_k)(z_k - z_{k-1}). \tag{2.6}$$

Proof: Let $\gamma: t \to \gamma(t)$, $a \leqq t \leqq b$, be a parameter representation of C. For each $\varepsilon > 0$ there exists a $\delta(\varepsilon) > 0$ satisfying:

$$|t - s| < \varepsilon \qquad \text{if } |\gamma(t) - \gamma(s)| < \delta(\varepsilon). \tag{2.7}$$

(For, suppose there exists an $\varepsilon > 0$ for which this is not true, then there are t_n and s_n for $n = 1, 2, 3, \ldots$ such that $|\gamma(t_n) - \gamma(s_n)| < 1/n$ and $|t_n - s_n| \geqq \varepsilon$. Choose subsequences $\{t_{n_j}\}$ and $\{s_{n_j}\}$, $n_1 < n_2 < \cdots < n_j < \cdots$, converging to t and s, respectively. Then $|t - s| \geqq \varepsilon$ and $\gamma(t) = \gamma(s)$, contradicting the fact that γ is one-to-one.)

Define t_k and τ_k by $\gamma(t_k) = z_k$ and $\gamma(\zeta_k) = \tau_k$, respectively. Then $a = t_0 < t_1 < \cdots < t_{m-1} < t_m = b$ and $\max_k |t_k - t_{k-1}| \to 0$ if $\delta[\Delta] = \max_k |z_k - z_{k-1}| \to 0$, by 2.7. Hence we have, by 1.33,

$$\lim_{\delta[\Delta] \to 0} \sum_{k=1}^m f(\zeta_k)(z_k - z_{k-1}) = \int_\gamma f(z)dz.$$

This proves (2.6).

Therefore (2.6) could be taken as the definition of $\int_C f(z)dz$. The Jordan curve obtained by reversing the orientation of C is denoted by $-C$. C and $-C$ have the same underlying point sets, but if z and w are points of C with $z \prec w$, then on $-C$ we have $w \prec z$. If $\gamma: t \to \gamma(t)$, $a \leqq t \leqq b$, is a parameter representation of C, then $\lambda: s \to \lambda(s) = \gamma(a+b-s)$, $a \leqq s \leqq b$, is a parameter representation of $-C$ (if s increases from a to b, $t = a+b-s$ decreases from b to a). If C is smooth, we have, by (2.4),

$$\int_{-C} f(z) = -\int_C f(z)dz. \tag{2.8}$$

We have given a new definition of Jordan arc, but we will not adhere to this definition too strictly. We adopt the following convention: If we talk about "a Jordan arc C" we have in mind a Jordan arc according to Definition 2.1, if we talk about "a Jordan arc γ" we have in mind the one-to-one continuous map used to define the actual Jordan arc. Furthermore, if $\gamma: t \to \gamma(t)$, $t \in I$, is a Jordan arc then $\gamma(I)$ is not just the point set $\{\gamma(t): t \in I\}$ but rather the Jordan arc C with parameter representation $\gamma: t \to \gamma(t)$, $t \in I$. Hence $C = \gamma(I)$. The point set $\{\gamma(t): t \in I\}$ will be denoted by $|C|$. Sometimes $|C|$ is also called a Jordan arc. In that case C is the Jordan arc obtained by defining an orientation on $|C|$. If C is smooth, $|C|$ is called a smooth Jordan arc. Let $C = \gamma(I)$, $I = [a, b]$, be a Jordan arc and let the $m-1$ points $a_1, a_2, \ldots, a_{m-1}$, $a < a_1 < a_2 < \cdots < a_{m-1} < b$, divide the interval I into m intervals $I_1 = [a, a_1], \ldots, I_k = [a_{k-1}, a_k], \ldots, I_m = [a_{m-1}, b]$. Then the Jordan arc C is also divided into m Jordan arcs $C_1 = \gamma(I_1), \ldots, C_k = \gamma(I_k), \ldots, C_m = \gamma(I_m)$. This is denoted by

$$C = C_1 + C_2 + \cdots + C_k + \cdots + C_m. \tag{2.9}$$

If the Jordan arcs C_k are all smooth, the Jordan arc C is called piecewise smooth. Thus C is piecewise smooth if $\gamma(t)$ is continuously differentiable with $\gamma'(t) \neq 0$ on each I_k.

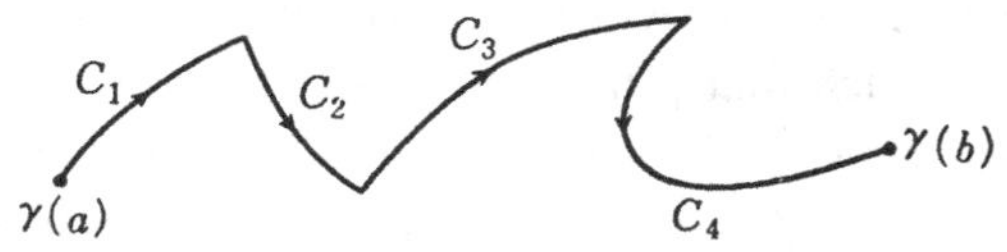

Fig. 2.1

If $f(z)$ is a continuous function defined on a region containing C, $\int_C f(z)dz$ is defined by

$$\int_C f(z)dz = \int_{C_1} f(z)dz + \int_{C_2} f(z)dz + \cdots + \int_{C_m} f(z)dz. \tag{2.10}$$

Since the + in (2.9) indicates a commutative operation, the sum $C_1+C_2+\cdots+C_m$ is independent of the order of $C_1, C_2, \ldots, C_m$. The Jordan curve C obtained by piecing together $C_1, C_2, \ldots, C_m$ in that order is represented as

$$C = C_1 \cdot C_2 \cdot \cdots \cdot C_k \cdot \cdots \cdot C_m, \tag{2.11}$$

If we use this notation, we write C^{-1} instead of $-C$. If $C = C_1 \cdot C_2 \cdot \cdots \cdot C_m$, then $C^{-1} = C_m^{-1} \cdot \cdots \cdot C_2^{-1} \cdot C_1^{-1}$, and $|C| = |C_1| \cup |C_2| \cup \cdots \cup |C_m|$.

The parameter representations of the arcs C_k occurring in the representations (2.9) or (2.11) of C can be chosen independent of the parameter representation of C. For example, we can choose representations $\gamma_k\colon t \to \gamma_k(t)$, $0 \leqq t \leqq 1$, for C_k. These representations of course satisfy $\gamma_k(1) = \gamma_{k+1}(0)$, $k = 1, 2, \ldots, m-1$. Conversely if parameter representations $\gamma_k\colon t \to \gamma_k(t)$, $a_k \leqq t \leqq b_k$, of arcs C_k are given, we put $l_k = b_k - a_k$ and we have the parameter representation

$$\gamma(t)\colon t \to \gamma(t), \quad 0 \leqq t \leqq \sum_{j=1}^{m} l_j;$$

$$\gamma(t) = \gamma_k\left(a + t - \sum_{j=1}^{k-1} l_j\right) \quad \text{for} \quad \sum_{j=1}^{k-1} l_j \leqq t \leqq \sum_{j=1}^{k} l_j$$

with $C = C_1 \cdot C_2 \cdot \cdots \cdot C_k \cdot \cdots \cdot C_m$.

If $\gamma\colon t \to \gamma(t)$, $t \in I = [a, b]$, is a Jordan curve, we have $\gamma(a) = \gamma(b)$ and therefore it is not possible to transfer the order $<$ on I to the point set $\gamma(I)$. However, if we split I into at least two intervals $I_1 = [a, a_1], \ldots, I_k = [a_{k-1}, a_k], \ldots, I_m = [a_{m-1}, b]$, it is possible to define a linear order $\prec$ on each point set $\gamma(I_k)$ using the order $\prec$ on each I_k. The Jordan arc

$$C = C_1 \cdot C_2 \cdot \cdots \cdot C_k \cdot \cdots \cdot C_m \tag{2.12}$$

obtained in this way by piecing together the Jordan arcs $C_1 = \gamma(I_1), \ldots, C_k = \gamma(I_k), \ldots, C_m = \gamma(I_m)$ in that order is called a Jordan curve and is denoted by $C = \gamma(I)$. Using the notation $|C| = \{\gamma(t)\colon t \in I\}$, we obviously have $|C| = |C_1| \cup |C_2| \cup \cdots \cup |C_m|$. Also, γ is called a parameter representation of the Jordan curve $C = \gamma(I)$. If all C_k occurring in (2.12) are smooth, C is called a piecewise smooth Jordan curve.

b. *Boundaries of bounded closed regions*

Let D be a region and let D be the interior of $[D]$, that is, the region consisting of all interior points of $[D]$. If $[D] - D = |C|$ for some Jordan curve C, we say that C is the *boundary* of $[D]$.

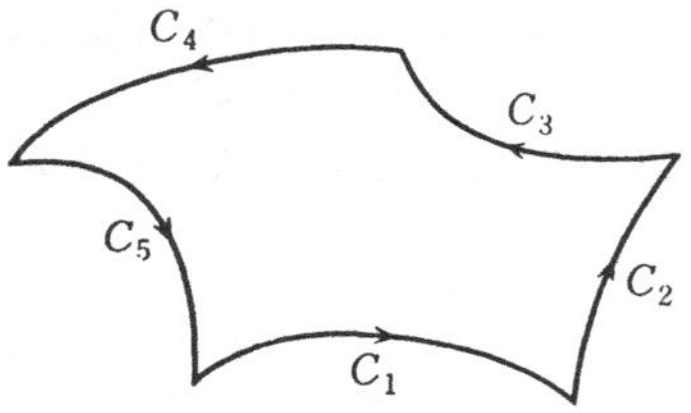

Fig. 2.2

Let the boundary $[D]-D$ of $[D]$ be a piecewise smooth Jordan curve $C = C_1 \cdot C_2 \cdot \cdots \cdot C_k \cdot \cdots \cdot C_m$ with parameter representation $\gamma: t \to \gamma(t)$, $t \in I = [a, b]$, and $C_k = \gamma(I_k)$ with $I_k = [a_{k-1}, a_k]$. Of course, we have $a = a_0 < a_1 < \cdots < a_{k-1} < a_k < \cdots < a_m = b$. We want to study the relation between D and the orientation of C. In order to do this, choose a point $\gamma(c) \in C$, $c \neq a_k$ for $k = 0, \ldots, m$. Since $c \neq a_k$, we have $a_{k-1} < c < a_k$ for some k, hence $\gamma(c) \in C_k$ and $\gamma'(c) \neq 0$. In order to facilitate our study of the neighborhood of $\gamma(c)$ we introduce a new coordinate $w = u + iv$ with origin at $\gamma(c)$ in $\mathbb{C}$ by

$$z = \gamma(c) + \gamma'(c)w. \tag{2.13}$$

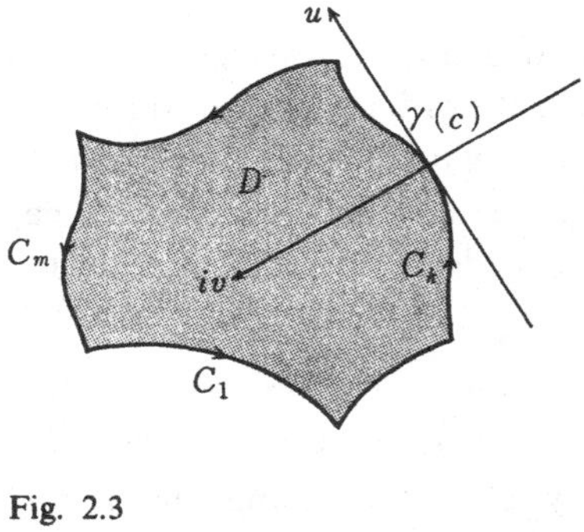

Fig. 2.3

Putting

$$\gamma(t) = \gamma(c) + \gamma'(c)(\rho(t) + i\sigma(t)), \tag{2.14}$$

$t \to w = \rho(t) + i\sigma(t)$ is the parameter representation of C in the new coordinate w.

We have $\rho(c) = \sigma(c) = 0$ and differentiating both sides of (2.14) with regard to t and substituting $t = c$ we get

$$\gamma'(c) = \gamma'(c)(\rho'(c) + i\sigma'(c)),$$

hence $\rho'(c) = 1$ and $\sigma'(c) = 0$. Since $\rho'(t)$ and $\sigma'(t)$ are continuous in a neighborhood of $t = c$, we have $\rho'(t) > 0$. Therefore $\rho(t)$ is a monotone increasing function and the inverse function $t = \tau(u)$ of $u = \rho(t)$ is a continuously differentiable monotone increasing function on a neighborhood of $u = 0$. Since $c = \tau(0)$, by applying the change of variable $t = \tau(u)$ to the parameter representation $t \to w = \rho(t) + i\sigma(t)$ we arrive at the parameter representation

$$u \to w = u + i\psi(u), \qquad \psi(u) = \sigma(\tau(u))$$

of C_k, valid on a neighborhood of the origin $w = 0$. Here, $\psi(u)$ is a continuously differentiable function of u, $\psi(0) = \sigma(c) = 0$ and $\psi'(0) = \sigma'(c)\tau'(0) = 0$. Therefore, if ε is small enough, we have $|\psi(u)| < \varepsilon$ if $|u| < \varepsilon$.

Let $U = \{w : w = u + iv, |u| < \varepsilon, |v| < 2\varepsilon\}$ be the rectangle with center 0, width 2ε, and height 4ε. Choose ε so small that $C \cap U = C \cap U_k$ and

$$|C| \cap U = \{w : w = u + i\psi(u), |u| < \varepsilon\}.$$

C divides U into two regions $U^{(+)}$ and $U^{(-)}$:

$$U^{(+)} = \{w : w = u + iv, |u| < \varepsilon, \psi(u) < v < 2\varepsilon\},$$
$$U^{(-)} = \{w : w = u + iv, |u| < \varepsilon, -2\varepsilon < v < \psi(u)\}.$$

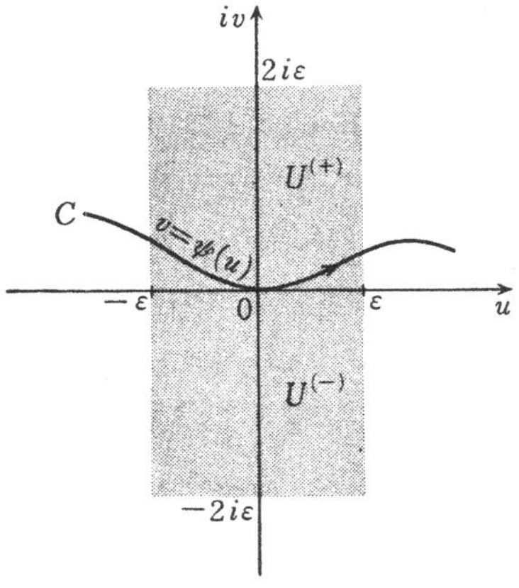

Fig. 2.4

If we remove $|C|$ from the complex plane, the remainder consists of two open sets D and $C-[D]$, that have no points in common.* Therefore, $U^{(+)}$ is also divided into two disjoint sets $U^{(+)} \cap D$ and $U^{+} \cap (\mathbb{C}-[D])$. Since U^{+} is connected, one of these sets has to be the empty set, i.e., either $U^{(+)} \subset D$ or $U^{(+)} \subset \mathbb{C}-[D]$ and similarly for $U^{(-)}$ either $U^{(-)} \subset D$ or $U^{(-)} \subset \mathbb{C}-[D]$ will hold. Therefore, one of the following possibilities will occur:

$$\text{(i)}\ U^{(+)} \subset D, \qquad U^{(-)} \subset \mathbb{C}-[D],$$
$$\text{(ii)}\ U^{(-)} \subset D, \qquad U^{(+)} \subset \mathbb{C}-[D].$$

(If $U^{(+)} \subset D$ and $U^{(-)} \subset D$, then $U \subset [D]$. Hence, all points of $C \cap U$ will be interior points of $[D]$, contradicting the fact that D is the interior of $[D]$. If $U^{(+)} \subset \mathbb{C}-[D]$ and $U^{(-)} \subset \mathbb{C}-[D]$, then $U \subset \mathbb{C}-D$, contradicting the fact that all points of $C \cap U$ are boundary points of D.)

If one looks at U moving along the curve C in the direction corresponding to increasing values of t, one sees $U^{(+)}$ on one's left and $U^{(-)}$ on one's right. In case (i) the region $U^{(+)}$ to the right of C will be in D, while in case (ii) the region $U^{(-)}$ to the left of C will be in D.

Now we return to the original coordinate system z. Remember that $w = u + iv$ was defined by

$$z = \gamma(c) + \gamma'(c)(u + iv).$$

Since $t = \tau(u)$ is a continuously differentiable and monotone increasing function of u and $\tau(0) = c$, we see, putting $\tau(-\varepsilon) = c_1$ and $\tau(\varepsilon) = c_2$, that $c_1 < c < c_2$ and that $\tau : u \to \tau(u)$ is a one-to-one map from the open interval $(-\varepsilon, \varepsilon)$ onto the open interval (c_1, c_2). By definition of $\psi(u)$ we have

$$\gamma(t) = \gamma(c) + \gamma'(c)(u + i\psi(u)), \qquad \psi(u) = \sigma(t).$$

Hence

$$z = \gamma(c) + \gamma'(c)(u + iv) = \gamma(t) + i\gamma'(c)(v - \psi(u)).$$

Putting

$$\eta = v - \psi(u) = v - \sigma(t)$$

we get

$$z = \gamma(t) + i\gamma'(c)\eta$$

and

$$U = \{z : z = \gamma(t) + i\gamma'(c)\eta, \qquad c_1 < t < c_2, \qquad |\eta + \sigma(t)| < 2\varepsilon\},$$
$$|C| \cap U = \{\gamma(t) : c_1 < t < c_2\},$$

* This assumes the validity of the Jordan Curve Theorem.

$$U^{(+)} = \{\gamma(t) + i\gamma'(c)\eta : c_1 < t < c_2, \qquad 0 < \eta < 2\varepsilon - \sigma(t)\},$$
$$U^{(-)} = \{\gamma(t) + i\gamma'(c)\eta : c_1 < t < c_2, \qquad -2\varepsilon - \sigma(t) < \eta < 0\}.$$

That $c_1 < t < c_2$ implies

$$2\varepsilon - \sigma(t) > \varepsilon, \qquad -2\varepsilon - \sigma(t) < -\varepsilon$$

follows from the fact that $|\psi(u)| < \varepsilon$ if $|u| < \varepsilon$.

The division of the neighborhood U of $\gamma(c)$ into a part to the left of C and a part to the right of C can be shown by introducing a smooth Jordan arc, which intersects $C \cap U$. Let $\gamma(t_0)$, $c_1 < t_0 < c_2$, be an arbitrary point of $C \cap U$ and let λ: $s \to \lambda(s)$, $-1 \leqq s \leqq 1$, be a smooth Jordan curve with $\lambda(0) = \gamma(t_0)$. If the angle between λ and C, that is the argument θ of $\lambda'(0)/\gamma'(t_0)$, is between 0 and π, then there exists a $\delta > 0$ such that $\lambda(s) \in U^{(+)}$ if $0 < s < \delta$ and $\lambda(s) \in U^{(-)}$ if $-\delta < s < 0$, that is, if $0 < s < \delta$ the point $\lambda(s)$ is to the left of C and if $-\delta < s < 0$, the point $\lambda(s)$ is to the right of C. A proof follows.

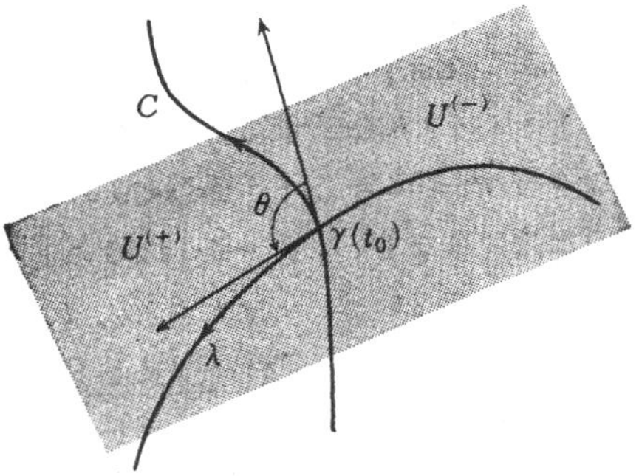

Fig. 2.5

For $\lambda(s) \in U$ we put

$$\lambda(s) = \gamma(c) + \gamma'(c)(u + iv) = \gamma(t) + i\gamma'(c)\eta. \tag{2.15}$$

Since $u = u(s)$ and $v = v(s)$ are continuously differentiable functions, $t = t(s) = \tau(u(s))$ and $\eta = \eta(s) = v(s) - \psi(u(s))$ are continuously differentiable functions of s with $t(0) = t_0$ and $\eta(0) = 0$. Differentiating both sides of (2.15) and substituting $s = 0$ yields

$$\lambda'(0) = \gamma'(t_0)t'(0) + i\gamma'(c)\eta'(0).$$

Hence

$$\frac{\lambda'(0)}{\gamma'(t_0)} = t'(0) + i\frac{\gamma'(c)}{\gamma'(t_0)}\eta'(0)$$
$$= A \cdot e^{i\theta} \qquad \text{with } A = \left|\frac{\lambda'(0)}{\gamma'(t_0)}\right| > 0.$$

Taking the imaginary parts of both sides of the above equality we get

$$A \sin\theta = \mathrm{Re}(\gamma'(c)/\gamma'(t_0)) \cdot \eta'(0).$$

From (2.14) we get $\gamma'(t_0) = \gamma'(c)(\rho'(t_0) + i\sigma'(t_0))$ with $\rho'(t_0) > 0$. Therefore, $\mathrm{Re}(\gamma'(c)/\gamma'(t_0)) > 0$. Since $0 < \theta < \pi$ by assumption, we conclude that $\eta'(0) > 0$. Therefore there exists a $\delta > 0$ such that $\eta(s) > 0$ if $0 < s < \delta$. Remember that $\eta(0) = 0$, i.e., $\lambda(s) = \gamma(t) + i\gamma'(c)\eta(s) \in U^{(+)}$. Similarly, $\lambda(s) \in U^{(-)}$ for $-\delta < s < 0$.

In general, if $\lambda: s \to \lambda(s)$, $-1 \leqq s \leqq 1$, is a smooth Jordan arc intersecting C in $\lambda(0) = \gamma(t), a < t < b, t \neq a_k$, such that the angle $\theta, 0 < \theta < 2\pi$, between λ and C at $\lambda(0) = \gamma(t)$ satisfies $\theta \neq 0$, $\theta \neq \pi$, then λ is said to *intersect C transversely* at $\lambda(0) = \gamma(t)$. In this book, we will say that λ *crosses C* at $\lambda(0) = \gamma(t)$ from right to left if $0 < \theta < \pi$ and from left to right if $\pi < \theta < 2\pi$.

Let λ be a Jordan arc crossing C from right to left at $\lambda(0) = \gamma(t)$. If there exists a $\delta > 0$ such that $\lambda(s) \in D$ for $0 < s < \delta$, then we say that D is on the left of C at the point $\gamma(t)$; if there exists a $\delta > 0$ such that $\lambda(s) \in D$ for $-\delta < s < 0$, then we say that D is on the right of C at the point $\gamma(t)$. The value of δ is dependent on λ, but D being on the left or right of C at $\gamma(t)$ is independent of the choice of λ. (To see this, put $c = t$, let U be the neighborhood of $\gamma(c)$ defined above. There are two possibilities: (1) $U^{(+)} \subset D$ and $U^{(-)} \subset \mathbb{C} - [D]$ or (2) $U^{(-)} \subset D$ and $U^{(+)} \subset \mathbb{C} - [D]$. In the first case, $\lambda(s) \in U^{(+)} \subset D$ for $0 < s < \delta$ and in the second case $\lambda(s) \in U^{(-)} \subset D$ for $-\delta < s < 0$ by the above.) If D is on the left of C at one point $\gamma(t_0)$, $t_0 \neq a_k$, of C, then D is on the left of all points $\gamma(t)$, $t \neq a_k$, of C. Similarly, if D is on the right of C at one point $\gamma(t_0)$, $t_0 \neq a_k$, of C, then D is on the right of all points $\gamma(t)$, $t \neq a_j$, of C.

To see this, choose c with $a_{k-1} < c < a_k$ and let U be the neighborhood of c as defined above. Let us assume that D is on the left of C at $\gamma(c)$, i.e., $U^{(+)} \subset D$. If λ is a Jordan curve crossing C from right to left at $\lambda(0) = \gamma(t) \in C \cap U$, $c_1 < t < c_2$, then, as shown above, there exists a $\delta > 0$ such that $\lambda(s) \in U^{(+)} \subset D$ for $0 < s < \delta$. Therefore, D is on the left of C at all points $\gamma(t)$ with $c_1 < t < c_2$. Let L be the set of all $t \in \mathbb{R}$, $a_{k-1} < t < a_k$, such that D is on the left of C at $\gamma(t)$. If $c \in L$, then the open interval (c_1, c_2), $c_1 < c < c_2$, is contained in L, that is, it is an open subset of $\mathbb{R}$. Similarly, let R be the set of all $t \in \mathbb{R}$, $a_{k-1} < t < a_k$ such that D is on the right of C at $\gamma(t)$. R is also an open subset of $\mathbb{R}$. Since $(a_{k-1}, a_k) = L \cup R$, $R \cap L = \varnothing$ and (a_{k-1}, a_k) is connected, we have either $L = (a_{k-1}, a_k)$ or $R = (a_{k-1}, a_k)$. If $L = (a_{k-1}, a_k)$, i.e., if D is on the left of C_k at all points $\gamma(t)$, $a_{k-1} < t < a_k$, then we say that D is on the left of C_k. Similarly, if $R = (a_{k-1}, a_k)$, we say that D is on the right of C_k.

Next we will prove that D is on the left of C_k if it is on the left of C_{k+1}. Choose ε sufficiently small and draw a circle λ: $s \to \lambda(s) = \gamma(a_k) + \varepsilon e^{is}$ with center $\gamma(a_k)$ and radius ε, such that λ intersects C_{k+1} at exactly one point $\lambda(\alpha)$, C_k at exactly one point $\lambda(\beta)$, $\alpha < \beta < \alpha + 2\pi$, and does not intersect $C_1, \ldots, C_{k-1}, C_{k+2}, \ldots, C_m$, that is λ intersects C at exactly two points $\lambda(\alpha)$ and $\lambda(\beta)$. The circle λ crosses C_{k+1} at $\lambda(\alpha)$ from right to left (we will prove this below). Since by assumption D is at the left of C_{k+1}, we have $\lambda(s) \in D$ for values of s, $\alpha < s < \beta$, sufficiently close to α. Since λ intersects the boundary C of D at $\lambda(\alpha)$ and $\lambda(\beta)$, we conclude that $\lambda(s) \in D$ for all s with $\alpha < s < \beta$ and that λ crosses C_k from left to right at $\lambda(\beta)$. Therefore, D is on the left of C_k at $\lambda(\beta) \in C_k$, hence D is on the left of C_k.

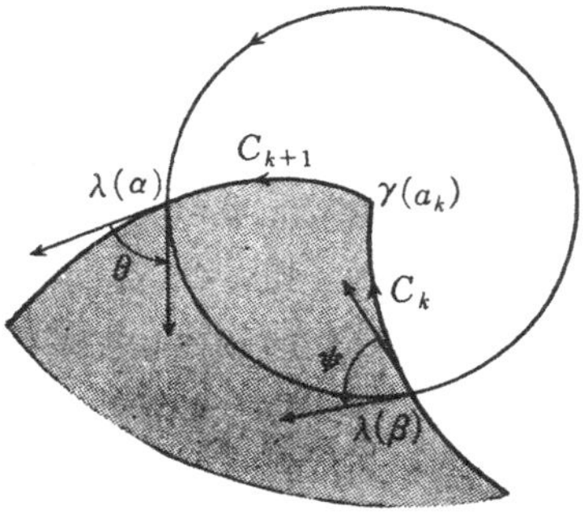

Fig. 2.6

We now want to prove that λ intersects C_{k+1} at exactly one point and that λ crosses C_{k+1} from right to left at $\lambda(\alpha)$. To simplify our notation, we may assume $\gamma(a_k) = 0$, $a_k = 0$. Let $\gamma: t \to \gamma(t)$, $t \in [0, a_{k+1}]$, be a parameter representation of the Jordan arc C_{k+1} (i.e., $\gamma(t)$ is a continuously differentiable function of t and $\gamma(0) = 0$). First we show that $|\gamma(t)|$ is a continuously differentiable function of t on $[0, a_{k+1}]$ and

$$\left(\frac{d}{dt}|\gamma(t)|\right)_{t=+0} = |\gamma'(0)| > 0. \tag{2.16}$$

Put $\gamma_1(t) = \gamma(t)/t$. From $\lim_{t\to+0} \gamma(t)/t = \gamma'(0)$ we conclude that $\gamma_1(t)$ is continuous on $[0, a_{k+1}]$ if we define $\gamma_1(0) = \gamma'(0)$.
Hence

$$\lim_{t\to+0} \frac{|\gamma(t)|}{t} = \lim_{t\to+0} |\gamma_1(t)| = |\gamma_1(0)| = |\gamma'(0)|,$$

that is, $|\gamma(t)|$ is differentiable at $t = 0$ and the differential coefficient is given by (2.16). On $(0, a_{k+1}]$, $|\gamma(t)|^2 = \gamma(t)\overline{\gamma(t)}$ is continuously differentiable. Since $|\gamma(t)|^2 > 0$, $|\gamma(t)| = \sqrt{(|\gamma(t)|^2)}$ is also continuously differentiable, while

$$\frac{d}{dt}|\gamma(t)| = \frac{\overline{\gamma(t)}\gamma'(t) + \gamma(t)\overline{\gamma'(t)}}{2|\gamma(t)|}.$$

Substitution of $t\gamma_1(t)$ for $\gamma(t)$ in the expression on the right yields

$$\lim_{t\to+0}\frac{d}{dt}|\gamma(t)| = \lim_{t\to+0}\frac{\overline{\gamma_1(t)}\gamma'(t) + \gamma_1(t)\overline{\gamma'(t)}}{2|\gamma_1(t)|} = |\gamma'(0)|$$

hence $d|\gamma(t)|/dt$ is continuous on $[0, a_{k+1}]$.

Therefore, taking a sufficiently small k, we have $d|\gamma(t)|/dt > 0$ if $0 \leqq t \leqq k$, hence $|\gamma(t)|$ is a monotone increasing function for $0 \leqq t \leqq k$. By (2.7) there exists a $\delta(k) > 0$ such that

$$0 \leqq t \leqq k \qquad \text{if } |\gamma(t)| < \delta(k).$$

Take an $\varepsilon > 0$ such that $\varepsilon < \delta(k)$ and $\varepsilon < |\gamma(k)|$. If $|\gamma(t)| = \varepsilon$, then $0 < t \leqq k$. If t increases from 0 till k, then $|\gamma(t)|$ increases from $|\gamma(0)|$ until $|\gamma(k)|$, so there is exactly one t $(0 < t \leqq a_{k+1})$ such that $|\gamma(t)| = \varepsilon$. We have proved that the circle λ: $s \to \lambda(s) = \varepsilon e^{is}$ intersects C_{k+1} at exactly one point:

$$\gamma(t) = \varepsilon e^{i\alpha} \qquad t > 0.$$

Let θ be the angle between C_{k+1} and λ at this point, then

$$\frac{\lambda'(\alpha)}{\gamma'(t)} = \left|\frac{\lambda'(\alpha)}{\gamma'(t)}\right| e^{i\theta}.$$

From: $\lambda'(\alpha) = i\varepsilon e^{i\alpha} = i\gamma(t) = it\gamma_1(t)$ we get

$$i\frac{\gamma_1(t)}{\gamma'(t)} = \left|\frac{\gamma_1(t)}{\gamma'(t)}\right| e^{i\theta}.$$

If $\varepsilon \to 0$, then $t \to 0$, hence: $\gamma_1(t)/\gamma'(t) \to 1$. Therefore, for sufficiently small ε, θ will be between, for example, $\pi/4$ and $3\pi/4$. Therefore, the circle λ crosses C_{k+1} from right to left at $\lambda(\alpha) = \gamma(t)$. The fact that λ intersects C_k at exactly one point $\lambda(\beta)$ and crosses C_k from left to right at that point is proved in the same way.

This finishes the proof of the fact that D is on the left of C_k if D is on the left of C_{k+1}. Therefore, if D is on the left of C_m, then D is on the left of C_{m-1}, $C_{m-2}, \ldots, C_2, C_1$, i.e., D is on the left of $C = C_1 \cdot C_2, \ldots, C_m$ at all points $\gamma(t)$, $t \neq a_k$. The fact that D is on the right of C at all points $\gamma(t)$, $t \neq a_k$, if D is on the right of C_m is proved similarly.

If D is on the left (right) of C at all points $\gamma(t)$, $t \neq a_k$, then D is said to be on the left (right) of C.

Remark: The above proof is based on local considerations. However complicated the form of a piecewise Jordan curve C may be, in a sufficiently small neighborhood of each point $\gamma(t) \in C$, the form of C is simple, as shown by the diagrams above. The proof is nothing but the confirmation by calculation of an intuitively clear fact. It is often said that Jordan's Theorem, stating that a Jordan curve C divides the complex plane in two parts, the interior and the exterior of C, is intuitively clear, but this is only based on the analogy with simple cases as circles and convex polygons. In fact this theorem is far from intuitively clear.

Till now we have assumed that the boundary of $[D]$ is a piecewise smooth Jordan curve C, i.e., that $|C| = [D] - D$. In this sense, the Jordan curve $-C = C^{-1}$ is also a boundary of $[D]$. However, C and $-C$ are different Jordan curves, and if D is on the left of C, then D is on the right of $-C$ and if D is on the right of C, then D is on the left of $-C$. We now define the boundary of the bounded, closed region as an oriented, piecewise smooth Jordan curve as follows.

Definition 2.2. If C is an oriented, piecewise smooth Jordan curve such that $|C| = [D] - D$ and D is on the left of C, then C is called the *boundary* of $[D]$, and denoted by $\partial[D]$. If $|C| = [D] - D$ and D is on the right of C, then we have $\partial[D] = -C$.

Till now, we have only considered the case that the boundary of $[D]$ consists of exactly one piecewise smooth Jordan curve, but the same considerations can be applied to a domain $[D]$, whose boundary consists of a finite number of mutually disjoint piecewise smooth Jordan curves.

Let $C_1, C_2, \ldots, C_\nu, \ldots, C_n$ be piecewise smooth Jordan curves such that

$$[D] - D = |C_1| \cup |C_2| \cup \cdots \cup |C_\nu| \cup \cdots \cup |C_n|, \quad |C_\lambda| \cap |C_\nu| = \emptyset, \ \lambda \neq \nu,$$

and such that D is either on the right or on the left of each C_ν.

Definition 2.3. If D is on the left of all C_ν, $\nu = 1, 2, \ldots, n$, then $C = C_1 + C_2 + \cdots + C_n$ is called the *boundary* of $[D]$ and denoted by $\partial[D]$:

$$\partial[D] = C = C_1 + C_2 + \cdots + C_n.$$

If $[D]-D=\bigcup_{\nu=1}^{n}|C_\nu|$, $|C_\lambda|\cap|C_\nu|=\emptyset$, $\lambda\neq\nu$, then

$$\partial[D]=\sum_{\nu=1}^{n}\pm C_\nu.$$

The sign of C_ν is $+$ if D is on the left of C_ν and $-$ if D is on the right of C_ν.

Example 2.1. Consider the smooth Jordan curve $C_r=\gamma_r(I)$ defined by γ_r: $\theta\to\gamma_r(\theta)=re^{i\theta}$, $\theta\in I=[0,\ 2\pi]$. If $D=\{z:\ \varepsilon<|z|<R\}$ then the boundary $\partial[D]$ of D is given by

$$\partial[D]=C_R-C_\varepsilon.$$

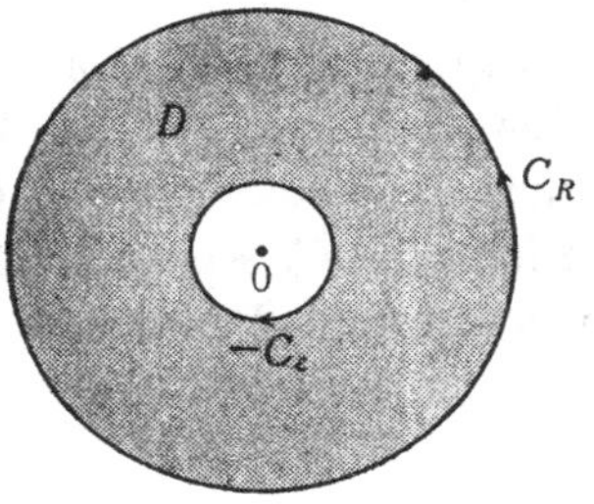

Fig. 2.7

2.2 Cellular decomposition

a. Cells

Let Γ: $(t,\ s)\to\Gamma(t,\ s)$ be a continuous map from the rectangle $K=\{(t,\ s):\ a\leqq t\leqq b,\ 0\leqq s\leqq 1\}$ into the complex plane $\mathbb{C}$, such that the partial derivatives $\Gamma_t(t,\ s)$, $\Gamma_s(t,\ s)$, and $\Gamma_{ts}(t,\ s)=\Gamma_{st}(t,\ s)$ exist and are continuous. Under certain conditions to be specified later, the image $\Gamma(K)$ of K under Γ is called a *cell*. We first want to give a few examples of cells.

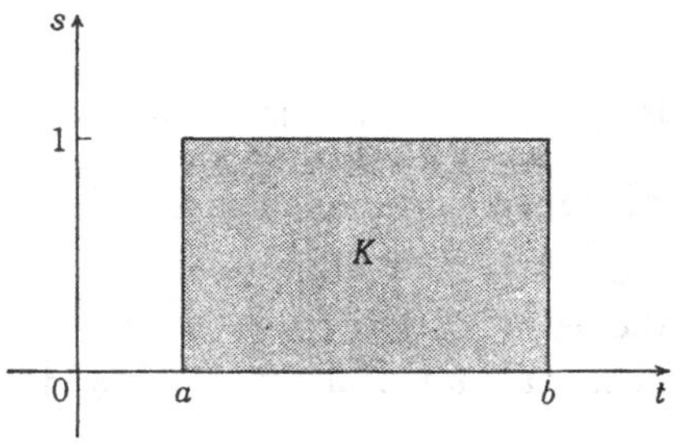

Fig. 2.8

Example 2.2. $K = \{(t, s): 0 < \varepsilon \leqq t \leqq R, 0 \leqq s \leqq \pi\}$ and $\Gamma(t, s) = te^{is}$. The cell $\Gamma(K)$ is the closed domain sketched in Fig. 2.9.

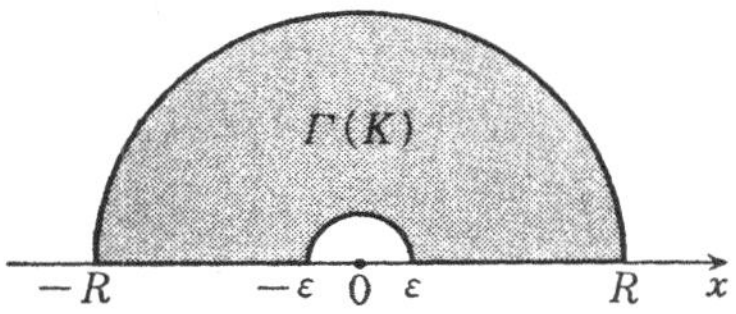

Fig. 2.9

Example 2.3. $K = \{(t, s): a \leqq t \leqq b, 0 \leqq s \leqq 1\}$ and

$$\Gamma(t, s) = t + (1 - s)i\phi(t) + si\psi(t),$$

where $\phi(t)$ and $\psi(t)$ are continuously differentiable functions, defined on $[a, b]$ and satisfying $\phi(t) < \psi(t)$ if $a < t < b$. We have $\phi(a) \leqq \psi(a)$ and $\phi(b) \leqq \psi(b)$. This gives four possibilities: the cell corresponding with each possibility is sketched in Figure 2.10 (a–d).

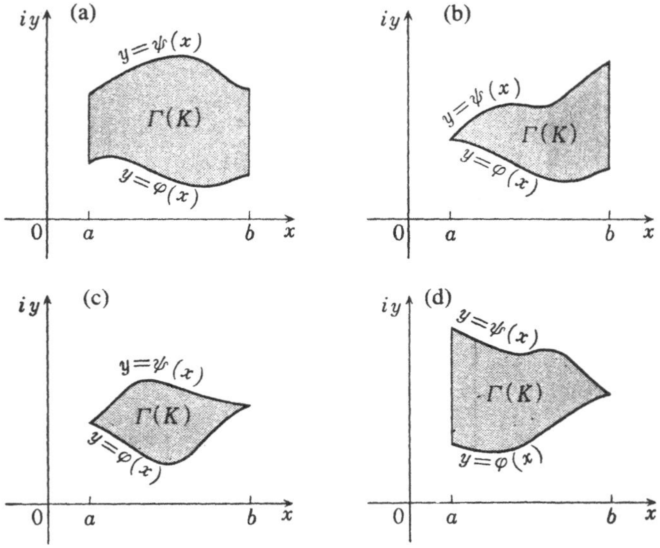

Fig. 2.10

Example 2.4. $K = \{(t, s): 0 \leqq t \leqq 1, 0 \leqq s \leqq 1\}$ and $\Gamma(t, s) = (t - 1)(1 + e^{-i\pi s})$.

The cell $\Gamma(K)$ is the closed domain enclosed by a semicircle and the line segment $[-2, 0]$ and is sketched in Fig. 2.11. Γ maps the side $\{(t, 0): 0 \leqq t \leqq 1\}$ of K on the segment $[-2, 0]$, the side $\{(0, s): 0 \leqq s \leqq 1\}$ on the semicircle, and the remaining two sides onto the origin 0.

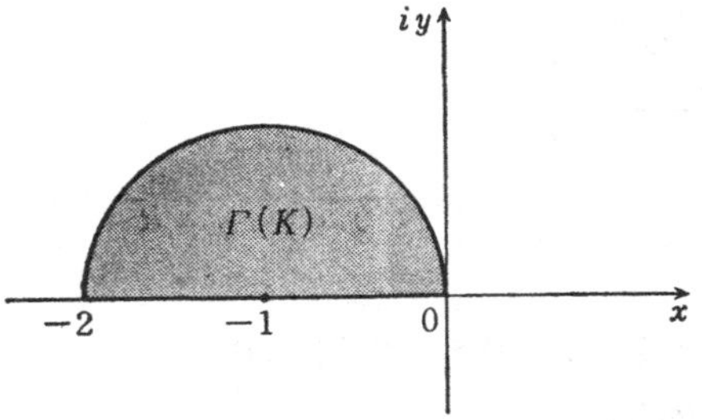

Fig. 2.11

Now let us introduce the following notations: $E = \{(t, s): a < t < b, 0 < s < 1\}$ is the interior of K and $L_1 = \{(t, 0: a \leqq t \leqq b\}$, $L_2 = \{(b, s): 0 \leqq s \leqq 1\}$, $L_3 = \{(t, 1: a \leqq t \leqq b\}$, and $L_4 = \{(a, s): 0 \leqq s \leqq 1\}$ are its sides. Of course, E is a region in $\mathbb{C}$ and $K = [E]$ is its closure.

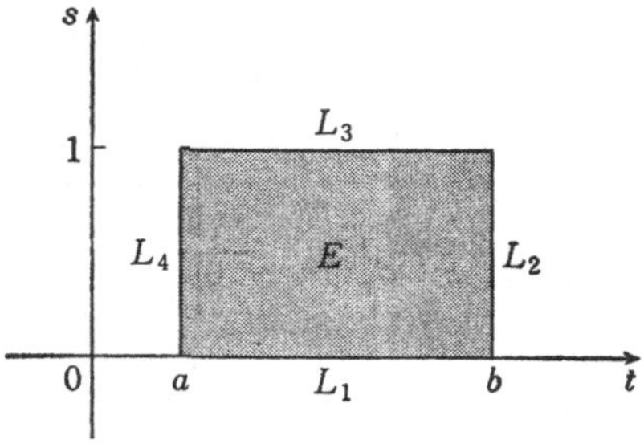

Fig. 2.12

We now want to examine the properties of the map Γ as defined in Example 2.3 under the assumption that $\phi(b) = \psi(b)$. From

$$\Gamma(t, s) = t + (1 - s) i\phi(t) + si\psi(t)$$

we get

$$\Gamma_t(t, s) = 1 + (1 - s) i\phi'(t) + si\psi'(t), \tag{2.17}$$

$$\Gamma_s(t, s) = i(\psi(t) - \phi(t)).$$

$\Gamma(L_1)$ is a smooth Jordan arc, with parameter representation $t \to \Gamma(t, 0) = t + i\phi(t)$, $a \leqq t \leqq b$, $\Gamma(L_2)$ is just one point $b + i\phi(b)$, $\Gamma(L_3)$ is a smooth Jordan arc with parameter representation $t \to \Gamma(t, 1) = t + i\psi(t), a \leqq t \leqq b$, and $\Gamma(L_4)$ is just one point $a + i\phi(a)$. Let $K'' = \{(t, s): a < t < b, 0 \leqq s \leqq 1\}$ be the set we get by taking L_2 and L_4 from K, then $\Gamma: K'' \to \Gamma(K'')$ is one-to-one on K'' and for each point $(t, s) \in K''$ we have by (2.17)

$$\operatorname{Im} \Gamma_s(t, s)\overline{\Gamma_t(t, s)} = \psi(t) - \phi(t) > 0.$$

$\Gamma(K)$ is a bounded, closed region, $\Gamma(E)$ is its interior and its boundary a piecewise smooth Jordan curve $C_1 \cdot C_3$, where $C_1 = \Gamma(L_1)$ and $C_3 = \Gamma(L_3)$, i.e., $\partial\Gamma(K) = C = C_1 \cdot C_3$. Redefining the Jordan arc $\Gamma(L_3)$ by the parameter representation $t \to \Gamma(b + a - t, 1)$, $a \leqq t \leqq b$, we get $C_3 = \Gamma(L_3)$.

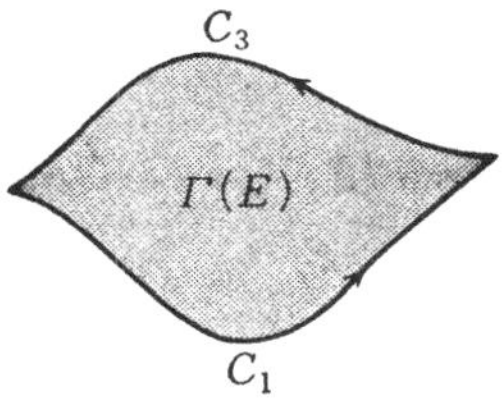

Fig. 2.13

We now define a cell as follows.

Definition 2.4. If Γ satisfies the following conditions (1)–(4), $\Gamma(K)$ is called a *smooth cell*, or more briefly, a *cell*:

(1) $\Gamma(K)$ is a closed domain, $\Gamma(E)$ is its interior, i.e., the set of all interior points of $\Gamma(K)$ and the boundary of $\Gamma(K)$ is a piecewise smooth Jordan curve.

(2) Each $\Gamma(L_K)$ is either a point or a smooth Jordan arc.

(3) If K'' is the set obtained from K by omitting those sides L_K such that $\Gamma(L_K)$ is a point, the map $\Gamma: K'' \to \Gamma(K'')$ is one-to-one and we have for all points $(t, s) \in K''$:

$$\operatorname{Im} \Gamma_s(t, s)\overline{\Gamma_t(t, s)} > 0. \tag{2.18}$$

(4) For fixed s the function $\Gamma(t, s)$ of t satisfies $\Gamma_t(t, s) \neq 0, a \leqq t \leqq b$, or is a constant. Similarly, for fixed t, the function $\Gamma(t, s)$ of s satisfies $\Gamma_s(t, s) \neq 0$ or is a constant.

The boundary $C = \partial\Gamma(K)$ of a cell $\Gamma(K)$ is a piecewise smooth Jordan curve by condition (1) and we have

$$|C| = \Gamma(K) - \Gamma(E) = |\Gamma(L_1)| \cup |\Gamma(L_2)| \cup |\Gamma(L_3)| \cup |\Gamma(L_4)|.$$

Each $C_k = \Gamma(L_k)$, $k = 1, 2, 3, 4$, is by condition (2) either a point or a smooth Jordan arc. If C_k is a smooth Jordan arc, by condition (4) parameter representations are given by:

$$\begin{aligned} &C_1 : t \to \Gamma(t, 0), && a \leqq t \leqq b, \\ &C_2 : s \to \Gamma(b, s), && 0 \leqq s \leqq 1, \\ &C_3 : t \to \Gamma(b + a - t, 1), && a \leqq t \leqq b, \\ &C_4 : s \to \Gamma(a, 1 - s), && 0 \leqq s \leqq 1. \end{aligned} \tag{2.19}$$

Since the orientation of each $\Gamma(L_k)$ is not implied by condition (2), we define the orientations for each $C_k = \Gamma(L_k)$ using the above parameter representations, i.e.,

$$C = \partial\Gamma(K) = C_1 \cdot C_2 \cdot C_3 \cdot C_4. \tag{2.20}$$

If C_1, C_2, C_3, and C_4 are all Jordan arcs, then $C_1 \cdot C_2 \cdot C_3 \cdot C_4$ is the Jordan curve obtained by piecing together C_1, C_2, C_3, and C_4 in that order. If among C_1, C_2, C_3, and C_4 there are one or more points, then $C_1 \cdot C_2 \cdot C_3 \cdot C_4$ is the Jordan curve obtained by piecing together only the Jordan arcs from among C_1, C_2, C_3, and C_4. If, for example, C_2 is a point and C_1, C_3, and C_4 are Jordan arcs, then $C_1 \cdot C_2 \cdot C_3 \cdot C_4 = C_1 \cdot C_3 \cdot C_4$ and if C_2 and C_3 are points, then $C_1 \cdot C_2 \cdot C_3 \cdot C_4 = C_1 \cdot C_4$.

Proof: Since $|C| = |C_1| \cup |C_2| \cup |C_3| \cup |C_4|$, it is sufficient to establish the fact that if C_k is a Jordan arc its orientation agrees with the orientation of C, i.e., that $\Gamma(E)$ is on the left of C_k. Choose $c \in (a, b)$ and consider the curve ${}_c\gamma$: $s \to {}_c\gamma(s) = \Gamma(c, s)$, $0 \leqq s \leqq 1$. By conditions (3) and (4), ${}_c\gamma$ is a smooth Jordan arc. If C_1 is a Jordan arc, then the angle between ${}_c\gamma$ and C_1 at ${}_c\gamma(0) = \Gamma(c, 0)$, i.e., the argument θ of ${}_c\gamma'(0)/\Gamma_t(c, 0)$, is equal to the argument of

$$\Gamma_s(c, 0)\overline{\Gamma_t(c, 0)} = {}_c\gamma'(0)/\Gamma_t(c, 0) \cdot |\Gamma_t(c, 0)|^2.$$

By (2.18) we have $\sin\theta > 0$, hence $0 < \theta < \pi$. Since ${}_c\gamma(s) = \Gamma(c, s) \in \Gamma(E)$ for $0 < s < 1$, $\Gamma(E)$ is on the right of C_1. Similarly, considering the Jordan arc λ: $s \to \lambda(s) = {}_c\gamma(1 - s)$, $0 \leqq s \leqq 1$, obtained from ${}_c\gamma$ by reversing the orientation, and observing that the angle ψ between λ and C_3 at $\Gamma(c, 1)$ is equal to the argument of $\Gamma_s(c, 1)\overline{\Gamma_t(c, 1)}$, we conclude $\sin\psi > 0$ by (2.18), hence $0 < \psi < \pi$. Therefore $\Gamma(E)$ is to the left of C_3.

Next consider $\gamma_{1/3}$: $t \to \gamma_{1/3}(t) = \Gamma(t, 1/3)$, $a \leqq t \leqq b$. $\gamma_{1/3}$ is a smooth Jordan arc and the angle ω between $\gamma_{1/3}$ and C_4 (if C_4 is not a point) at the point $\Gamma(a, \frac{1}{3})$ is equal to the argument of $-\Gamma_t(a, \frac{1}{3})\overline{\Gamma_s(a, \frac{1}{3})}$. Therefore, $\sin\omega > 0$ by (2.18), i.e., $0 < \omega < \pi$. Hence $\Gamma(E)$ is on the left of C_4 since

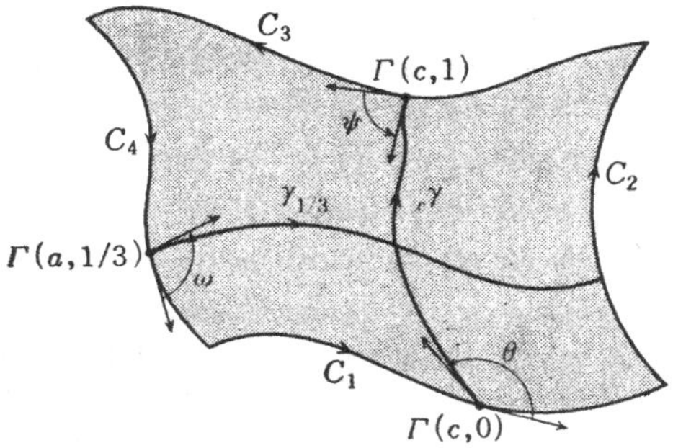

Fig. 2.14

$\gamma_{1/3}(t) \in \Gamma(E)$ for $a < t < b$. Similarly, one proves that $\Gamma(E)$ is on the left of C_2 if C_2 is not a point.

To illustrate what happens if some of the C_1, C_2, C_3, and C_4 are points, we have sketched in Fig. 2.15 what the curves ${}_c\gamma$ and $\gamma_{1/3}$ look like for the cell of Example 2.4. In this case, $C_2 = C_3 = 0$ and $C = \partial\Gamma(K) = C_1 \cdot C_4$.

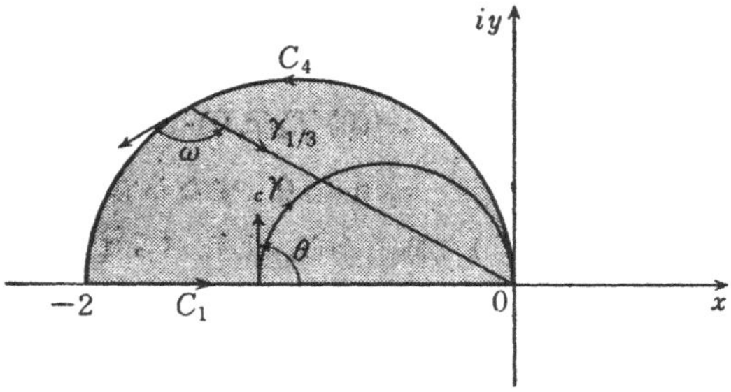

Fig. 2.15

For notational convenience we have defined K as the rectangle $\{(t, s): a \leqq t \leqq b, 0 \leqq s \leqq 1\}$, but it is obvious that we could have used an arbitrary rectangle $\{(t, s): a \leqq t \leqq b, c \leqq s \leqq d\}$. An arbitrary rectangle is transformed into the square $K' = \{(t, s): 0 \leqq t \leqq 1, 0 \leqq s \leqq 1\}$ by the change of coordinates $t = a + (b - a)\tau$, $s = c + (d - c)\sigma$, so we could just as well have started with K'.

b. Cellular decomposition

Let $[D]$ be a closed region in $\mathbb{C}$ and let D be its interior.

Definition 2.5. A collection $\mathscr{K} = \{\Gamma_1(K_1) \ldots \Gamma_\lambda(K_\lambda), \ldots, \Gamma_\mu(K_\mu)\}$ of cells is called a *cellular decomposition* of $[D]$ if

(1) an arbitrary pair of cells has no interior points in common.

(2) $[D] = \Gamma_1(K_1) \cup \Gamma_2(K_2) \cup \cdots \cup \Gamma_\lambda(K_\lambda) \cup \cdots \cup \Gamma_\mu(K_\mu)$. (2.21)

(3) If $\Gamma_\lambda(K_\lambda)$ and $\Gamma_\nu(K_\nu)$, $\lambda \neq \nu$, have a nonempty intersection, then $\Gamma_\lambda(K_\lambda) \cap \Gamma_\nu(K_\nu)$ is either one point or a smooth Jordan arc $|C_{\lambda\nu}|$.

If such a cellular decomposition exists, $[D]$ is called *cellularly decomposable.*

Example 2.5. Figure 2.16 shows a cellular decomposition of the closed disk. In this example, $|C_{12}| = \Gamma_1(K_1) \cap \Gamma_2(K_2)$ is the segment $\{x: 0 \leqq x \leqq 1\}$, $|C_{31}| = \Gamma_3(K_3) \cap \Gamma_1(K_1)$ is the segment $\{iy: 0 \leqq y \leqq 1\}$, and so on.

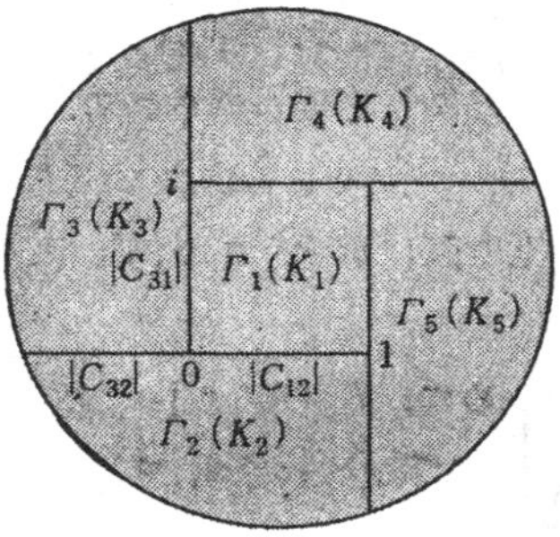

Fig. 2.16

Theorem 2.1. If the boundary of a bounded closed region $[D]$ consists of a finite number of mutually disjoint, piecewise smooth Jordan curves, then $[D]$ is cellularly decomposable.

The proof is in three parts.

Part (I) of proof: We first assume that the boundary of $[D]$ consists of exactly one smooth Jordan curve C. The proof, which is basically straightforward, is based on the fact that a smooth Jordan curve is "almost a straight line" on a sufficiently small neighborhood of each of its points.

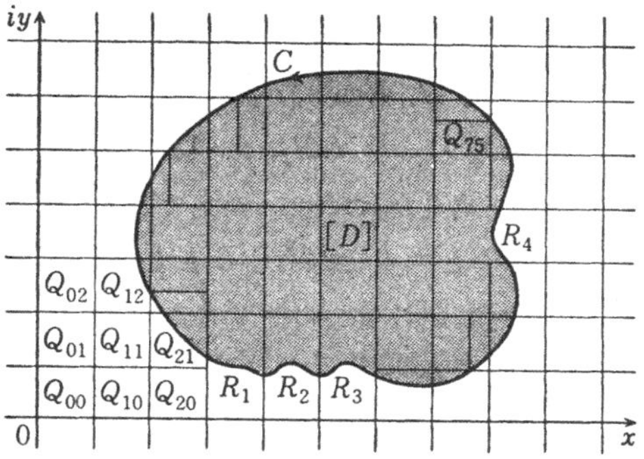

Fig. 2.17

Let

$$Q_{hk} = \{x + iy: h\delta \leqq x \leqq h\delta + \delta,\ k\delta \leqq y \leqq k\delta + \delta\},$$
$$h, k = 0, \pm 1, \pm 2, \ldots,$$

define an infinite number of squares with sides equal to $\delta > 0$ and covering the whole complex plane $\mathbb{C}$

$$\mathbb{C} = \bigcup_{h,k} Q_{h,k}.$$

Then $[D]$ is covered by a finite number of closed subsets $[D] \cap Q_{hk} \neq \emptyset$

$$[D] = \bigcup_{h,k} ([D] \cap Q_{h,k}).$$

If this is a cellular decomposition, we are finished. If it is not, we make adjustments as shown in Fig. 2.17 to change this decomposition into a cellular decomposition.

The details of this adjustment procedure are as follows. First, let Q_{hk} be a square such that one of its sides intersects C in at least two points or is tangent to C. Let $Q_{h'k'}$ be the square that has this side in common with Q_{hk} (i.e., $(h', k') = (h, k+1), (h, k-1), (h+1, k)$, or $(h-1, k)$) and let the rectangle R be defined by $R = Q_{hk} \cup Q_{h'k'}$. Denote the rectangles obtained in this way by $R_1, R_2, \ldots, R_p, \ldots, R_m$. If $p \neq q$, R_p and R_q have no interior points in common and $[D]$ is covered by the sets $[D] \cap R_p$, $p = 1, 2, \ldots, m$, and the remaining $[D] \cap Q_{hk}$, where $Q_{hk} \not\subset \bigcup_p R_p$. All $[D] \cap R_p$ are cells. Let $Q_{hk} \not\subset \bigcup_p R_p$ be such that Q_{hk} and $[D]$ have interior points in common, then a side of Q_{hk} intersects C not at all or transversely in exactly one point. If $[D] \cap Q_{hk}$ is not a cell, then one of the cases sketched in Fig. 2.19 must occur. By dividing $[D] \cap Q_{hk}$ into two cells, as shown above, we obtain a cellular

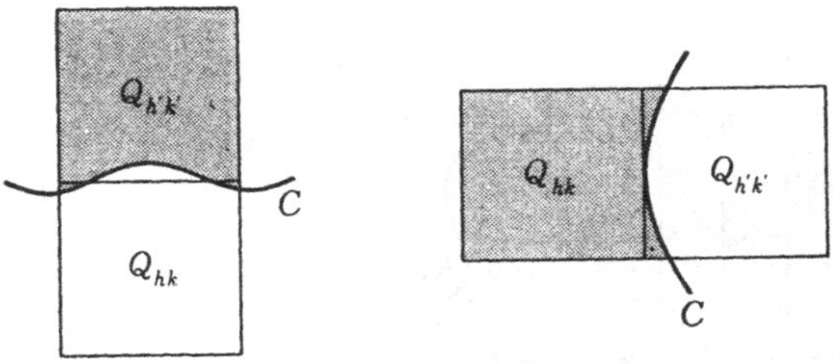

Fig. 2.18

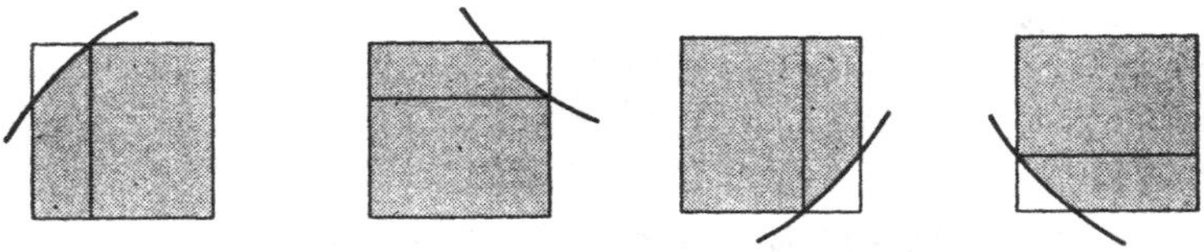

Fig. 2.19

decomposition of $[D]$. The cellular decomposition shown above was obtained in this way.

We now proceed to prove that the decomposition defined in this way is really a cellular decomposition.* Let $\gamma: t \to \gamma(t)$, $0 \leqq t \leqq 1$, be a parameter representation of C. By assumption, $\gamma(t)$ is a continuously differentiable function of t, such that $\gamma'(t) \neq 0$ for $t \in [0, 1]$, $\gamma(0) = \gamma(1)$, and $\gamma(s) \neq \gamma(t)$ for $0 \leqq s < t < 1$. In order to be able to treat the point $\gamma(0) = \gamma(1)$ just as all other points $\gamma(t)$, we extend $\gamma(t)$ to a function with period one defined on the real line $\mathbb{R}$ by putting $\gamma(t) = \gamma(t-k)$, where t is an arbitrary real number such that $k \leqq t \leqq k+1$, k an integer.

Splitting $\gamma(t)$ into its real and imaginary part, we write

$$\gamma(t) = \rho(t) + i\sigma(t).$$

Let $\zeta = \gamma(\tau) = \xi + i\eta = \rho(\tau) + i\sigma(\tau)$, $0 \leqq \tau \leqq 1$ be an arbitrary point on C. Since $\gamma'(\tau) \neq 0$, at least one of $\rho'(\tau)$ and $\sigma'(\tau)$ is not equal to 0. Let us assume $\rho'(\tau) \neq 0$; then $|C|$ can be represented by an equation $y = \phi(x)$, where $\phi(x)$ is a continuously differentiable function of x on a neighborhood of ζ.

If $\rho'(\tau) > 0$, then $\rho'(t) > 0$ on a neighborhood of ζ, hence $x = \rho(t)$ is a monotone increasing function of t. Therefore the inverse function

* This fact is intuitively clear from the diagrams. The proof that follows is nothing but a verification through calculation of this intuitively obvious fact.

$t = \rho^{-1}(x)$ is a continuously differentiable, monotone increasing function on a neighborhood of $\xi = \rho(\tau)$. By applying the change of coordinate $t = \rho^{-1}(x)$ to $\gamma(t)$ we obtain the parameter representation

$$x \to z = x + i\phi(x), \qquad \phi(x) = \sigma(\rho^{-1}(x))$$

on a neighborhood of ζ. If $\rho'(\tau) < 0$, then $t = \rho^{-1}(x)$ is a continuously differentiable, monotone decreasing function on a neighborhood of ξ, and $x \to z = x + i\phi(x)$ is a parameter representation of the Jordan curve $-C$ obtained from C by reversing the orientation on a neighborhood of ζ. In both cases, $|C|$ is represented by the equation $y = \phi(x)$ on a neighborhood of ζ.

Since $\phi'(x)$ is a continuous function of x, there exists for each $\varepsilon > 0$ a $\delta(\varepsilon) > 0$ such that

$$|\phi'(x) - \phi'(\xi)| < \varepsilon \qquad \text{if } |x - \xi| < \delta(\varepsilon). \tag{2.22}$$

By the intermediate value theorem there exist a θ, $0 < \theta < 1$, such that

$$\phi(x) = \phi(\xi) + \phi'(\xi + \theta(x - \xi))(x - \xi).$$

Since $\phi(\xi) = \eta$ we get from (2.22)

$$|\phi(x) - \eta - \phi'(\xi)(x - \xi)| \leqq \varepsilon|x - \xi| \qquad \text{if } |x - \xi| < \delta(\varepsilon). \tag{2.23}$$

We have proved that if $\rho'(\tau) \neq 0$, then $|C|$ can be represented on a neighborhood of $\zeta = \gamma(\tau)$ by the equation $y = \phi(x)$ and that inequality (2.23) is satisfied on this neighborhood. Next, we want to prove the following. If the condition

$$3|\rho'(\tau)| \geqq |\sigma'(\tau)| \tag{2.24}$$

is satisfied, then the "size" of the neighborhood and the value of $\delta(\varepsilon)$ occurring in (2.23) can be determined independent of the point $\zeta = \gamma(\tau)$.

Since $\gamma'(t)$ is a continuous function of t and since $\gamma'(t+1) = \gamma'(t)$, $\gamma'(t)$ is uniformly continuous. Therefore, for each $\varepsilon > 0$, there exists a $\beta(\varepsilon) > 0$ such that

$$|\gamma'(t) - \gamma'(s)| < \varepsilon \qquad \text{if } |t - s| < \beta(\varepsilon). \tag{2.25}$$

Since $\gamma'(t) \neq 0$ for all t, there exists a $\kappa > 0$ such that

$$|\gamma'(t)| \geqq 8\kappa > 0. \tag{2.26}$$

It is sufficient to prove that $\beta(\varepsilon)$ is a monotone nondecreasing function of ε. (For under this assumption $\alpha(\varepsilon) = \sup_{0 < u \leqq \varepsilon} \beta(u)$ is a monotone nondecreasing function of ε. Hence there exists a u with $0 < u \leqq \varepsilon$ such that $|t - s| < \beta(u)$ if $|t - s| < \alpha(\varepsilon)$, so $|\gamma'(t) - \gamma'(s)| < u \leqq \varepsilon$ if $|t - s| < \alpha(\varepsilon)$.) From (2.24) and (2.26) we get

$$10|\rho'(\tau)|^2 \geqq |\rho'(\tau)|^2 + |\sigma'(\tau)|^2 = |\gamma'(\tau)|^2 \geqq 64\kappa^2,$$

hence

$$|\rho'(\tau)| > 2\kappa,$$

that is, $\rho'(\tau) > 2\kappa$ or $\rho'(\tau) < -2\kappa$.

First, let us assume $\rho'(\tau) > 2\kappa$. By (2.25)

$$\rho'(t) > \kappa \qquad \text{if } |t-\tau| < \beta(\kappa). \tag{2.27}$$

Hence, $x = \rho(t)$ is a continuously differentiable, monotone increasing function of t on the interval $[\tau - \beta(\kappa), \tau + \beta(\kappa)]$, so its inverse function $t = \rho^{-1}(x)$ is a continuously differentiable, monotone increasing function of x on the interval $[\rho(\tau - \beta(\kappa)),\ \rho(\tau + \beta(\kappa))]$. Therefore, $\phi(x) = \sigma(\rho^{-1}(x))$ is also continuously differentiable on $[\rho(\tau - \beta(\kappa)), \rho(\tau + \beta(\kappa))]$. By (2.27) we have

$$\rho(\tau + \beta(\kappa)) - \rho(\tau) = \int_{\tau}^{\tau+\beta(\kappa)} \rho'(t)\,dt > \kappa\beta(\kappa)$$

and similarly $\rho(\tau) - \rho(\tau - \beta(\kappa)) > \kappa\beta(\kappa)$. Since $\rho(\tau) = \xi$ we conclude that the interval $[\xi - \kappa\beta(\kappa),\ \xi + \kappa\beta(\kappa)]$ is contained in the interval $[\rho(\tau - \beta(\kappa)), \rho(\tau + \beta(\kappa))]$. From now on, we regard $\phi(x)$ as a function with domain $[\xi - \kappa\beta(\kappa),\ \xi + \kappa\beta(\kappa)]$. From

$$\phi'(x) = \frac{\phi'(t)}{\rho'(t)}, \qquad t = \rho^{-1}(x),$$

we get

$$|\phi'(x) - \phi'(\xi)| \leqq \left|\frac{\sigma'(t)}{\rho'(t)} - \frac{\sigma'(\tau)}{\rho'(t)}\right| + \left|\frac{\sigma'(\tau)}{\rho'(t)} - \frac{\sigma'(\tau)}{\rho'(\tau)}\right|$$

$$\leqq \frac{|\sigma'(t) - \sigma'(\tau)|}{\rho'(t)} + |\sigma'(\tau)|\,\frac{|\rho'(t) - \rho'(\tau)|}{\rho'(t)\rho'(\tau)}.$$

Therefore, by (2.27) and (2.24)

$$|\phi'(x) - \phi'(\xi)| \leqq \frac{4}{\kappa}|\gamma'(t) - \gamma'(\tau)|;$$

hence, by (2.25)

$$|\phi'(x) - \phi'(\xi)| < \varepsilon \qquad \text{if } |t-\tau| < \beta(\kappa\varepsilon/4).$$

Since $dt/dx = 1/\rho'(t) < 1/\kappa$ by (2.27) we have

$$|t-\tau| = \left|\int_{\xi}^{x} \frac{dt}{dx}\,dx\right| \leqq \frac{1}{\kappa}|x-\xi|.$$

Hence

$$|\phi'(x) - \phi'(\xi)| < \varepsilon \qquad \text{if } |x-\xi| < \kappa\beta(\kappa\varepsilon/4).$$

Note that x belongs to the domain of $\phi(x)$ since $|x-\xi| \leqq \kappa\beta(\kappa)$. Putting

$$\delta(\varepsilon) = \begin{cases} \kappa\beta(\kappa\varepsilon/4), & 0 < \varepsilon \leqq 4, \\ \kappa\beta(\kappa), & \varepsilon > 4, \end{cases} \tag{2.28}$$

we have $\delta(\varepsilon) \leqq \kappa\beta(\kappa)$ for all ε since $\beta(\varepsilon)$ is a monotone nondecreasing function of ε. Therefore

$$|\phi'(x) - \phi'(\xi)| < \varepsilon \qquad \text{if } |x-\xi| < \delta(\varepsilon),$$

hence

$$|\phi(x) - \eta - \phi'(\xi)(x-\xi)| \leqq \varepsilon|x-\xi| \qquad \text{if } |x-\xi| < \delta(\varepsilon).$$

Putting $\varepsilon = 1$ and using $|\phi'(\xi)| = |\sigma'(\tau)/\rho'(\tau)| \leqq 3$ (by 2.24), we get

$$|\phi(x) - \eta| < 4|x-\xi| \qquad \text{if } |x-\xi| < \delta(1).$$

Defining a neighborhood $U(\zeta)$ of $\zeta = \xi + i\eta$ by

$$U(\zeta) = \{x+iy \colon |x-\xi| < \delta(1), |y-\eta| < 4\delta(1)\}$$

the points $x + i\phi(x) \in |C|$ with $\xi - \delta(1) < x < \xi + \delta(1)$ belong to $U(\zeta)$.

Since $\delta(1) = \kappa\beta(\kappa/4)$, we get for sufficiently small κ by (2.7)

$$|C| \cap U(\zeta) = \{x + i\phi(x) \colon \xi - \delta(1) < x < \xi + \delta(1)\}.$$

The case $\rho'(\tau) < 2\kappa$ is treated similarly, the only difference being that now $\rho(t)$ and $\rho^{-1}(x)$ are monotone decreasing functions.

To summarize, if condition (2.24),

$$3|\rho'(\tau)| \geqq |\sigma'(\tau)|,$$

is satisfied at a point $\zeta = \xi + i\eta = \gamma(\tau) = \rho(\tau) + i\sigma(\tau)$ of C, then $|C|$ can be represented on the neighborhood $U(\zeta)$ by

$$|C| \cap U(\zeta) = \{x + i\phi(x) \colon \xi - \delta(1) < x < \xi + \delta(1)\}. \tag{2.29}$$

Here, $\phi(x)$ is a continuously differentiable function of x on the closed interval $[\xi - \kappa\beta(\kappa), \xi + \kappa\beta(\kappa)]$, where $\kappa\beta(\kappa) \geqq \delta(1)$, satisfying

$$|\phi(x) - \eta - \phi'(\xi)(x-\xi)| \leqq \varepsilon|x-\xi| \qquad \text{if } |x-\xi| < \delta(\varepsilon). \tag{2.30}$$

$U(\zeta)$ is divided by $|C|$ into two subregions $U^+(\zeta)$ and $U^-(\zeta)$ defined by the inequalities $y > \phi(x)$ and $y < \phi(x)$, respectively:

$$U(\zeta) - |C| = U^+(\zeta) \cup U^-(\zeta).$$

Let $l_x \colon y \to l_x(y) = x + iy$ represent the line through the point $z = x + i\phi(x)$ on C and parallel to the imaginary axis.

If $\rho'(\tau) > 2\kappa$, then $x = \rho(\tau)$ is a monotone increasing function. Since

$$x \to z = x + i\phi(x)$$

is a parameter representation of C on $U(\zeta)$, we see that the line l_x crosses C from right to left at $l_x(\phi(x)) = x + i\phi(x)$. Therefore, if

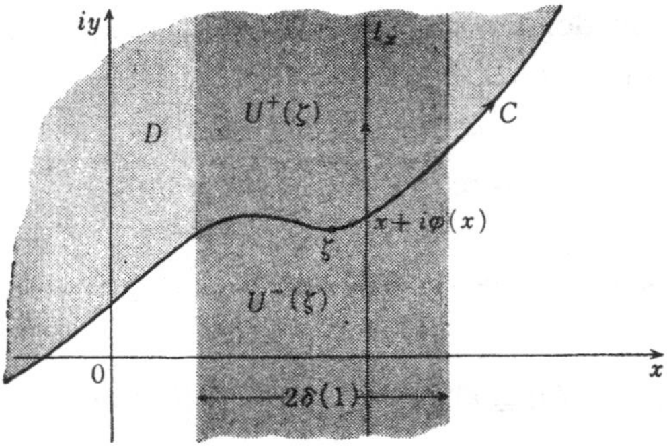

Fig. 2.20

$\phi(x) < y < \eta + 4\delta(1)$ then $x + iy \in D$, that is, $U^+(\zeta) \subset D$ and $U^-(\zeta) \subset \mathbb{C} - [D]$. If $\rho'(\tau) < -2\kappa$, then $x \to z = x + i\phi(x)$ is a parameter representation of $-C$, hence $U^-(\zeta) \subset D$ and $U^+(\zeta) \subset \mathbb{C} - [D]$. Similarly, if the condition

$$3|\sigma'(\tau)| \geqq |\rho'(\tau)| \tag{2.31}$$

is satisfied at $\zeta = \xi + i\eta = \gamma(\tau)$ and if we put

$$V(\zeta) = \{x + iy: |x - \xi| < 4\delta(1), |y - \eta| < \delta(1)\},$$

then

$$|C| \cap V(\zeta) = \{\psi(y) + iy: \eta - \delta(1) < y < \eta + \delta(1)\}. \tag{2.32}$$

$\psi(y)$ is a continuously differentiable function of y defined on the closed interval $[\eta - \kappa\beta(\kappa), \eta + \kappa\beta(\kappa)]$ and

$$|\psi(y) - \xi - \psi'(\eta)(y - \eta)| \leqq \varepsilon(y - \eta) \qquad \text{if } |y - \eta| < \delta(\varepsilon). \tag{2.33}$$

If a point $x_1 + iy_1$, on C with $0 < |x_1 - \xi| < \delta(\varepsilon)$ and $0 < |y_1 - \eta| < \delta(\varepsilon)$ is such that

$$\left|\frac{y_1 - \eta}{x_1 - \xi}\right| \leqq \frac{3}{3\varepsilon + 1} \tag{2.34}$$

holds, then condition (2.24), $3|\rho'(\tau)| \geqq |\sigma'(\tau)|$, is satisfied. For, assuming $3|\sigma'(\tau)| \geqq |\rho'(\tau)|$, we have $x_1 = \psi(y_1)$ by (2.32), hence

$$|x_1 - \xi - \psi'(\eta)(y_1 - \eta)| \leqq \varepsilon|y_1 - \eta|$$

by (2.33), that is,

$$\left|\frac{x_1 - \xi}{y_1 - \eta} - \psi'(\eta)\right| \leqq \varepsilon.$$

Using (2.34) we get

$$\left|\frac{\rho'(\tau)}{\sigma'(\tau)}\right| = |\psi'(\eta)| \geqq \left|\frac{x_1-\xi}{y_1-\eta}\right| - \varepsilon \geqq \frac{3\varepsilon+1}{3} - \varepsilon = \frac{1}{3}.$$

Similarly, if

$$\left|\frac{x_1-\xi}{y_1-\eta}\right| \leqq \frac{3}{3\varepsilon+1} \tag{2.35}$$

holds, then we have $3|\sigma'(\tau)| \geqq |\rho'(\tau)|$.

Using these results we can easily ascertain that a cellular decomposition of $[D]$ can be obtained by the procedure described in the beginning of this proof. First select ε with $0 < \varepsilon < 1/3$, next select $\delta > 0$ such that

$$3\delta \leqq \delta(\varepsilon)$$

and cover the complex plane $\mathbb{C}$ with an infinite number of squares Q_{hk} with width δ, as described above.

We consider three cases:

Case 1: A side of Q_{hk} intersects C in at least two points or is tangent to C. Let us put $Q_{hk} = Q_{00}$ to simplify the notation. Let

$$L = \{x + i\delta : 0 \leqq x \leqq \delta\}$$

represent the side of Q_{00} that intersects C in at least two points or is tangent to C and let $\zeta = \xi + i\delta = \gamma(\tau)$ be one of the points of intersection or the point of tangency. The rectangle $R = Q_{00} \cup Q_{01}$ is given by

$$R = \{x + iy : 0 \leqq x \leqq \delta, 0 \leqq y \leqq 2\delta\}.$$

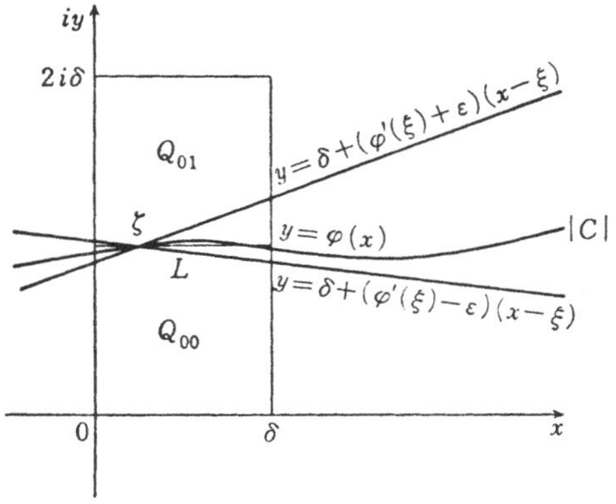

Fig. 2.21

Assuming that $3|\sigma'(\tau)| \geqq |\rho'(\tau)|$ holds at $\zeta = \gamma(\tau)$, C is given by the equation $x = \psi(y)$ on $V(\zeta)$ by (2.32). Hence L intersects C transversely in exactly one point, contrary to the assumption. (Since $3\delta \leqq \delta(\varepsilon) \leqq \delta(1)$, we have $V(\zeta) \supset R$.) Therefore $|C|$ is given by the equation $y = \phi(x)$ on $U(\zeta)$ by (2.29). By (2.30) we have

$$|\phi(x) - \delta - \phi'(\xi)(x-\xi)| \leqq \varepsilon|x-\xi| \qquad \text{if } |x-\xi| < \delta(\varepsilon), \tag{2.36}$$

that is, on the neighborhood $\{x+iy\colon |x-\xi| < \delta(\varepsilon),\ |y-\delta| < 4\delta(1)\}$ of $\zeta = \xi + i\delta$, the curve C is between the lines $y = \delta + (\phi'(\xi)+\varepsilon)(x-\xi)$ and $y = \delta + (\phi'(\xi)-\varepsilon)(x-\xi)$, as shown in Fig. 2.21. If L intersects C in at least two points, we put one such other point of intersection $\zeta_1 = \xi_1 + i\delta$, $\xi_1 \neq \xi$. Since $\phi(\xi_1) = \delta$ and $|\xi_1 - \xi| \leqq \delta < \delta(\varepsilon)$, we have $\phi'(\xi) \leqq \varepsilon$ by (2.36). If L is tangent to C at ζ, then $\phi'(\xi) = 0$. So, in both cases, we have $|\phi'(\xi)| \leqq \varepsilon$. Hence, by (2.36)

$$|\phi(x) - \delta| \leqq 2\varepsilon|x-\xi| \leqq 2\varepsilon\delta \qquad \text{if } 0 \leqq x \leqq \delta,$$

therefore

$$0 < \phi(x) < 2\delta \qquad \text{if } 0 \leqq x \leqq \delta$$

since $\varepsilon \leqq \frac{1}{3}$.

Therefore, assuming $\rho'(\tau) > 0$, we have $U^+(\zeta) \subset D$ and

$$[D] \cap R = \{x+iy\colon 0 \leqq x \leqq \delta,\ \phi(x) \leqq y \leqq 2\delta\}.$$

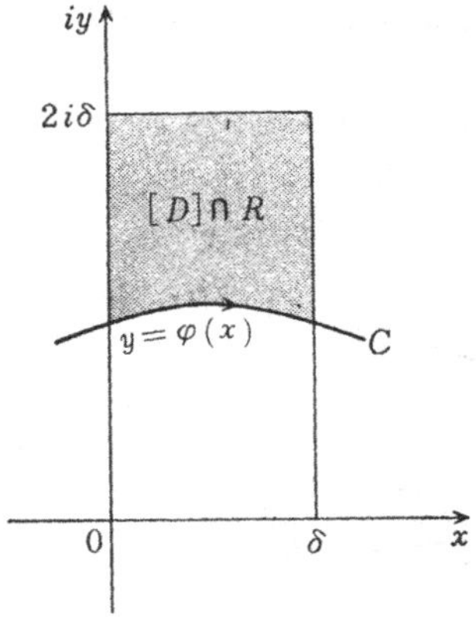

Fig. 2.22

If $\rho'(\tau) < 0$, we have

$$[D] \cap R = \{x+iy\colon 0 \leqq x \leqq \delta,\ 0 \leqq y \leqq \phi(x)\}.$$

In both cases $[D] \cap R$ is a cell as described in Example 2.3. Let $L_1, L_2, \ldots, L_p, \ldots, L_m$ represent all sides from among the sides of the squares Q_{hk}, $h, k = 0, \pm 1, \pm 2, \ldots$, which intersect C in at least two points or are tangent to C. For each L_p let $R_p = Q_{hk} \cup Q_{h'k'}$ be the union of the two squares Q_{hk} and $Q_{h'k'}$ which have L_p as one of their sides. We have shown that all $[D] \cap R_p$ are cells. If $p \neq q$, then R_p and R_q do not have interior points in common. (For, suppose $R_p = R = Q_{00} \cup Q_{01}$. If R_q has interior points in common with R, then one of the squares making up R, say, Q_{00} is contained in R. Hence L_q has to be a side of Q_{00}, but all sides of Q_{00} except $L = L_p$ either do not intersect C at all or intersect C transversely in exactly one point. Hence $L_q = L = L_p$, i.e., $p = q$.)

Case 2: $Q_{hk} \not\subset \bigcup_p R_p$ and $Q_{h,k}$ has interior points in common with $[D]$. As before, we put $Q_{hk} = Q_{00}$. If the boundary C of $[D]$ does not pass through the interior of Q_{00}, we have $Q_{00} \subset [D]$ and $[D] \cap Q_{00} = Q_{00}$ is a cell. If C passes through the interior of Q_{00}, we conclude from (2.29) that C intersects the boundary of Q_{00} in at least two points. Since $Q_{00} \not\subset \bigcup_p R_p$, each side of Q_{00} either does not intersect C at all or intersects C transversely in exactly one point. Therefore C intersects at least two sides of Q_{00} in one point each. We distinguish two cases:

(a) C intersects two sides of Q_{00} which are parallel. Let us assume that C intersects the side on the imaginary axis at $i\eta = \gamma(\tau)$, $0 \leqq \eta \leqq \delta$, and the other side at $\zeta_1 = \delta + i\eta_1$, $0 \leqq \eta_1 \leqq \delta$. At $i\eta$ we have $3|\rho'(\tau)| \geqq |\sigma'(\tau)|$. (For, since $\varepsilon < \frac{1}{3}$, we have $3/(3\varepsilon + 1) > \frac{3}{2}$. Hence, the inequality (2.34) $|\eta_1 - \eta|/\delta \leqq 3(3\varepsilon + 1)$, holds at the point $\delta + i\eta_1 \neq i\eta = \gamma(\tau)$ of C.) Therefore, by (2.29), $|C|$ is given by the equation $y = \phi(x)$ in a neighborhood $U(i\eta)$ of $i\eta$. The side $\{x: 0 \leqq x \leqq \delta\}$ of Q_{00} either does not intersect C at all or intersects C transversely in exactly one point. If C intersects this side transversely in $\xi, 0 < \xi < \delta$, then $\phi(\xi) = 0$ and $\phi'(\xi) \neq 0$. If $\phi'(\xi) > 0$, then $\eta = \phi(0) < 0$ and if $\phi'(\xi) < 0$, then $\eta_1 = \phi(\delta) < 0$ by the intermediate value theorem. This contradicts our assumption, hence $\phi(x) > 0$ if $0 < x < \delta$. Similarly, $\phi(x) < \delta$ if $0 < x < \delta$, hence

$$0 < \phi(x) < \delta \qquad \text{if } 0 < x < \delta.$$

Therefore if $\rho'(\tau) > 0$ then $U^+(i\eta) \subset D$ and

$$[D] \cap Q_{00} = \{x + iy: 0 \leqq x \leqq \delta,\ \phi(x) \leqq y \leqq \delta\},$$

If $\rho'(\tau) < 0$, then

$$[D] \cap Q_{00} = \{x + iy: 0 \leqq x \leqq \delta,\ 0 \leqq y \leqq \phi(x)\}.$$

In both cases $[D] \cap Q_{00}$ is a cell as described in Example 2.3.

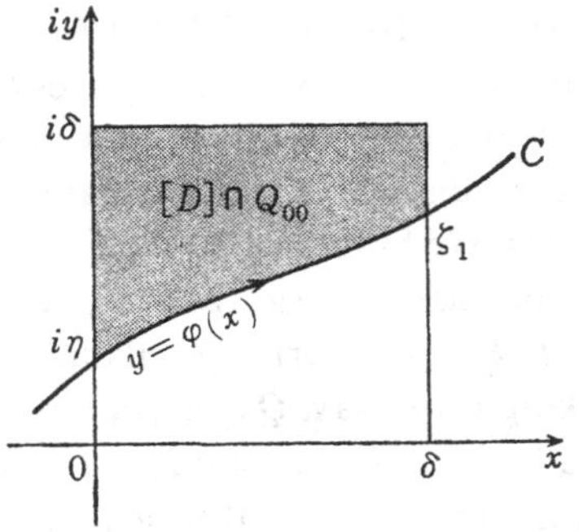

Fig. 2.23

(b) C intersects two adjacent sides of Q_{00} in one point each. Let us assume that C intersects the side $\{iy: 0 \leqq y \leqq \delta\}$ at $i\eta$, $0 < \eta \leqq \delta$, and the side $\{x: 0 \leqq x \leqq \delta\}$ at $\xi, 0 < \xi \leqq \delta$. $|C|$ is given by the equation $y = \phi(x)$ in a neighborhood $U(i\eta)$ of $i\eta$ or by the equation $x = \psi(y)$ in a neighborhood $V(i\eta)$ of $i\eta$. Let us assume that $|C|$ is given by an equation $y = \phi(x)$. If $\phi'(\xi) < 0$, then $0 < \phi(x) < \delta$ for $0 \leqq x < \xi$ and $\phi(x) < 0$ for $\xi < x \leqq \delta$. Therefore, if $\rho'(\tau) > 0$, we have

$$[D] \cap Q_{00} = \{x + iy: 0 \leqq x \leqq \delta,\ \phi(x) \leqq y \leqq \delta,\ 0 \leqq y \leqq \delta\}.$$

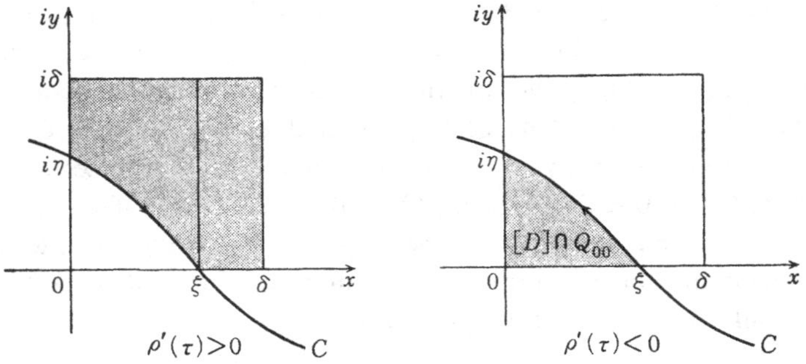

Fig. 2.24

In this case $[D] \cap Q_{00}$ is the union of a cell, as described in Example 2.3,

$$\{x + iy: 0 \leqq x \leqq \xi,\ \phi(x) \leqq y \leqq \delta\},$$

and a rectangle,

$$\{x + iy: \xi \leqq x \leqq \delta,\ 0 \leqq y \leqq \delta\}.$$

If $\rho'(\tau) < 0$ then

$$[D] \cap Q_{00} = \{x+iy: 0 \leqq x \leqq \xi, 0 \leqq y \leqq \phi(x)\},$$

which is a cell as described in Example 2.3.

Case 3: $Q_{hk} \not\subset \bigcup_p R_p$ and Q_{hk} has no interior points in common with $[D]$. In this case, $[D] \cap Q_{hk}$ is either empty or consists only of the points where C intersects the sides of Q_{hk}. Since C intersects each side of Q_{hk} in at most one point, $[D] \cap Q_{hk}$ consists of at most two points. Since $[D]$ has no isolated points, we can omit these sets $[D] \cap Q_{hk}$ without changing the validity of the equality

$$[D] = \bigcup([D] \cap Q_{hk}).$$

Hence, $[D]$ is decomposed by the procedure above into a finite number of cells, no two of which have interior points in common.

Part (II) of proof: The boundary of $[D]$ is a piecewise smooth Jordan curve C. Let $C = C_1 \cdot C_2 \cdot \cdots \cdot C_j \cdot \cdots \cdot C_m$, with all C_j smooth Jordan arcs and let $\gamma_j: t \to \gamma_j(t), 0 \leqq t \leqq 1$, be a parameter representation of C_j. We have

$$\gamma_1(1) = \gamma_2(0), \gamma_2(1) = \gamma_3(0), \ldots, \gamma_j(1) = \gamma_{j+1}(0), \ldots, \gamma_m(1) = \gamma_1(0).$$

Just as in Part (I) we cover the complex plane $\mathbb{C}$ with an infinite number of squares Q_{hk} with width δ. Then $[D]$ is covered by a finite number of closed subsets $[D] \cap Q_{hk} \neq \varnothing$:

$$[D] = \bigcup([D] \cap Q_{hk}).$$

By adjusting this decomposition, we arrive at a cellular decomposition, but now the points $\gamma_j(0)$, $j = 1, \ldots, m$, need special consideration. Put $\zeta_1 = \gamma_1(0) = \gamma_m(1), \ldots, \zeta_j = \gamma_j(0) = \gamma_{j-1}(1), \ldots, \zeta_m = \gamma_m(0) = \gamma_{m-1}(1)$ and pick squares $Q_{h_j k_j}$ such that $\zeta_j \in Q_{h_j k_j}$. Let Q_j be the union of the nine squares $Q_{h'' k''}$ such that $Q_{h''k''} \cap Q_{h_j k_j} \neq \varnothing$ (i.e., $h'' = h_{j-1}, h_j, h_{j+1}$, $k'' = k_{j-1}, k_j, k_{j+1}$). Q_j is a square of width 3δ, its center coinciding with that of Q_{hk_j}. Next we divide all squares Q_{hk}, which are not contained in $\bigcup_{j=1}^m Q_j$, into n^2 squares Q_{hkpq} with width δ/n (where n is a natural number) defined by

$$Q_{hkpq} = \{x+iy: (p-1)\delta/n \leqq x - h\delta \leqq p\delta/n, (q-1)\delta/n \leqq y - k\delta \leqq q\delta/n\}.$$

We have

$$Q_{hk} = \bigcup_{p,q=1}^{n} Q_{hkpq}$$

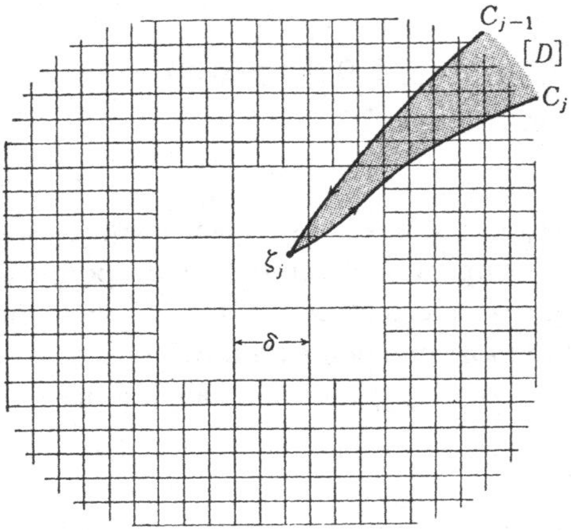

Fig. 2.25

and the decomposition of $[D]$

$$[D] = \bigcup_{h,k,p,q} ([D] \cap Q_{hkpq}) \cup ([D] \cap Q_1) \cup \cdots \cup ([D] \cap Q_m).$$

$\bigcup_{h,k,p,q}([D] \cap Q_{hkpq})$ can be changed into a cellular decomposition by the procedure described in Part (I). (Figure 2.25 illustrates why it is necessary to divide the squares $Q_{hk} \not\subset \bigcup_j Q_j$ into n^2 smaller squares for some sufficiently large n.) Therefore it suffices to show that each $[D] \cap Q_j$ can be decomposed into a finite number (in fact, at most three) of cells, as illustrated in Fig. 2.26.*

Define $\beta(\varepsilon), \kappa, \delta(\varepsilon)$ for $\gamma_j(t), j = 1, 2, \ldots, m$, as they were defined in Part (I) for $\gamma(t)$, that is, $\beta(\varepsilon)$ is a monotone nondecreasing function of ε, $\varepsilon > 0$, such that $\beta(\varepsilon) > 0$ and $|\gamma_j'(t) - \gamma_j'(s)| < \varepsilon, j = 1, 2, \ldots, m$, if $|t - s| < \beta(\varepsilon)$, $0 \leqq t < s \leqq 1$, κ is a constant satisfying

$$|\gamma_j'(t)| \geqq 8\kappa > 0, \qquad j = 1, 2, \ldots, m$$

and $\delta(\varepsilon)$ is defined by

$$\delta(\varepsilon) = \begin{cases} \kappa\beta(\kappa\varepsilon/4), & 0 < \varepsilon \leqq 4, \\ \kappa\beta(\kappa), & \varepsilon > 4. \end{cases}$$

Splitting $\gamma_j(t)$ into its real and imaginary part, we write $\gamma_j(t) = \rho_j(t) + i\sigma_j(t)$. For $\zeta = \xi + i\eta$ we define neighborhoods $U(\zeta)$ and $V(\zeta)$ as in Part (I) by

$$U(\zeta) = \{x + iy : |x - \xi| < \delta(1), |y - \eta| < 4\delta(1)\},$$

$$V(\zeta) = \{x + iy : |y - \eta| < \delta(1), |x - \xi| < 4\delta(1)\}.$$

* This is also intuitively clear: if C_{j-1} and C_j are both segments, the proof is trivial. By taking Q_j sufficiently small the parts of C_{j-1} and C_j contained in Q_j are almost segments.

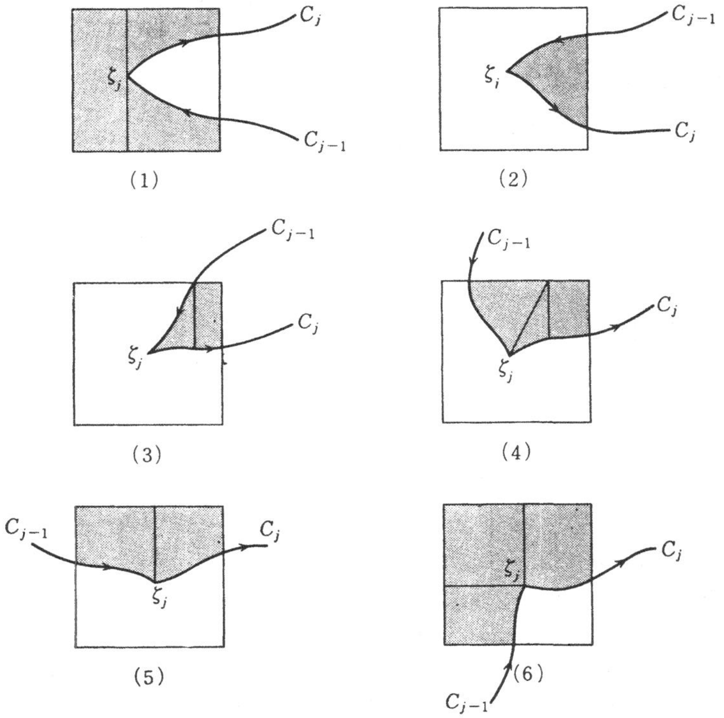

Fig. 2.26

By selecting a sufficiently small κ, we may assume that similar results to (2.29) and (2.30) or (2.32) and (2.33) are valid for $|C_j|$ in the neighborhood

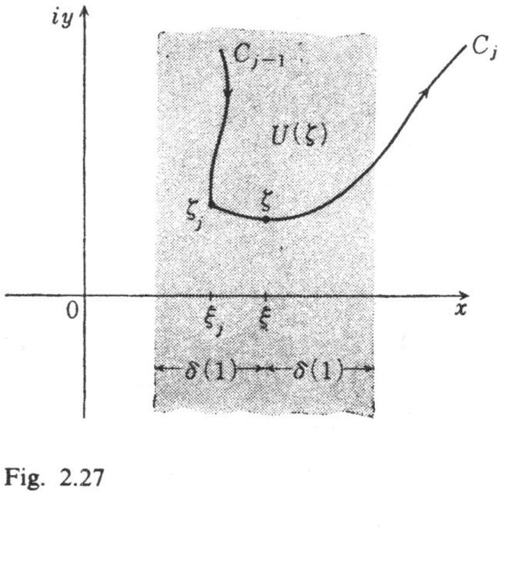

Fig. 2.27

$U(\zeta)$ or $V(\zeta)$ of the point $\zeta = \xi + i\eta = \gamma_j(t)$ of C_j. However, since C_j is not a Jordan curve, we have to adjust these results if ζ is close to $\gamma_j(0)$ or $\gamma_j(1)$. For example, if condition (2.24), $3\,|\rho'_j(\tau)| \geqq |\sigma'_j(\tau)|$, is satisfied and if $\rho'_j(\tau) > 0$ and $\xi_j = \rho_j(0) > \xi - \delta(1)$, then

$$|C_j| \cap U(\zeta) = \{x + i\phi_j(x) \colon \xi_j \leqq x < \xi + \delta(1)\}.$$

If $\zeta = \zeta_j = \xi_j + in_j = \gamma_j(0)$ this result becomes (if $3\,|\rho'_j(0)| \geqq |\sigma'_j(0)|$ and if $\rho'_j(0) > 0$)

$$|C_j| \cap U(\zeta_j) = \{x + i\phi_j(x) \colon \xi_j \leqq x < \xi_j + \delta(1)\}, \tag{2.37}$$

and if $\rho'_j(0) < 0$,

$$|C_j| \cap U(\zeta_j) = \{x + i\phi_j(x) \colon \xi_j - \delta(1) < x \leqq \xi_j\}. \tag{2.38}$$

If $3\,|\sigma'_j(0)| \geqq |\rho'_j(0)|$ and $\sigma'_j(0) > 0$, then

$$|C_j| \cap V(\zeta_j) = \{\psi_j(y) + iy \colon \eta_j \leqq y < \eta_j + \delta(1)\}, \tag{2.39}$$

and if $\sigma'_j(0) < 0$, then

$$|C_j| \cap V(\zeta_j) = \{\psi_j(y) + iy \colon \eta_j - \delta(1) < y \leqq \eta_j\}. \tag{2.40}$$

The functions $\phi_j(x)$ and $\psi_j(y)$ are continuously differentiable functions of x and y, respectively, satisfying the inequalities (2.30) and (2.33), respectively, that is, we have for ϕ_j

$$|\phi_j(x) - \eta_j - \phi'_j(\xi_j)\,(x - \xi_j)| \leqq \varepsilon\,(x - \xi_j) \qquad \text{if } \xi_j \leqq x < \xi_j + \delta(\varepsilon)$$

and for $\psi_j(y)$

$$|\psi_j(y) - \xi_j - \psi'_j(\eta_j)\,(y - \eta_j)| \leqq \varepsilon\,(y - \eta_j) \qquad \text{if } \eta_j \leqq y < \eta_j + \delta(\varepsilon).$$

Hence, if the point $x_1 + iy_1$ of C_j, $\xi_1 < x_1 < \xi_1 + \delta(1)$, $|y_1 - \eta_j| < \delta(1)$, is such that the inequality (2.34)

$$\left|\frac{y_1 - \eta_j}{x_1 - \xi_j}\right| \leqq \frac{3}{3\varepsilon + 1}$$

is valid, then $3|\rho'_j(0)| \geqq |\sigma'_j(0)|$. If the point $x_1 + iy$ of C_j, $\eta_1 < y_1 < \eta_1 + \delta(1)$, $|x_1 - \xi_j| < \delta(1)$, is such that the inequality,

$$\left|\frac{x_1 - \xi_j}{y_1 - \eta_j}\right| \leqq \frac{3}{3\varepsilon + 1}$$

is valid, then $3\,|\sigma'_j(0)| \geqq |\rho'_j(0)|$.

Since $\zeta_j = \gamma_{j-1}(1)$ is the terminal end point of the arc $-C_{j-1}$ with parameter representation $t \to \gamma_{j-1}(1 - t)$, $0 \leqq t \leqq 1$, we can replace C_j, $\rho'_j(0)$, and $\sigma'_j(0)$ in the above results by $-C_{j-1}$, $-\rho'_{j-1}(1)$, and $-\sigma'_{j-1}(1)$. For example, if $3\,|\rho'_{j-1}(1)| \geqq |\sigma'_{j-1}(1)|$ and $\rho'_{j-1}(1) < 0$, then

$$|C_{j-1}| \cap U(\zeta_j) = \{x + i\phi_{j-1}(x) \colon \xi_j \leqq x < \xi_j + \delta(1)\}, \tag{2.41}$$

and if $3|\sigma'_{j-1}(1)| \geqq |\rho'_{j-1}(1)|$ and $\sigma'_{j-1}(1) < 0$, then

$$|C_{j-1}| \cap V(\zeta_j) = \{\psi_{j-1}(y) + iy: \eta_j \leqq y < \eta_j + \delta(1)\}. \quad (2.42)$$

Choose $\varepsilon < \frac{1}{9}$ and $3\delta \leqq \delta(\varepsilon)$ and consider $Q_1 \cap [D]$. If $\zeta_1 \in Q_{11}$, we have

$$Q_1 = \{x + iy: 0 \leqq x \leqq 3\delta, 0 \leqq y \leqq 3\delta\}.$$

Let $L_1 = \{3\delta + iy: \ 0 \leqq y \leqq 3\delta\}$, $L_2 = \{x + 3i\delta: \ 0 \leqq x \leqq 3\delta\}$, $L_3 = \{iy: 0 \leqq y \leqq 3\delta\}$, and $L_4 = \{x: 0 \leqq x \leqq 3\delta\}$ be the four sides of the square Q_{11}. The point $\zeta_1(0) = \gamma_m(1)$ is the beginning end point of C_1 and the terminal end point of C_m.

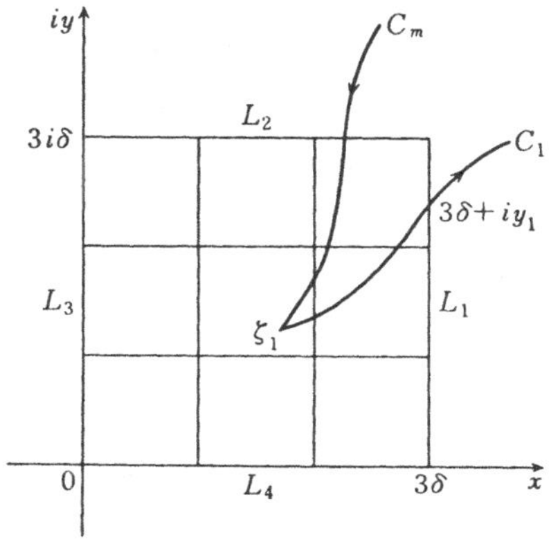

Fig. 2.28

Since $3\delta \leqq \delta(\varepsilon) \leqq \delta(1)$ we have $Q_1 \subset U(\zeta_1)$. Since $\delta(1) = \kappa\beta(\kappa/4)$ and since we may assume that κ is sufficiently small, $U(\zeta_1)$ does not intersect $C_2, C_3, \ldots, C_{m-1}$, that is, $|C| \cap Q_1 = (|C_1| \cap Q_1) \cup (|C_m| \cap Q_1)$. Hence, if z_1 and z_2 are two points in the interior of Q_1, but not on $|C_1| \cup |C_m|$, that can be connected by a polygonal line in the interior of Q_1 not intersecting C_1 and C_m, then z_1 and z_2 either belong to D or to the exterior of $[D]$.

C_1 intersects the boundary $L_1 \cup L_2 \cup L_3 \cup L_4$ of Q_1. We assume that C_1 intersects L_1 and let $3\delta + iy_1, 0 \leqq y_1 \leqq 3\delta$, be the point of intersection of C_1 and L_1. Since $\zeta_1 = \xi_1 + i\eta_1 \in Q_{11}$, we have

$$\left|\frac{y_1 - \eta_1}{3\delta - \xi_1}\right| < \frac{2\delta}{\delta} = 2 < \frac{3}{3\varepsilon + 1},$$

hence $3|\rho'_1(0)| \geqq |\sigma'_1(0)|$ and $|C_1|$ is represented on $U(\zeta_1)$ by (2.37) or (2.38). Since C_1 intersects L_1 we conclude that $|C_1|$ is represented by (2.37), that is,

C_1 is given by the parameter representation

$$x \to x + i\phi_1(x), \qquad \xi_1 \leqq x < \xi_1 + \delta(1),$$

on $U(\zeta_1)$.

We have

$$\text{if } \xi_1 \leqq x < 3\delta, \qquad \text{then } 0 < \phi_1(x) < 3\delta. \tag{2.43}$$

To see this let x_2 be such that $\phi_1(x_2) = 3\delta$, $\xi_1 < x_2 < 3\delta$, then C_1 intersects L_2 in the point $x_2 + 3i\delta$, hence C_1 has the parameter representation

$$y \to \psi_1(y) + iy, \qquad \eta_1 \leqq y < \eta_1 + \delta(1)$$

on $V(\zeta_1)$. Hence $3\delta + iy_1 = \psi(y_1) + iy_1$, i.e., $\psi_1(y_1) = 3\delta, \eta_1 < y_1 \leqq 3\delta$. Since $\phi_1(\xi_1) = \eta_1$ we conclude from the intermediate value theorem that there exists an x_3 with $\phi_1(x_3) = y_1$ and $\xi_1 < x_3 \leqq x_2$. Hence $\psi_1(y_1) = x_3 \leqq x_2 < 3\delta$, contradicting the fact that $\psi_1(y_1) = 3\delta$.

As shown by (2.43), C_1 intersects the boundary of Q_1 in exactly one point $3\delta + iy_1$. Similarly, we prove that C_m intersects the boundary $L_1 \cup L_2 \cup L_3 \cup L_4$ of Q_1 in exactly one point. There are four possibilities to consider:

(1) C_m intersects L_1. In this case,

$$x \to x + i\phi_m(x), \qquad \xi_1 \leqq x < \xi_1 + \delta(1)$$

is a parameter representation of $-C_m$ on $U(\zeta_1)$. C_1 and C_m have no other points of intersection in $U(\zeta_1)$ but ζ_1. Hence, $\phi_m(x) > \phi_1(x)$ for all x with $\xi_1 < x \leqq 3\delta$ or $\phi_1(x) > \phi_m(x)$ for all x with $\xi_1 < x \leqq 3\delta$. Let l_x: $y \to l_x(y) = x + iy$ be the line through x parallel to the imaginary axis for $\xi_1 < x \leqq 3\delta$. The line l_x crosses C_1 from right to left at $x + i\phi_1(x)$ and C_m from left to right at $x + i\phi_m(x)$. Furthermore, l_x intersects $C = \partial[D]$ in no other points on $U(\zeta_1)$. Hence, since $Q_1 \subset U(\zeta_1)$ we have, if $\phi_m(x) < \phi_1(x)$,

$$\begin{aligned} x + iy &\in [D] && \text{if } 0 \leqq y \leqq \phi_m(x), \\ x + iy &\notin [D] && \text{if } \phi_m(x) < y < \phi_1(x), \\ x + iy &\in [D] && \text{if } \phi_1(x) \leqq y \leqq 3\delta. \end{aligned}$$

Hence $[D] \cap Q_1$ is decomposed into three cells:

$$\{x + iy: \xi_1 \leqq x \leqq 3\delta,\ 0 \leqq y \leqq \phi_m(x)\},$$
$$\{x + iy: \xi_1 \leqq x \leqq 3\delta,\ \phi_1(x) \leqq y \leqq 3\delta\},$$
$$\{x + iy: 0 \leqq x \leqq \xi_1,\ 0 \leqq y \leqq 3\delta\}.$$

That the third cell, which is a rectangle, is contained in $[D]$ follows from the fact that a point z in the rectangle can be connected with the point x in D by a polygonal line not intersecting C_1 and C_m. This case corresponds with Fig. 2.26 (1).

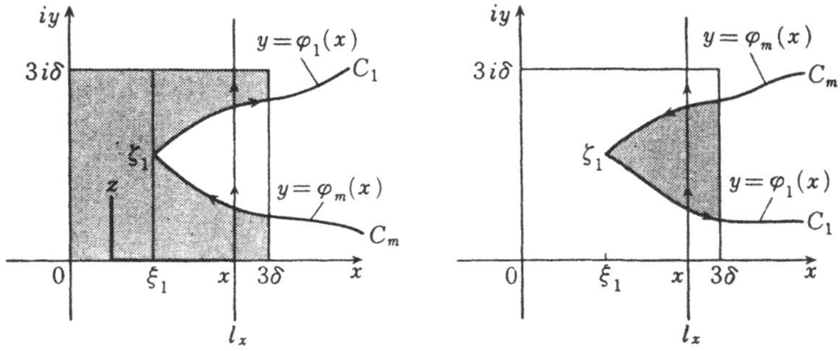

Fig. 2.29

If $\phi_1(x) < \phi_m(x)$, then

$$x + iy \in [D] \qquad \text{if } \phi_1(x) \leqq y < \phi_m(x),$$

$$x + iy \notin [D] \qquad \text{if } 0 \leqq y < \phi_1(x) \text{ or } \phi_m(x) < y \leqq 3\delta.$$

Hence

$$[D] \cap Q_1 = \{x + iy\colon \xi_1 \leqq x \leqq 3\delta,\ \phi_1(x) \leqq y \leqq \phi_m(x)\}.$$

This case corresponds with Fig. 2.26 (2) and $[D] \cap Q_1$ is a cell as described in Example 2.3.

(2) C_m intersects L_2. Let C_m intersect L_2 in the point $x_2 + 3i\delta$, $0 \leqq x_2 \leqq 3\delta$. On $V(\zeta_1)$, $-C_m$ is given by he parameter representation

$$y \to \psi_m(y) + iy, \qquad \eta_1 \leqq y < \delta(1). \tag{2.44}$$

If $-C_m$ is also given by the parameter representation

$$x \to x + i\phi_m(x), \qquad \xi_1 \leqq x < \delta(1), \tag{2.45}$$

then $\phi_1(x) < \phi_m(x)$ for $\xi_1 < x \leqq 3\delta$ and, by 2.44,

$$\phi_m(x) < 3\delta \text{ for } \xi_1 \leqq x < x_2 \quad \text{and} \quad \phi_m(x) > 3\delta \text{ for } x_2 < x \leqq 3\delta.$$

Hence, $[D] \cap Q_1$ is decomposed into two cells:

$$\{x + iy\colon \xi_1 \leqq x \leqq x_2, \phi_1(x) \leqq y \leqq \phi_m(x)\},$$

$$\{x + iy\colon x_2 \leqq x \leqq 3\delta, \phi_1(x) \leqq y \leqq 3\delta\}.$$

This case corresponds to Fig. 2.26 (3). The equation

$$\frac{y - \eta_1}{x - \xi_1} = \frac{3}{1 + 3\varepsilon}$$

represents a line l through $\zeta_1 = \xi_1 + i\eta_1$. Solving for x and y we get

$$x = \lambda(y) = \xi_1 + [\tfrac{1}{3} + \varepsilon](y - \eta_1),$$

$$y = \mu(x) = \eta_1 + \frac{3}{1 + 3\varepsilon}(x - \xi_1).$$

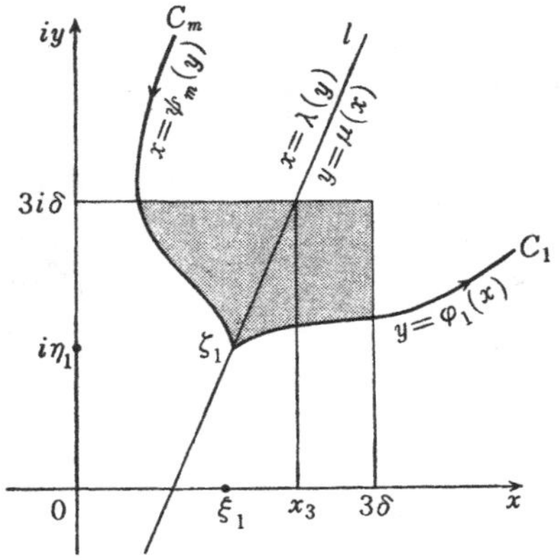

Fig. 2.30

If there exists a point $x+iy=\psi_m(y)+iy$ with $\eta_1 < y \leqq 3\delta$ and $x \geqq \lambda(y)$ on C_m, then

$$\left|\frac{y-\eta_1}{x-\xi_1}\right| \leqq \frac{3}{1+3\varepsilon},$$

hence $3|\rho'_m(1)| \geqq |\sigma'_m(1)|$ and there exists another parameter representation (2.45) for $-C_m$. Hence, if a parameter representation (2.45) does not exist, then

$$\psi_m(y) < \lambda(y) \qquad \text{if } \eta_1 < y \leqq 3\delta. \tag{2.46}$$

In this case

$$\phi_1(x) < \mu(x) \qquad \text{if } \xi_1 < x \leqq 3\delta. \tag{2.47}$$

For, if $\xi_1 < x < \xi_1 + \delta(\varepsilon)$, then

$$\left|\frac{\phi_1(x)-\eta_1}{x-\xi_1} - \phi'_1(\xi_1)\right| \leqq \varepsilon,$$

and since $y_1 = \phi_1(3\delta) \leqq 3\delta$ we have

$$\phi'_1(\xi_1) \leqq \frac{3\delta-\eta_1}{3\delta-\xi_1} + \varepsilon \leqq \frac{2\delta}{\delta} + \varepsilon = 2+\varepsilon.$$

We assumed $\varepsilon < \frac{1}{9}$, hence $2+2\varepsilon < 3/(1+3\varepsilon)$ and we conclude

$$\phi_1(x) \leqq \eta_1 + (2+2\varepsilon)(x-\xi_1) < \mu(x)$$

if $\xi_1 < x \leqq 3\delta$.

Putting $x_3 = \lambda(3\delta)$ we conclude from (2.46) and (2.47) that $[D] \cap Q_1$ can be decomposed into three cells

$$\{x+iy\colon x_3 \leqq x \leqq 3\delta,\ \phi_1(x) \leqq y \leqq 3\delta\},$$
$$\{x+iy\colon \xi_1 \leqq x \leqq x_3,\ \phi_1(x) \leqq y \leqq \mu(x)\},$$
$$\{x+iy\colon \eta_1 \leqq y \leqq 3\delta,\ \psi_m(y) \leqq x \leqq \lambda(y)\}.$$

This case corresponds with Fig. 2.26 (4).

(3) C_m intersects L_3. On $U(\zeta_1)$ the arc C_m is given by the parameter representation

$$x \to x + i\phi_m(x), \qquad 0 \leqq x \leqq \xi_1,$$

and $[D] \cap Q_1$ is decomposed into two cells

$$\{x+iy\colon 0 \leqq x \leqq \xi_1,\ \phi_m(x) \leqq y \leqq 3\delta\},$$
$$\{x+iy\colon \xi_1 \leqq x \leqq 3\delta,\ \phi_1(x) \leqq y \leqq 3\delta\}$$

as shown in Fig. 2.26 (5).

(4) C_m intersects L_3. On $V(\zeta_1)$ the arc C_m is given by a parameter representation

$$y \to \psi_m(y) + iy, \qquad 0 \leqq y \leqq \eta_1,$$

and $[D] \cap Q_1$ is decomposed into three cells

$$\{x+iy\colon \xi_1 \leqq x \leqq 3\delta,\ \phi_1(x) \leqq y \leqq 3\delta\},$$
$$\{x+iy\colon 0 \leqq x \leqq \xi_1,\ \eta_1 \leqq y \leqq 3\delta\},$$
$$\{x+iy\colon 0 \leqq y \leqq \eta_1,\ 0 \leqq x \leqq \psi_m(y)\}$$

as shown in Fig. 2.26 (6).

Part (III) of proof: Since all considerations above were local (which means that it was possible to restrict our attention to sufficiently small neighborhoods of points of C), the same considerations go through if the boundary of $[D]$ consists of a finite number of mutually disjoint piecewise smooth Jordan curves.

Theorem 2.1 is now proved.

Let $\{\Gamma_\lambda(K)\colon \lambda = 1, 2, \ldots, \mu\}$ be a cellular decomposition of $[D]$ as obtained above:

$$[D] = \Gamma_1(K) \cup \Gamma_2(K) \cup \cdots \cup \Gamma_\lambda(K) \cup \cdots \cup \Gamma_\mu(K), \tag{2.48}$$

where $K = \{(t,s)\colon 0 \leqq t \leqq 1,\ 0 < s < 1\}$. Let $E = \{(t,s);\ 0 < t < 1,\ 0 < s < 1\}$ be K's interior. It is clear from the above proof that the cells that have no point or exactly one point in common with the boundary C of $[D]$ are squares or rectangles. For all other cells $\Gamma_\lambda(K)$ the intersection

$$C_\lambda = C \cap \Gamma_\lambda(K)$$

is a Jordan arc and the boundary $|\partial\Gamma_\lambda(K)| = \Gamma_\lambda(K) - \Gamma_\lambda(E)$ of such a $\Gamma_\lambda(K)$ consists of C_λ and at most three segments.

If $\Gamma_\lambda(K) \cap \Gamma_\nu(K)$, $\lambda \neq \nu$, is neither empty nor a set consisting of one point, then $|C_{\lambda\nu}| = \Gamma_\lambda(K) \cap \Gamma_\nu(K)$ is a segment and $C_{\lambda\nu}$ is defined by assigning an orientation to $|C_{\lambda\nu}|$ in such a way that $\Gamma_\lambda(E)$ is on the left of $C_{\lambda\nu}$. Now, if $C \cap \Gamma_\lambda(K)$ is the empty set or a set consisting of one point, we have

$$\partial\Gamma_\lambda(K) = \sum_\nu C_{\lambda\nu}. \tag{2.49}$$

For example, in Fig. 2.31, $\partial\Gamma_5(K)$ is given by

$$\partial\Gamma_5(K) = C_{54} + C_{59} + C_{57} + C_{56} + C_{53},$$

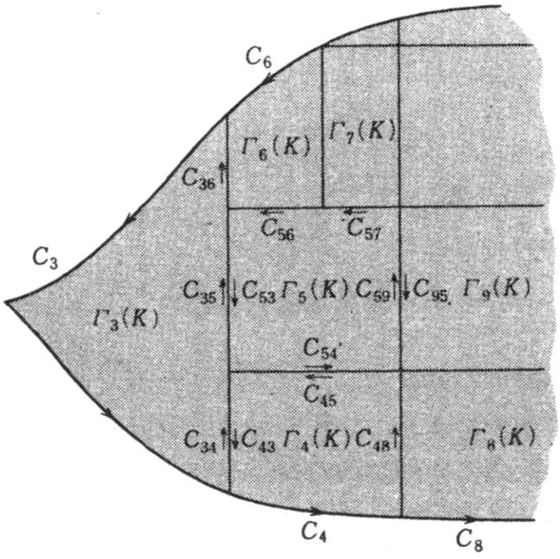

Fig. 2.31

If $C_\lambda = C \cap \Gamma_\lambda(K)$ is a Jordan arc, we have

$$\partial\Gamma_\lambda(K) = C_\lambda + \sum_\nu C_{\lambda\nu}. \tag{2.50}$$

For example,

$$\partial\Gamma_3(K) = C_3 + C_{34} + C_{35} + C_{36},$$
$$\partial\Gamma_4(K) = C_4 + C_{48} + C_{45} + C_{43}$$

in Fig. 2.31. Obviously

$$C_{\lambda\nu} = -C_{\nu\lambda} \tag{2.51}$$

and

$$C = \sum_{\lambda} C_{\lambda}. \tag{2.52}$$

Each Jordan arc C_λ is smooth or piecewise smooth. If C_λ is piecewise smooth, then C_λ consists of two smooth Jordan arcs. For example, in Fig. 2.31, C_4 and C_6 are smooth while C_3 is piecewise smooth.

2.3 Cauchy's Theorem

In this section $[D]$ is a bounded, closed region in the complex plane and its boundary $\partial[D]$ consists of piecewise smooth Jordan curves. D is the interior of $[D]$.

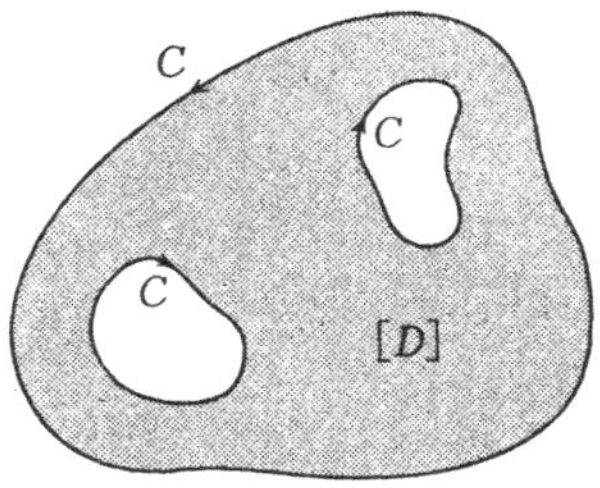

Fig. 2.32

a. Cauchy's Theorem

As agreed upon already in Section 1.1c, a function $f(z)$ is called holomorphic on $[D]$ if $f(z)$ is holomorphic on a region containing $[D]$.

Theorem 2.2 (Cauchy's Theorem). If $f(z)$ is holomorphic on $[D]$, then

$$\int_C f(z)\,dz = 0, \qquad C = \partial[D]. \tag{2.53}$$

Proof: Let $\{\Gamma_\lambda(K): \lambda = 1, 2, \ldots, \mu\}$ be the cellular decomposition of $[D]$ obtained in the proof of Theorem 2.1. The equality

$$\sum_{\lambda=1}^{\mu} \int_{\partial\Gamma_\lambda(K)} f(z)\,dz = \int_C f(z)\,dz \tag{2.54}$$

follows from (2.49), (2.50), (2.51), and (2.52).

Since $\partial\Gamma_\lambda(K) = \sum_\nu C_{\lambda\nu}$ or $\partial\Gamma_\lambda(K) = C_\lambda + \sum_\nu C_{\lambda\nu}$ by (2.49) and

(2.50), we have

$$\int_{\partial\Gamma_\lambda(K)} f(z)\,dz = \sum_\nu \int_{C_{\lambda\nu}} f(z)\,dz$$

or

$$\int_{\partial\Gamma_\lambda(K)} f(z)\,dz = \int_{C_\lambda} f(z)\,dz + \sum_\nu \int_{C_{\lambda\nu}} f(z)dz.$$

Hence

$$\sum_{\lambda=1}^{\mu} \int_{\partial\Gamma_\lambda(K)} f(z)\,dz = \sum_\lambda \int_{C_\lambda} f(z)\,dz + \sum_\lambda \sum_\nu \int_{C_{\lambda\nu}} f(z)\,dz.$$

Here, the sum $\Sigma_\lambda \Sigma_\nu$ is extended over all pairs $(\Gamma_\lambda(K), \Gamma_\nu(K))$ that have a segment in common. For each segment $|C_{\lambda\nu}| = |C_{\nu\lambda}|$ we have $C_{\nu\lambda} = -C_{\lambda\nu}$ by (2.51), hence by (2.8)

$$\int_{C_{\lambda\nu}} f(z)\,dz + \int_{C_{\nu\lambda}} f(z)\,dz = 0,$$

and therefore

$$\sum_\lambda \sum_\nu \int_{C_{\lambda\nu}} f(z)\,dz = 0.$$

Further, $\sum_\lambda C_\lambda = C$ by (2.52), hence

$$\sum_\lambda \int_{C_\lambda} f(z)\,dz = \int_C f(z)\,dz.$$

This proves (2.54).

Therefore, it suffices to prove $\int_{\partial\Gamma_\lambda(K)} f(z)\,dz = 0$ for each cell $\Gamma_\lambda(K)$.

Lemma 2.2. If $f(z)$ is holomorphic on the cell $\Gamma(K)$, then

$$\int_{\partial\Gamma(K)} f(z)\,dz = 0. \tag{2.55}$$

Proof: This lemma is a consequence of Theorem 1.14 of Section 1.3. Let $a \leqq t \leqq b, 0 \leqq s \leqq 1$. Let $K = \{(t, s)\colon\ a \leqq t \leqq b,\ 0 \leqq s \leqq 1\}$ and put $\gamma_s(t) = {}_t\gamma(s) = \Gamma(t, s)$, thereby defining smooth arcs $\gamma_s\colon t \to \gamma_s(t)$ and ${}_t\gamma\colon s \to {}_t\gamma(s)$. By (1.44) we have

$$\int_{\gamma_0} f(z)\,dz + \int_{{}_b\gamma} f(z)\,dz - \int_{\gamma_1} f(z)\,dz - \int_{{}_a\gamma} f(z)\,dz = 0. \tag{2.56}$$

By (2.20) we have $\partial\Gamma(K) = C_1 \cdot C_2 \cdot C_3 \cdot C_4$, where each C_i, $i = 1, 2, 3, 4$, is a smooth Jordan arc or a single point. First assume that all C_i are smooth Jordan arcs. By (2.19) parameter representations are given by $C_1\colon t \to \gamma_0(t)$,

$C_2: s \to {}_b\gamma(s)$, $C_3: t \to \gamma_1(b+a-t)$, $C_4: s \to {}_a\gamma(1-s)$. Substituting t for $a+b-t$ in $\gamma_1(b+a-t)$ and s for $1-s$ in ${}_a\gamma(1-s)$, we obtain parameter representations $t \to \gamma_1(t)$ for $-C_3$ and $s \to {}_a\gamma(s)$ for $-C_4$. So, we get from (2.56)

$$\int_{C_1} f(z)\,dz + \int_{C_2} f(z)\,dz + \int_{C_3} f(z)\,dz + \int_{C_4} f(z)dz = 0$$

and this proves (2.55).

If one or more of the C_i are points, the corresponding integrals equal 0. If, for example, C_1 is a point, then $\gamma_0(t)$ is a constant, hence $\int_{\gamma_0} f(z)\,dz = \int_a^b f(\gamma_0(t))\,\gamma_0'(t)\,dt = 0$. Hence (2.55) is also valid in this case.

Cauchy's Theorem is among the most beautiful theorems of mathematics and has many applications.

Theorem 2.3 (*Strong form of Cauchy's Theorem*). If $f(z)$ is holomorphic on D and continuous on $[D]$, then

$$\int_C f(z)\,dz = 0, \qquad C = \partial[D].$$

Proof: Since equality (2.54) is valid for any continuous function on $[D]$, it suffices to show that $\int_{\partial\Gamma_\lambda(K)} f(z)\,dz = 0$.

Lemma 2.3. Let E be the interior of the rectangle K and let $f(z)$ be continuous on the cell $\Gamma(K)$ and holomorphic on $\Gamma(E)$. Then

$$\int_{\partial\Gamma(K)} f(z)\,dz = 0.$$

Proof: If we put $K = \{(t, s)\colon a \leqq t \leqq b,\ 0 \leqq s \leqq 1\}$, then $E = \{(t, s)\colon a < t < b, 0 < s < 1\}$ and $\Gamma(E)$ is the interior of $\Gamma(K)$ by Definition 2.4. For a sufficiently small ε, put

$$K^\varepsilon = \{(t, s)\colon a+\varepsilon \leqq t \leqq b-\varepsilon, \varepsilon \leqq s \leqq 1-\varepsilon\}.$$

Since $\Gamma(K^\varepsilon) \subset \Gamma(E)$ and $f(z)$ is holomorphic on $\Gamma(E)$, Theorem 1.14 is true for the rectangle K^ε and the map: $\Gamma\colon (t, s) \to \Gamma(t, s)$. Hence, defining smooth arcs by

$$\begin{aligned}
\gamma_0^\varepsilon &: t \to \Gamma(t, \varepsilon), & a+\varepsilon &\leqq t \leqq b-\varepsilon,\\
\gamma_1^\varepsilon &: t \to \Gamma(t, 1-\varepsilon), & a+\varepsilon &\leqq t \leqq b-\varepsilon,\\
{}_a\gamma^\varepsilon &: s \to \Gamma(a+\varepsilon, s), & \varepsilon &\leqq s \leqq 1-\varepsilon,\\
{}_b\gamma^\varepsilon &: s \to \Gamma(b-\varepsilon, s), & \varepsilon &\leqq s \leqq 1-\varepsilon,
\end{aligned}$$

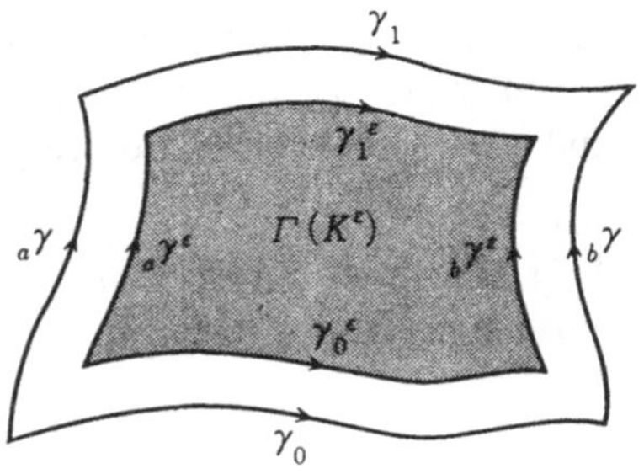

Fig. 2.33

we have

$$\int_{\gamma_0^\varepsilon} f(z)\,dz + \int_{{}_b\gamma^\varepsilon} f(z)\,dz - \int_{\gamma_1\varepsilon} f(z)dz - \int_{{}_a\gamma^\varepsilon} f(z)dz = 0.$$

Letting ε tend to $+0$ in this equality, we obtain (2.56). Consider, for example

$$\int_{\gamma_0^\varepsilon} f(z)\,dz = \int_{a+\varepsilon}^{b-\varepsilon} f(\Gamma(t,\varepsilon))\,\Gamma_t(t,\varepsilon)\,dt.$$

Since $f(z)$ is continuous on $\Gamma(K)$ by assumption, $f(\Gamma(t,s))\,\Gamma_t(t,s)$ is a continuous function of t and s on K. Hence

$$\lim_{\varepsilon\to+0}\int_{\gamma_0^\varepsilon} f(z)\,dz = \int_a^b f(\Gamma(t,0))\,\Gamma_t(t,0)\,dt = \int_{\gamma_0} f(z)\,dz.$$

We have already seen in the proof of Lemma 2.2, that

$$\int_{\partial\Gamma(K)} f(z)\,dz = 0$$

follows from equality (2.56).

b. *Cauchy's integral formula*

Theorem 2.4 (*Cauchy's integral formula*). If $f(z)$ is holomorphic on D and continuous on $[D]$, then the value of $f(z)$ at a point $w\in D$ is given by

$$f(w) = \frac{1}{2\pi i}\int_C \frac{f(z)}{z-w}\,dz, \qquad C = \partial[D]. \tag{2.57}$$

Proof: Let $U_\varepsilon(w) = \{z: |z-w| < \varepsilon\}$ be a disk with center w and radius $\varepsilon > 0$, contained in D and let $C_\varepsilon = \gamma_\varepsilon(I)$ with $\gamma_\varepsilon\colon\ \theta \to \gamma_\varepsilon(\theta) = w + \varepsilon e^{i\theta}$, $\theta\in I = [0, 2\pi]$, represent the circle with center w and radius ε. The boundary

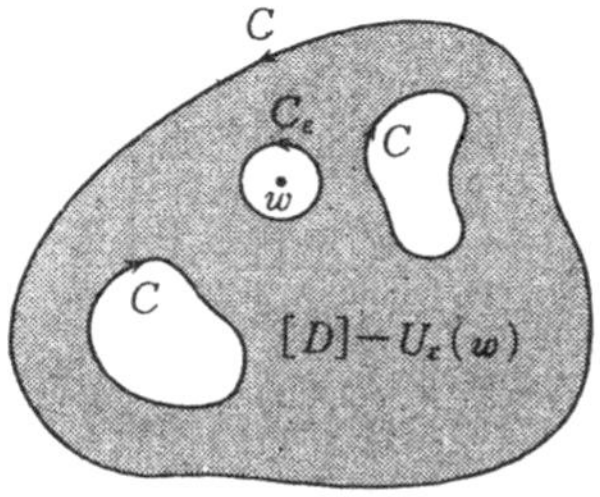

Fig. 2.34

of the closed region $[D]-U_\varepsilon(w)$ is given by

$$\partial([D]-U_\varepsilon(w)) = C - C_\varepsilon.$$

The function $f(z)/(z-w)$ of z is holomorphic on $D-\{w\}$ and continuous on $[D]-\{w\}$; therefore we can apply Cauchy's Theorem (Theorem 2.3) to this function and the closed region $[D]-U_\varepsilon(w)$,

$$\int_{C-C_\varepsilon} \frac{f(z)}{z-w}\, dz = 0,$$

hence

$$\int_C \frac{f(z)}{z-w}\, dz = \int_{C_\varepsilon} \frac{f(z)}{z-w}\, dz.$$

Since $\gamma_\varepsilon(\theta) - w = \varepsilon e^{i\theta}$ and $\gamma'_\varepsilon(\theta) = i\varepsilon e^{i\theta}$, we get

$$\frac{1}{2\pi i}\int_{C_\varepsilon} \frac{f(z)}{z-w}\, dz = \frac{1}{2\pi i}\int_0^{2\pi} \frac{f(\gamma_\varepsilon(\theta))}{\gamma_\varepsilon(\theta)-w}\, \gamma'_\varepsilon(\theta)\, d\theta = \frac{1}{2\pi}\int_0^{2\pi} f(w+\varepsilon e^{i\theta})\, d\theta$$

hence

$$\lim_{\varepsilon\to 0} \frac{1}{2\pi i}\int_{C_\varepsilon} \frac{f(z)}{z-w}\, dz = f(w),$$

proving (2.57).

Formula (2.57) is called *Cauchy's integral formula*. Formula (1.34) is a special case of this result.

Replacing w by z and z by ζ we get

$$f(z) = \frac{1}{2\pi i}\int_C \frac{f(\zeta)}{\zeta - z}\, d\zeta, \qquad C = \partial[D]. \tag{2.58}$$

Corollary. If $f(z)$ is holomorphic on D and continuous on $[D]$, then the

mth derivative $f^{(m)}(z)$, $m = 1, 2, \ldots$, of $f(z)$ on D is given by

$$f^{(m)}(z) = \frac{m!}{2\pi i} \int_C \frac{f(\zeta)}{(\zeta - z)^{m+1}}\, d\zeta, \qquad C = \partial[D]. \tag{2.59}$$

Proof: The result follows immediately from (2.58) and Theorem 1.17.

In Section 1.4d we deduced the existence of the Laurent expansion of $f(z)$ at an isolated singularity from (1.58), which is nothing but Cauchy's integral formula for the closed region $[D] = \{z\colon \varepsilon \leqq |z| \leqq r\}$. (Indeed, $\partial[D] = C_r - C_\varepsilon$, where C_r and C_ε are circles with center 0 and radius r and ε, respectively. According to (2.58) we have

$$f(z) = \frac{1}{2\pi i} \int_{C_\gamma} \frac{f(\zeta)}{\zeta - z}\, d\zeta - \frac{1}{2\pi i} \int_{C_\varepsilon} \frac{f(\zeta)}{\zeta - z}\, d\zeta$$

and this is nothing but (1.58).)

c. Residues

If the point c is an isolated singularity of $f(z)$, i.e., if $f(z)$ is holomorphic on a neighborhood of c from which the point c has been removed, then $f(z)$ can be expanded in a Laurent series (Theorem 1.22)

$$f(z) = \sum_{n=-\infty}^{\infty} a_n (z - c)^n.$$

The coefficient a_{-1} of $(z - c)^{-1}$ is called the *residue* of $f(z)$ at c and denoted by $\mathrm{Res}_{z=c}[f(z)]$:

$$\mathrm{Res}_{z=c}[f(z)] = a_{-1}. \tag{2.60}$$

If C_ε is a circle with center c and radius ε, where ε is a sufficiently small positive number, we have by (1.56)

$$\int_{C_\varepsilon} f(z)\, dz = 2\pi i\, \mathrm{Res}_{z=c}[f(z)]. \tag{2.61}$$

If c is a pole of the mth order of $f(z)$ we have

$$(z - c)^m f(z) = a_{-m} + a_{-m+1}(z - c) + \cdots + a_{-1}(z - c)^{m-1} + \cdots$$

hence

$$\frac{d^{m-1}}{dz^{m-1}}((z - c)^m f(z)) = (m - 1)!\, a_{-1} + \frac{m!}{1!} a_0 (z - c) + \cdots.$$

Therefore $\mathrm{Res}_{z=c}[f(z)] = a_{-1}$ is given by

$$\mathrm{Res}_{z=c}[f(z)] = \frac{1}{(m - 1)!} \lim_{z \to c} \frac{d^{m-1}}{dz^{m-1}}((z - c)^m f(z)). \tag{2.62}$$

In particular, if c is a pole of the first order, we have

$$\mathrm{Res}_{z=c}[f(z)] = \lim_{z \to c} (z-c)f(z). \tag{2.63}$$

Theorem 2.5 (*Residue Theorem*). If $f(z)$ is holomorphic on D with the exception of a finite number of isolated singularities $c_1, c_2, \ldots, c_j, \ldots, c_m$ and continuous on $[D]-\{c_1, c_2, \ldots, c_m\}$, then

$$\int_C f(z)\,dz = 2\pi i \sum_{j=1}^{m} \mathrm{Res}_{z=c_j}[f(z)], \qquad C = \partial[D]. \tag{2.64}$$

Proof: Let $U_\varepsilon(c_j)$ be disks with center c_j and sufficiently small radius ε and put $C_\varepsilon(c_j) = \partial[U_\varepsilon(c_j)]$. $[D] - \bigcup_{j=1}^{m} U_\varepsilon(c_j)$ is a closed region and

$$\partial\left[[D] - \bigcup_{j=1}^{m} U_\varepsilon(c_j)\right] = C - \sum_{j=1}^{m} C_\varepsilon(c_j).$$

Furthermore, $f(z)$ is holomorphic on $D - \bigcup_{j=1}^{m} [U_\varepsilon(c_j)]$ and continuous on $[D] - \bigcup_{j=1}^{m} U_\varepsilon(c_j)$. By Cauchy's Theorem (Theorem 2.3) we get

$$\int_{C - \sum_j C_\varepsilon(c_j)} f(z)\,dz = 0$$

hence

$$\int_C f(z)\,dz = \sum_{j=1}^{m} \int_{C_\varepsilon(c_j)} f(z)\,dz.$$

Therefore, by (2.61)

$$\int_C f(z)\,dz = 2\pi i \sum_{j=1}^{m} \mathrm{Res}_{z=c_j}[f(z)].$$

Now, let $f(z)$ be holomorphic on $[D]$ and $f(z) \neq 0$ for $z \in C = \partial[D]$. Under these circumstances, $f(z)$ has at most a finite number of zeros in D.

Proof: Suppose $f(z)$ has an infinite number of zeros in D. Then the collection of zeros of $f(z)$ has an accumulation point $c \in [D]$. Since $f(z)$ is continuous on $[D]$, we conclude $f(c) = 0$, hence $c \in D$. Since $f(z)$ is holomorphic on D, there exists by (1.60) an $\varepsilon > 0$ such that $f(z) \neq 0$ for $0 < |z - c| < \varepsilon$. This contradicts the fact that c is an accumulation point of the collection of zeros of $f(z)$.

Theorem 2.6. Let $c_1, c_2, \ldots, c_k$ be the zeros of $f(z)$ in D and let m_j be the order of $c_j, j = 1, 2, \ldots, k$. If $\phi(z)$ is an arbitrary holomorphic function of z on $[D]$, we have

$$\frac{1}{2\pi i} \int_C \frac{f'(z)}{f(z)} \phi(z)\,dz = \sum_{j=1}^{k} m_j \phi(c_j). \tag{2.65}$$

Proof: Each c_j has a neighborhood on which

$$f(z) = (z - c_j)^{m_j} g_j(z), \tag{2.66}$$

where $g_j(z)$ is holomorphic and $g_j(z) \neq 0$ on that neighborhood. From

$$f'(z) = m_j (z - c_j)^{m_j - 1} g_j(z) + (z - c_j)^{m_j} g'_j(z),$$

we have

$$\frac{f'(z)}{f(z)} \phi(z) = \frac{m_j \phi(z)}{z - c_j} + \frac{g'_j(z)}{g_j(z)} \phi(z)$$

and $g'_j(z) \cdot \phi(z)/g_j(z)$ is holomorphic on the selected neighborhood of c_j. Hence c_j is a pole of first order of $f'(z) \cdot \phi(z)/f(z)$ (if $\phi(c_j) = 0$ then $f'(z) \cdot \phi(z)/f(z)$ is holomorphic on a neighborhood of c_j) and the residue at c_j equals $m_j \phi(c_j)$ by (2.63). Since $f'(z) \cdot \phi(z)/f(z)$ is holomorphic on $[D] - \{c_1, c_2, \ldots, c_k\}$, we have by the residue theorem:

$$\int_C \frac{f'(z)}{f(z)} \phi(z)\, dz = 2\pi i \sum_{j=1}^{k} m_j \phi(c_j).$$

Corollary 1. The number of zeros of $f(z)$ in D is given by

$$\sum_{j=1}^{k} m_j = \frac{1}{2\pi i} \int_C \frac{f'(z)}{f(z)}\, dz. \tag{2.67}$$

Corollary 2. The sum of the nth powers of the zeros of $f(z)$ in D (where n is a natural number) is given by

$$\sum_{j=1}^{k} m_j c_j^n = \frac{1}{2\pi i} \int_C \frac{f'(z)}{f(z)} z^n\, dz. \tag{2.68}$$

The above is also valid if some of the m_j are negative integers. If $m_j = -|m_j|$ is a negative integer, (2.66) tells us that c_j is a pole of $f(z)$ of the order $|m_j|$. Therefore we have

Corollary 3. If $f(z)$ is holomorphic and not equal to zero on $[D] - \{c_1, c_2, \ldots, c_k\}$, where $c_j \in D$ is either a zero of order m_j or a pole of order $|m_j| = -m_j$ of $f(z)$, then (2.65) is valid without change.

In particular, if N denotes the number of zeros of $f(z)$ on D and P the number of poles of $f(z)$ on D, then $N - P = \sum_{j=1}^{k} m_j$ and hence

$$N - P = \frac{1}{2\pi i} \int_C \frac{f'(z)}{f(z)}\, dz. \tag{2.69}$$

d. Evaluation of definite integrals

Cauchy's Theorem can be used to evaluate certain definite integrals of analytic functions of a real variable.* We give a few examples.

Example 2.6

$$\int_0^{+\infty} \frac{\sin x}{x}\, dx = \frac{\pi}{2}.$$

This is a standard example treated in most books on complex analysis. Since $(\sin x)/x = (e^{ix} - e^{-ix})/2ix$ we are led to consider the function e^{iz}/z. Choosing $0 < \varepsilon < R$ we put

$$D = \{re^{i\theta}\colon \varepsilon < r < R, 0 < \theta < \pi\}.$$

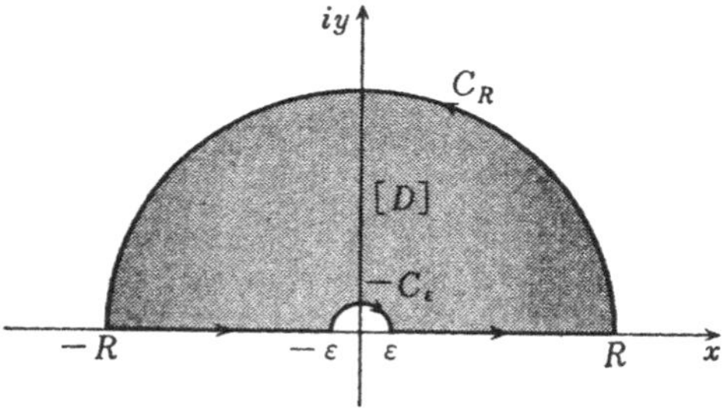

Fig. 2.35

Let C_R be the semicircle with parameter representation $\theta \to Re^{i\theta}$, $0 \leqq \theta \leqq \pi$; then the boundary of the closed region $[D]$ is given by

$$\partial [D] = [\varepsilon, R] + C_R + [-R, -\varepsilon] - C_\varepsilon.$$

Since e^{iz}/z is holomorphic on $\mathbb{C} - \{0\}$, we have by Cauchy's Theorem (Theorem 2.2)

$$\int_\varepsilon^R \frac{e^{ix}}{x}\, dx + \int_{C_R} \frac{e^{iz}}{z}\, dz + \int_{-R}^{-\varepsilon} \frac{e^{ix}}{x}\, dx - \int_{C_\varepsilon} \frac{e^{iz}}{z}\, dz = 0.$$

Replacing x by $-x$ we obtain

$$\int_{-R}^{-\varepsilon} \frac{e^{ix}}{x}\, dx = \int_R^\varepsilon \frac{e^{-ix}}{x}\, dx = -\int_\varepsilon^R \frac{e^{-ix}}{x}\, dx$$

* It seems that Cauchy first studied integrals of complex functions with the purpose of finding a uniform method of evaluating a greater number of real definite integrals known at that time.

hence

$$\int_{\varepsilon}^{R} \frac{e^{ix}}{x}\,dx + \int_{-R}^{-\varepsilon} \frac{e^{ix}}{x}\,dx = 2i\int_{\varepsilon}^{R} \frac{\sin x}{x}\,dx.$$

Therefore, it suffices to evaluate the limits

$$\lim_{R\to+\infty}\int_{C_R} \frac{e^{iz}}{z}\,dz, \qquad \lim_{\varepsilon\to+0}\int_{C_\varepsilon} \frac{e^{iz}}{z}\,dz.$$

Since $z = Re^{i\theta} = R(\cos\theta + i\sin\theta)$ and $dz = iRe^{i\theta}\,d\theta$, we have

$$\int_{C_R} \frac{e^{iz}}{z}\,dz = i\int_0^{\pi} e^{iR\cos\theta - R\sin\theta}\,d\theta$$

hence

$$\left|\int_{C_R} \frac{e^{iz}}{z}\,dz\right| \leqq \int_0^{\pi} e^{-R\sin\theta}\,d\theta.$$

We will show that the right-hand side of this inequality tends to 0 if $R \to +\infty$. By replacing the variable θ by $\pi - \theta$ on the interval $[\pi/2, \pi]$ we get

$$\int_0^{\pi} e^{-R\sin\theta}\,d\theta = 2\int_0^{\pi/2} e^{-R\sin\theta}\,d\theta.$$

Since $\sin\theta/\theta$ is continuous and positive on $[0, \pi/2]$, its minimum value is also positive. Let μ denote this minimum value, then $\sin\theta \geqq \mu\theta$, $\mu > 0$, and

$$\int_0^{\pi/2} e^{-R\sin\theta}\,d\theta \leqq \int_0^{\pi/2} e^{-R\mu\theta}\,d\theta = \frac{1}{R\mu}(1 - e^{-R\mu\pi/2}) \to 0, \quad \text{as } R \to +\infty.$$

Hence

$$\lim_{R\to+\infty}\int_{C_R} \frac{e^{iz}}{z}\,dz = 0.$$

Since $z = \varepsilon e^{i\theta}$ on C_ε we have

$$\int_{C_\varepsilon} \frac{e^{iz}}{z}\,dz = i\int_0^{\pi} e^{i\varepsilon(\cos\theta + i\sin\theta)}\,d\theta.$$

Since

$$|e^z - 1| \leqq |z| + \frac{|z|^2}{2!} + \frac{|z|^3}{3!} + \cdots = e^{|z|} - 1 \to 0, \quad \text{as } |z| \to 0$$

$e^{i\varepsilon(\cos\theta + i\sin\theta)}$ converges to 1 uniformly in θ if $\varepsilon \to 0$. Hence

$$\lim_{\varepsilon\to+0}\int_{C_\varepsilon} \frac{e^{iz}}{z}\,dz = i\int_0^{\pi} 1\cdot d\theta = i\pi$$

and therefore

$$\int_0^{+\infty} \frac{\sin x}{x}\,dx = \frac{\pi}{2}.$$

Example 2.7

$$\int_0^{+\infty} \cos(x^2)\,dx = \int_0^{+\infty} \sin(x^2)\,dx = \tfrac{1}{2}\sqrt{\frac{\pi}{2}}.$$

These integrals are called *Fresnel's integrals*. Since $\cos(x^2) - i\sin(x^2) = e^{-ix^2}$ we are led to consider the function e^{-z^2}.

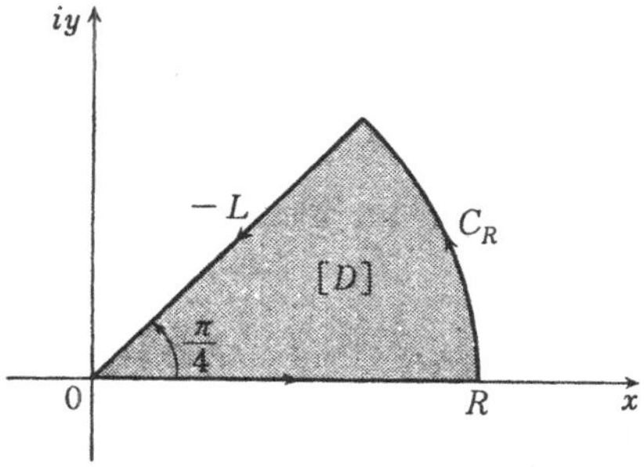

Fig. 2.36

$$[D] = \left\{re^{i\theta} : 0 \leqq r \leqq R, 0 \leqq \theta \leqq \frac{\pi}{4}\right\}$$

represents the fan-shaped closed region shown in Fig. 2.36. The boundary of $[D]$ is given by

$$\partial[D] = [0, R] + C_R - L,$$

where C_R is given by the parameter representation $\theta \to Re^{i\theta}, 0 \leqq \theta \leqq \pi/4$, and L by the parameter representation $r \to re^{i\pi/4}, 0 \leqq r \leqq R$.

Since e^{-z^2} is holomorphic on $\mathbb{C}$, we have by Cauchy's Theorem

$$\int_0^R e^{-x^2}\,dx + \int_{C_R} e^{-z^2}\,dz - \int_L e^{-z^2}\,dz = 0.$$

Since $z = re^{i\pi/4}$, $z^2 = ir^2$ and $dz = e^{i\pi/4}\,dr$ on L, we have

$$\int_L e^{-z^2}\,dz = e^{i\pi/4}\int_0^R e^{-ir^2}\,dr = e^{i\pi/4}\int_0^R (\cos r^2 - i\sin r^2)\,dr.$$

Since $z = Re^{i\theta}$, $z^2 = R^2\cos 2\theta + iR^2\sin 2\theta$ and $dz = iRe^{i\theta}\,d\theta$, we have

$$\left|\int_{C_R} e^{-z^2}\,dz\right| \leqq \int_0^{\pi/4} e^{-R^2\cos 2\theta}\,R\,d\theta = \frac{R}{2}\int_0^{\pi/2} e^{-R^2\cos\theta}\,d\theta$$

$$= \frac{R}{2}\int_0^{\pi/2} e^{-R^2\sin\theta}\,d\theta < \frac{R}{2}\cdot\frac{1}{R^2\mu} = \frac{1}{2R\mu} \to 0, \text{ as } R \to +\infty.$$

Since, as is known from real analysis, $\int_0^\infty e^{-x^2}\,dx = \sqrt{\pi}/2$, we have

$$\int_0^{+\infty} (\cos r^2 - i \sin r^2)\,dr = e^{-i\pi/4} \int_0^{+\infty} e^{-x^2}\,dx = \frac{(1-i)\sqrt{\pi}}{2\sqrt{2}},$$

and therefore

$$\int_0^{+\infty} \cos(x^2)\,dx = \int_0^{+\infty} \sin(x^2)\,dx = \tfrac{1}{2}\sqrt{\frac{\pi}{2}}.$$

Example 2.8

$$\int_{-\infty}^{+\infty} \frac{\cos x}{(x^2+1)^2}\,dx = \frac{\pi}{e}.$$

Let $R > 1$ and $[D] = \{re^{i\theta}: 0 \leqq r \leqq R, 0 \leqq \theta \leqq \pi\}$.

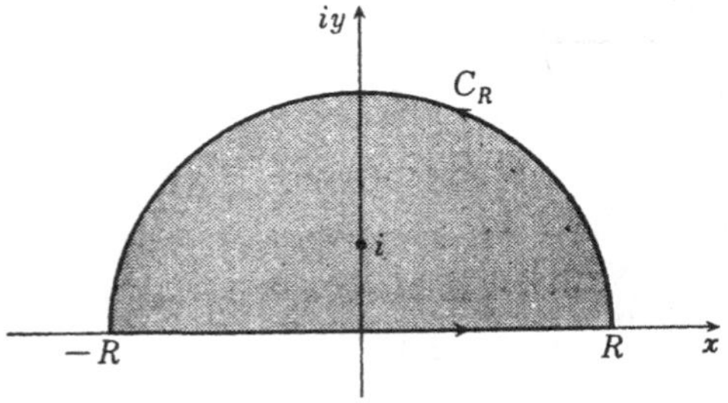

Fig. 2.37

Now, the boundary of $[D]$ is given by

$$\partial[D] = [-R, R] + C_R,$$

where C_R is given by the parameter representation $\theta \to Re^{i\theta}$, $0 \leqq \theta \leqq \pi$. The function of z

$$\frac{e^{iz}}{(z^2+1)^2} = \frac{e^{iz}}{(z-i)^2\,(z+i)^2}$$

is holomorphic on $[D] - \{i\}$ and has a pole of order two at i. The residue at i is by (2.62)

$$\lim_{z\to i} \frac{d}{dz}\left[\frac{(z-i)^2 e^{iz}}{(z^2+1)^2}\right] = \lim_{z\to i} \frac{d}{dz}\left[\frac{e^{iz}}{(z+i)^2}\right] = \frac{e^{-1}}{2i}.$$

Hence by the residue theorem (Theorem 2.5)

$$\int_{-R}^{R} \frac{e^{ix}}{(x^2+1)^2}\,dx + \int_{C_R} \frac{e^{iz}}{(z^2+1)^2}\,dz = \frac{\pi}{e}.$$

Since $z = Re^{i\theta}$ on C_R we have

$$\left|\frac{e^{iz}}{(z^2+1)^2}\right| = \left|\frac{e^{iR\cos\theta - R\sin\theta}}{(R^2e^{2i\theta}+1)^2}\right| \leqq \frac{e^{-R\sin\theta}}{(R^2-1)^2} \leqq \frac{1}{(R^2-1)^2}.$$

Therefore

$$\left|\int_{C_R} \frac{e^{iz}}{(z^2+1)^2}\,dz\right| \leqq \int_0^{\pi} \frac{1}{(R^2-1)^2}\,Rd\theta = \frac{\pi R}{(R^2-1)^2} \to 0, \text{ as } R \to +\infty,$$

so

$$\int_{-\infty}^{\infty} \frac{e^{ix}}{(x^2+1)^2}\,dx = \frac{\pi}{e}$$

from which the desired result follows.

2.4 Differentiability and homology

As we remarked in connection with Definition 1.3 of a holomorphic function, usually, in complex analysis, a function $f(z)$ of a complex variable z is called holomorphic on a region D if it is differentiable at each point of D. In this section we will prove that both definitions are equivalent, i.e., that the derivative $f'(z)$ of a function $f(z)$ differentiable at each point of a region D is continuous on D.

Let $f(z)$ be differentiable at each point of the region D, and let

$$K = \{t + is: a \leqq t \leqq x, b \leqq t \leqq y\}, \qquad a < x, b < y,$$

be a rectangle such that $K \subset D$, then we have

$$\int_{\partial K} f(z)\,dz = 0. \tag{2.70}$$

To see this, put $S(K) = \int_{\partial K} f(z)\,dz$. We have

$$\begin{aligned} S(K) = &\int_a^x f(t+ib)\,dt + \int_b^y f(x+is)\,i\,ds \\ &- \int_a^x f(t+iy)\,dt - \int_b^y f(a+is)\,i\,ds. \end{aligned} \tag{2.71}$$

We will prove that $S(K) = 0$ by subdividing K into smaller rectangles. First divide K into four congruent rectangles K', K'', K''', and K'''', as shown in Fig. 2.38. By writing $S(K')$, $S(K'')$, $S(K''')$, and $S(K'''')$ in the form of (2.71) it is clear that

$$S(K) = S(K') + S(K'') + S(K''') + S(K'''').$$

Hence $|S(K)| \leqq |S(K')| + |S(K'')| + |S(K''')| + |S(K'''')|$. Therefore there is at least one rectangle, say, K''', among the rectangles K', K'', K''', and K''''

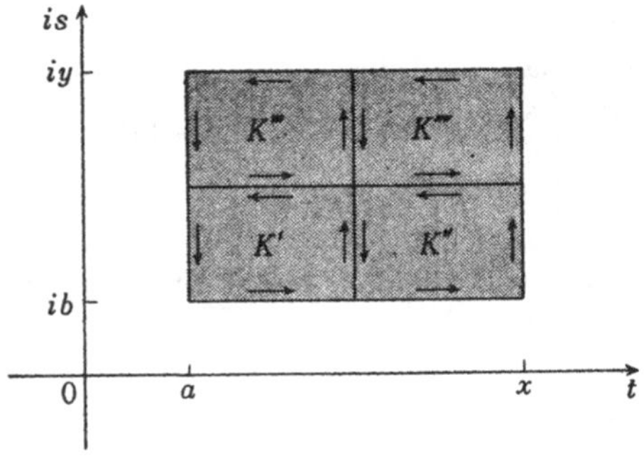

Fig. 2.38

satisfying

$$\frac{|S(K)|}{4} \leqq |S(K''')|.$$

We write $K_1 = K'''$ and next we subdivide K_1 into four congruent rectangles K_1', K_1'', K_1''', and K_1''''. Again, among the rectangles K_1', K_1'', K_1''', and K_1'''' there is at least one rectangle, say, K_1'' such that

$$\frac{|S(K_1)|}{4} \leqq |S(K_1'')|.$$

We write $K_2 = K_1''$ and $K_3, K_4, \ldots, K_m, \ldots$ are defined similarly. So we obtain a sequence of rectangles satisfying

$$K \supset K_1 \supset K_2 \supset \cdots \supset K_m \supset \cdots$$

and $|S(K_{m-1})|/4 \leqq |S(K_m)|$. Therefore

$$\frac{|S(K)|}{4^m} \leqq |S(K_m)|. \tag{2.72}$$

The length of the side of K_m parallel with the real axis is $(x-a)/2^m$, that of the side parallel with the imaginary axis is $(y-b)/2^m$ and its diameter $\delta(K_m)$ is given by $\delta(K_m) = \sqrt{[(x-a)^2 + (y-b)^2]}/2^m$, hence $\delta(K_m) \to 0$ if $m \to \infty$. Therefore, there exists exactly one point c which belongs to all K_m: $\{c\} = \bigcap_{m=1}^{\infty} K_m$. Since, by assumption, $f(z)$ is differentiable at c, we have, by (1.9),

$$f(z) = f(c) + f'(c)(z-c) + O(z-c).$$

A direct computation based on (2.71) yields $\int_{\partial K_m} (f(c) + f'(c))(z-c)\,dz = 0$. So we have

$$S(K_m) = \int_{\partial K_m} f(z)\,dz = \int_{\partial K_m} O(z-c)\,dz.$$

For each $\varepsilon > 0$, there exists a $\delta(\varepsilon) > 0$ such that

$$|O(z-c)| \leqq \varepsilon|z-c| \qquad \text{if } |z-c| < \delta(\varepsilon).$$

Choosing m such that $\delta(K_m) < \delta(\varepsilon)$ we have for $z \in K_m$

$$|z-c| \leqq \delta(K_m) < \delta(\varepsilon),$$

hence

$$|O(z-c)| = \leqq \varepsilon|z-c| \leqq \varepsilon\delta(K_m).$$

Observing that ∂K_m consists of two segments of length $(x-a)/2^m$ and of two segments of length $(y-b)/2^m$, we have

$$|S(K_m)| = \left|\int_{\partial K_m} O(z-c)dz\right| \leqq \frac{\varepsilon\delta(K_m)l}{2^m} = \frac{\varepsilon\delta(K)l}{2^{2m}},$$

where $l = 2(x-a) + 2(y-b)$.

Combining this inequality with (2.72) we get

$$|S(K)| \leqq \varepsilon\delta(K)l$$

and since ε is an arbitrary positive number, we conclude $|S(K)| = 0$.

Define a function $F(x+iy)$ of $z = x+iy$ by

$$F(x+iy) = \int_a^x f(t+iy)\,dt + i\int_b^y f(a+is)\,ds.$$

$F(x+iy)$ is partially differentiable with respect to x and

$$\frac{\partial}{\partial x}F(x+iy) = f(x+iy).$$

By (2.70) we have $S(K) = 0$, hence by (2.71)

$$F(x+iy) = \int_a^x f(t+ib)\,dt + i\int_b^y f(x+is)\,ds.$$

Therefore $F(x+iy)$ is also partially differentiable with respect to y and

$$\frac{\partial}{\partial y}F(x+iy) = if(x+iy).$$

Writing $F(x+iy) = U(x,y) + iV(x,y)$, $f(x+iy) = u(x,y) + iv(x,y)$, we have

$$U_x(x,y) = V_y(x,y) = u(x,y),$$
$$V_x(x,y) = -U_y(x,y) = v(x,y).$$

Therefore, $U(x,y)$ and $V(x,y)$ are continuously differentiable functions with respect to x and y satisfying the Cauchy–Riemann equations. So $F(z)$ is a holomorphic function in the sense of Definition 1.3 by Theorem 1.4, and

$$f(z) = F'(z)$$

by (1.11). Therefore, $f(z)$ is also holomorphic in the sense of Definition (1.3) by the corollary to Theorem 1.16.

Remark: The above ingenious proof by repeated subdivision was first given by Goursat. In modern treatments of complex analysis it is customary to prove Cauchy's Theorem by Goursat's method. We proved Cauchy's Theorem (Theorems 2.2 and 2.3), by first reducing it by cellular decomposition to the case of cells, and by next deducing Cauchy's Theorem for a cell $\Gamma(K)$ (Lemmas 2.2 and 2.3) from equality (1.44) in Theorem 1.14:

$$\int_{\gamma_0} f(z)\,dz + \int_{b\gamma} f(z)\,dz - \int_{\gamma_1} f(z)\,dz - \int_{a\gamma} f(z)\,dz = 0.$$

Writing the left-hand side of this equality in explicit form we get

$$\int_a^b f(\Gamma(t,0))\,\Gamma_t(t,0)\,dt + \int_0^1 f(\Gamma(b,s))\,\Gamma_s(b,s)\,ds$$
$$- \int_a^b f(\Gamma(t,1))\,\Gamma_t(t,1)\,dt - \int_0^1 f(\Gamma(a,s))\,\Gamma_s(a,s)\,ds$$

and representing this expression by $S(\Gamma(K))$, (1.44) can be written as

$$S(\Gamma(K)) = 0. \tag{2.73}$$

We proved (1.44) under the assumption that $f(z)$ is holomorphic in the sense of Definition 1.3, i.e., that the derivative $f'(z)$ exists and is continuous.

If we assume only that $f'(z)$ exists, but is not necessarily continuous, we first reduce Cauchy's Theorem to (2.73) and proceed from there by subdividing the rectangle K into four congruent rectangles K', K'', K''', and K'''' and observing that

$$S(\Gamma(K)) = S(\Gamma(K')) + S(\Gamma(K'')) + S(\Gamma(K''')) + S(\Gamma(K''''))$$

It is possible to prove (2.73) by repeated subdivision of rectangles, just as we proved $S(K) = 0$ above. The reason that we did not adopt this approach in this book is, as already stated earlier, that it is more natural to assume not only existence but also continuity of the derivative $f'(z)$ and that we really wanted to use the continuity of $f'(z)$ in the proof.

3

Conformal mappings

3.1 Conformal mappings

Let $w = f(z)$ be a holomorphic function defined on a region $D \subset \mathbb{C}$. Let us call the complex plane, the points of which are denoted by the letter z, the z-plane and the complex plane, the points of which are denoted by the letter w, the w-plane. The function f then becomes a mapping, assigning to each point z belonging to the region D of the z-plane a point $w = f(z)$ of the w-plane. The complex function f is of course the same thing as the mapping f, but it is customary to call f a mapping if one wants to emphasize the geometric aspects of f.

If S is an arbitrary subset of D, then $f(S) = \{f(z): z \in S\}$ is called the *image* of S under f. If W is an arbitrary subset of the w-plane, the set of all points $z \in D$ such that $f(z) \in W$ is called the *inverse image* of W and denoted by $f^{-1}(W)$: $f^{-1}(W) = \{z \in D: f(z) \in W\}$. If $f(D) \cap W = \emptyset$, then $f^{-1}(W)$ is also the empty set. If W is an open set, so is its inverse image.

To prove this, select an arbitrary point $c \in f^{-1}(W)$. Since a holomorphic function is continuous, there exists for each $\varepsilon > 0$ a $\delta(\varepsilon) > 0$ satisfying

$$|f(z) - f(c)| < \varepsilon \qquad \text{if } |z - c| < \delta(\varepsilon).$$

Since $f(c) \in W$ and W is open, we can find a (sufficiently small) ε such that $f(z) \in W$ if $|z - c| < \delta(\varepsilon)$, i.e., $U_{\delta(\varepsilon)}(c) \subset f^{-1}(W)$. Hence $f^{-1}(W)$ is open.

Theorem 3.1. Let $f(z)$ be a holomorphic function on the region D, which is not identically equal to zero, then the set of zeros of $f(z)$ has no accumulation point in D.

Proof: Let D_0 be the set of all points $z \in D$ such that $f(z) = f'(z) = f''(z) = \cdots = f^{(m)}(z) = \cdots = 0$ and let D_1 be the set of all points $z \in D$ such that at least one of $f(z), f'(z), \ldots, f^{(m)}(z), \ldots$ is not equal to 0, then

$$D = D_0 \cup D_1, \qquad D_0 \cap D_1 = \emptyset.$$

Since $f(z), f'(z), \ldots, f^{(m)}(z), \ldots$ are all holomorphic functions of z (by the corollary to Theorem 1.16), we have that if $f^{(m)}(c) \neq 0$ for some $c \in D$, then $f^{(m)}(z) \neq 0$ for all z in a sufficiently small ε-neighborhood $U_\varepsilon(c) \subset D$ of c. Hence, if $c \in D_1$ then $U_\varepsilon(c) \subset D_1$, i.e., D_1 is open.

D_0 is also open. To prove this, we consider the power series expansion

$$f(z) = \sum_{n=0}^{\infty} a_n(z-c)^n$$

around $c \in D$, which is valid on a certain neighborhood $U_\varepsilon(c) \subset D$ of c (Theorem 1.16). Since $a_n = f^{(n)}(c)/n!$ by (1.23) we conclude from $c \in D_0$ that $f(z)$ is identically equal to zero on $U_\varepsilon(c)$, hence that $U_\varepsilon(c) \subset D_0$.

Since D is a region (a connected open set), we have either $D = D_0$ or $D = D_1$. If $D = D_0$, the function $f(z)$ is identically equal to 0 on D. Since we excluded this possibility, we conclude $D = D_1$. Therefore, if $c \in D$ is a zero of $f(z)$, then there exists a natural number m such that $a_m = f^{(m)}(c)/m! \neq 0$. Hence, by (1.60) we have

$$f(z) \neq 0 \qquad \text{if } 0 < |z-c| < \varepsilon$$

for some sufficiently small $\varepsilon > 0$, i.e., $f(z)$ has no other zeros on $U_\varepsilon(c)$ than c.

Now if c is an accumulation point of the set of zeros of $f(z)$ in D, then c is also a zero of $f(z)$, contradicting the above.

Corollary. Let $f(z)$ and $g(z)$ be holomorphic functions on the region D.

(1) If $f(z) = g(z)$ for all z from some nonempty open subset U of D, then $f(z) = g(z)$ for all $z \in D$.

(2) If $f(z) = g(z)$ for all z on some curve $C \subset D$ (we exclude the possibility that C is the constant curve), then $f(z) = g(z)$ for all $z \in D$.

Proof: If $f(z)$ is not identically equal to $g(z)$ on D, the set $\{z \in D: f(z) = g(z)\}$ has no accumulation points in D. Therefore, $\{z \in D: f(z) = g(z)\}$ cannot contain an open set U or a curve C.

Let $f(z)$ be a holomorphic, nonconstant function on a region $D \subset \mathbb{C}$, then $f(z)$ is nonconstant on all subregions U of D. (Suppose $f(z) = a$ for all z in some subregion U of D, then $f(z) = a$ for all $z \in D$ by the above corollary.) Further, for all $c \in D$ at least one of the coefficients $a_1, a_2, \ldots, a_n, \ldots$ appearing in the power series expansion of $f(z)$ around c

$$f(z) = a_0 + \sum_{n=1}^{\infty} a_n(z-c)^n$$

is not equal to 0. (If $a_1 = a_2 = \cdots = a_n = \cdots = 0$, then $f(z) = a_0$ for all z in some ε-neighborhood $U_\varepsilon(c) \subset D$ of c.) Just as we called a point c such that $f(c) = 0$ a zero of $f(z)$, we will call a point c such that $f(c) = a$ an *a-point* of $f(z)$. The a-points of $f(z)$ are just the zeros of $f(z) = a$. Therefore, we have by Theorem 3.1 that if $f(z)$ is not identically equal to a on the region D, then the set of a-points of $f(z)$ has no accumulation points in D. If c is a zero of the mth order of $f(z) - a$, c is called an a-point of $f(z)$ of the mth order. In this case, the power series expansion of $f(z)$ around c takes the form

$$f(z) = a + a_m (z-c)^m + a_{m+1}(z-c)^{m+1} + \cdots, \qquad a_m \neq 0.$$

Theorem 3.2. Let $f(z)$ be a holomorphic function on the region D. If $c \in D$ is such that $f'(c) \neq 0$, then there is a neighborhood $U \subset D$ of c, such that $W = f(U)$ is a neighborhood of $a = f(c)$ and f is a one-to-one mapping from U onto W. The inverse mapping f_U^{-1}: $w = f(z) \to z = f_U^{-1}(w)$ of the restriction f_U of f to U is holomorphic on D and

$$\frac{d}{dw} f_U^{-1}(w) = \frac{1}{f'(z)} \neq 0, \qquad z = f_U^{-1}(w). \tag{3.1}$$

Proof: Expanding $f(z) - a$ around c in a power series, we get

$$f(z) - a = a_1(z-c) + a_2(z-c)^2 + \cdots, \qquad a_1 = f'(c) \neq 0.$$

Therefore $z = c$ is a zero of the first order of $f(z) - a$, hence, by (1.59),

$$f(z) - a = (z-c)\, g(z),$$

with $g(z) \neq 0$ on a sufficiently small neighborhood of c. Therefore

$$f(z) - a \neq 0 \qquad \text{if } 0 < |z-c| \leqq \varepsilon \tag{3.2}$$

for some sufficiently small $\varepsilon > 0$. The continuous function $|f(z) - a|$ assumes a minimum on the circle $C = \partial[U_\varepsilon(c)]$ with center c and radius ε. Denoting this minimum by μ, we have, by (3.2),

$$|f(z) - a| \geqq \mu > 0 \qquad \text{for } z \in C.$$

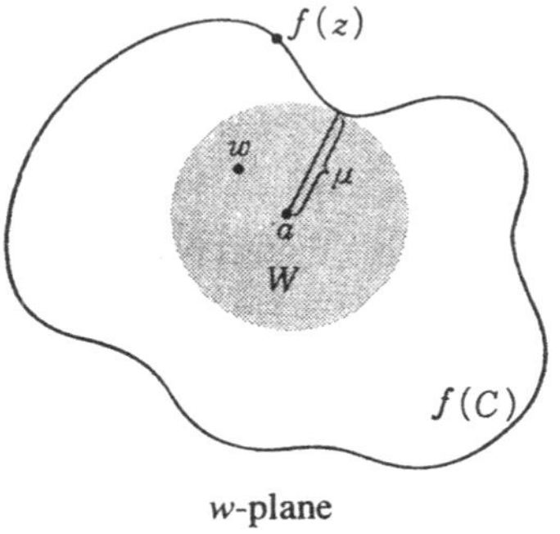

Fig. 3.1

Putting $W = \{w: |w-a| < \mu\}$, we have

$$|f(z)-w| > 0 \qquad \text{if } z \in C \text{ and } w \in W.$$

Put

$$m(w) = \frac{1}{2\pi i}\int_C \frac{f'(z)}{f(z)-w}\,dz, \qquad w \in W.$$

By Corollary 1 of Theorem 2.6, $m(w)$ equals the number of zeros of $f(z)-w$, i.e., the number of w-points of $f(z)$ on $U_\varepsilon(c)$. Since $W \subset \mathbb{C} - f(c)$, we know by Theorem 1.17 that $m(w)$ is a holomorphic function of w on W. Since a holomorphic function which only assumes integer values is obviously a constant, we conclude that $m(w)$ is constant on W. The point c is an a-point of the first order of $f(z)$ and $f(z)$ has no other a-points than c on $U_\varepsilon(c)$ by (3.2). Therefore, $m(a) = 1$ and hence

$$m(w) = m(a) = 1.$$

Therefore, there exists exactly one point $z \in U_\varepsilon(c)$ such that $f(z) = w$. Putting

$$U = f^{-1}(W) \cap U_\varepsilon(c) = \{z \in U_\varepsilon(c): f(z) \in W\}$$

U is a neighborhood of c and the point $z \in U_\varepsilon(c)$ such that $f(z) = w$ can be represented by $z = f_U^{-1}(w)$. Since $m(w) = 1$ we have by Corollary 2 of Theorem 2.6

$$f_U^{-1}(w) = \frac{1}{2\pi i}\int_C \frac{f'(z)}{f(z)-w}\,z\,dz, \qquad w \in W.$$

Therefore, $f_U^{-1}(w)$ is a holomorphic function of w on W by Theorem 1.17 and it is clear that f maps U one-to-one onto W. Differentiating both sides of the equality $f_U^{-1}(f(z)) = z$, we get, by the chain rule (1.15),

$$\frac{d}{dw} f_U^{-1}(w) \cdot f'(z) = 1, \qquad w = f(z),\ z \in U,$$

from which (3.1) follows at once.

Theorem 3.3. Let $f(z)$ be a holomorphic function on D and $c \in D$ such that

$$f'(c) = f''(c) = \cdots = f^{(m-1)}(c) = 0, \qquad f^{(m)}(c) \neq 0,\ m \geqq 2.$$

Let ε be a positive real number, then there exists a neighborhood $U \subset U_\varepsilon(c)$ of c such that $W = f(U)$ is a neighborhood of $a = f(c)$ and the mapping $f: z \to w = f(z)$ is an m-to-1 mapping of $U - \{c\}$ onto $W - \{a\}$.

Proof: Let

$$f(z) - a = a_m(z-c)^m + a_{m+1}(z-c)^{m+1} + \cdots, \qquad a_m = \frac{f^{(m)}(c)}{m!} \neq 0,$$

be the power series expansion of $f(z)-a$ around c. We have by (1.59)

$$f(z)-a=(z-c)^m g(z)$$

and $g(z)\neq 0$ on a sufficiently small neighborhood of c. Hence

$$f(z)-a\neq 0 \qquad \text{if } 0<|z-c|\leqq \varepsilon$$

for a sufficiently small ε. Just as in the proof of Theorem 3.2, the minimum μ of $|f(z)-a|$ on the circle $C=\partial[U_\varepsilon(c)]$ is positive and putting $W=\{w:|w-a|<\mu\}$, for $w\in W$ the number of w-points of $f(z)$ in $U_\varepsilon(c)$ is given by

$$m(w)=\frac{1}{2\pi i}\int_C \frac{f'(z)}{f(z)-w}\,dz,$$

hence is independent of w.

The point c is an a-point of the mth order of $f(z)$ and by (3.2) there are no other a-points of $f(z)$ in $U_\varepsilon(c)$. Therefore, $m(a)=m$ hence

$$m(w)=m(a)=m, \qquad w\in W.$$

So, for $w\in W$ there are m w-points in $U_\varepsilon(c)$. Assuming that ε has been chosen sufficiently small and $w\neq a$, these m w-points are all w-points of the first order, hence there are m different w-points of $f(z)$ in $U_\varepsilon(c)$. For, since $f'(c)=0$ we have by (1.60).

$$f'(z)\neq 0 \qquad \text{if } 0<|z-c|<\varepsilon,$$

hence if $z\in U_\varepsilon(c)$ and $f(z)=w\neq a$, then $f'(z)\neq 0$. Therefore, putting

$$U=f^{-1}(W)\cap U_\varepsilon(c)=\{z\in U_\varepsilon(c):f(z)\in W\}$$

we see that $U\subset U_\varepsilon(c)$ is a neighborhood of c such that $f(U)=W$ and for each $w\in W-\{a\}$, $f_U^{-1}(w)$ consists of m points. Hence f is an m-to-1 mapping between $U-\{c\}$ and $W-\{a\}$.

Corollary. If $f(z)$ is holomorphic on D and the map $f\colon z\to w=f(z)$ is one-to-one on a neighborhood of $c\in D$, then $f'(c)\neq 0$.

If $f(z)$ is a holomorphic function of z on a region D, then, by Theorems 3.2 and 3.3, there exists for each point $c\in D$ a neighborhood $U\subset D$ of c such that $W=f(U)$ is a neighborhood of the point $a=f(c)$. Therefore, if $V\subset D$ is open, $f(V)\subset f(D)$ is open too. A mapping which maps open sets onto open sets is called an *open mapping*.

We conclude: Holomorphic mappings are open mappings.

Furthermore, $f(D)$ is a region in the w-plane. To see this, let W_1 and W_2 be two disjoint open subsets of $f(D)$ such that $f(D)=W_1\cup W_2$. Then $D=f^{-1}(W_1)\cup f^{-1}(W_2)$, $f^{-1}(W_1)\cap f^{-1}(W_2)=\varnothing$, and $f^{-1}(W_1)$ and

$f^{-1}(W_2)$ are both nonempty open subsets of D, contradicting the connectedness of D.

Let $f(z)$ be a holomorphic function on the region D such that $f'(z) \neq 0$ for $z \in D$ and consider the mapping $f: z \to w = f(z)$. Let $\gamma: t \to \gamma(t), 0 \leqq t \leqq 1$, be a smooth curve in D and put $\lambda(t) = f(\gamma(t))$. By the chain rule (Theorem 1.12), $\lambda(t)$ is a continuously differentiable function of t and since $f'(\gamma(t)) \neq 0$ by assumption, we have

$$\lambda'(t) = f'(\gamma(t)) \cdot \gamma'(t) \neq 0. \tag{3.3}$$

Therefore, λ: $t \to w = \lambda(t)$ is a smooth curve in the w-plane. Since $\lambda(t) = f(\gamma(t))$, the curve λ is the image of the curve γ: $\lambda = f(\gamma)$.

Let γ_1: $t \to \gamma_1(t)$, γ_2: $t \to \gamma_2(t)$, $0 \leqq t \leqq 1$, be smooth curves such that $c = \gamma_1(0) = \gamma_2(0)$. We defined the angle θ between γ_1 and γ_2 at c as the argument of $\gamma_2'(0)/\gamma_1'(0)$, i.e.,

$$\frac{\gamma_2'(0)}{\gamma_1'(0)} = \left|\frac{\gamma_2'(0)}{\gamma_1'(0)}\right| e^{i\theta}.$$

The mapping $f: z \to w = f(z)$ maps the curves γ_1 and γ_2 onto smooth curves $\lambda_1 = f(\gamma_1): t \to \lambda_1(t)$ and $\lambda_2 = f(\gamma_2)$: $t \to \lambda_2(t)$, $0 \leqq t \leqq 1$, with common initial point $a = f(c)$. By (3.3)

$$\frac{\lambda_2'(0)}{\lambda_1'(0)} = \frac{\gamma_2'(0)}{\gamma_1'(0)}$$

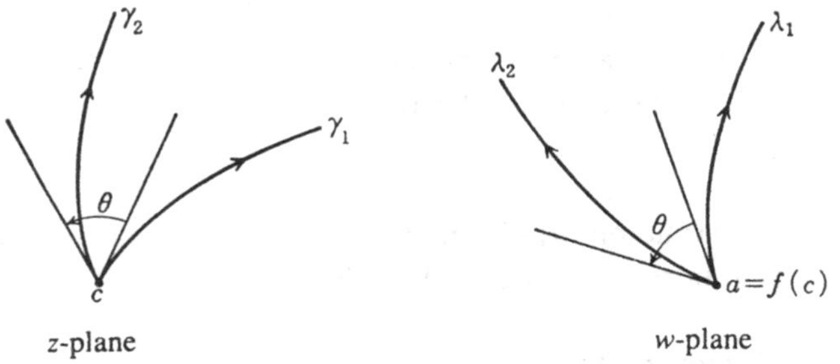

Fig. 3.2

therefore the angle between the curves λ_1 and λ_2 at $a = \lambda_1(0) = \lambda_2(0)$ equals the angle θ between the curves γ_1 and γ_2 at c. We conclude that if $f(z)$ is holomorphic and $f'(z) \neq 0$ everywhere, then the angle between two curves at a point of intersection is invariant under the mapping $f: z \to w = f(z)$.

A mapping preserving angles of intersection at points of intersection of two curves is called a *conformal mapping*. Next we want to show that the definition of conformal mapping also makes sense in a somewhat more general situation.

Let $f(z)$ be a continuous function of z defined on the region D in the z-plane. Splitting $f(z)$ into its real and imaginary part yields

$$f(z) = f(x+iy) = u(x, y) + iv(x, y), \qquad z = x+iy.$$

We assume that $u(x, y)$ and $v(x, y)$ are continuously differentiable functions of x and y. Consider the mapping $f\colon z \to w = f(z)$. The mapping f maps the smooth curve $\gamma\colon t \to z = \gamma(t)$, $0 \leqq t \leqq 1$, in D with initial point $\gamma(0) = c \in D$ onto the smooth curve $f(\gamma)\colon t \to w = f(\gamma(t))$, $0 \leqq t \leqq 1$, in the w-plane with initial point $a = f(c)$. Writing $\gamma(t) = \rho(t) + i\sigma(t)$, we get

$$f(\gamma(t)) = u(\rho(t), \sigma(t)) + iv(\rho(t), \sigma(t))$$

and remembering that $f(\gamma(t))$ is continuously differentiable

$$\frac{d}{dt} f(\gamma(t)) = u_x \rho'(t) + u_y \sigma'(t) + i(v_x \rho(t) + v_y \sigma'(t)). \tag{3.4}$$

Assuming that at c

$$\frac{\partial(u, v)}{\partial(x, y)} = \begin{vmatrix} u_x(x, y) & u_y(x, y) \\ v_x(x, y) & v_y(x, y) \end{vmatrix} \neq 0$$

we have $df(\gamma(t))/dt \neq 0$ at $t = 0$ since $\rho'(t) + i\sigma'(t) = \gamma'(t) \neq 0$, i.e., the curve $f(\gamma)$ is smooth in a neighborhood of its initial point $a = f(\gamma(0))$. Therefore, if γ_1 and γ_2 are smooth curves with common initial point c, the angle between $f(\gamma_1)$ and $f(\gamma_2)$ at $a = f(c)$ is well-defined.

The mapping f is called *conformal* at c if $\partial(u, v)/\partial(x, y) \neq 0$ at c and if for an arbitrary pair of smooth curves γ_1 and γ_2 in D with common initial point c the angle between $f(\gamma_1)$ and $f(\gamma_2)$ at $a = f(c)$ equals the angle between γ_1 and γ_2 at c. If f is conformal at each point of its domain, f is called conformal.

If $f(z)$ is a holomorphic function of z, we deduce from equation (1.11),

$$f'(z) = u_x(x, y) + iv_x(x, y) = v_y(x, y) - iu_y(x, y),$$

that

$$\frac{\partial(u, v)}{\partial(x, y)} = \begin{vmatrix} u_x(x, y) & u_y(x, y) \\ v_x(x, y) & v_y(x, y) \end{vmatrix} = |f'(z)|^2. \tag{3.5}$$

Therefore, if $f(z)$ is holomorphic on the region D in the z-plane and $f'(z) \neq 0$ for $z \in D$, then the mapping $z \to w = f(z)$ is a conformal mapping. Conversely, if $f\colon z \to w = f(z)$ is a conformal mapping defined on the region D, then $f(z)$ is holomorphic on D and $f'(z) \neq 0$ for $z \in D$.

To see this, let z be an arbitrary point from D. Choose $\varepsilon > 0$ such that $[U_\varepsilon(z)] \subset D$ and put $\gamma_\theta(t) = z + te^{i\theta}$, $0 \leqq t \leqq \varepsilon$, for arbitrary real θ.

Obviously γ_θ is a smooth curve with initial point z and the angle between γ_0 and γ_θ at z equals θ. Put $\lambda_\theta(t) = f(\gamma_\theta(t))$. Then $\lambda_\theta: t \to \lambda_\theta(t)$, $0 \leqq t \leqq \varepsilon$, is a smooth curve in the w-plane with initial point $w = f(z)$ and $\lambda_0 = f(\gamma_0)$ and $\lambda_\theta = f(\gamma_\theta)$. Since f is conformal, the angle between λ_0 and λ_θ, i.e., the argument of $\lambda'_\theta(0)/\lambda'_0(0)$, is equal to θ. Writing the argument of $\lambda'_0(0)$ as α we have

$$e^{-i(\theta+\alpha)}\lambda'_\theta(0) = |\lambda'_\theta(0)| > 0. \tag{3.6}$$

By (3.4) we have

$$\begin{aligned}\lambda'_\theta(0) &= u_x \cos\theta + u_y \sin\theta + i(v_x \cos\theta + v_y \sin\theta)\\ &= (u_x + iv_x)\cos\theta + (u_y + iv_y)\sin\theta = f_x \cos\theta + f_y \sin\theta,\end{aligned}$$

hence

$$2\lambda'_\theta(0) = (f_x - if_y)e^{i\theta} + (f_x + if_y)e^{-i\theta}.$$

Therefore, by (3.6)

$$(f_x - if_y)e^{-i\alpha} + (f_x + if_y)e^{-2i\theta - i\alpha} = 2|\lambda'_\theta(0)|.$$

Differentiating both sides with respect of θ we get

$$-2i(f_x + if_y)e^{-2i\theta - i\alpha} = \frac{2d}{d\theta}|\lambda'_\theta(0)|$$

Since θ is a real variable in this equality and since the right-hand side only takes real values, we conclude

$$f_x + if_y = 0,$$

hence

$$u_x - v_y + i(v_x + u_y) = 0.$$

Since $z = x + iy$ is an arbitrary point of D, we see that $u(x, y)$ and $v(x, y)$ satisfy the Cauchy–Riemann equations on D. Hence $f(z) = u(x, y) + iv(x, y)$ is a holomorphic function of $z = x + iy$ on D by Theorem 1.4 and $f'(z) = 0$ by (3.5).

Example 3.1. Consider the function $w = f(z) = z^m$ (m a natural number such that $m \geqq 2$), holomorphic on the z-plane. Obviously, $f(0) = f'(0) = \cdots = f^{(m-1)}(0) = 0$, $f^{(m)}(0) = m! \neq 0$, $f(z) \neq 0$ if $z \neq 0$ and $f'(z) = mz^{m-1} \neq 0$ if $z \neq 0$. To find all z satisfying $z^m = w$ for a certain $w \neq 0$ that is, to find all mth roots of w, we put $w = |w|e^{i\omega}$ and $z = re^{i\theta}$. From $r^m e^{im\theta} = |w|e^{i\omega}$ we get $r^m = |w|$ and $m\theta = \omega + 2k\pi$, k an integer. Hence

$$r = |w|^{1/m}, \qquad \theta = \frac{\omega}{m} + \frac{2k\pi}{m}.$$

Therefore

$$z = \rho^k |w|^{1/m} e^{i\omega/m}, \qquad \rho = e^{2\pi i/m}, \qquad k = 0, 1, 2, \ldots, m-1. \tag{3.7}$$

We conclude that there are m different mth roots z of $w \neq 0$, given by (3.7). Since $f'(z) \neq 0$ for $z \neq 0, f: z \to w = z^m$ is an m-to-one conformal mapping from the region $\{z: 0 < |z| < +\infty\}$ (i.e., the z-plane minus $\{0\}$) onto the region $\{w: 0 < |w| < +\infty\}$ (i.e., the w-plane minus $\{0\}$).

The mapping $f: z \to w = z^m$ maps the ray $r \to z = re^{i\theta}$, $0 \leqq r < \infty$, of the z-plane onto the ray: $t \to w = te^{im\theta}$, $0 \leqq t = r^m < +\infty$. Hence the angle between two rays with initial point 0 is multiplied by m under the mapping f, and f is not conformal at 0.

We now examine the case $m = 2$, $f: z = x + iy \to w = u + iv = z^2$ in more detail. The mapping f maps the circle with center 0 and radius r of the z-plane onto the circle with center 0 and radius r^2 of the w-plane and the ray $r \to z = re^{i\theta}$ onto the ray $t \to w = te^{2i\theta}$.

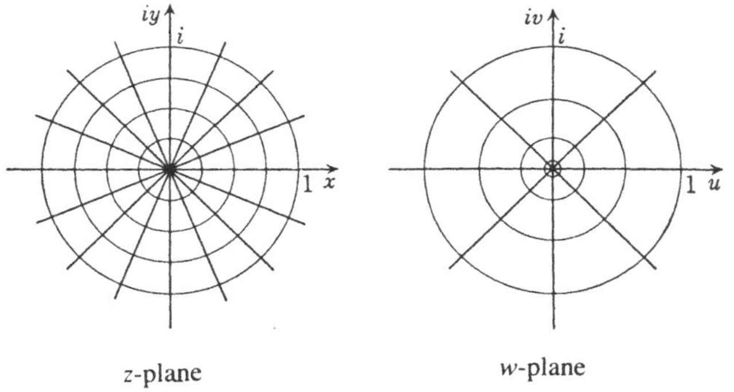

Fig. 3.3

Put $x = a \neq 0$. Since $u = x^2 - y^2$ and $v = 2xy$, we get $y = v/2a$ and $u = (a^2 - v^2)/ya^2$. Therefore the line $x = a$, which runs parallel to the imaginary axis in the z-plane, is mapped onto the parabola u. Similarly, f maps the line $y = b \neq 0$, which runs parallel to the real axis in the z-plane, onto the parabola $u = v^2/4b^2 - b^2$ of the w-plane. Since f is conformal except at $z = 0$, these two kinds of parabolas intersect each other orthogonally. Furthermore, f maps the imaginary axis of the z-plane onto the nonpositive part $\{u: u \leqq 0\}$ of the real axis of the w-plane and the real axis on the nonnegative part $\{u: u \geqq 0\}$ of the real axis of the w-plane.

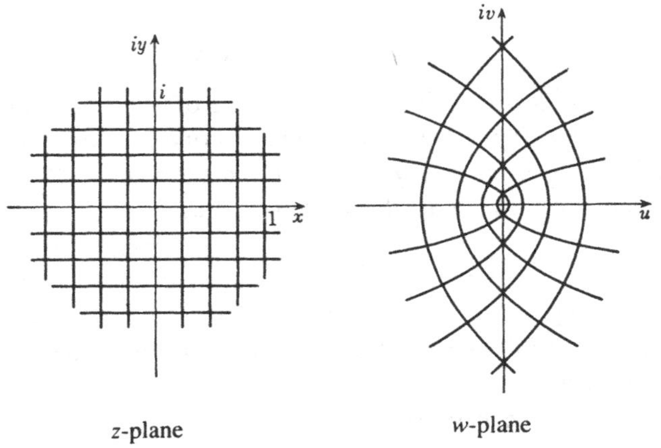

Fig. 3.4

Each of the equations $x^2 - y^2 = \alpha \neq 0$ and $2xy = \beta \neq 0$, where α and β are constants, defines a hyperbola in the z-plane. The mapping f maps the hyperbola $x^2 - y^2 = \alpha \neq 0$ onto the line $u = \alpha$ of the w-plane and the hyperbola $2xy = \beta$ onto the line $v = \beta$ of the w-plane.

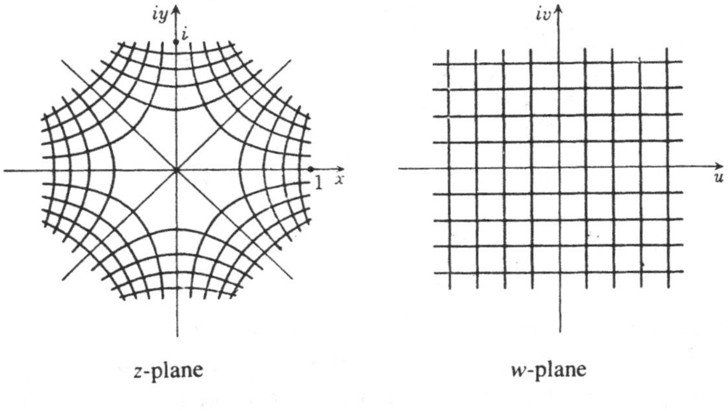

Fig. 3.5

In complex analysis an important role is played by conformal mappings which map a certain region one-to-one onto another region. From now on, a conformal mapping which maps a region D onto a region E will always be a conformal mapping that maps D one-to-one onto E.

Theorem 3.4. (1) If $f\colon z \to w = f(z)$ maps the region D of the z-plane onto the region E of the w-plane conformally, then the inverse mapping: f^{-1}: $w \to z = f^{-1}(w)$ maps E onto D conformally.

(2) Given such a mapping f and if $g\colon w \to \zeta = g(w)$ is a conformal mapping from D onto the region Ω of the ζ-plane, then the composite mapping $g \circ f\colon z \to \zeta = g(f(z))$ is a conformal mapping from D onto Ω.

Proof. (1) Since $f(z)$ is holomorphic on D and $f'(z) \neq 0$ on D, the inverse mapping $f^{-1}\colon w \to z = f^{-1}(w)$ is holomorphic on E and $df^{-1}(w)/dw \neq 0$ by Theorem 3.2. Hence f^{-1} is a conformal mapping between E and D.

(2) By the chain rule for holomorphic functions (Theorem 1.6) the function $g(f(z))$ is holomorphic on D and

$$\frac{d}{dz} g(f(z)) = g'(f(z)) \cdot f'(z) \neq 0.$$

Therefore, the one-to-one mapping $g \circ f\colon z \to \zeta = g(f(z))$, which maps E onto Ω, is conformal.

If $f(z)$ is a holomorphic function on a region D, which maps two different points of D onto different points (i.e., if $z_1, z_2 \in D$ and $z_1 \neq z_2$ then $f(z_1) \neq f(z_2)$), then $f(z)$ is called *univalent* on D. If $f\colon z \to w = f(z)$ is a conformal mapping which maps the region D onto the region E, then, of course, D is univalent. Conversely, we have the following result.

Theorem 3.5. If $f(z)$ is a univalent holomorphic function on a region D, then $f\colon z \to w = f(z)$ maps D conformally onto the region $E = f(D)$.

Proof: We have already shown that $E = f(D)$ is a region. Since f is one-to-one on D by assumption, $f'(z) \neq 0$ at all points $z \in D$, hence f is a conformal mapping which maps D onto E.

Corollary. The inverse function of a univalent holomorphic function defined on some region is a univalent holomorphic function.

If $w = f(z)$ is a holomorphic function defined on a region, such that its inverse function $z = f^{-1}(w)$ exists and is holomorphic, then $f(z)$ is called *biholomorphic*. If the function $f(z)$, defined on the region D, is biholomorphic, we conclude from the equality $f^{-1}(f(z)) = 1$:

$$\frac{d}{dw} f^{-1}(w) \cdot f'(z) = 1, \qquad w = f(z).$$

Therefore $f'(z) \neq 0$, hence $f\colon z \to w = f(z)$ is a conformal mapping which maps D onto the region $E = f(D)$. Conversely, if $f\colon z \to w = f(z)$ is a conformal mapping which maps the region D onto the region E, then $f(z)$ is biholomorphic by Theorem 3.4(1). Biholomorphic functions are obviously univalent and we have by the above corollary that a univalent holomorphic function is biholomorphic.

The disk with center 0 and radius 1 in the complex plane is called the *unit disk*. Conformal mappings that map the unit disk onto itself are of fundamental importance. For $\alpha \in \mathbb{C}$ such that $0 < |\alpha| < 1$ we put $\alpha^* = 1/\bar{\alpha} = \alpha/|\alpha|^2$. The points α and α^* are on the same ray with initial point 0 and $|\alpha^*| = 1/|\alpha| > 1$. Put

$$f(z) = \frac{z-\alpha}{1-\bar{\alpha}z} = -\frac{1}{\bar{\alpha}}\cdot\frac{z-\alpha}{z-\alpha^*}, \qquad z \neq \alpha^*.$$

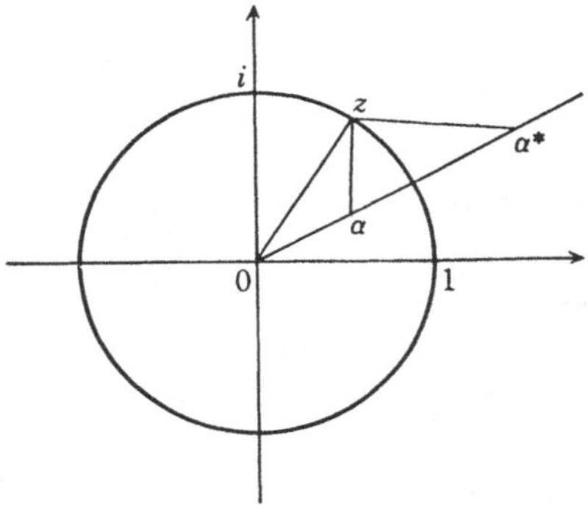

Fig. 3.6

The function $f(z)$ is holomorphic on the z-plane with the exception of α^*. If $|z| = 1$, then $|z|/|\alpha| = |\alpha^*|/|z|$ hence the triangles $z\alpha 0$ and $\alpha^* z 0$ are similar and therefore

$$\frac{|z-\alpha|}{|\alpha^* - z|} = \frac{|\alpha|}{|z|} = |\alpha|,$$

that is, $|f(z)| = 1$. Therefore, $|f(z)| = 1$ for all points z on the unit circle $\{z \in \mathbb{C}\colon |z| = 1\}$. Hence by the maximum principle (corollary to Theorem 1.21), $|f(z)| < 1$ if $|z| < 1$. This can also be verified directly:

$$\begin{aligned} 1 - |f(z)|^2 &= 1 - \frac{(z-\alpha)(\bar{z}-\bar{\alpha})}{(1-\bar{\alpha}z)(1-\alpha\bar{z})} \\ &= \frac{1 - z\bar{z} - \alpha\bar{\alpha} + \bar{\alpha}z\alpha\bar{z}}{(1-\bar{z}\alpha)(1-\alpha\bar{z})} = \frac{(1-|\alpha|^2)(1-|z|^2)}{|1-\bar{\alpha}z|^2} \end{aligned}$$

and as $1-|\alpha|^2 > 0$ by assumption, $|f(z)| < 1$ if $|z| < 1$. Conversely, if $|f(z)| < 1$ then $|z| < 1$. In order to find the inverse function of $f(z)$ we solve the equation $f(z) = w$ for some arbitrary w. From $f(z) = w$ we get $(1-\bar{\alpha}z)\,w = z-\alpha$ and, conversely, if $(1-\bar{\alpha}z)\,w = z-\alpha$, then $1-\bar{\alpha}z \neq 0$ and $f(z) = w$. (For if $1-\bar{\alpha}z = 0$, then $z-\alpha = 0$, hence $0 = 1-\bar{\alpha}z = 1-\bar{\alpha}\alpha$ and so $|\alpha|^2 = 1$, contradicting our assumption.) Therefore, in order to solve the equation $f(z) = w$ it suffices to solve the linear equation $(1-\bar{\alpha}z)\,w = z-\alpha$. This equation is equivalent to

$$(1+\bar{\alpha}w)z = w+\alpha.$$

If $1+\bar{\alpha}w \neq 0$ the solution of this equation is given by

$$z = \frac{w+\alpha}{1+\bar{\alpha}w}.$$

If $1+\bar{\alpha}w = 0$, then $w+\alpha = 0$, hence $0 = 1+\bar{\alpha}w = 1-|\alpha|^2$, contradicting our assumption. Therefore, the inverse function of $w = f(z)$ is given by

$$z = f^{-1}(w) = \frac{w+\alpha}{1+\bar{\alpha}w}, \qquad w \neq -\alpha^*.$$

Since $f^{-1}(w)$ is holomorphic on the w-plane minus $\{-\alpha^*\}$, $f\colon z \to w = f(z)$ is a conformal mapping which maps the region $\{z \in \mathbb{C}\colon z \neq \alpha^*\}$ onto the region $\{w \in \mathbb{C}\colon w \neq -\alpha^*\}$. As seen above, $|f(z)| < 1$ if $|z| < 1$ and $|z| < 1$ if $|f(z)| < 1$; therefore f maps the unit disk $\{z \in \mathbb{C}\colon |z| < 1\}$ onto the unit disk $\{w \in \mathbb{C}\colon |w| < 1\}$. Restricting the domain of the function $f(z)$ to the former we get the conformal mapping

$$f\colon z \to w = f(z) = \frac{z-\alpha}{1-\bar{\alpha}z}, \qquad |z| < 1, \tag{3.8}$$

which maps onto the unit disk $\{w \in \mathbb{C}\colon |w| < 1\}$. The conformal mapping f maps α onto the unit disk $\{z \in \mathbb{C}\colon |z| < 1\}$ $0\colon f(\alpha) = 0$. The inverse mapping of f is given by

$$f^{-1}\colon w \to z = f^{-1}(w) = \frac{w+\alpha}{1+\bar{\alpha}w}, \qquad |w| < 1. \tag{3.9}$$

Theorem 3.6. All conformal mappings which map the unit disk $\{z \in \mathbb{C}\colon |z| < 1\}$ onto the unit disk $\{w \in \mathbb{C}\colon |w| < 1\}$ and the point α, $|\alpha| < 1$, onto 0, can be represented as

$$f\colon z \to w = f(z) = e^{i\theta} \cdot \frac{z-\alpha}{1-\bar{\alpha}z}, \qquad \theta \in \mathbb{R}. \tag{3.10}$$

The mapping $w \to e^{i\theta}w$ is a rotation through an angle θ, and therefore leaves the unit disk unchanged. Hence, by the above result it is clear that the mapping (3.10) is a conformal mapping which maps the unit disk onto itself.

The converse is an easy consequence of the following lemma.

Schwarz's Lemma. If $f(z)$ is a holomorphic function defined on the unit disk $\{z \in \mathbb{C}: |z| < 1\}$ satisfying:

(1) $|f(z)| \leqq 1$ for all z with $|z| < 1$,

(2) $f(0) = 0$,

then

$$|f(z)| \leqq |z|, \tag{3.11}$$

$$|f'(0)| \leqq 1. \tag{3.12}$$

Unless

$$f(z) = c \cdot z, \quad \text{with } c \text{ a constant satisfying } |c| = 1,$$

we have

$$|f(z)| < |z| \quad \text{if } 0 < |z| < 1, \tag{3.13}$$

$$|f'(0)| < 1. \tag{3.14}$$

Proof: By Theorem (1.16), $f(z)$ can be expanded into a power series

$$f(z) = a_0 + a_1 z + a_2 z^2 + \cdots + a_n z^n + \cdots, |z| < 1.$$

Since $a_0 = f(0) = 0$ we have

$$f(z) = zg(z), \qquad g(z) = a_1 + a_2 z + \cdots + a_n z^{n-1} + \cdots,$$

and $g(z)$ is holomorphic for $|z| < 1$. To prove (3.11) it suffices to prove that $|g(z)| \leqq 1$ for $|z| < 1$. Let $M(r)$ be the maximum of $|g(z)|$ on the closed disk $[U_r(0)] = \{z: |z| \leqq r\}$, $0 < r < 1$. By the maximum principle (corollary to Theorem 1.21), there is a point z on the circumference of $[U_r(0)]$ such that $|g(z)| = M(r)$. By condition (1)

$$|g(z)| = \left|\frac{f(z)}{z}\right| = \frac{|f(z)|}{r} \leqq \frac{1}{r} \qquad \text{if } |z| = r.$$

Hence $M(r) \leqq 1/r$ and

$$|g(z)| \leqq 1/r \qquad \text{if } |z| \leqq r < 1.$$

If $|z| < 1$, then $|g(z)| \leqq 1/r$ for some r satisfying $|z| \leqq r < 1$, therefore $|g(z)| \leqq \lim_{r \to 1-0} 1/r = 1$, hence

$$|g(z)| \leqq 1 \qquad \text{if } |z| < 1.$$

Since $f(z) = zg(z)$, we have $|f(z)| \leqq z$ if $|z| \leqq 1$, proving (3.11). Since $f'(0) = g(0)$ we have $|f'(0)| = |g(0)| < 1$, proving (3.12).

The function $g(z)$ is holomorphic on the unit disk $U_1(0) = \{z: |z| < 1\}$ and $|g(z)|, \leqq 1$, does not assume a maximum on $U_1(0)$ unless $g(z)$ is a constant, by the maximum principle. Therefore $|g(z)| < 1$. If $g(z) = c$, a constant, then $|c| \leqq 1$ and $|g(z)| < 1$ unless $|c| = 1$. So we have proved that $|g(z)| < 1$

if $|z| < 1$ unless $g(z) = c$ with c a constant satisfying $|c| = 1$, whence (3.13) and (3.14) follow directly.

Corollary. All conformal mappings which map the unit disk onto itself and 0 onto 0 are represented by

$$z \to w = e^{i\theta} z, \qquad \theta \in \mathbb{R}.$$

Proof: Let $f\colon z \to w = f(z)$ be a conformal mapping which maps the unit disk onto itself and the point 0 onto 0. Since $f(z)$ is holomorphic and satisfies $|f(z)| < 1$ on the unit disk, we have $|f(z)| \leqq |z|$ by Schwarz's Lemma. Similarly, $|f^{-1}(w)| \leqq w$ for $|w| \leqq 1$. Therefore

$$|z| = |f^{-1}(f(z))| \leqq |f(z)| \leqq |z|,$$

that is, $|f(z)| = |z|$. Hence, by Schwarz's Lemma $f(z) = c \cdot z$ with c a constant satisfying $|c| = 1$.

Proof of Theorem 3.6: Let $f\colon z \to w = f(z)$ be a conformal mapping which maps the unit disk onto itself.

Putting

$$g(z) = \frac{z - \alpha}{1 - \bar{\alpha} z}, \qquad \alpha = f^{-1}(0),$$

we have proved that $g\colon z \to w = g(z)$ is a conformal mapping which maps the unit disk onto itself, the inverse of which is given by

$$g^{-1}\colon w \to z = g^{-1}(w) = \frac{w + \alpha}{1 + \bar{\alpha} w}.$$

The composite mapping $\phi = f \circ g^{-1}\colon w \to \phi(w) = f(g^{-1}(w))$ is a conformal mapping that maps the unit disk onto itself by Theorem 3.4(2). Since $g^{-1}(0) = \alpha$ and $f(\alpha) = 0$ we have $\phi(0) = 0$. Hence

$$f(g^{-1}(w)) = \phi(w) = e^{i\theta} w, \qquad \theta \in \mathbb{R},$$

by the above corollary.

Substituting $w = g(z)$ we get

$$f(z) = e^{i\theta} g(z) = e^{i\theta} \cdot \frac{z - \alpha}{1 - \bar{\alpha} z}.$$

Theorem 3.7. All conformal mappings that map the complex plane $\mathbb{C}$ onto itself can be represented as

$$f\colon z \to w = f(z) = \alpha z + \beta, \qquad \alpha,\ \beta \text{ constants } \alpha \neq 0.$$

Proof: Let $z \to w = f(z)$ be a conformal mapping which maps the z-plane onto the w-plane, then $f(z)$ is an entire function of z and its inverse $z = f^{-1}(w)$ is an entire function of w. Entire functions are either polynomials or transcendental functions. Suppose $f(z)$ is transcendental, then there exists a sequence $\{z_n\}$ such that $\lim_{n\to\infty} z_n = \infty$ and $\lim_{n\to\infty} f(z_n) = 0$. From $\lim_{n\to\infty} f(z_n) = 0$ we get $\lim_{n\to\infty} z_n = \lim_{n\to\infty} f^{-1}(f(z_n)) = f^{-1}(0) \neq \infty$. This is a contradiction and, hence, $f(z)$ is a polynomial

$$f(z) = a_0 + a_1 z + a_2 z^2 + \cdots + a_m z^m, \qquad a_m \neq 0.$$

By assumption $f'(z) \neq 0$ for all z, but by the Fundamental Theorem of Algebra, the equation

$$f'(z) = a_1 + 2a_2 z + \cdots + m a_m z^{m-1} = 0$$

has roots if $m \geqq 2$. Therefore, $m = 1$ and $f(z) = a_0 + a_1 z$, where $a_1 \neq 0$.

3.2 The Riemann sphere

a. The Riemann sphere

Let S be the sphere with center 0 and radius 1 in three-dimensional Euclidean space $\mathbb{R}^3$, that is, $S = \{\xi = (\xi_1, \xi_2, \xi_3) \in \mathbb{R}^3 : \xi_1^2 + \xi_2^2 + \xi_3^2 = 1\}$. Let us identify the coordinate plane $\xi_3 = 0$ in $\mathbb{R}^3$ with the complex plane, that is, we put

$$z = x + iy = (x, y, 0).$$

In this way $\mathbb{R}^3$ becomes the direct product of the complex plane $\mathbb{C}$ and the real line $\mathbb{R}$: $\mathbb{R}^3 = \mathbb{C} \times \mathbb{R}$. The point $(0, 0, 1)$ of S is called the *north pole* and is denoted by N, the point $(0, 0, -1)$ is called the *south pole*. For every point $\xi = (\xi_1, \xi_2, \xi_3) \neq (0, 0, 1)$ of S, the line through N and ξ intersects the coordinate plane $\xi_3 = 0$ in exactly one point $z = (x, y, 0)$.

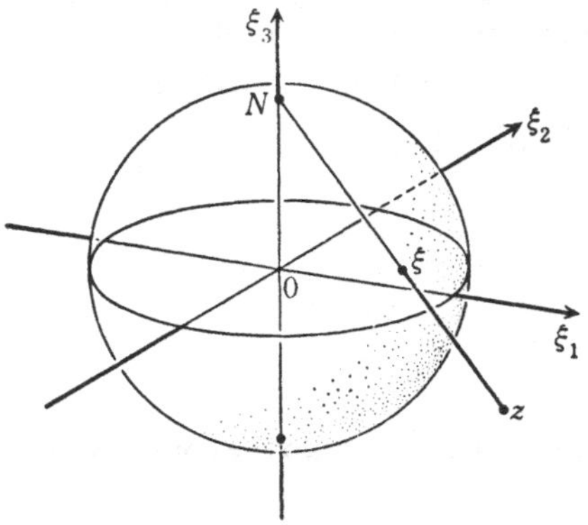

Fig. 3.7

From

$$x : y : -1 = \xi_1 : \xi_2 : \xi_3 - 1$$

we get

$$x = \frac{\xi_1}{1 - \xi_3} \qquad y = \frac{\xi_2}{1 - \xi_3}.$$

Hence

$$z = \frac{\xi_1 + i\xi_2}{1 - \xi_3}. \tag{3.15}$$

The mapping associating with each point $\xi \neq N$ of S the point $z = (x, y, 0)$ is called the *stereographic projection.* The stereographic projection $\xi \to z$ gives a one-to-one mapping from $S - \{N\}$ onto $\mathbb{C}$. Given $z = x + iy$ it is possible to solve (3.15) and find ξ_1, ξ_2, and ξ_3. From

$$|z|^2 = \frac{\xi_1^2 + \xi_2^2}{(1 - \xi_3)^2} = \frac{1 - \xi_3^2}{(1 - \xi_3)^2} = \frac{1 + \xi_3}{1 - \xi_3}$$

we get

$$\xi_3 = \frac{|z|^2 - 1}{|z|^2 + 1}. \tag{3.16}$$

Further

$$\xi_1 + i\xi_2 = z(1 - \xi_3) = \frac{2z}{|z|^2 + 1},$$

hence

$$\xi_1 = \frac{2x}{|z|^2 + 1}, \qquad \xi_2 = \frac{2y}{|z|^2 + 1}. \tag{3.17}$$

Since by means of the stereographic projection there is a one-to-one mapping between the points $\xi \neq N$ of the sphere S and the points $z = (\xi_1 + i\xi_2)/(1 - \xi_3)$ of the complex plane $\mathbb{C}$, the complex number z can be represented by its corresponding point $\xi \neq N$ of the sphere S. There is no complex number corresponding with the north pole. As is clear from the above figure and from (3.16) and (3.17) we have $\xi \to N$ as $|z| \to +\infty$. Therefore, we extend the complex plane with a point ∞, called the *point at infinity*, which corresponds with N. $\mathbb{C} \cup \{\infty\}$ is called the *extended complex plane.* The sphere S, the points of which are used to represent the points of $\mathbb{C}$, is called the *complex sphere* or the *Riemann sphere.* If we extend $\mathbb{C}$ to become the extended complex plane $\mathbb{C} \cup \{\infty\}$, we also extend the lines l of $\mathbb{C}$ to become lines $l \cup \{\infty\}$ of the extended complex plane. The plane through l and N intersects S in a circle. Hence the lines $l \cup \{\infty\}$ are represented by circles through N on the Riemann sphere S. By identifying the point $\xi = (\xi_1, \xi_2, \xi_3) \neq N$ of S with the complex number $z = (\xi_1 + i\xi_2)/(1 - \xi_3)$ and $N = (0, 0, 1)$ with the point at infinity, S is identified with the extended complex plane $\mathbb{C} \cup \{\infty\}$. Notions such as limits of point sequences, neighborhoods of a point and so on are defined for the extended complex plane $\mathbb{C} \cup \{\infty\}$ through this identification with S. For example, an ε-neighborhood of N on S is defined by

$$\{\xi\colon \xi_1^2 + \xi_2^2 + (\xi_3 - 1)^2 < \varepsilon^2\}.$$

If $\xi \neq N$, we have, by (3.16) and (3.17),

$$\xi_1^2 + \xi_2^2 + (\xi_3 - 1)^2 = \frac{4}{|z|^2 + 1}$$

and since $4/(|z|^2 + 1) < \varepsilon^2$ is equivalent to $|z|^2 > (4 - \varepsilon^2)/\varepsilon^2$ we find, putting

$$U_R(\infty) = \{z \in \mathbb{C} : |z| > R\} \cup \{\infty\},$$

that

$$\{\xi \in S\colon \xi_1^2 + \xi_2^2 + (\xi_3 - 1)^2 < \varepsilon^2\} = \mathrm{U}_{\rho(\varepsilon)}(\infty), \qquad \rho(\varepsilon) = \sqrt{}(4 - \varepsilon^2)/\varepsilon.$$

Therefore an ε-neighborhood of ∞ is given by $U_{\rho(\varepsilon)}(\infty)$. Obviously, $\rho(\varepsilon) \to +\infty$ if $\varepsilon \to +0$.

Now, consider a sequence $\{z_n\}$ in $\mathbb{C}$. If $\lim_{n \to \infty} |z_n| = +\infty$ then $\lim_{n \to \infty} z_n = \infty$. For suppose $\varepsilon > 0$ be given, then there exists a natural number $n_0(\varepsilon)$ such that $|z_n| > \rho(\varepsilon)$ if $n > n_0(\varepsilon)$, i.e., $z_n \in U_{\rho(\varepsilon)}(\infty)$ if $n > n_0(\varepsilon)$. Similarly, if $f(z)$ is a function defined on a neighborhood of $c \in \mathbb{C}$ minus $\{c\}$ such that $\lim_{z \to c} |f(z)| = +\infty$, then $\lim_{z \to c} f(z) = \infty$. Because, for every $\varepsilon > 0$, there exists a $\delta(\varepsilon)$ such that $|f(z)| > \rho(\varepsilon)$, i.e., $f(z) \in U_{\rho(\varepsilon)}(\infty)$, if $0 < |z - c| < \delta(\varepsilon)$. If $f(z)$ is holomorphic for $0 < |z - c| < r$ and c is a pole of $f(z)$, then $\lim_{z \to c} |f(z)| = +\infty$, hence $\lim_{z \to c} f(z) = \infty$. We now define the value of $f(z)$ at $z = c$ by

$$f(c) = \lim_{z \to c} f(z) = \infty.$$

Thereby $z \to f(z)$ becomes a continuous mapping from the neighborhood $U_r(c)$ of c into S. However, $f(z)$ is not considered to be continuous at c as a function which assumes the value ∞ at c. If we put $f(z) = a/z$, a a constant, $a \neq 0$, then

$$\frac{a}{0} = \lim_{z \to 0} \frac{a}{z} = \infty, \qquad a \neq 0.$$

b. Holomorphic functions with an isolated singularity at ∞

Let D be a region in $\mathbb{C}$ such that $\{z \in \mathbb{C}: |z| > R_0\} \subset D$ for some $R_0 > 0$, let $f(z)$ be a function defined on D, and let a be a complex number. If for every $\varepsilon > 0$ there exists an $R(\varepsilon) \geqq R_0$ such that

$$|f(z) - a| < \varepsilon \quad \text{if} \quad |z| > R(\varepsilon), \text{ i.e., if } z \in U_{R(\varepsilon)}(\infty),$$

then we say that the limit of $f(z)$ for z tending to ∞ equals a. This is written: $\lim_{z \to \infty} f(z) = a$ or $f(z) \to a$ as $z \to \infty$. If $f(z)$ is defined on $D \cup \{\infty\}$ and $\lim_{z \to \infty} f(z) = f(\infty)$, then $f(z)$ is said to be continuous at ∞. If $f(z)$ is defined and continuous on D and if $\lim_{z \to \infty} f(z) = a$, then $f(z)$ can be extended to a continuous function on $D \cup \{\infty\}$ by putting $f(\infty) = a$. Let $f(z)$ be a nonconstant holomorphic function defined on $\{z \in \mathbb{C}: |z| > R\} = U_R(\infty) - \{\infty\}$, $R > 0$. Putting $\hat{z} = 1/z$, the mapping $z \to \hat{z}$ is a biholomorphic mapping that maps $U_R(\infty) - \{\infty\}$ onto $U_r(0) - \{0\} = \{\hat{z} \in \mathbb{C}: 0 < |\hat{z}| < r\}$, where $r = 1/R$. Hence the function $\hat{f}(\hat{z})$ defined by

$$\hat{f}(\hat{z}) = f(z) = f(1/\hat{z})$$

is holomorphic on $U_r(0) - \{0\}$. Since $\hat{z} = 0$ is an isolated singularity of $\hat{f}(\hat{z})$, the following three cases can occur (see Section 1.4d):

Case 1: $\hat{z} = 0$ is a removable singularity of $\hat{f}(\hat{z})$. $\hat{f}(\hat{z})$ can be expanded in a power series that converges absolutely for $0 < |\hat{z}| < r$

$$\hat{f}(\hat{z}) = a_0 + a_1 \hat{z} + a_2 z^2 + \cdots + a_n \hat{z}^n + \cdots.$$

Hence

$$f(z) = \hat{f}(1/z)$$

can be expanded in a power series which converges absolutely for $|z| > R$,

$$f(z) = a_0 + \frac{a_1}{z} + \frac{a_2}{z^2} + \cdots + \frac{a_n}{z^n} + \cdots. \tag{3.18}$$

Since $\lim_{z \to \infty} f(z) = \lim_{\hat{z} \to \infty} \hat{f}(\hat{z}) = a_0$, we put

$$f(\infty) = \lim_{z \to \infty} f(z) = a_0$$

and we say that $f(z)$ is holomorphic on $U_R(\infty)$. In particular, taking $f(z) = a/z$, $a \neq 0$, we get

$$a/\infty = 0, \qquad a \neq 0.$$

Hence, substituting ∞ for z in (3.18) yields $f(\infty) = a_0$. If $f(\infty) = a_0 = 0$, (3.18) takes the form

$$f(z) = \frac{1}{z^m}\left(a_m + \frac{a_{m+1}}{z} + \frac{a_{m+2}}{z^2} + \cdots\right), \qquad a_m \neq 0.$$

In this case, ∞ is called a *zero of the mth order of* $f(z)$.

Case 2: $\hat{z} = 0$ is a pole of $\hat{f}(\hat{z})$. Since for $0 < |\hat{z}| < r$ we have

$$\hat{f}(\hat{z}) = \frac{a_{-m}}{\hat{z}^m} + \cdots + \frac{a_{-1}}{\hat{z}} + \sum_{n=0}^{+\infty} a_n \hat{z}^n, \qquad a_{-m} \neq 0,$$

we get for $|z| > R$

$$f(z) = a_{-m} z^m + \cdots + a_{-1} z + a_0 + \sum_{n=1}^{+\infty} \frac{a_n}{z^n}, \qquad a_{-m} \neq 0, \tag{3.19}$$

where the power series $\sum_{n=1}^{+\infty} (a_n/z^n)$ converges absolutely for $|z| > R$. In this case, ∞ is called a *pole of the mth order of* $f(z)$. Putting

$$g(z) = a_{-m} + \frac{a_{-m+1}}{z} + \cdots + \frac{a_0}{z^m} + \frac{a_1}{z^{m+1}} + \cdots$$

$g(z)$ is holomorphic on $U_R(\infty)$ and

$$f(z) = z^m g(z), \qquad g(\infty) = a_{-m} \neq 0.$$

Hence $\lim_{z\to\infty} f(z) = \infty$ and we put

$$f(\infty) = \lim_{z\to\infty} f(z) = \infty.$$

In this case, f: $z \to f(z)$ is a continuous mapping from $U_R(\infty)$ into the Riemann sphere S, but we do not say that $f(z)$ is a continuous function which assumes the value ∞ at ∞. As a point of the domain of the function $f(z)$, the point at infinity, ∞, is treated in the same way as a point $c \neq \infty$, but as a value assumed by the function $f(z)$, ∞ is considered as a special "number." By taking $f(z) = az = za$, $a \neq 0$ or $f(z) = z^2$ we get

$$a \cdot \infty = \infty \cdot a = \infty, \qquad a \neq 0, \qquad \infty \cdot \infty = \infty.$$

Case 3: $\hat{z} = 0$ is an essential singularity of $\hat{f}(\hat{z})$. In this case

$$\hat{f}(\hat{z}) = \sum_{n=-\infty}^{+\infty} a_n \hat{z}^n, \qquad 0 < |\hat{z}| < r,$$

where an infinite number of the coefficients $a_{-1}, a_{-2}, \ldots, a_{-n}, \ldots$ are not equal to zero. Hence

$$f(z) = \cdots + a_{-n} z^n + \cdots + a_{-1} z + a_0 + \sum_{n=1}^{+\infty} \frac{n}{z^n} \tag{3.20}$$

for $|z| > R$, where an infinite number of the coefficients a_{-1}, $a_{-2}, \ldots, a_{-n}, \ldots$ are not equal to zero. In this case, ∞ is called an *essential singularity* of $f(z)$. For example, if $f(z)$ is a transcendental entire function, then ∞ is an essential singularity of $f(z)$. If ∞ is an essential singularity of $f(z)$ then, by Weierstrass' Theorem (Theorem 1.23) it is possible, for each complex number w, to find a sequence $\{z_n\}$ such that $\lim_{n\to\infty} z_n = \infty$ and $\lim_{n\to\infty} f(z_n) = w$ and to find a sequence $\{z_n\}$ such that $\lim_{n\to\infty} z_n = \infty$ and $\lim_{n\to\infty} f(z) = \infty$.

If ∞ is an essential singularity of $f(z)$, $\lim_{z\to\infty} f(z)$ does not exist. Hence, if $f(\infty) = \lim_{z\to\infty} f(z) \in \mathbb{C}$ exists, then $f(z)$ is holomorphic at $z = \infty$ and if $f(\infty) = \lim_{z\to\infty} f(z) = \infty$, then $f(z)$ has a pole at ∞.

Summarizing the above computational rules for ∞ we have for $a \neq 0$,

$$\frac{a}{0} = \infty, \quad \frac{a}{\infty} = 0, \quad a \cdot \infty = \infty \cdot a = \infty, \quad \infty \cdot \infty = \infty. \tag{3.21}$$

c. Local coordinates

In the above we have seen how the points $\xi \neq N$ of the Riemann sphere S represent complex numbers $z = (\xi_1 + i\xi_2)/(1 - \xi_3)$. Conversely, the complex numbers z can be used to represent the points ξ of S. In this case the correspondence

$$\xi = (\xi_1, \xi_2, \xi_3) \to z = \frac{\xi_1 + i\xi_2}{1 - \xi_3}, \qquad \xi \neq N$$

is called a *complex coordinate* defined on $S - \{N\}$ and the complex number z corresponding with ξ is called the *coordinate* of the point ξ. This complex coordinate does not associate a coordinate with N. In order to define a complex coordinate which can be applied to N, we consider $\hat{z} = 1/z$. Since $\xi_1^2 + \xi_2^2 + \xi_3^2 = 1$ we have, for $-1 < \xi_3 < 1$,

$$\frac{\xi_1 + i\xi_2}{1 - \xi_3} \cdot \frac{\xi_1 - i\xi_2}{1 + \xi_3} = \frac{\xi_1^2 + \xi_2^2}{1 - \xi_3^2} = 1,$$

hence, for $z \neq 0$,

$$\hat{z} = \frac{1}{z} = \frac{1 - \xi_3}{\xi_1 + i\xi_2} = \frac{\xi_1 - i\xi_2}{1 + \xi_3}. \tag{3.22}$$

Therefore we define a complex coordinate on $S - \{(0, 0, -1)\}$ by

$$\xi = (\xi_1, \xi_2, \xi_3) \to \hat{z} = \frac{\xi_1 - i\xi_2}{1 + \xi_3}, \qquad \xi \neq (0, 0, -1).$$

With respect to this complex coordinate, the coordinate of $N = (0, 0, 1)$ is $\hat{z} = 0$. The complex coordinate $\xi \to \hat{z}$ can be obtained from the complex coordinate $\xi \to z$ by a rotation through the angle π about the ξ_1-axis

$$(\xi_1, \xi_2, \xi_3) \to (\xi_1, -\xi_2, -\xi_3).$$

Since $\xi \to z$ is a complex coordinate defined on a subregion $S - \{N\}$ of S, we call it a *local complex coordinate* on S. Of course, $\xi \to \hat{z}$ is also a local complex coordinate. Each point $\xi \in S - \{(0, 0, 1), (0, 0, -1)\}$ has two complex coordinates, related by

$$z \to \hat{z} = 1/z, \qquad \hat{z} \to z = 1/\hat{z}.$$

This correspondence is called a *coordinate transformation* between the local complex coordinates $\xi \to z$ and $\xi \to \hat{z}$. With the north pole and the south pole, $(0, 0, -1)$, only one complex coordinate is associated: $\hat{z} = 0$ and $z = 0$, respectively. The range of the complex coordinates is the $\hat{z}$-plane, which we denote by $\hat{\mathbb{C}}$. Extending $\hat{\mathbb{C}}$ with the point at infinity $\hat{\infty}$, which corresponds with $(0, 0, -1) \in S$, we can identify S and $\hat{\mathbb{C}} \cup \{\hat{\infty}\}$, just as we identified S and $\mathbb{C} \cup \{\infty\}$. With these identifications, $z \in \mathbb{C}$, $z \neq 0$, coincides with $\hat{z} = 1/z \in \hat{\mathbb{C}}$, $0 \in \mathbb{C}$ with $\hat{\infty}$, and ∞ with $0 \in \mathbb{C}$. Hence

$$S = \mathbb{C} \cup \hat{\mathbb{C}}$$

and S is the surface obtained by "pasting together" $\mathbb{C}$ and $\hat{\mathbb{C}}$ such that $z \in \mathbb{C}$, $z \neq 0$, coincides with $\hat{z} = 1/z \in \hat{\mathbb{C}}$. The coordinate transformation $z \to \hat{z} = 1/z$ determines the way the pasting should be done.

In Section 3.2b we have considered the function $\hat{f}(\hat{z}) = f(1/\hat{z})$ corresponding with a holomorphic function $f(z)$ defined on $U_R(\infty) - \{\infty\}$. This function is nothing but $f(z)$ considered as a function of the local coordinate $\hat{z}$.

d. *Homogeneous coordinates*

The Riemann sphere can also be considered as a one-dimensional complex projective space. Consider the direct product $\mathbb{C}^2 = \mathbb{C} \times \mathbb{C} = \{(\zeta_1, \zeta_2): \zeta_1 \in \mathbb{C}, \zeta_2 \in \mathbb{C}\}$ and put

$$\zeta = \{(\lambda\zeta_1, \lambda\xi_2): \lambda \in \mathbb{C}\}$$

for $(\zeta_1, \zeta_2) \neq (0, 0)$. Now, ζ is a complex line in $\mathbb{C}_2$ passing through the origin $(0, 0)$. The collection of all complex lines of $\mathbb{C}^2$ passing through $(0, 0)$ is called the *one-dimensional complex projective space* and is denoted by $\mathbb{P}^1$. The complex lines $\zeta \in \mathbb{P}^1$ are called the *points* of $\mathbb{P}^1$ and if

$$\zeta = \{(\lambda\zeta_1, \lambda\zeta_2): \lambda \in \mathbb{C}\}, \qquad (\zeta_1, \zeta_2) \neq (0,0),$$

then (ζ_1, ζ_2) are called *homogeneous coordinates* of the point ζ. $(\zeta'_1, \zeta'_2) \neq (0,0)$ and $(\zeta_1, \zeta_2) \neq (0,0)$ are homogeneous coordinates of the same point $\zeta \in \mathbb{P}^1$ if and only if $\zeta'_1 = \lambda\zeta_1$ and $\zeta'_2 = \lambda\zeta_2$ for some complex number $\lambda \neq 0$. If (ζ_1, ζ_2) are homogeneous coordinates of the point $\zeta \in \mathbb{P}^1$, we write

$$\zeta = (\zeta_1, \zeta_2).$$

Hence, $(\zeta'_1, \zeta'_2) = (\zeta_1, \zeta_2)$ if and only if $\zeta'_1 = \lambda\zeta_1$ and $\zeta'_2 = \lambda\zeta_2$ for some complex number $\lambda \neq 0$. The equality sign $=$, used for homogeneous coordinates, will always have this meaning. We want to show that there is a natural mapping from $\mathbb{P}^1$ onto the Riemann sphere $S = \mathbb{C} \cup \hat{\mathbb{C}}$. For that purpose put

$$U_1 = \{\zeta \in \mathbb{P}^1 : \zeta_2 \neq 0\}, \qquad U_2 = \{\zeta \in \mathbb{P}^1 : \zeta_1 \neq 0\}.$$

Obviously,

$$\mathbb{P}^1 = U_1 \cup U_2.$$

For $\zeta_2 \neq 0$ we have

$$\zeta = (\zeta_1, \zeta_2) = (\zeta_1/\zeta_2, 1) = (z, 1), \qquad z = \zeta_1/\zeta_2,$$

hence the mapping $\zeta \to z = \zeta_1/\zeta_2$ maps U_1 one-to-one onto $\mathbb{C} \subset S$. Similarly, for $\zeta_1 \neq 0$ we have

$$\zeta = (\zeta_1, \zeta_2) = (1, \hat{z}), \qquad \hat{z} = \zeta_2/\zeta_1,$$

hence the mapping $\zeta \to \hat{z} = \zeta_2/\zeta_1$ maps U_2 one-to-one onto $\hat{\mathbb{C}} \subset S$. Since $\hat{z} = \zeta_2/\zeta_1 = 1/z$ for $\zeta_1 \neq 0$ and $\zeta_2 \neq 0$, the two mappings $\zeta \to z$ and $\zeta \to \hat{z}$ coincide on $U_1 \cap U_2$. Therefore, a one-to-one mapping is obtained between $\mathbb{P}^1$ and $S = \mathbb{C} \cup \hat{\mathbb{C}}$ by combining these two mappings into one. Using this mapping, we can identify the points of S with the points of $\mathbb{P}^1$ and thereby the Riemann sphere S becomes a one-dimensional complex projective space. Under this identification, the point $z \in \mathbb{C}$ gets the homogeneous coordinates $(z, 1)$ and ∞ gets the homogeneous coordinates $(1, 0)$, i.e., $z = (z, 1)$ and $\infty = (1, 0)$. Now z and ∞ are called the *inhomogeneous coordinates* of the points $(z, 1)$ and $(1, 0)$ of $\mathbb{P}^1$, respectively. Since $\zeta_1/0 = \infty$ if $\zeta_1 \neq 0$ by (3.21), the inhomogeneous coordinate of $\zeta = (\zeta_1, \zeta_2)$ is always given by ζ_1/ζ_2.

3.3 Linear fractional transformations

a. Linear fractional transformations

A function of the form

$$f(z) = \frac{\alpha z + \beta}{\gamma z + \delta}, \qquad \alpha, \beta, \gamma, \delta \text{ constants}, \qquad \alpha\delta - \beta\gamma \neq 0,$$

is called a *linear fractional function* or a *linear function*. The condition $\alpha\delta - \beta\gamma \neq 0$ is included in order to exclude the case that $f(z)$ reduces to a constant. If $\gamma = 0$, we write α instead of α/δ and β instead of β/δ, obtaining

$$f(z) = \alpha z + \beta, \qquad \alpha \neq 0.$$

We distinguish two cases:

Case 1: $\gamma \neq 0$. In this case $f(z)$ is a holomorphic function on the z-plane excluding $\{-\delta/\gamma\}$ and since

$$f(\infty) = \lim_{z \to \infty} \frac{\alpha z + \beta}{\gamma z + \delta} = \lim_{z \to \infty} \frac{\alpha + \beta/z}{\gamma + \delta/z} = \frac{\alpha}{\gamma},$$

$f(z)$ is holomorphic at ∞ too. Writing

$$f(z) = \frac{\beta\gamma - \alpha\delta}{\gamma^2(z+\delta/\gamma)} + \frac{\alpha}{\gamma} \tag{3.23}$$

it is clear that $-\delta/\gamma$ is a pole of the first order of $f(z)$, i.e.,

$$f(-\delta/\gamma) = \infty.$$

Hence, in order to prove that $f: z \to w = f(z)$ is a one-to-one mapping from the extended plane $\mathbb{C} \cup \{\infty\}$ onto itself, it suffices to show that for each $w \in \mathbb{C}$, $w \neq \alpha/\gamma$, the equation $f(z) = w$ has exactly one solution $z \in \mathbb{C}$, $z \neq -\delta/\gamma$. From $f(z) = w$ we get $(\gamma z + \delta)w = \alpha z + \beta$. Conversely, assume $(\gamma z + \delta)w = \alpha z + \beta$. From $\alpha\delta - \beta\gamma \neq 0$ it follows that $\gamma z + \delta$ and $\alpha z + \beta$ are not simultaneously equal to 0, hence $\gamma z + \delta \neq 0$ and $f(z) = w$. The equation $(\gamma z + \delta)w = \alpha z + \beta$ is equivalent to the equation

$$(-\gamma w + \alpha)z = \delta w - \beta$$

and since $w \neq \alpha/\gamma$ we have $-\gamma w + \alpha \neq 0$, hence the equation $f(z) = w$ has exactly one solution

$$z = \frac{\delta w - \beta}{-\gamma w + \alpha}, \qquad z \neq -\delta/\gamma.$$

Therefore, f is a one-to-one mapping from $\mathbb{C} \cup \{\infty\}$ onto $\mathbb{C} \cup \{\infty\}$ and its inverse mapping is given by

$$f^{-1}: w \to z = f^{-1}(w) = \frac{\delta w - \beta}{-\gamma w + \alpha}. \tag{3.24}$$

Since $f(z)$ is holomorphic if $z \neq -\delta/\gamma$ and $f^{-1}(w)$ if $w \neq \alpha/\delta$, $f(z)$ is biholomorphic on the region $\mathbb{C} - \{-\delta/\gamma\}$. Hence the restriction of $f(z)$ to $\mathbb{C} - \{-\delta/\gamma\}$ is a conformal mapping which maps $\mathbb{C} - \{-\delta/\gamma\}$ onto $\mathbb{C} - \{\alpha/\gamma\}$. The function $f(z)$, considered as a function on the Riemann sphere $S = \mathbb{C} \cup \{\infty\}$, gives a one-to-one continuous mapping from S onto itself. For by the results of Section 3.2a, f, as a mapping from S into S, is also continuous at $z = -\delta/\gamma$.

Case 2: $\gamma = 0$. The function $f(z) = \alpha z + \beta$ is of course holomorphic on the z-plane, ∞ is a pole of the first order of $f(z)$, and $f(\infty) = \infty$. Solving the equation $w = \alpha z + \beta$ we get $z = (w - \beta)/\alpha$. Hence, $f: z \to w = f(z)$ is a one-to-one mapping from $\mathbb{C} \cup \{\infty\}$ onto $\mathbb{C} \cup \{\infty\}$ and its inverse is given by

$$f^{-1}: w \to z = f^{-1}(w) = \frac{w - \beta}{\alpha}.$$

This formula can also be obtained from (3.24) by putting $\gamma = 0$ and $\delta = 1$. Restricting $f(z)$ to $\mathbb{C}$ we get a mapping from $\mathbb{C}$ into $\mathbb{C}$.

Considering $f(z)$ as a function defined on the Riemann sphere $S = \mathbb{C} \cup \{\infty\}$ we get a one-to-one continuous mapping from S onto itself by Section 3.2b, (2).

A one-to-one mapping from the extended plane onto itself of the form

$$f: z \to w = f(z) = \frac{\alpha z + \beta}{\gamma z + \delta}, \qquad \alpha\delta - \beta\gamma \neq 0, \tag{3.25}$$

is called a *linear fractional transformation* or simply a *linear transformation*. The inverse transformation f^{-1} of a linear fractional transformation is given by (3.24). Let (ζ_1, ζ_2) and (ω_1, ω_2) be homogeneous coordinates of z and w, respectively. From $z = \zeta_1/\zeta_2$, $w = \omega_1/\omega_2$, we get

$$w = \frac{\omega_1}{\omega_2} = \frac{\alpha(\zeta_1/\zeta_2) + \beta}{\gamma(\zeta_1/\zeta_2) + \delta} = \frac{\alpha\zeta_1 + \beta\zeta_2}{\gamma\zeta_1 + \delta\zeta_2}.$$

Hence, (3.25) can be represented as

$$f: \begin{bmatrix} \zeta_1 \\ \zeta_2 \end{bmatrix} \to \begin{bmatrix} \omega_1 \\ \omega_2 \end{bmatrix} = \begin{bmatrix} \alpha & \beta \\ \gamma & \delta \end{bmatrix} \begin{bmatrix} \zeta_1 \\ \zeta_2 \end{bmatrix}. \tag{3.26}$$

Strictly, the transformation (3.26) only determines the matrix to within a scalar multiple but, as is clear from the following discussion, we can select any matrix satisfying (3.26). Since

$$\begin{bmatrix} \alpha & \beta \\ \gamma & \delta \end{bmatrix}^{-1} = \frac{1}{\lambda} \begin{bmatrix} \delta & -\beta \\ -\gamma & \alpha \end{bmatrix}, \qquad \lambda = \alpha\delta - \beta\gamma,$$

we get from (3.26)

$$\begin{bmatrix} \lambda\zeta_1 \\ \lambda\zeta_2 \end{bmatrix} = \begin{bmatrix} \delta & -\beta \\ -\gamma & \alpha \end{bmatrix} \begin{bmatrix} \omega_1 \\ \omega_2 \end{bmatrix}.$$

Hence

$$z = \frac{\zeta_1}{\zeta_2} = \frac{\delta\omega_1 - \beta\omega_2}{-\gamma\omega_1 + \alpha\omega_2} = \frac{\delta w - \beta}{-\gamma w + \alpha},$$

the same formula as (3.24). If

$$f_1(z) = \frac{\alpha_1 z + \beta_1}{\gamma_1 z + \delta_1}, \qquad f_2(z) = \frac{\alpha_2 z + \beta_2}{\gamma_2 z + \delta_2}$$

are linear fractional transformations, their product $f_1 \cdot f_2$ (i.e., the composite transformation) in homogeneous coordinates is given by

$$f_1 \cdot f_2: \begin{bmatrix} \zeta_1 \\ \zeta_2 \end{bmatrix} \to \begin{bmatrix} \omega_1 \\ \omega_2 \end{bmatrix} = \begin{bmatrix} \alpha_1 & \beta_1 \\ \gamma_1 & \delta_1 \end{bmatrix} \begin{bmatrix} \alpha_2 & \beta_2 \\ \gamma_2 & \delta_2 \end{bmatrix} \begin{bmatrix} \zeta_1 \\ \zeta_2 \end{bmatrix}.$$

Since

$$\begin{bmatrix} \alpha_1 & \beta_1 \\ \gamma_1 & \delta_1 \end{bmatrix} \begin{bmatrix} \alpha_2 & \beta_2 \\ \gamma_2 & \delta_2 \end{bmatrix} = \begin{bmatrix} \alpha_1\alpha_2 + \beta_1\gamma_1 & \alpha_1\beta_2 + \beta_1\delta_2 \\ \gamma_1\alpha_2 + \delta_1\gamma_2 & \gamma_1\beta_2 + \delta_1\delta_2 \end{bmatrix}$$

we get the inhomogeneous formula

$$f_1 \cdot f_2 : z \to w = f_1(f_2(z)) = \frac{(\alpha_1\alpha_2 + \beta_1\gamma_2)z + \alpha_1\beta_2 + \beta_1\delta_2}{(\gamma_1\alpha_2 + \delta_1\gamma_2)z + \gamma_1\beta_2 + \delta_1\delta_2}.$$

We see from this that the product of two linear fractional transformations is again a linear fractional transformation. Since the inverse of a linear fractional transformation is also a linear fractional transformation, we conclude that the collection of all linear fractional transformations constitutes a group.

Theorem 3.8. There is exactly one linear fractional function $f(z)$ such that

$$f(z_1) = 0, \qquad f(z_2) = 1, \qquad f(z_3) = \infty,$$

where z_1, z_2, and z_3 are three different points of $\mathbb{C} \cup \{\infty\}$ and $f(z)$ is given by

$$f(z) = \frac{z - z_1}{z - z_3} \bigg/ \frac{z_2 - z_1}{z_2 - z_3}. \tag{3.27}$$

and where the formula for $f(z)$, if one of the z_j, say z_1, equals ∞, is found by letting z_1 tend to ∞ in (3.27).

Proof: Since a function given by (3.27) clearly satisfies the requirements, it suffices to show that if $g(z)$ is a linear fractional function satisfying $g(z_1) = 0$, $g(z_2) = 1$, and $g(z_3) = \infty$, then $f(z)$ and $g(z)$ are identical. For this purpose consider the linear fractional function $g(f^{-1}(w))$. Putting

$$g(f^{-1}(w)) = \frac{\alpha w + \beta}{\gamma w + \delta}, \qquad \alpha\delta - \beta\gamma \neq 0,$$

we have

$$\frac{\beta}{\delta} = g(f^{-1}(0)) = 0, \qquad \frac{\alpha + \beta}{\gamma + \delta} = g(f^{-1}(1)) = 1, \qquad \frac{\alpha}{\gamma} = g(f^{-1}(\infty)) = \infty$$

that is, $\beta = 0$, $\gamma = 0$, and $\alpha/\delta = 1$, hence $g(f^{-1}(w)) = w$. Substitution of $f(z)$ for w yields $f(z) = g(z)$.

Corollary. Given three distinct points w_1, w_2, w_3 of $\mathbb{C} \cup \{\infty\}$ there is, for each triple of distinct points z_1, z_2, and z_3, exactly one linear fractional function $f(z)$ such that $f(z_1) = w_1$, $f(z_2) = w_2$, and $f(z_3) = w_3$.

b. *Cross ratio*

As explained in Section 3.2a, a line in the extended plane $\mathbb{C} \cup \{\infty\}$ is obtained by extending a line l of the complex plane $\mathbb{C}$ with $\{\infty\}$. Circles of $\mathbb{C}$ will also be considered to be circles of $\mathbb{C} \cup \{\infty\}$.

For three distinct points z_1, z_2, and z_3 of $\mathbb{C}\cup\{\infty\}$ the expression $((z-z_1)/(z-z_3))/((z_2-z_1)/(z_2-z_3))$ is called the *cross ratio* of the four points z, z_1, z_2, and z_3 and is denoted by (z, z_1, z_2, z_3)

$$(z, z_1, z_2, z_3) = \frac{z-z_1}{z-z_3}\Bigg/\frac{z_2-z_1}{z_2-z_3} = \frac{(z-z_1)(z_2-z_3)}{(z-z_3)(z_2-z_1)}.$$

Considering z as a variable, the cross ratio (z, z_1, z_2, z_3) is a linear fractional function of z. Putting $z=z_1$, $z=z_2$, and $z=z_3$ we get the values 0, 1, and ∞, respectively for the cross ratio (z, z_1, z_2, z_3).

Theorem 3.9. If z, z_1, z_2, and z_3 are four distinct points of $\mathbb{C}\cup\{\infty\}$, then z, z_1, z_2, and z_3 are on one circle or one line if and only if their cross ratio (z, z_1, z_2, z_3) is a real number.

Proof: (1) Assume that z_1, z_2, and z_3 are on a circle C. Let θ, $-\pi<\theta\leqq\pi$, be the argument of $(z-z_1)/(z-z_3)$ and let ψ, $-\pi<\psi\leqq\psi$, be the argument of $(z_2-z_1)/(z_2-z_3)$. Since the argument of (z, z_1, z_2, z_3) equals $\theta-\psi$, $-2\pi<\theta-\psi<2\pi$, (z, z_1, z_2, z_3) is a real number if and only if

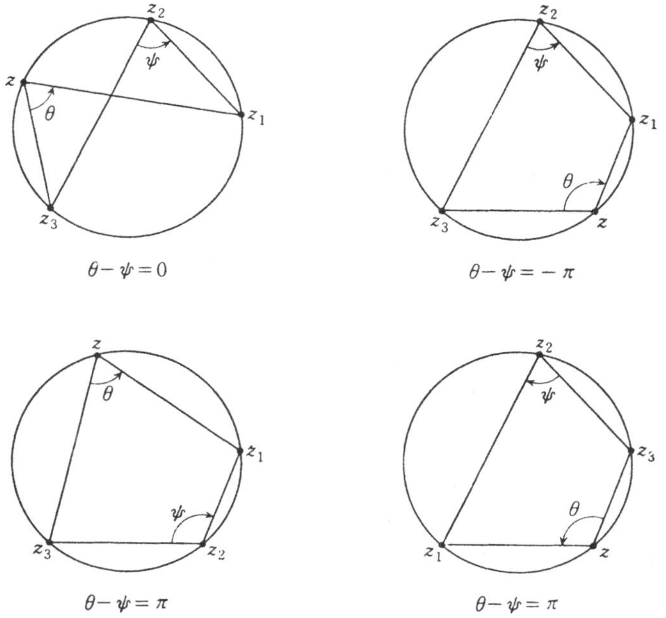

Fig. 3.8

$\theta-\psi=0$ or $\theta-\psi=\pm\pi$ and it is clear from simple geometric considerations that z is on the circle C if and only if $\theta-\psi=0$ or $\theta-\psi=\pm\pi$, hence, if z_1, z_2, and z_3 are on a circle, the points z, z_1, z_2, and z_3 are on a circle if and only if (z, z_1, z_2, z_3) is a real number.

(2) Assume that z_1, z_2, and z_3 are on a line $l \subset \mathbb{C}$. Since $(z_2-z_1)/(z_2-z_3)$ is a real number, (z, z_1, z_2, z_3) is a real number if and only if $(z-z_1)/(z-z_3)$ is a real number, that is, if and only if z is on l or $z=\infty$.

(3) Assume that one of the points z_1, z_2, and z_3 is ∞. Let us assume that $z_1=\infty$ (the other cases are similar). Since

$$(z, \infty, z_2, z_3)=\frac{z_2-z_3}{z-z_3},$$

(z, ∞, z_2, z_3) is real if and only if z, z_2, and z_3 are on a line l, that is, if and only if z, $z_1=\infty$, z_2 and z_3 are on a line $l \cup \{\infty\}$.

Theorem 3.10. Let $f(z)$ be a linear fractional function and let z, z_1, z_2, and z_3 be four distinct points in $\mathbb{C} \cup \{\infty\}$. Then

$$(f(z), f(z_1), f(z_2), f(z_3))=(z, z_1, z_2, z_3). \tag{3.28}$$

Proof: Putting $f(z)=(\alpha z+\beta)/(\gamma z+\delta)$, $\alpha\delta-\beta\gamma \neq 0$, we have, for $v=1,3$,

$$f(z)-f(z_v)=\frac{\alpha z+\beta}{\gamma z+\delta}-\frac{\alpha z_v+\beta}{\gamma z_v+\delta}=\frac{(\alpha\delta-\beta\gamma)(z-z_v)}{(\gamma z+\delta)(\gamma z_v+\delta)},$$

hence

$$\frac{f(z)-f(z_1)}{f(z)-f(z_3)}=\frac{(\gamma z_3+\delta)(z-z_1)}{(\gamma z_1+\delta)(z-z_3)}.$$

Therefore

$$\frac{f(z)-f(z_1)}{f(z)-f(z_3)}\Big/\frac{f(z_2)-f(z_1)}{f(z_2)-f(z_3)}=\frac{z-z_1}{z-z_3}\Big/\frac{z_2-z_1}{z_2-z_3}.$$

This theorem shows that the cross ratio is invariant under linear fractional transformations. As an immediate consequence of this theorem and Theorem 3.9 we have

Theorem 3.11. A fractional linear transformation maps circles or lines of $\mathbb{C} \cup \{\infty\}$ onto circles or lines (i.e., a circle is mapped onto a circle or a line and a line is mapped onto a circle or a line).

Let C be a circle in $\mathbb{C}$ with center c and radius r and let $\theta \to z=c+re^{i\theta}$, $0 \leqq \theta \leqq 2\pi$ be a parameter representation of C. C is a Jordan curve on which an orientation is defined by this parameter representation and C divides the Riemann sphere $S = \mathbb{C} \cup \{\infty\}$ into the interior $U = \{z \in \mathbb{C}: |z - c| < r\}$ and the exterior $V = (\mathbb{C} - [U]) \cup \{\infty\}$ of C:

$$S - C = U \cup V, \qquad U \cap V = \emptyset.$$

The circle C is the boundary of the closed disk $[U]$. The interior U of C is on the left of C and the exterior of V is on the right of C (Section 2.1b). Now, let z_1, z_2, and z_3 be three points on C, let C_1 be the arc with initial point z_1 and terminal point z_3, and let C_2 be the arc with initial point z_3 and terminal point z_1; then $C = C_1 \cdot C_2$. On C_1 and C_2 a linear order $\prec$ is induced by the given orientation (Section 2.1a). If $z_2 \in C_1$, then $z_1 \prec z_2 \prec z_3$ and if $z_2 \in C_2$, then $z_3 \prec z_2 \prec z_1$. If $z_1 \prec z_2 \prec z_3$ then $z_2 \prec z_3 \prec z_1$ and $z_3 \prec z_1 \prec z_2$, and if $z_3 \prec z_2 \prec z_1$, then $z_2 \prec z_1 \prec z_3$ and $z_1 \prec z_3 \prec z_2$.

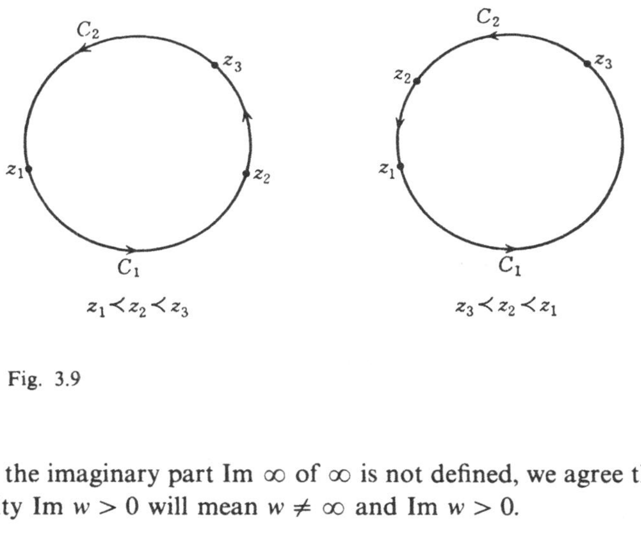

Fig. 3.9

Since the imaginary part Im ∞ of ∞ is not defined, we agree that as an inequality Im $w > 0$ will mean $w \neq \infty$ and Im $w > 0$.

Theorem 3.12. Let the interior of the circle C be denoted by U, its exterior by V. If z_1, z_2, and z_3 are points on C such that $z_1 \prec z_2 \prec z_3$, then $z \in U$ if and only if $\operatorname{Im}(z, z_1, z_2, z_3) > 0$ and $z \in V$ if and only if $\operatorname{Im}(z, z_1, z_2, z_3) < 0$.

Proof: Consider the cross ratio

$$h(z) = (z, z_1, z_2, z_3) = \frac{(z - z_1)(z_2 - z_3)}{(z - z_3)(z_2 - z_1)}$$

as a linear fractional function of z. As $h(z)$ is holomorphic on $S - \{z_3\}$, Im $h(z)$ is a continuous function on $S - \{z_3\}$. Since $h(z) \neq 0$ if $z \in C$ by

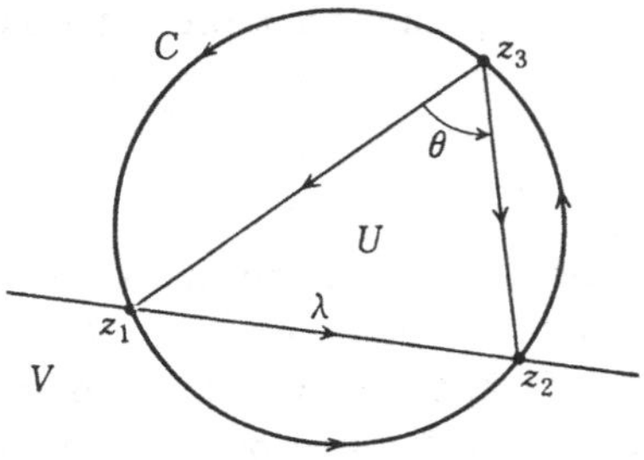

Fig. 3.10

Theorem 3.9, $\operatorname{Im} h(z)$ does not change sign on the region U or on the region V. In order to determine which sign is valid in which region we consider the line through z_1 and z_2. $\lambda(0) = z_1 \in C$ and $\lambda(\varepsilon) \in U$ and $\lambda(-\varepsilon) \in V$ for sufficiently small ε, so it suffices to prove that $\operatorname{Im} h(\lambda(\varepsilon)) > 0$ and $\operatorname{Im} h(\lambda(-\varepsilon)) < 0$. Since $\operatorname{Im} h(\lambda(0)) = \operatorname{Im} h(z_1) = 0$ it suffices to prove

$$\left| \frac{d}{dt} \operatorname{Im} h(\lambda(t)) \right]_{t=0} > 0.$$

From

$$h'(z) = \frac{(z_1 - z_3)(z_2 - z_3)}{(z - z_3)^2 (z_2 - z_1)}, \tag{3.29}$$

we conclude

$$\left[\frac{d}{dt} h(\lambda(t)) \right]_{t=0} = h'(z_1)\lambda'(0) = \frac{z_2 - z_3}{z_1 - z_3}.$$

Putting $\theta = \arg((z_2 - z_3)/(z_1 - z_3))$, we have $0 < \theta < \pi$ since $z_1 \prec z_2 \prec z_3$, hence

$$\left[\frac{d}{dt} \operatorname{Im} h(\lambda(t)) \right]_{t=0} = \operatorname{Im} \frac{z_2 - z_3}{z_1 - z_3} = \left| \frac{z_2 - z_3}{z_1 - z_3} \right| \sin\theta > 0.$$

Therefore, $\operatorname{Im} h(z) > 0$ if $z \in U$ and $\operatorname{Im} h(z) < 0$ if $z \in V$. Since $S = U \cup V \cup C$, $h(z_3) = \infty$ and $\operatorname{Im} h(z) = 0$ if $z \in C, z \neq z_3$, by Theorem 3.9, we see that, conversely $z \in U$ if $\operatorname{Im} h(z) > 0$ and $z \in V$ if $\operatorname{Im} h(z) < 0$.

Let $f(z)$ be a linear fractional function of z. Then the linear fractional transformation $f: z \to z' = f(z)$ transforms the circle C into a circle or a line of the z'-plane by Theorem 3.11.

Case 1: $C' = f(C)$ is a circle. C' is a Jordan curve on which an orientation can be defined in the same way as on C. The parameter representation

$\theta \to z = c + re^{i\theta}$, $0 \leqq \theta \leqq 2\pi$, of C induces a parameter representation,

$$\theta \to z' = f(c + re^{i\theta}), \qquad 0 \leqq \theta \leqq 2\pi,$$

of $f(C)$. The orientation on $f(C)$, defined by the parameter representation, is not necessarily the same as the original orientation of C'. Neglecting orientations we have $C' = f(C)$, taking them into account we have

$$f(C) = \pm C', \tag{3.30}$$

where C' represents the circle $|C'|$ with reversed orientation (Section 2.1a). In order to determine the sign of the right-hand side of (3.30), we take three different points z_1, z_2, and z_3 on C such that $z_1 \prec z_2 \prec z_3$. Putting $z'_1 = f(z_1)$, $z'_2 = f(z_2)$, and $z'_3 = f(z_3)$, we have only to determine if $z'_1 \prec z'_2 \prec z'_3$ or $z'_3 \prec z'_2 \prec z'_1$ on C': if the former, then $f(C) = +C'$; if the latter then $f(C) = -C'$.

Let U and U' denote the interiors and V and V' the exteriors of C representing C'. If $f(C) = +C'$, then $f(U) = U'$ and $f(V) = V'$, while if $f(C) = -C'$, then $f(U) = V'$ and $f(V) = U'$. We prove this as follows: If $f(C) = +C'$, then $z'_1 < z'_2 < z'_3$ on C'. Since

$$\mathrm{Im}\,(f(z), z'_1, z'_2, z'_3) = \mathrm{Im}\,(z, z_1, z_2, z_3)$$

by Theorem 3.10, we have $f(z) \in U'$ if $z \in U$ and $f(z) \in V'$ if $z \in V$ by Theorem 3.12. If $f(C) = -C'$, then $z'_3 \prec z'_2 \prec z'_1$, hence $f(z) \in U'$ if $\mathrm{Im}\,(f(z), z'_3, z'_2, z'_1) > 0$ and $f(z) \in V'$ if $\mathrm{Im}(f(z), z'_3, z'_2, z'_1) < 0$ and since

$$(f(z), z'_3, z'_2, z'_1) = (z, z_3, z_2, z_1) = 1/(z, z_1, z_2, z_3)$$

by Theorem 3.10, we see that the signs of $\mathrm{Im}(f(z), z'_3, z'_2, z'_1)$ and $\mathrm{Im}(z, z_1, z_2, z_3)$ are opposite. Hence, if $z \in U$, then $f(z) \in V'$ and if $z \in V$, then $f(z) \in U'$.

Case 2: $C' = f(C)$ is a line. Put $C' = l' \cup \{\infty\}$, where l' is a line in the z'-plane. Putting $z_3 = f^{-1}(\infty)$, the point z_3 is on C. Let z_1 and z_2 be points on C with $z_1 \prec z_2 \prec z_3$, and put $z'_1 = f(z_1)$ and $z'_2 = f(z_2)$, then z'_1 and z'_2 are two different points of l'. Let $\prec$ be the orientation on l', under which $z'_1 \prec z'_2$. Then l' divides the z'-plane into a region U' to the left of l' and a region V' to the right of l'

$$S - C' = \mathbb{C} - l' = U' \cup V' \qquad U' \cap V' = \emptyset.$$

For an arbitrary $z \neq z_3$ we have by Theorem 3.10

$$(z, z_1, z_2, z_3) = (f(z), z'_1, z'_2, \infty) = \frac{f(z) - z'_1}{z'_2 - z'_1}.$$

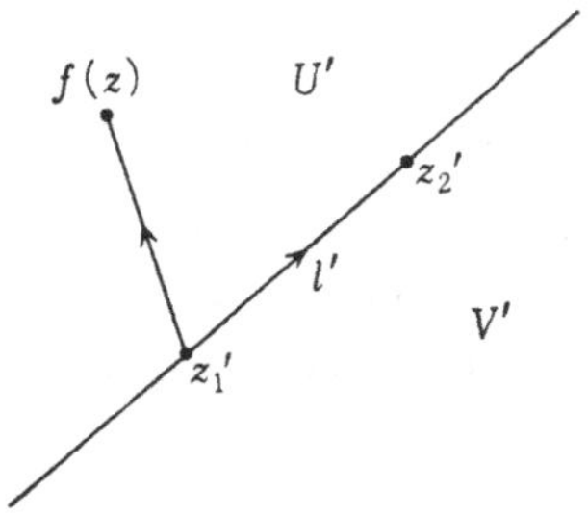

Fig. 3.11

Hence, putting $\theta = \arg((f(z) - z'_1)/(z'_2 - z'_1))$,

$$\left|\frac{f(z) - z'_1}{z'_2 - z'_1}\right| \cdot \sin\theta = \operatorname{Im}(z, z_1, z_2, z_3).$$

Therefore, $f(z) \in U'$ if $z \in U$ and $f(z) \in V'$ if $z \in V$ by Theorem 3.12, hence $f(U) = U'$ and $f(V) = V'$. The transformation f maps the left-hand side region of C onto the left-hand side region of l' and the right-hand side region of C on the right-hand side region of l'.

Assume that the linear fractional transformation $f: z \to z' = f(z)$ maps the line $l \cup \{\infty\}$ onto a circle $C' = f(l \cup \{\infty\})$, and denote the interior of C' by U' and its exterior by V'. Define an orientation $\prec$ on l such that if z_1, and z_2 are different points of l with $z_1 \prec z_2$, then $f(z_1) \prec f(z_2) \prec f(\infty)$ on C'. Letting U be the region to the left of l and V be the region to the right of l, then $f(U) = U'$ and $f(V) = V'$. This follows from case (2) above by considering the inverse transformation f^{-1}.

c. Elementary conformal mappings

In this section we want to give a few examples of elementary conformal mappings.

First consider the linear fractional transformation

$$f: z \to w = f(z) = (z - \beta)/(z - \bar{\beta}), \tag{3.31}$$

where $\beta = a + ib$ is a complex number with $b > 0$. Let l be the real axis $\mathbb{R}$ and denote the region $\{z \in \mathbb{C}: \operatorname{Im} z > 0\}$ by H^+ and the region $\{z \in \mathbb{C}: \operatorname{Im} z < 0\}$ by H^-. H^+ is called the *upper half-plane* and H^- the *lower half-plane*. a, $a + b$, and ∞ are three points on $\mathbb{R} \cup \{\infty\}$ and

$$f(a) = -1, \qquad f(a+b) = -i, \qquad f(\infty) = 1$$

hence $C = f(\mathbb{R} \cup \{\infty\})$ is the circle passing through -1, $-i$ and 1, that is, the unit circle. Let U denote the interior of C and V its exterior. Put on $\mathbb{R}$ the orientation $\prec$ induced by the usual linear order $<$. Since $a \prec a+b$ on $\mathbb{R}$ and $f(a) \prec f(a+b) \prec f(\infty)$ on C and H^+ is on the left of $\mathbb{R}$ and H^- on the right of $\mathbb{R}$, we have by the above result

$$f(H^+) = U, \qquad f(H^-) = V.$$

Obviously, $f(\beta) = 0$, $f(\overline{\beta}) = \infty$. Therefore, if we restrict the domain of the linear fractional function $f(z)$ to H^+, the transformation

$$f: z \to w = f(z) = \frac{z-\beta}{z-\overline{\beta}}, \qquad \operatorname{Im} z > 0,$$

is a conformal mapping, mapping the upper half-plane H^+ onto the unit disk U and the point β onto 0. The following theorem is a direct consequence of this result and the corollary to Schwarz's Lemma.

Theorem 3.13. All conformal mappings which map the upper half-plane H^+ onto the unit disk $U = \{w \in \mathbb{C}: |w| < 1\}$ and the point $\beta \in H^+$ onto 0 are given by

$$f: z \to w = f(z) = e^{i\theta}\frac{z-\beta}{z-\overline{\beta}}, \qquad \theta \in \mathbb{R}.$$

Example 3.2. Let C_1 and C_2 be two arcs in the z-plane with initial point β and the terminal point δ, let the angle between C_1 and C_2 at β equal $\pi/m, m$ a natural number, $m \geqq 2$. We want to determine a conformal mapping which maps the area D enclosed between C_1 and C_2 onto the unit disk. Let L be the line segment connecting β and δ, ψ be the angle between C_1 and L at β and put

$$f: z \to \zeta = f(z) = -e^{i\psi}\frac{z-\beta}{z-\delta}$$

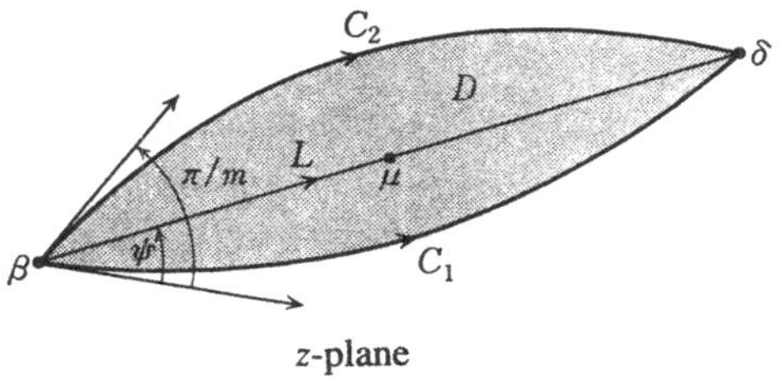

Fig. 3.12

Since $f(\beta) = 0$ and $f(\delta) = \infty$, f maps C_1, L, and C_2 onto rays $\Gamma_1 = f(C_1)$, $\Gamma = f(L)$, and $\Gamma_2 = f(C_2)$ with initial points 0 in the ex-

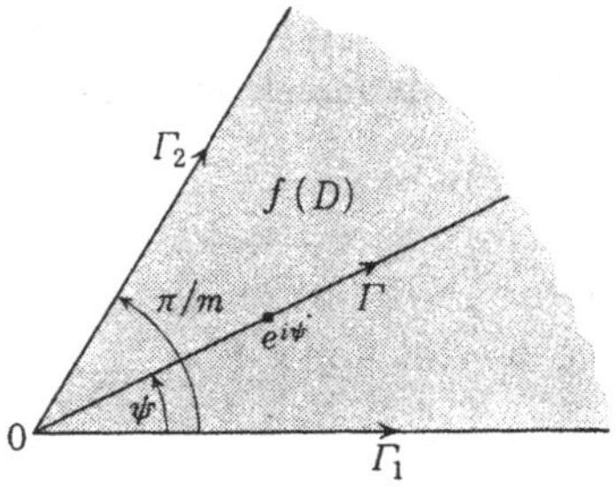

Fig. 3.13

tended ζ-plane. Let $\mu = (\beta + \delta)/2$ be the midpoint of L. Since $f(\mu) = e^{i\psi}$, the ray Γ passes through $e^{i\psi}$, hence the angle between the positive real axis of the ζ-plane and Γ equals ψ. Since f is a conformal mapping on the z-plane minus the point δ, the angle between Γ_1 and Γ_2 at 0 equals π/m and the angle between Γ_1 and Γ equals ψ. Hence Γ_1 coincides with the nonnegative part of the real axis of the ζ-plane. Since D is to the left of C_1 and to the right of C_2, $f(D)$ is to the left of Γ_1 and to the right of Γ_2 by (2) of Section 3.3b, that is,

$$f(D) = \{\zeta \colon \zeta = \rho e^{i\theta}, 0 < \rho < +\infty, 0 < \theta < \pi/m\}.$$

$\omega = \zeta^m = \rho^m e^{im\theta}$ is a holomorphic function of $\zeta = \rho e^{i\theta}$, which is univalent on $f(D)$ and $\{\omega \colon \omega = \zeta^m, \zeta \in f(D)\}$ is the upper half-plane H^+ of the ω-plane. Therefore, by Theorem 3.5, $\zeta \to \omega = \zeta^m$ is a conformal mapping which maps $f(D)$ onto H^+. Putting

$$g(z) = f(z)^m = e^{im\psi}\left[\frac{z-\beta}{\delta - z}\right]^m, \qquad z \in D,$$

the mapping $z \to \omega = g(z)$ is a conformal mapping which maps D onto H^+ by Theorem 3.4 (2). Therefore, the mapping $h \colon z \to w = h(z)$, given by

$$h(z) = \frac{g(z) - g(\mu)}{g(z) - \overline{g(\mu)}}$$

is a conformal mapping which maps the region D onto the unit disk, by Theorem 3.13. Obviously, $h(\mu) = 0$. Since $g(\mu) = e^{im\psi}$, we have

$$h(z) = \frac{(z-\beta)^m - (\delta - z)^m}{(z-\beta)^m - e^{-2im\psi}(\delta - z)^m}.$$

If D is symmetric with respect to the segment L, i.e., if $2\psi = \pi/m$, then

$$h(z) = \frac{(z-\beta)^m - (\delta - z)^m}{(z-\beta)^m + (\delta - z)^m}.$$

Example 3.3. Let C_1 be the circle with center $i/2$ passing through 0, let C_2 be the circle with center $i/4$ passing through 0, and let D be the region enclosed by C_1 and C_2, that is, D is the collection of all points that are inside C_1 and outside C_2. We want to find a conformal mapping which maps D onto the unit disk. First, let us put

$$\zeta = f(z) = 1/z.$$

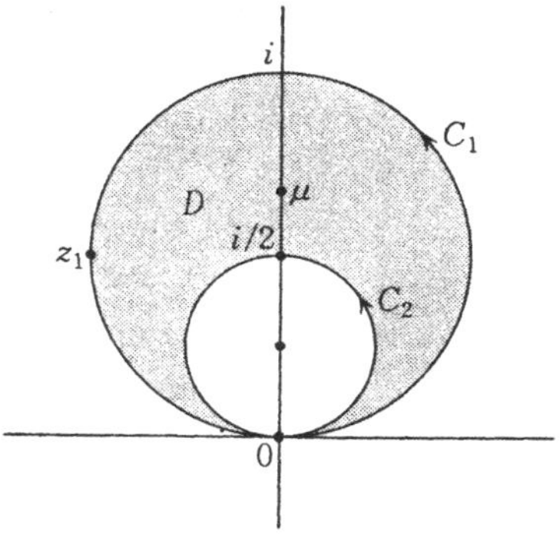

Fig. 3.14

The points i, $z_1 = (i-1)/2$, and 0 are points of C_1 and $f(i) = -i, f(z_1) = -i-1$, and $f(0) = \infty$, hence $f(C_1) = l_1 \cup \{\infty\}$, where l_1 is the line through $-i$ and $-i-1$ in the ζ-plane. Similarly, $f(C_2) = l_2 \cup \{\infty\}$, where l_2 is the line through $-2i$ and $-2i-2$ in the ζ-plane. Define orientations on l_1 and l_2 as indicated in Fig. 3.15. Since f maps the interior of C_1 onto the region to the left of l_1 and the exterior of C_2 onto the region to the right of l_2, $f(D)$ is the

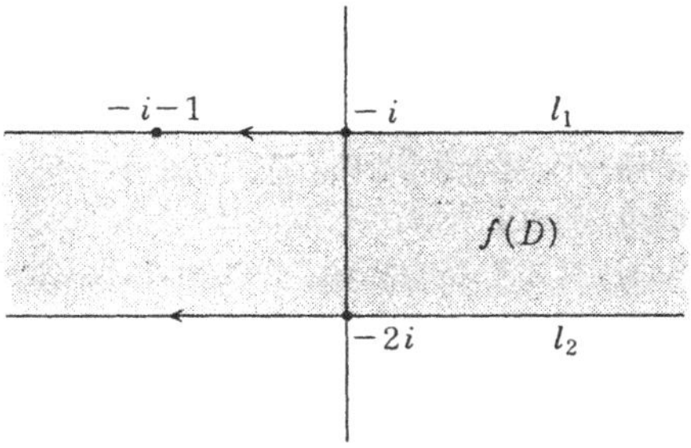

Fig. 3.15

region of the ζ-plane between l_1 and l_2. $\omega = e^{\pi\zeta}$ is a holomorphic function of ζ, which is univalent on $f(D)$ and $\{\omega: \omega = e^{\pi\zeta}, \zeta \in f(D)\}$ is the upper half-plane H^+ of the ω-plane. Hence $\zeta \to \omega = e^{\pi\zeta}$ is a conformal mapping that maps $f(D)$ onto H^+ and therefore $z \to \omega = e^{\pi f(z)} = e^{\pi/z}$ is a conformal mapping that maps D onto H^+. Therefore,

$$h: z \to w = h(z) = \frac{e^{\pi/z} - i}{e^{\pi/z} + i}, \qquad z \in D,$$

is a conformal mapping that maps the region D onto the unit disk; by Theorem 3.13, $\mu = h^{-1}(0) = 2i/3$.

4

Analytic continuation

4.1 Analytic continuation

a. Analytic continuation

Let $f_0(z)$ be a holomorphic function of z, defined on the region D_0 of the complex plane and let D_1 be a region such that $D_0 \cap D_1 \neq \emptyset$. A holomorphic function $f_1(z)$ defined on D_1 which coincides with $f_0(z)$ on $D_1 \cap D_0$ is called an *analytic continuation* of $f_0(z)$ into D_1. (The meaning of "coinciding" is, of course, that $f_0(z) = f_1(z)$ for $z \in D_0 \cap D_1$.) Given a holomorphic function $f_0(z)$ on D_0, there does not always exist an analytic continuation of $f_0(z)$ into D_1, but if one does, then it is unique by the corollary to Theorem 3.1. In this case, $f_0(z)$ is said to be *analytically continuable into* D_1. Then, putting $f(z) = f_0(z)$ if $z \in D_0$ and $f(z) = f_1(z)$ if $z \in D_1$, we obtain a holomorphic function $f(z)$ defined on the region $D_0 \cup D_1$, called the *analytic continuation* of $f_0(z)$ into D.

If the analytic continuation $f_1(z)$ of $f_0(z)$ into D_1 exists and if the analytic continuation $f_2(z)$ of $f_1(z)$ into the region D_2 exists, then $f_2(z)$ is called the analytic continuation of $f_0(z)$ into D_2. Generally, if a sequence of regions,

$$D_0, D_1, D_2, \ldots, D_{n-1}, D_n, \qquad D_k \cap D_{k-1} \neq \emptyset, \qquad k = 1, 2, \ldots, n,$$

is given and if the analytic continuation $f_1(z)$ of $f_0(z)$ into D_1, the analytic continuation $f_2(z)$ of $f_1(z)$ into $D_2, \ldots,$ the analytic continuation $f_n(z)$ of $f_{n-1}(z)$ into D_n exist, then $f_0(z)$ is said to be analytically continuable along the sequence $D_1, D_2, \ldots, D_n$ and the functions $f_1(z), f_2(z), \ldots, f_n(z)$ are called analytic continuations of $f_0(z)$. The analytic continuations $f_1(z)$, $f_2(z), \ldots, f_n(z)$ are uniquely determined by $f_0(z), D_0, D_1, \ldots, D_n$. If the analytic continuations $f_1(z), f_2(z), \ldots, f_n(z)$ of $f_0(z)$ exist then a holomorphic function $f(z)$ is defined on the region

$$D = D_0 \cup D_1 \cup D_2 \cup \cdots \cup D_n$$

by putting $f(z) = f_k(z)$ if $z \in D_k$. However, if, as in Fig. 4.1, $D_0 \cap D_n \neq \emptyset$, then it is not necessarily true that $f_0(z)$ and $f_n(z)$ coincide on $D_0 \cap D_n$. If, for a

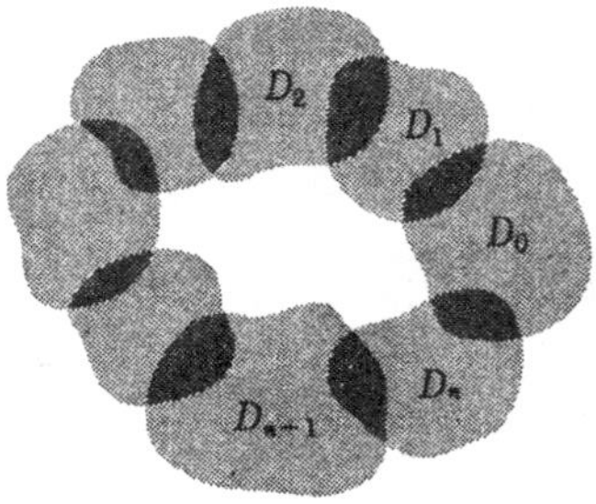

Fig. 4.1

point $\alpha \in D_0 \cap D_n$, $f_0(\alpha) \neq f_n(\alpha)$, then the function f assumes two different values at $z = \alpha$. Moreover, if analytic continuations $f_{n+1}(z)$ into D_{n+1}, $f_{n+2}(z)$ into $D_{n+2}, \ldots, f_m(z)$ into D_m exist and if we put again $f(z) = f_k(z)$ for $z \in D_k$, $k = n+1, n+2, \ldots, m$, we arrive at a holomorphic function $f(z)$ defined on a region

$$D = D_0 \cup D_1 \cup D_2 \cup \cdots \cup D_n \cup D_{n+1} \cup \cdots \cup D_m.$$

If $D_0 \cap D_n \cap D_m \neq \varnothing$ then $f(z)$ could assume three different values $f_0(\alpha)$, $f_n(\alpha)$, and $f_m(\alpha)$ at $\alpha \in D_0 \cap D_n \cap D_m$. In this way a multivalued holomorphic function $f(z)$ is defined. The analytic continuations $f_1(z)$, $f_2(z)$, $f_3(z), \ldots$ are uniquely determined by $f_0(z)$, hence $f(z)$ is also uniquely determined by $f_0(z)$.

A possibly multivalued holomorphic function is called an analytic function, while a holomorphic function will always be understood to be one-valued.

The function $f_1(z)$, in contradistinction to $f_2(z)$, $f_3(z), \ldots$ is called a *direct analytic continuation*. Hence, each $f_k(z)$ is a direct analytic continuation of $f_{k-1}(z)$ and $f(z)$ is obtained by a repeated process of direct analytic continuation starting from $f_0(z)$. The function $f(z)$, obtained by repeating the process of direct analytic continuation starting from $f_0(z)$, as many times as possible, is called a *complete analytic function* (see L. Ahlfors, 1966, Chapter 8, Section 1.1).

"Analytic relations" between holomorphic functions are preserved by analytic continuation. For example, if $f_1(z)$, $f_2(z)$, $f_3(z), \ldots$ are analytic continuations of $f_0(z)$, then $f'_1(z)$, $f'_2(z)$, $f'_3(z), \ldots$ are analytic continuations of $f'_0(z)$. Further, if $\phi(w)$ is a holomorphic function defined on a region Ω of the w-plane and if the ranges of $f_0(z)$ and its analytic continuations $f_1(z)$, $f_2(z)$, $f_3(z)$, ... are all included in Ω, then $\phi(f_1(z))$, $\phi(f_2(z))$, $\phi(f_3(z))$, ... are analytic continuations of $\phi(f_0(z))$. Therefore, if $f_0(z)$ and $g_0(z)$ are holomorphic functions defined on the region D_0, if $f_1(z)$, $f_2(z)$, ..., $f_k(z)$, ... and $g_1(z)$, $g_2(z)$, ..., $g_k(z)$, ... are analytic continuations along $D_1, D_2, \ldots, D_k, \ldots$ of $f_0(z)$ and $g_0(z)$ respectively, and if $g_0(z) = f_0'(z)$, then $g_k(z) = f_k'(z)$; if $g_0(z) = f_0''(z)$, then $g_k(z) = f_k''(z)$; ...; and if $g_0(z) = \phi(f_0(z))$, then $g_k(z) = \phi(f_k(z))$. Up to now, we have considered holomorphic functions $f_0(z)$, $f_1(z)$, $f_2(z)$, ... defined on regions of the complex plane but it is also possible to consider analytic continuations for holomorphic functions defined on regions of the Riemann sphere S.

As a simple example of analytic continuation let us consider the logarithmic function. The exponential function e^w is a holomorphic function of w on the whole w-plane (Example 1.2 of Section 1.2b). Writing $w = u + iv$ we get $e^w = e^u e^{iv}$, hence the range of e^w is the region $\mathbb{C}^* = \mathbb{C} - \{0\}$. For $z \in \mathbb{C}^*$ a solution w of the equation $e^w = z$ is denoted by $w = \log z$ and $\log z$ is called the *logarithmic function* of z. Putting $z = |z|e^{i\theta}$, $\theta = \arg z$, we conclude from $e^w = e^u e^{iv} = z$ that $e^u = |z|$ and $e^{iv} = e^{i\theta}$, i.e., $u = \log|z|$ and $v = \theta + 2n\pi$, n an integer. Therefore, $\log z$ is a multivalued function taking an infinite number of values:

$$w = \log z = \log|z| + i\theta + 2n\pi i, \qquad n = 0, \pm 1, \pm 2, \ldots \tag{4.1}$$

at each $z = |z|e^{i\theta}$, $|z| > 0$. The logarithmic function $w = \log z$ is the "inverse function" of the exponential function $z = e^w$, but since e^w is not univalent, we cannot use the corollary to Theorem 3.5 to infer immediately that $w = \log z$ is a holomorphic function. However, if we restrict the domain of e^w to a region W on which e^w is univalent, the inverse function of the restriction is holomorphic. For each integer k define W_k by

$$W_k = \left\{u + iv: \ -\infty < u < +\infty, \frac{2k\pi}{3} - \frac{\pi}{2} < v < \frac{2k\pi}{3} + \frac{\pi}{2}\right\}$$

and let $e_k(w)$ denote the restriction of e^w to W_k, that is, $e_k(w) = e^w$, $w \in W_k$, then $e_k(w)$ is a univalent holomorphic function on W_k. Putting

$$k = 3n + \nu, \qquad n \text{ an integer and } \nu = 0, 1, \text{ or } 2,$$

and $v = \theta + 2n\pi$, the range of $e_k(w)$ is given by

$$D_\nu = \left\{z : z = re^{i\theta}, r > 0, \frac{2\nu\pi}{3} - \frac{\pi}{2} < \theta < \frac{2\nu\pi}{3} + \frac{\pi}{2}\right\}.$$

Hence, the inverse $w = f_k(z)$ of $z = e_k(w)$ is holomorphic on D_ν by the corollary to Theorem 3.5 and given by

$$f_k(z) = \log|z| + i\theta + 2n\pi i, \quad z = |z|e^{i\theta}, \quad \frac{2\nu\pi}{3} - \frac{\pi}{2} < \theta < \frac{2\nu\pi}{3} + \frac{\pi}{2}. \tag{4.2}$$

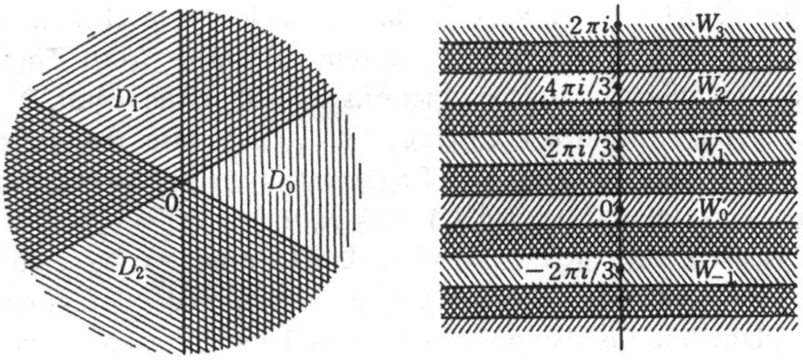

Fig. 4.2

Obviously,

$$C^* = D_0 \cup D_1 \cup D_2.$$

$f_0(z)$, $f_1(z)$, and $f_2(z)$ are holomorphic functions defined on the regions D_0, D_1, and D_2, respectively, and $f_0(z)$ and $f_1(z)$ coincide on $D_0 \cap D_1$, and $f_1(z)$ and $f_2(z)$ coincide on $D_1 \cap D_2$ by (4.2), that is, $f_1(z)$ is the direct analytic continuation of $f_0(z)$ and $f_2(z)$ is the direct analytic continuation of $f_1(z)$. The function $f_3(z)$, which is holomorphic on D_0, coincides with $f_2(z)$ on $D_0 \cap D_2$. For, writing an arbitrary point $z \in D_0 \cap D_2$ as

$$z = |z|e^{i\theta} = |z|e^{i\psi}, \qquad -\frac{\pi}{2} < \theta < -\frac{\pi}{6}, \qquad \frac{3\pi}{2} < \psi < \frac{11\pi}{6},$$

we have

$$f_3(z) = \log|z| + i\theta + 2\pi i = \log|z| + i\psi = f_2(z)$$

by (4.2). Hence $f_3(z)$ is the direct analytic continuation of $f_2(z)$. Similarly, it is proved that the functions $f_1(z), f_2(z), f_3(z), \ldots, f_{k-1}(z), f_k(z), \ldots$ are analytic continuations of $f_0(z)$ along the sequence of regions

$$D_1, D_2, D_0, D_1, D_2, D_0, D_1, D_2, \ldots$$

and that the functions $f_{-1}(z), f_{-2}(z), f_{-3}(z), \ldots$ are analytic continuations of $f_0(z)$ along the sequence of domains

$$D_2, D_1, D_0, D_2, D_1, D_0, D_2, \ldots.$$

Hence $f_k(z)$ is the analytic continuation of $f_0(z)$ for all integers k. By comparing (4.1) and (4.2) it is obvious that $\log z$ coincides with one of the $f_k(z)$, $k = 3n + \nu$, on each region D_ν. Hence the logarithmic function $\log z$ is a holomorphic function obtained from $f_0(z)$ by analytic continuation. Since

$D_0 \cup D_1 \cup D_2 = \mathbb{C}^*$ $\log z$ is holomorphic on $\mathbb{C}^*$, 0 is called a *logarithmic singularity* of $\log z$. From the definition of $\log z$ we have

$$e^{\log z} = z, \qquad z \neq 0. \tag{4.3}$$

Since $de^u/du = e^u$ for the real variable u, we have, by the corollary to Theorem 3.1,

$$\frac{de^w}{dw} = e^w.$$

Therefore, differentiating both sides of (4.3) yields

$$\frac{d\log z}{dz} = \frac{1}{z}. \tag{4.4}$$

For an arbitrary fixed complex number α the power function z^α is defined for $z \neq 0$ by

$$z^\alpha = e^\alpha \log z. \tag{4.5}$$

In general, z^α is a multivalued holomorphic function on $\mathbb{C}^*$. Since $de^{\alpha \log z}/dz = e^{\alpha \log z}\alpha d \log z/dz = z^\alpha \alpha/z$ by (4.4), we have

$$\frac{dz^\alpha}{dz} = \alpha z^{\alpha-1}. \tag{4.6}$$

If $\alpha = m$ is an integer, then $z^\alpha = e^{m \log z} = (e^{\log z})^m = z^m$ is the usual mth power of z. If $\alpha = 1/m$, m a natural number such that $m \geqq 2$, then we have

$$(z^{1/m})^m = [\exp(1/m)\log z]^m = e^{\log z} = z,$$

that is, $z^{1/m}$ is an mth root of z. Since $0^m = 0$, we define $0^{1/m} = 0$. Putting $z = |z|e^{i\theta}$, $|z| > 0$, we have, by (4.1),

$$z^{1/m} = \exp[(1/m)\log|z| + i\theta/m + 2n\pi i/m],$$

that is,

$$z^{1/m} = \rho^n |z|^{1/m} e^{i\theta/m}, \qquad \rho = e^{2\pi i/m}, \qquad n = 0, 1, 2, \ldots, m-1.$$

(Compare (3.7).) The function $z^{1/m}$ is an m-valued holomorphic function. Since $\lim_{z\to 0}|z^{1/m}| = \lim_{z\to 0}|z|^{1/m} = 0$, $z^{1/m}$ is continuous at $z = 0$, but $z^{1/m}$ is of course not holomorphic at $z = 0$. The point 0 is called a *branch point* of the power function $z^{1/m}$.

b. Analytic continuation by expansion in power series

Let r_0, $0 < r_0 < +\infty$, be the radius of convergence of the power series

$$f_0(z) = \sum_{n=0}^{\infty} a_{0n}(z - c_0)^n.$$

Then $f_0(z)$ is a holomorphic function on the interior $U_0 = U_{r_0}(c_0)$ of the circle of convergence. In order to find the analytic continuation of $f_0(z)$, we pick a point $c_1 \in U_0$, $c_1 \neq c_0$, and we expand $f_0(z)$ into a power series about c_1

$$f_0(z) = \sum_{n=0}^{\infty} a_{1n}(z-c_1)^n, \qquad a_{1n} = \frac{1}{n!} f_0^{(n)}(c_1). \tag{4.7}$$

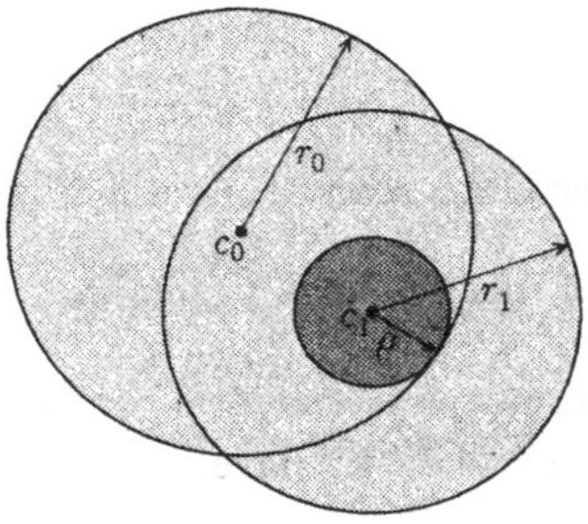

Fig. 4.3

The power series in the right-hand side of (4.7) converges on the interior of a circle with center c_1 and radius $\rho = r_0 - |c_1 - c_0|$ by Theorem 1.16. Therefore, if r_1 represents the radius of convergence of the power series, we have

$$r_1 \geqq r_0 - |c_1 - c_0| \tag{4.8}$$

and the function $f_1(z)$ defined by

$$f_1(z) = \sum_{n=0}^{\infty} a_{1n}(z-c_1)^n, \qquad |z - c_1| < r_1,$$

is a holomorphic function on the disk $U_1 = U_{r_1}(c_1)$ with center c_1 and radius r_1. Since $f_1(z) = f_0(z)$ on a ρ-neighborhood $U_\rho(c_1)$ of c_1, $f_1(z) = f_0(z)$ on $U_1 \cap U_0$ by the corollary to Theorem 3.1. Therefore, if $U_1 \not\subset U_0$, i.e., if $r_1 > \rho = r_0 - |c_1 - c_0|$, $f_1(z)$ is the direct analytic continuation of $f_0(z)$ into U_1.

We have

$$f_0^{(n)}(z) = \sum_{m=n}^{\infty} \frac{m!}{(m-n)!} a_{0m}(z-c_0)^{m-n}$$

by Theorem 1.11, hence the coefficients a_{1n} of the power series for $f_1(z)$ are given by

$$a_{1n} = \sum_{m=n}^{\infty} \binom{m}{n} a_{0m}(c_1 - c_0)^{m-n}. \tag{4.9}$$

Repeating the above procedure, we pick a point $c_2 \in U_1$, $c_2 \neq c_1$, and we expand the holomorphic function into a power series about c_2

$$f_1(z) = \sum_{n=0}^{\infty} a_{2n}(z - c_2)^n, \qquad a_{2n} = \frac{1}{n!} f_1^{(n)}(c_2).$$

Putting

$$f_2(z = \sum_{n=0}^{\infty} a_{2n}(z - c_2)^n, \qquad |z - c_2| < r_2,$$

where r_2 denotes the radius of convergence of the power series under consideration, and assuming $r_2 > r_1 - |c_2 - c_1|$, $f_2(z)$ is the direct analytic continuation of $f_1(z)$ into $U_2 = U_{r_2}(c_2) \not\subset U_1$. Continuing in this way, we arrive at the direct analytic continuation $f_3(z)$ of $f_2(z), \ldots$, the direct analytic continuation $f_k(z)$ of $f_{k-1}(z), \ldots$. Here we know that

$$r_k \geqq r_{k-1} - |c_k - c_{k-1}|,$$

where c_k denotes the center of each power expansion $f_k(z)$ and r_k its radius of convergence. If for each choice of $c_1 \in U_0$, the equality $r_1 = r_0 - |c_1 - c_0|$ holds, $f_0(z)$ cannot be continued analytically outside its circle of convergence.

Example 4.1. Consider

$$f_0(z) = \sum_{m=0}^{\infty} z^{m!} = 1 + z + z^2 + z^6 + z^{24} + \ldots.$$

If $|z| < 1$, then $\sum_{m=0}^{\infty} |z|^{m!} \leqq \sum_{n=0}^{\infty} |z|^n < +\infty$ and if $|z| > 1$, then $\sum_{m=0}^{\infty} |z|^{m!} = +\infty$, hence the radius of convergence of this power series is 1 and its circle of convergence the unit circle. For $z = re^{i\theta}$, $0 < r < 1$, with $\theta = 2\pi k/n$, k,n natural numbers, we have $(e^{i\theta})^{m!} = 1$ if $m \geqq n$, hence

$$f_0(re^{i\theta}) = \sum_{m=0}^{n-1} (re^{i\theta})^{m!} + \sum_{m=n}^{\infty} r^{m!}.$$

Therefore

$$\lim_{r \to 1-0} |f_0(re^{i\theta})| = +\infty. \tag{4.10}$$

Suppose that for some $c_1 \in U_0$ we have $r_1 > r_0 - |c_1 - c_0|$, i.e., $U_1 \not\subset U_0$: then for a $\theta = 2\pi k/n$ such that $e^{i\theta} \leqq U_1$ we would have

$$\lim_{r \to 1-0} f_0(re^{i\theta}) = \lim_{r \to 1-0} f_1(re^{i\theta}) = f_1(e^{i\theta})$$

contradicting (4.10). Hence, for all $c_1 \in U_0$ the equality $r_1 = r_0 - |c_1 - c_0|$ must hold and the function $f_0(z)$ cannot be continued analytically across the circle of convergence. In general, if a holomorphic function $f(z)$ defined on some region D cannot be continued analytically across the boundary of D, then the boundary of D is called the *natural boundary* of $f(z)$. In the above example, the unit circle is the natural boundary of $f_0(z)$.

4.2 Analytic continuation along curves

Let γ: $t \to \gamma(t)$, $a \leqq t \leqq b$, be a curve in the complex plane. As explained in Section 2.1a for the case of Jordan arcs, the curve $\lambda: \tau \to \lambda(\tau) = \gamma(\phi(\tau))$ obtained from γ by a change of variables $t = \phi(\tau)$, $\alpha \leqq \tau \leqq \beta$, where $\phi(\tau)$ is a continuous nondecreasing function of τ with $\phi(\alpha) = a$ and $\phi(\beta) = b$, is considered to be the same curve as γ, i.e., γ and λ are considered as two different parameter representations of the same curve. A Jordan arc was defined to be the point set C together with an orientation $\prec$ (Definition 2.1), but this definition cannot be used in the general case. If γ is a smooth curve, then we assume that $\phi(\tau)$ is a continuously differentiable function with $\phi'(\tau) > 0$ for all τ. For example, if $\phi(\tau) = a + (b-a)\tau$, $0 \leqq \tau \leqq 1$, then γ is the same curve as λ: $\tau \to \lambda(\tau) = \gamma(a + (b-a)\tau)$ hence each curve can be written in the form: γ: $t \to \gamma(t)$, $0 \leqq t \leqq 1$.

If γ_1 is a curve connecting c_0 and c_1 and if γ_2 is a curve connecting c_1 and c_2, then the curve obtained by piecing together γ_1 and γ_2 is denoted by

$$\gamma = \gamma_1 \cdot \gamma_2.$$

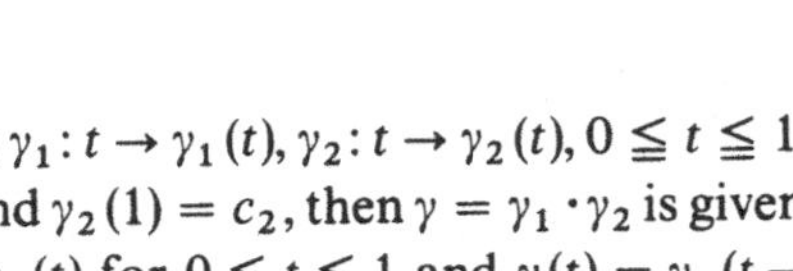

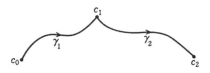

Fig. 4.4

Putting $\gamma_1: t \to \gamma_1(t)$, $\gamma_2: t \to \gamma_2(t)$, $0 \leqq t \leqq 1$, with $\gamma_1(0) = c_0$, $\gamma_1(1) = \gamma_2(0) = c_1$, and $\gamma_2(1) = c_2$, then $\gamma = \gamma_1 \cdot \gamma_2$ is given by $\gamma: t \to \gamma(t)$, $0 \leqq t \leqq 2$, where $\gamma(t) = \gamma_1(t)$ for $0 \leqq t \leqq 1$ and $\gamma(t) = \gamma_2(t-1)$ for $1 \leqq t \leqq 2$.

In case of three or more curves γ_k: $t \to \gamma_k(t)$, $0 \leqq t \leqq 1$, with $\gamma_k(0) = c_{k-1}$ and $\gamma_k(1) = c_k$, $k = 1, 2, \ldots, n$, we put

$$\gamma(t) = \gamma_k(t - k + 1) \qquad \text{if } k - 1 \leqq t \leqq k, \quad k = 1, 2, \ldots, n$$

and γ: $t \to \gamma(t)$, $0 \leqq t \leqq n$, is the curve obtained by piecing together

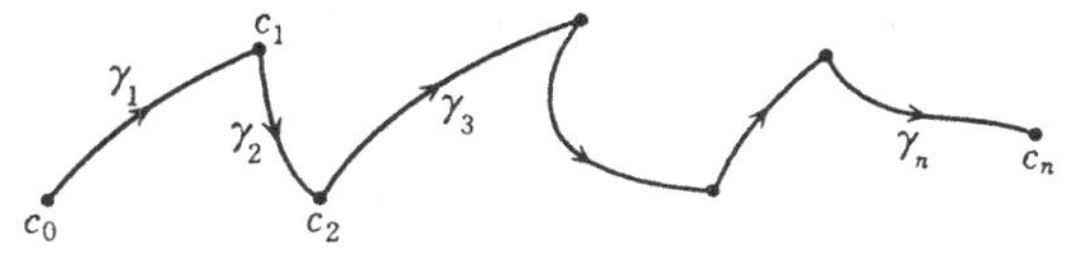

Fig. 4.5

$\gamma_1, \gamma_2, \ldots, \gamma_n$ in that order

$$\gamma = \gamma_1 \cdot \gamma_2 \cdot \cdots \cdot \gamma_k \cdot \cdots \cdot \gamma_n.$$

Given a curve $\gamma: t \to \gamma(t), a \leqq t \leqq b$, the oppositely oriented curve is denoted by γ^{-1}

$$\gamma^{-1}: t \to \gamma^{-1}(t) = \gamma(b+a-t), \qquad a \leqq t \leqq b.$$

If $\gamma = \gamma_1 \cdot \gamma_2 \cdot \cdots \cdot \gamma_n$, then $\gamma^{-1} = \gamma_n^{-1} \cdot \cdots \cdot \gamma_2^{-1} \cdot \gamma_1^{-1}$.

The radius of convergence of the power series $\sum_{n=0}^{\infty} a_n(z-c)^n$ is given by the Cauchy–Hadamard formula (1.19)

$$r = \frac{1}{\limsup_{n \to \infty} |a_n|^{1/n}}.$$

If $r = 0$, $f(z) = \sum_{n=0}^{\infty} a_n(z-c)^n$ is reduced to a function defined at only one point. Since this is an uninteresting case, from now on we will always assume that our power series have a radius of convergence r such that $0 < r \leqq +\infty$, unless the contrary is explicitly stated. The sum of a power series $\sum_{n=0}^{\infty} a_n(z-c)^n$ is a holomorphic function of z on the interior $U_r(c)$ of the circle of convergence.

Let γ: $t \to \gamma(t)$, $0 \leqq t \leqq b$, be a curve with initial point c_0 and let us consider the analytic continuation by power series expansion along the curve γ of

$$f_0(z) = \sum_{n=0}^{\infty} a_n(z-c_0)^n.$$

Definition 4.1. If for all t in the closed interval $[0, b]$, the power series with center $\gamma(t)$

$$f(z, t) = \sum_{n=0}^{\infty} a_n(t)\,(z-\gamma(t))^n$$

exists such that the following conditions are satisfied:

(1) $f(z, 0) = f_0(z)$,

(2) for each $s \in [0, b]$ there is a positive $\varepsilon(s)$ such that $f(z, t)$ is the direct analytic continuation of $f(z, s)$ if $|t-s| < \varepsilon(s)$,

then $f_0(z)$ is said to be *analytically continuable along* γ and the family of power series

$$F = \{f(z, t): 0 \leqq t \leqq b\}$$

is called the *analytic continuation of* $f_0(z)$ *along* γ. Let $r(t)$ denote the radius of convergence for each power series $f(z, t)$ belonging to this family F. Then $r(t) > 0$ by the above assumption and $f(z, t)$ is a holomorphic function defined on the disk $U_t = U_{r(t)}(\gamma(t))$ with center $\gamma(t)$ and radius $r(t)$. By condition (2) we have if $|t-s| < \varepsilon(s)$, then $U_t \cap U_s \neq \varnothing$ and

$$f(z, t) = f(z, s) \qquad \text{if } z \in U_t \cap U_s. \tag{4.11}$$

Theorem 4.1. If the power series $f_0(z)$ with center c_0 is analytically continuable along a curve $\gamma: t \to \gamma(t)$, $0 \leqq t \leqq b$, with initial point c_0, then the analytic continuation of $f_0(z)$ along γ

$$F = \{f(z, t): \ 0 \leqq t \leqq b\}$$

is uniquely determined.

Proof: Let $F = \{f(z, t): 0 \leqq t \leqq b\}$ and $G = \{g(z, t): 0 \leqq t \leqq b\}$ be two analytic continuations of $f_0(z)$ along γ, and let $r(t)$ denote the smallest of the radii of convergence of the power series $f(z, t)$ and $g(z, t)$ with center t. If $s > 0$ is sufficiently small, then $U_t \cap U_0 \neq \varnothing$ if $0 \leqq t \leqq s$ and $f(z, t) = f_0(z)$ and $g(z, t) = g_0(z)$ on $U_t \cap U_0$ by (4.11), hence $f(z, t) = g(z, t)$ by the corollary to Theorem 3.1. Let u denote the supremum of all s such that $f(z, t) = g(z, t)$ if $0 \leqq t \leqq s$. Suppose that $u < b$, then $f(z, t) = g(z, t)$ if $0 \leqq t \leqq u$, and for a sufficiently small $\varepsilon > 0$ we have $U_t \cap U_u \neq \varnothing$ if $u - \varepsilon < t < u + \varepsilon$, while $f(z, t) = f(z, u)$ and $g(z, t) = g(z, u)$ on $U_t \cap U_u$ by (4.11). Hence we conclude that $f(z, t) = g(z, t)$ for $0 \leqq t < u + \varepsilon$, contradicting the definition of u. Therefore $u = b$, that is, $f(z, t) = g(z, t)$ for $0 \leqq t < b$. The same argument now shows that $f(z, t) = g(z, t)$ for $0 \leqq t \leqq b$.

Corollary. If $f(z)$ is a holomorphic function defined on a region D, if $f_0(z)$ is the power series expansion of $f(z)$ about a point $c_0 \in D$, and if $F = \{f(z, t): 0 \leqq t \leqq b\}$ is the analytic continuation of $f_0(z)$ along the curve γ: $t \to \gamma(t)$, $0 \leqq t \leqq b$, with $\gamma(0) = c_0$ and $\gamma(t) \in D$ for all t, then all $f(z, t)$ are power series expansions of $f(z)$ about $\gamma(t)$.

Let $F = f(z, t): 0 \leqq t \leqq b\}$ be the analytic continuation of $f_0(z)$ along γ, let $r(t)$ denote the circle of convergence of $f(z, t)$ and put $U_t = U_{r(t)}(\gamma(t))$. If $r(s) = +\infty$ for some s, then $f(z, s)$ is holomorphic on the whole z-plane and since by the above corollary $f(z,t) = f(z,s)$ for all t, $0 \leqq t \leqq b$, we are reduced to a trivial case. Therefore, we assume

$$r(t) < +\infty, \qquad 0 \leqq t \leqq b.$$

The radius of convergence $r(t)$ of $f(z,t)$ is a continuous function of t, $0 \leqq t \leqq b$. For, since $\gamma(t)$ is a continuous function of t, there exists for arbitrary s, $0 \leqq s \leqq b$, a $\delta > 0$ such that $\gamma(t) \in U_s$ if $|t-s| < \delta$. Hence, $f(z,t)$ is the power series expansion of the holomorphic function $f(z,s)$ about $z = \gamma(t)$ by the corollary to Theorem 4.1. Therefore, by (4.8)

$$r(t) \geqq r(s) - |\gamma(t) - \gamma(s)|.$$

If $r(t) > r(s)$ then $\gamma(s) \in U_t$ and

$$r(s) \geqq r(t) - |\gamma(s) - \gamma(t)|.$$

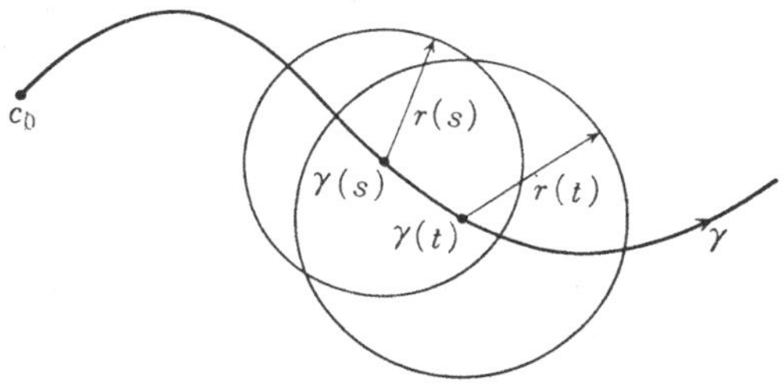

Fig. 4.6

Hence, if $|t-s| < \delta$, then

$$|r(t) - r(s)| \leqq |\gamma(t) - \gamma(s)|, \tag{4.12}$$

and therefore $\lim_{t \to s} r(t) = r(s)$.

The continuous function $r(t)$ assumes a minimum value ρ on $[0,b]$:

$$r(t) \geqq \rho > 0. \tag{4.13}$$

Since $\gamma(t)$ is uniformly continuous on $[0,b]$, there exists a $\delta > 0$ such that

$$|\gamma(t) - \gamma(s)| < \rho \qquad \text{if } |t-s| < \delta. \tag{4.14}$$

$r(s) > \rho$, hence, if $|t-s| < \delta$, $f(z,t)$ is the power series expansion of the holomorphic function $f(z,s)$ about $\gamma(t)$ by the corollary to Theorem 4.1:

$$f(z,t) = \sum_{n=0}^{\infty} a_n(t)(z-\gamma(t))^n, \qquad a_n(t) = \frac{1}{n!} f^{(n)}(\gamma(t), s).$$

Therefore, by (4.9),

$$a_n(t) = \sum_{m=n}^{\infty} \binom{m}{n} a_m(s)(\gamma(t)-\gamma(s))^{m-n}, \qquad |t-s|<\delta, \tag{4.15}$$

and if $|t-s|<\delta$, then $f(z,t)$ is the direct analytic continuation of $f(z,s)$ by power series expansion. Let $\gamma\colon t \to \gamma(t)$, $0 \leqq t \leqq b$, be a curve connecting $c_0 = \gamma(0)$ and $c_* = \gamma(b)$, $f_0(z)$ be a power series with center c_0 and $\{f(z,t)\colon 0 \leqq t \leqq b\}$ be the analytic continuation of $f_0(z)$ along the curve γ and put $f_*(z) = f(z,b)$. We say that the power series $f_*(z)$ with center c_* has been obtained from $f_0(z)$ by analytic continuation along γ, and $f_*(z)$ is called the *result of the analytic continuation* of $f_0(z)$ along the curve γ.

Let $\gamma_1\colon t \to \gamma_1(t)$, $0 \leqq t \leqq a$, be a curve connecting $c_0 = \gamma_1(0)$ and $c_1 = \gamma_1(a)$ and let $\gamma_2\colon t \to \gamma_2(t)$, $a \leqq t \leqq b$, be a curve connecting $c_1 = \gamma_2(a)$ and $c_* = \gamma_2(b)$. Let $F_1 = \{f(z,t)\colon 0 \leqq t \leqq a\}$ be the analytic continuation of the power series $f_0(z)$ with center c_0 along γ_1 and let $F_2 = \{f(z,t)\colon a \leqq t \leqq b\}$ be the analytic continuation of the power series $f(z,a)$ with center $c_1 = \gamma_1(a) = \gamma_2(a)$ along γ_2. Then

$$F = F_1 \cup F_2 = \{f(z,t)\colon 0 \leqq t \leqq b\}$$

is the analytic continuation of $f_0(z)$ along $\gamma = \gamma_1 \cdot \gamma_2$. Hence, if $f_1(z)$ is the result of the analytic continuation of $f_0(z)$ along γ_1, and $f_*(z)$ the result of the analytic continuation of $f_1(z)$ along γ_2, then $f_*(z)$ is the result of the analytic continuation of $f_0(z)$ along $\gamma = \gamma_1 \cdot \gamma_2$. A similar result holds if three or more curves $\gamma_1, \gamma_2, \gamma_3, \ldots, \gamma_n$ are pieced together to form $\gamma = \gamma_1 \cdot \gamma_2 \cdot \gamma_3 \cdot \ldots \cdot \gamma_n$.

Next we want to study the connection between analytic continuation along a curve and analytic continuation along a sequence of regions, as discussed in Section 4.1a. Let $g_0(z)$ be a holomorphic function defined on a region D_0 and let $g_1(z), g_2(z), \ldots, g_n(z)$ be the analytic continuation of $g_0(z)$ along the sequence of regions

$$D_1, D_2, D_3, \ldots, D_n, \qquad D_k \cap D_{k-1} \neq \emptyset, \qquad k = 1, 2, \ldots, n.$$

Let $f_0(z)$ be the power series expansion of $g_0(z)$ about the point $c_0 \in D_0$ and let $f_*(z)$ be the power series expansion of $g_n(z)$ about the point $c_* \in D_n$. In order to obtain $f_*(z)$ as the result of analytic continuation of $f_0(z)$ along a curve, we select for each $k = 1, 2, \ldots, n$ a point $c_k \in D_k \cap D_{k-1}$ and we put $c_{n+1} = c_*$. Next, let

$$\gamma_k\colon t \to \gamma_k(t) \in D_k, \qquad 0 \leqq t \leqq 1,\ \gamma_k(0) = c_k,\ \gamma_k(1) = c_{k+1},$$

be the curves connecting c_k with c_{k+1} in D_k, and let γ be the curve obtained by piecing together $\gamma_0, \gamma_1, \ldots, \gamma_k, \ldots, \gamma_n$:

$$\gamma = \gamma_0 \cdot \gamma_1 \cdot \gamma_2 \cdot \ldots \cdot \gamma_n.$$

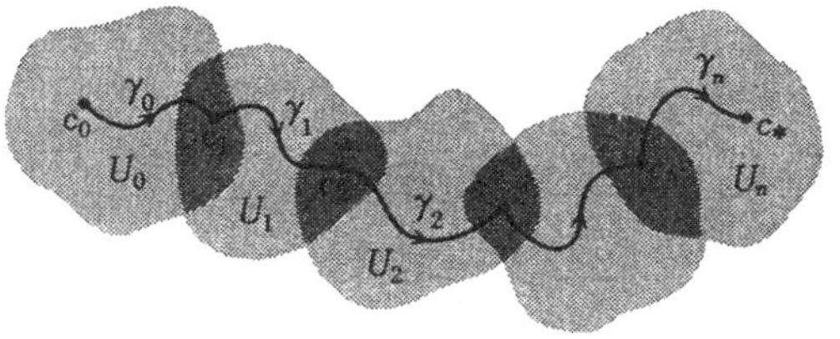

Fig. 4.7

Hence γ is given by

$$\gamma\colon t \to \gamma(t) = \gamma_k(t-k), \qquad k \leqq t \leqq k+1,\ k = 0, 1, 2, \ldots, n.$$

Since $g_k(z)$ is a holomorphic function on D_k, $F_k = \{f_k(z,t)\colon 0 \leqq t \leqq 1\}$, where $f_k(z,t)$ is the power series expansion of $g_k(z)$ with center $\gamma_k(t)$, $0 \leqq t \leqq 1$, is the analytic continuation of $f_k(z,0)$ along γ_k. Since $g_k(z) = g_{k-1}(z)$ on $D_k \cap D_{k-1}$, we have $f_k(z,0) = f_{k-1}(z,1)$. Hence, defining $f(z,t)$ by

$$f(z,t) = f_k(z, t-k) \qquad \text{if } k \leqq t \leqq k+1,\ k = 0, 1, 2, \ldots, n,$$

$f(z,t)$ is a power series with center $\gamma(t)$ and

$$F = F_1 \cup F_2 \cup \cdots \cup F_n = \{f(z,t)\colon 0 \leqq t \leqq n+1\}$$

is the analytic continuation of $f_0(z)$ along γ. Obviously $f(z, n+1) = f_*(z)$. We have proved.

Theorem 4.2. Let $g_1(z), g_2(z), \ldots, g_n(z)$ be the analytic continuation of the holomorphic function $g_0(z)$ defined on the region D_0 along the sequence of regions $D_1, D_2, \ldots, D_n$ with $D_k \cap D_{k-1} \neq \varnothing$, $k = 1, 2, \ldots, n$, let $f_0(z)$ be the power expansion of $g_0(z)$ with center $c_0 \in D_0$ and let $f_*(z)$ be the power series expansion of $g_n(z)$ with center $c_* \in D_n$. If a curve $\gamma\colon t \to \gamma(t)$, $0 \leqq t \leqq b$, connecting c_0 and c_* is selected such that for suitably chosen points $t_0, t_1, t_2, \ldots, t_k, \ldots, t_n, t_{n+1}$, $t_0 = 0 < t_1 < t_2 < \cdots < t_k < \cdots < t_n < t_{n+1} = b$, $\gamma(t) \in D_k$ if $t_k \leqq t \leqq t_{k+1}$, $k = 0, 1, 2, \ldots, n$, then $f_*(z)$ is the result of the analytic continuation of $f_0(z)$ along γ.

Let Ω be a region in the complex plane and let

$$f_0(z) = \sum_{n=0}^{\infty} a_n (z - c_0)^n$$

be a power series with center $c_0 \in \Omega$. If $f_0(z)$ is analytically continuable along all curves in Ω with initial point c_0, then $f_0(z)$ is said to be *freely analytically continuable* in Ω.

Let $f_0(z)$ be freely analytically continuable in Ω. Let c_1 be an arbitrary point of Ω and let $\gamma: t \to \gamma(t)$, $a \leqq t \leqq b$, connect c_0 and c_1, i.e., $\gamma(a) = c_0$ and $\gamma(b) = c_1$. Let the power series $f_1(z)$ with center c_1 be the result of the analytic continuation of $f_0(z)$ along γ. In general, $f_1(z)$ depends on the choice of γ, but $f_1(z)$ does not change if the curve γ is transformed continuously in Ω. Before proving this important fact, we have to give an exact definition of the phrase "is transformed continuously in Ω."

Definition 4.2. Let $\gamma_0: t \to \gamma_0(t)$ and $\gamma_1: t \to \gamma_1(t)$, $a \leqq t \leqq b$, be two curves in Ω connecting c_0 and c_1. The curves γ_1 and γ_0 are called *homotopic* in Ω (denoted $\gamma_1 \simeq \gamma_0$) if there exists a continuous function $\Gamma(t, s)$ from the rectangle $K = \{(t, s): a \leqq t \leqq b, 0 \leqq s \leqq 1\}$ into the complex plane satisfying

(i) $\Gamma(t, s) \in \Omega$ for all t, s,
(ii) $\Gamma(t, 0) = \gamma_0(t)$, $\Gamma(t, 1) = \gamma_1(t)$, $a \leqq t \leqq b$,
(iii) $\Gamma(a, s) = c_0$, $\Gamma(b, s) = c_1$, $0 \leqq s \leqq 1$.

If a continuous function $\Gamma(t, s)$ satisfying these three conditions exists, then for each $s \in [0, 1]$, the curve $\gamma_s: t \to \gamma_s(t) = \Gamma(t, s)$, $a \leqq t \leqq b$, connects c_0 and c_1 in Ω and γ_s varies continuously with the variable s. Hence we can say that γ_1 is obtained by *varying* γ_0 *continuously* in Ω. The initial point $c_0 = \gamma_s(a)$ and the terminal point $c_1 = \gamma_s(b)$ of γ_s remain fixed by condition (iii). We have already considered the case that $\Gamma_t(t, s)$, $\Gamma_s(t, s)$, and $\Gamma_{ts}(t, s)$ exist and are continuous in Section 1.3c, without, however, assuming that condition (ii) is satisfied. Since "γ_1 and γ_0 are homotopic in Ω" means "γ_0 can be transformed continuously into γ_1 in Ω," it is obvious that $\gamma_0 \simeq \gamma_1$ if $\gamma_1 \simeq \gamma_0$ and that $\gamma_2 \simeq \gamma_0$ if $\gamma_1 \simeq \gamma_0$ and $\gamma_2 \simeq \gamma_1$. It is easy to derive these facts formally from Definition 4.2. In order to prove that $\gamma_1 \simeq \gamma_0$ implies $\gamma_0 \simeq \gamma_1$ it suffices to consider the function $\Gamma(t, 1 - s)$. If $\gamma_1 \simeq \gamma_0$ and $\gamma_2 \simeq \gamma_1$ in Ω

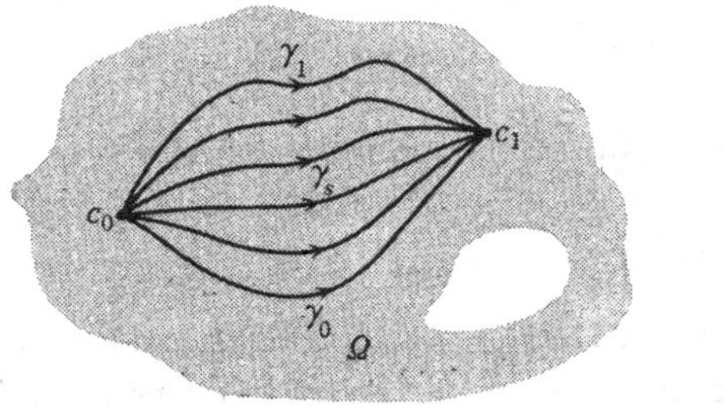

Fig. 4.8

then there exist continuous functions $\Gamma_1(t, s)$ and $\Gamma_2(t, s)$ such that

$$\Gamma_1(t, 0) = \gamma_0(t), \quad \Gamma_1(t, 1) = \Gamma_2(t, 0) = \gamma_1(t), \quad \Gamma_2(t, 1) = \gamma_2(t).$$

Hence, the function $\Gamma(t, s)$ defined by

$$\Gamma(t, s) = \begin{cases} \Gamma_1(t, 2s), & 0 \leqq s \leqq 1/2, \\ \Gamma_2(t, 2s-1), & 1/2 \leqq s \leqq 1, \end{cases}$$

is a continuous function satisfying the conditions (i) and (iii), while $\Gamma(t, 0) = \gamma_0(t)$ and $\Gamma(t, 1) = \gamma_2(t)$, that is, $\gamma_2 \simeq \gamma_0$ in Ω. Finally, we note that $\gamma_0 \simeq \gamma_0$ in Ω. We have proved that the homotopy relation $\simeq$ in Ω is an equivalence relation on the set of all curves connecting two fixed points $c_0, c_1 \in \Omega$.

Definition 4.3. Let Ω be a region and let $c_0, c_1 \in \Omega$. The equivalence classes under $\simeq$ of curves connecting c_0 and c_1 are called *homotopy classes.* The homotopy class to which γ belongs is denoted by $[\gamma]$.

If γ_0 and γ_1 are two curves connecting c_0 and c_1, then $[\gamma_0] = [\gamma_1]$ if and only if $\gamma_1 \simeq \gamma_0$ in Ω.

We have given a definition of homotopy for two curves $\gamma_0: t \to \gamma_0(t)$ and $\gamma_1: t \to \gamma_1(t)$ connecting c_0 and c_1 with the same domain $[a, b]$ for the variable t. If the two curves have different domains, for example, γ_0: $t \to \gamma_0(t)$, $a \leqq t \leqq b$, and $\gamma_1: \tau \to \gamma_1(\tau)$, $\alpha \leqq \tau \leqq \beta$, with $\gamma_0(a) = \gamma_1(\alpha) = c_0$ and $\gamma_0(b) = \gamma_1(\beta) = c_1$, we first apply a linear transformation

$$\tau = l(t) = \frac{\beta - \alpha}{b - a}(t - a) + a \tag{4.16}$$

before applying the definition of homotopy. That is, we say that γ_1 and γ_0 are homotopic in Ω (written as $\gamma_1 \simeq \gamma_0$) if $\gamma_1(l) \simeq \gamma_0$ in Ω in the sense of Definition 4.2, where $\gamma_1(l): t \to \gamma_1(l(t))$, $a \leqq t \leqq b$. Rather than transforming τ into t through $\tau = l(t)$, we could have transformed t into τ through $t = l^{-1}(\tau)$ with the same result.

We have considered the curve $\lambda: \tau \to \lambda(\tau) = \gamma(\phi(\tau))$, $\alpha \leqq \tau \leqq \beta$, obtained from $\gamma: t \to \gamma(t)$, $a \leqq t \leqq b$, by the change of variables $t = \phi(\tau)$ to be the same curve as γ. Therefore we have to prove $\lambda \simeq \gamma$ in Ω. By, if necessary, transforming the variable τ into the variable t by the linear transformation (4.16), we may assume that λ is given by

$$\lambda: t \to \lambda(t) = \gamma(\psi(t)), \qquad \psi(t) = \phi(l(t)) \qquad a \leqq t \leqq b.$$

Since $\psi(t)$ is a continuous, monotone increasing function of t such that $\psi(a) = a$ and $\psi(b) = b$, the function

$$\psi_s(t) = s\psi(t) + (1-s)t, \qquad a \leqq t \leqq b,$$

is also a monotone increasing function of t such that $\psi_s(a) = a$ and $\psi_s(b) = b$ for each $s \in [0, 1]$. Putting

$$\Gamma(t, s) = \gamma(s\psi(t) + (1 - s)t)$$

$\Gamma(t, s)$ is a continuous function defined on the rectangle $K = \{(t, s); a \leqq t \leqq b, 0 \leqq s \leqq 1\}$ such that: $\Gamma(t, s) \in \Omega$, $\Gamma(t, 0) = \gamma(t)$, $\Gamma(t, 1) = \lambda(t)$, $\Gamma(a, s) = \gamma(a)$, and $\Gamma(b, s) = \gamma(b)$. Therefore, $\lambda \simeq \gamma$ in Ω.

Theorem 4.3 (*Monodromy Theorem*). Let Ω be a region in the complex plane, $f_0(z)$ a power series with center $c_0 \in \Omega$, and $c_1 \in \Omega$. Further, let $f_0(z)$ be freely analytically continuable in Ω and let the power series $f_1(z)$ with center c_1 be the result of the analytic continuation of $f_0(z)$ along a curve γ_0 connecting c_0 and c_1 in Ω. If γ_1 is another curve connecting c_0 and c_1 in Ω and if $\gamma_1 \simeq \gamma_0$, the result of the analytic continuation of $f_0(z)$ along γ_1 equals $f_1(z)$.

In other words, the result of the analytic continuation of $f_0(z)$ along a curve γ connecting c_0 and c_1 in Ω is uniquely determined by the homotopy class $[\gamma]$ of γ.

Proof: Putting γ_0: $t \to \gamma_0(t)$ and γ_1: $t \to \gamma_1(t)$, $a \leqq t \leqq b$, there exists a function $\Gamma(t, s)$ defined on the rectangle $K = \{(t, s)\colon a \leqq t \leqq b, 0 \leqq s \leqq 1\}$ such that $\Gamma(t, s) \in \Omega$, $\Gamma(t, 0) = \gamma_0(t)$, $\Gamma(t, 1) = \gamma_1(t)$, $\Gamma(a, s) = c_0$, and $\Gamma(b, s) = c_1$. Putting $\gamma_s(t) = \Gamma(t, s)$, γ_s: $t \to \gamma_s(t)$, $a \leqq t \leqq b$, is a curve connecting c_0 and c_1 in Ω, hence there exists an analytic continuation

$$F_s = \{f_s(z, t)\colon a \leqq t \leqq b\}$$

of $f_0(z)$ along γ_s. In order to prove that the result $f_1(z, b)$ of the analytic continuation of $f_0(z)$ along γ_1 coincides with the result $f_1(z) = f_0(z, b)$ of the analytic continuation of $f_0(z)$ along γ_0 it suffices to prove that $f_s(z, b)$ is independent of s, i.e.,

$$f_s(z, b) = f_1(z), \qquad 0 \leqq s \leqq 1.$$

In order to prove this, it suffices to prove that for each $s \in [0, 1]$ there exists an $\varepsilon(s) > 0$ such that

$$f_u(z, b) = f_s(z, b) \qquad \text{if } |u - s| < \varepsilon. \tag{4.17}$$

(Because, if $f_u(z, b)$ is independent of u on a neighborhood of each point $s \in [0, 1]$, then $f_u(z, b)$ is independent of u on $[0, 1]$.) In order to prove (4.17) we will use Theorem 4.2.

Fix s and consider F_s, defined above. Let $r(t)$ denote the radius of convergence of the power series $f_s(z, t)$ with center $\gamma_s(t)$ and let U_t denote the interior of its circle of convergence. By (4.13)

$$r(t) \geqq \rho > 0$$

and for δ from (4.14)

$$|\gamma_s(t') - \gamma_s(t)| < \rho \qquad \text{if } |t' - t| < \delta$$

hence $f_s(z, t')$ is the direct analytic continuation by power series expansion of $f_s(z, t)$. Since $\Gamma(t, s)$ is uniformly continuous on K, there exists an ε with $0 < \varepsilon < \delta$ such that

$$|\Gamma(t', u) - \Gamma(t, s)| < \rho \qquad \text{if } |t' - t| < \varepsilon \text{ and } |u - s| < \varepsilon. \tag{4.18}$$

Pick $t_0 = a < t_1 < \cdots < t_k < \cdots < t_{n+1} = b$ such that

$$|t_{k+1} - t_k| < \varepsilon, \qquad k = 0, 1, 2, \ldots, n.$$

We have $f_s(z, t_0) = f_0(z)$. Putting $f_k(z) = f_s(z, t_k)$, $f_1(z), f_2(z), \ldots, f_n(z)$ is an analytic continuation of $f_0(z)$ along the sequence of regions $U_{t_1}, U_{t_2}, \ldots, U_{t_n}$ and $f_s(z, b)$ is the power series expansion with center $c_1 = \gamma_s(b)$ of $f_n(z)$. Let $\gamma_u: t \to \gamma_u(t)$, $0 \leqq t \leqq b$, be a curve connecting c_0 and c_1, where u satisfies $|u - s| < \varepsilon$ and let us apply Theorem 4.2 to this curve. If $t \in [t_k, t_{k+1}]$, then $|t - t_k| < \varepsilon$, hence by (4.18):

$$|\gamma_u(t) - \gamma_s(t_k)| < \rho.$$

Since U_{t_k} is the disk with center $\gamma_s(t_k)$ and radius $r(t_k) \geqq \rho$ we have

$$\gamma_u(t) \in U_{t_k} \qquad \text{if } t \in [t_k, t_{k+1}].$$

By Theorem 4.2, $f_s(z, b)$ equals the result $f_u(z, b)$ of the analytic continuation of $f_0(z)$ along γ_u. Hence (4.17) is valid.

If γ: $t \to \gamma(t)$, $a \leqq t \leqq b$, is a closed curve, that is, if $c_0 = \gamma(a) = \gamma(b)$, then c_0 is called the *base point* of γ. If the curve γ_0: $t \to \gamma_0(t)$, $a \leqq t \leqq b$, occurring in Theorem 4.3 is a closed curve with base point c_0, that is, if $\gamma_0(b) = \gamma_0(a) = c_0$, then the result $f_1(z)$ of the analytic continuation of $f_0(z)$ along γ_0 is a power series with center c_0, which in general will be different from $f_0(z)$.

If $\delta(t)$ maps each $t \in [a, b]$ onto the same point c_0, the curve δ: $t \to \delta(t) = c_0$ is a closed curve. If γ is a closed curve with base point c_0, then γ is said to be *homotopic to* 0 *in* Ω (written as $\gamma \simeq 0$) if $\gamma \simeq \delta$ in Ω. If $\gamma \simeq 0$, the curve γ can be contracted onto c_0 while the point c_0 remains fixed.

Since the result of the analytic continuation of $f_0(z)$ along δ is of course $f_0(z)$, we have

Corollary. If γ is a closed curve with base point c_0 in Ω and if $\gamma \simeq 0$ in Ω, then the result of the analytic continuation of $f_0(z)$ along γ is $f_0(z)$.

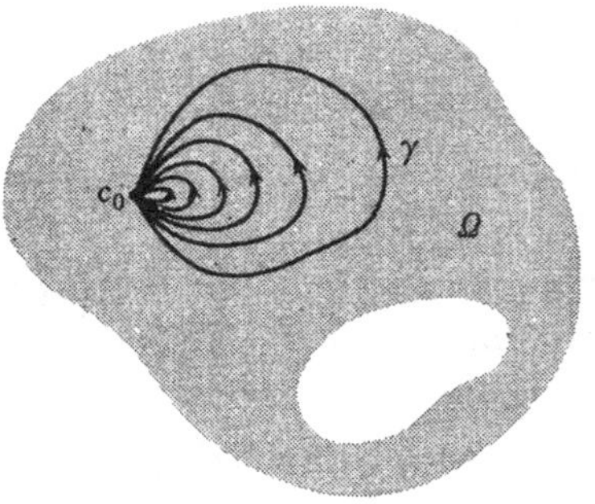

Fig. 4.9

Let C be the unit circle in the ζ-plane, and let $z = \gamma(\zeta)$ be a continuous function defined on C. Then $t \to z = \gamma(e^{it})$, $-\pi \leqq t \leqq \pi$, is a closed curve with base point $c_0 = \gamma(-1)$ in the z-plane. It is clear that each closed curve can be written in the form $t \to \gamma(e^{it})$, $-\pi \leqq t \leqq \pi$, for some continuous function $\gamma(\zeta)$ defined on C. Hence, we can define closed curves in the z-plane as continuous mappings $\gamma: \zeta \to z = \gamma(\zeta)$, which map the unit circle C into the z-plane. Let Δ be the closed unit disk $\{\zeta: |\zeta| \leqq 1\}$ in the ζ-plane, then C is the boundary of Δ. Let $\gamma: \zeta \to \gamma(\zeta) \in \Omega$, $\zeta \in C$, be a closed curve with base point $c_0 = \gamma(-1)$ in Ω.

Theorem 4.4. $\gamma \simeq 0$ in Ω if and only if the continuous mapping $\gamma: C \to \Omega$ can be extended to a continuous mapping $\Gamma: \Delta \to \Omega$.

Proof: Consider the continuous mapping

$$\Phi: (t, s) \to \zeta = \Phi(t, s) = se^{it} + s - 1$$

which maps the rectangle $K = \{(t, s): -\pi \leqq t \leqq \pi, 0 \leqq s \leqq 1\}$ onto Δ. (That $\Phi(K) = \Delta$ follows from the fact that for each s, $0 < s \leqq 1$, the mapping $t \to \Phi(t, s)$ maps $[-\pi, \pi]$ onto the circle with center $-1+s$ passing through -1.) We have $\Phi(t, 1) = e^{it} \in C$ and $\Phi(t, 0) = \Phi(-\pi, s) = \Phi(\pi, s) = -1$ and conversely, if $\Phi(t, s) = -1$, then $s = 0$ or $t = \pm\pi$. If $\zeta \neq -1$, the inverse mapping

$$\Phi^{-1}: \zeta = se^{it} + s - 1 \to (t, s)$$

is continuous. Since $\gamma \simeq 0$, there exists a continuous function $\Gamma(t, s)$ defined on K satisfying:

(i) $\Gamma(t, s) \in \Omega$,

(ii) $\Gamma(t, 0) = c_0, \qquad \Gamma(t, 1) = \gamma(e^{it})$,

(iii) $\Gamma(-\pi, s) = \Gamma(\pi, s) = c_0$.

Let $\Gamma: \zeta \to \Gamma(\zeta)$ be a continuous mapping from Δ into Ω extending the continuous mapping $\gamma: \zeta \to \gamma(\zeta)$ from C into Ω. Putting $\Gamma(t, s) = \Gamma(\Phi(t, s))$, $\Gamma(t, s)$ is clearly a continuous function defined on K satisfying conditions (i), (ii), and (iii). Conversely, if there exists a continuous function $\Gamma(t, s)$ defined on K satisfying the conditions (i), (ii), and (iii) we put

$$\Gamma(\zeta) = \Gamma(t, s) \quad \text{if} \quad \zeta = \Phi(t, s).$$

If $\Phi(t, s) = -1$, then $s = 0$ or $t = \pm\pi$, hence $\Gamma(-1) = c_0$ by conditions (ii) and (iii). Since $\Phi(t, 1) = e^{it}$, $\Gamma(e^{it}) = \Gamma(t, 1) = \gamma(e^{it})$ by (ii). Furthermore, $\Gamma(\zeta) \in \Omega$ by (i). Hence, $\Gamma: \zeta \to \Gamma(\zeta) \in \Omega$ is an extension of $\gamma: \zeta \to \gamma(\zeta)$. If $\zeta \neq -1$, $\Gamma(\zeta) = \Gamma(t, s)$ and $(t, s) = \Phi^{-1}(\zeta)$ are continuous functions of ζ. To show that $\Gamma(\zeta)$ is also continuous at $\zeta = -1$, let us assume that there exists a sequence of points $\zeta_n \in \Delta$ with $\lim_{n\to\infty} \zeta_n = -1$ and $|\Gamma(\zeta_n) - c_0| \geqq \varepsilon$ for some $\varepsilon > 0$. Put $\zeta_n = \Phi(t_n, s_n)$. The point sequence $\{(t_n, s_n)\}$ of points in K possesses a subsequence which converges to a point $(t_*, s_*) \in K$. Hence, we may just as well assume that $\lim_{n\to\infty} (t_n, s_n) = (t_*, s_*)$. Since

$$\Gamma(t_*, s_*) = \lim_{n\to\infty} \Gamma(t_n, s_n) = \lim_{n\to\infty} \Gamma(\zeta_n) \neq c_0$$

we have $s_* \neq 0$, $t_* \neq \pm\pi$, hence $\Phi(t_*, s_*) \neq -1$. This contradicts $\Phi(t_*, s_*) = \lim_{n\to\infty} \Phi(t_n, s_n) = \lim_{n\to\infty} \zeta_n = -1$. Therefore, $\Gamma(\zeta)$ is also continuous at $\zeta = -1$.

If the continuous mapping $\gamma: \zeta \to \gamma(\zeta)$ from $C = \partial\Delta$ into Ω is extended to a continuous mapping $\Gamma: \zeta \to \Gamma(\zeta)$ from Δ into Ω, for each $r \in [0, 1]$ a closed curve in Ω is given by $\gamma_r: \zeta \to \gamma_r(\zeta) = \Gamma(r\zeta)$, $\zeta \in C$. If r decreases from 1 to 0, the curve $\gamma_1 = \gamma$ shrinks continuously via the curves γ_r to become the point $C = \gamma_0(C) = \Gamma(0)$. Therefore, we say that the closed curve γ is *contractible* onto a point. The closed curve γ is contractible onto a point if and only if $\gamma \simeq 0$ by Theorem 4.4.

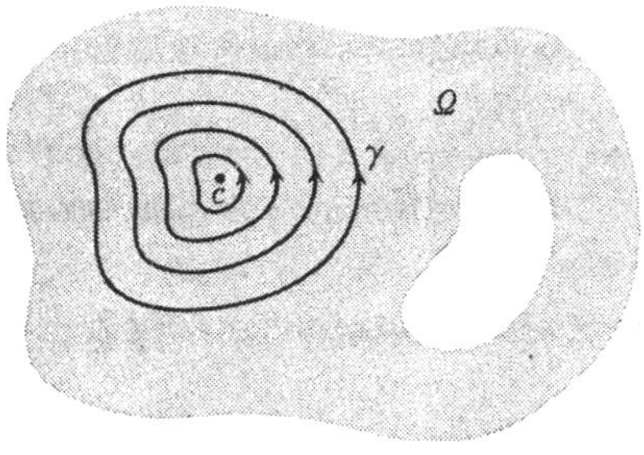

Fig. 4.10

Theorem 4.5. Let γ_0 and γ_1 be curves connecting c_0 and c_1 in Ω, then $\gamma_1 \simeq \gamma_0$ in Ω if and only if $\gamma_0 \cdot \gamma_1^{-1} \simeq 0$ in Ω.

Proof: Put $\gamma_0 : t \to \gamma_0(t)$ and $\gamma_1 : t \to \gamma_1(t)$, $0 \leqq t \leqq \pi$, with $\gamma_0(0) = \gamma_1(0) = c_0$ and $\gamma_0(\pi) = \gamma_1(\pi) = c_1$. Then $\gamma = \gamma_0 \cdot \gamma_1^{-1}$ is given by the mapping

$$\gamma : e^{it} \to \gamma(e^{it}) = \begin{cases} \gamma_0(\pi + t), & -\pi \leqq t \leqq 0, \\ \gamma_1(\pi - t), & 0 \leqq t \leqq \pi, \end{cases}$$

from the unit circle C into Ω. Consider the mapping

$$\Psi : (t, s) \to \zeta = \Psi(t, s) = (1 - s)e^{i(t-\pi)} + se^{i(\pi - t)}$$

which maps the rectangle $K = \{(t, s) : 0 \leqq t \leqq \pi, 0 \leqq s \leqq 1\}$ onto the closed unit disk Δ. If $\gamma \simeq 0$, then the continuous function $\gamma(\zeta)$ defined on $C = \partial\Delta$, can be extended to a continuous function from Δ into Ω, by Theorem 4.4. Putting $\Gamma(t, s) = \Gamma(\Psi(t, s))$, $\Gamma(t, s)$ is a continuous function on K which is easily verified to satisfy the conditions (i), (ii), and (iii) of Definition 4.2. Hence, $\gamma_1 \simeq \gamma_0$ in Ω. Conversely, if $\gamma_1 \simeq \gamma_0$ in Ω, then there exists a continuous function $\Gamma(t, s)$ on K satisfying conditions (i), (ii), and (iii). Putting

$$\Gamma(\zeta) = \Gamma(t, s) \qquad \text{if } \zeta = \Psi(t, s)$$

it can be shown just as in the proof of Theorem 4.4 that $\Gamma(\zeta)$ is a continuous function on Δ, which extends $\gamma(\zeta)$. Since obviously $\Gamma(\zeta) \in \Omega$, we conclude $\gamma \simeq 0$ in Ω.

Definition 4.4. A region Ω is called *simply connected* if all closed curves in Ω are contractible onto one point. If Ω is simply connected, then all closed curves are homotopic to 0 in Ω. Hence, if c_0 and c_1 are two points in Ω, and γ_0 and γ two curves in Ω connecting c_0 and c_1, then $\gamma_0 \cdot \gamma^{-1} \simeq 0$, i.e., $\gamma_0 \simeq \gamma_1$ in Ω by Theorem 4.5.

Example 4.2. A region Ω is called *convex* if any pair of points can be connected by a line segment in Ω. Disks and the interior of rectangles are examples of convex domains. Convex domains are simply connected. Let Δ be the closed unit disk and let the mapping $\gamma : e^{i\theta} \to z = \gamma(e^{i\theta})$, $0 \leqq \theta \leqq 2\pi$, from $C = \partial\Delta$ into Ω define a closed curve in Ω. Let z_0 be an arbitrary point of Ω. Since Ω is convex, $r\gamma(e^{i\theta}) + (1 - r)z_0 \in \Omega$ for $0 \leqq r \leqq 1$, hence γ can be extended to a continuous mapping

$$\Gamma : re^{i\theta} \to z = \Gamma(re^{i\theta}) = r\gamma(e^{i\theta}) + (1 - r)z_0, \qquad re^{i\theta} \in \Delta,$$

from Δ into Ω.

Example 4.3. Let Ω be the region obtained by deleting from $(Q) = \{z : z = x + iy,\ 0 < x < 1,\ 0 < y < 1\}$ an infinite number of segments

$L_k = \{z : z = 2^{-k} + iy,\ \frac{1}{2} \leqq y < 1,\ k = 1, 2, 3, \ldots\}$. Then Ω is simply connected. To prove this, it suffices to show that a continuous mapping $\gamma: e^{i\theta} \to z = \gamma(e^{i\theta})$, $0 \leqq \theta \leqq 2\pi$, from the unit circle $C = \partial\Delta$ into Ω can be extended to a continuous mapping from the closed unit disk Δ into Ω. Writing

$$\gamma(e^{i\theta}) = \rho(e^{i\theta}) + i\sigma(e^{i\theta})$$

we define a function $\Gamma(re^{i\theta})$ by

$$\Gamma(re^{i\theta}) = \begin{cases} \rho(e^{i\theta}) + (2r-1)i\sigma(e^{i\theta}) + \dfrac{(1-r)i}{2}, & \frac{1}{2} \leqq r \leqq 1, \\[2ex] 2r\rho(e^{i\theta}) + \dfrac{1-2r}{2} + \dfrac{i}{4}, & 0 \leqq r \leqq \frac{1}{2}, \end{cases}$$

which is a continuous mapping from Δ into Ω such that $\Gamma(e^{i\theta}) = \gamma(e^{i\theta})$.

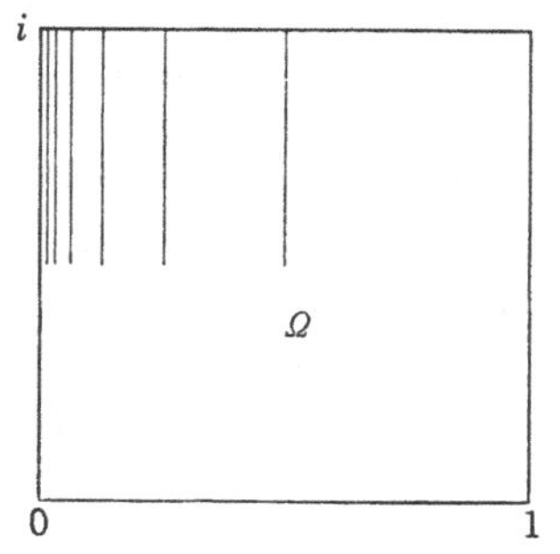

Fig. 4.11

Theorem 4.6. Let Ω be a region in the complex plane.

(1) If β_1 and γ_1 are curves in Ω connecting c_0 and c_1 and if β_2 and γ_2 are curves in Ω connecting c_1 and c_2, such that $\beta_1 \simeq \gamma_1$ and $\beta_2 \simeq \gamma_2$ in Ω then $\beta_1 \cdot \beta_2 = \gamma_1 \cdot \gamma_2$ in Ω.

(2) If β and γ are curves in Ω, such that $\beta \simeq \gamma$ in Ω, then $\beta^{-1} \simeq \gamma^{-1}$ in Ω.

(3) Let γ be a curve connecting c_0 and c_1 in Ω, α a closed curve in Ω with c_0 as base point, and β a closed curve in Ω with c_1 as base point. If $\alpha \simeq 0$, then $\alpha \cdot \gamma \simeq \gamma$ and if $\beta \simeq 0$, then $\gamma \cdot \beta \simeq \gamma$.

Proof: Parts (1) and (2) follow directly from Definition 4.2. For part (3), let $\delta: t \to \delta(t) = c_0$, $0 \leqq t \leqq 1$, be the curve consisting only of the point c_0. If $\alpha \simeq 0$ then $\alpha \simeq \delta$, hence $\alpha \cdot \gamma \simeq \delta \cdot \gamma$ by (1). Therefore, in order to prove

$\alpha \cdot \gamma \simeq \gamma$ it suffices to prove $\delta\gamma \simeq \gamma$. Let γ be given by: $t \to \gamma(t)$, $0 \leqq t \leqq 1$, $\gamma(0) = c_0$ and $\gamma(1) = c_1$, and define a continuous function $\Gamma(t, s)$ on the rectangle $K = \{(t, s): 0 \leqq t \leqq 2, 0 \leqq s \leqq 1\}$ by

$$\Gamma(t, s) = \begin{cases} c_0, & 0 \leqq t \leqq s, \\ \gamma\left(\dfrac{t-s}{2-s}\right), & s \leqq t \leqq 2, \end{cases}$$

(i) Obviously, $\Gamma(t, s) \in \Omega$ for all $(t, s) \in K$.

(ii) Put $\gamma_0(t) = \Gamma(t, 0)$ and $\gamma_1(t) = \Gamma(t, 1)$, $0 \leqq t \leqq 2$. Since $\gamma_0(t) = \gamma(t/2)$ the curve γ_0: $t \to \gamma_0(t)$, $0 \leqq t \leqq 2$ is identical with γ: $\gamma_0 = \gamma$. Since $\gamma_1(t) = c_0 = \delta(t)$ if $0 \leqq t \leqq 1$ and $\gamma_1(t) = \gamma(t-1)$ if $1 \leqq t \leqq 2$, we have $\gamma_1 = \delta \cdot \gamma$.

(iii) $\Gamma(0, s) = c_0$ and $\Gamma(2, s) = \gamma(1) = c_1$. Hence $\delta \cdot \gamma \simeq \gamma$.

The second half of part (3) is proved similarly.

Corollary. If $\beta = \beta_1 \cdot \beta_2 \cdot \cdots \cdot \beta_m$ and $\gamma = \gamma_1 \cdot \gamma_2 \cdot \cdots \cdot \gamma_m$ are curves in Ω such that $\beta_1 \simeq \gamma_1, \beta_2 \simeq \gamma_2, \ldots, \beta_m \simeq \gamma_m$ in Ω, then $\beta \simeq \gamma$ in Ω.

Next, we want to discuss briefly the so-called *fundamental group* of a region Ω. Let c_0 be a fixed point in Ω and let us consider the homotopy classes $[\gamma]$ of closed curves γ with base point c_0 in Ω. If γ_1 and γ_2 are closed curves with base point c_0 in Ω, then $\gamma_1 \cdot \gamma_2$ is also a closed curve with base point c_0 in Ω, and by Theorem 4.6 (1), the homotopy class $[\gamma_1 \cdot \gamma_2]$ is uniquely determined by $[\gamma_1]$ and $[\gamma_2]$. Therefore, we define the product of $[\gamma_1]$ and $[\gamma_2]$ by

$$[\gamma_1][\gamma_2] = [\gamma_1 \cdot \gamma_2].$$

The homotopy class of the curve δ: $t \to \delta(t) = c_0, 0 \leqq t \leqq 1$, is denoted by **1**: $[\delta] = \mathbf{1}$. By Theorem 4.6(3) we have

$$\mathbf{1}[\gamma] = [\gamma]\mathbf{1} = [\gamma].$$

By Theorem 4.6(2) we have that the homotopy class $[\gamma^{-1}]$ of γ^{-1} is uniquely determined by $[\gamma]$. Since $\gamma \cdot \gamma^{-1} \simeq 0$ by Theorem 4.5, we have

$$[\gamma][\gamma^{-1}] = \mathbf{1}.$$

Since $(\gamma^{-1})^{-1} = \gamma$ we also have by replacing γ by γ^{-1}

$$[\gamma^{-1}][\gamma] = \mathbf{1}.$$

Since both $[\gamma_1]([\gamma_2] \cdot [\gamma_3])$ and $([\gamma_1] \cdot [\gamma_2])\, [\gamma_3]$ equal $[\gamma_1 \cdot \gamma_2 \cdot \gamma_3]$, the associative law is satisfied. Therefore, the collection of homotopy classes of closed curves with base point c_0 in Ω forms a group. This group is called the *fundamental group* of Ω and is denoted by $\pi_1(\Omega, c_0)$. The identity of the fundamental group $\pi_1(\Omega, c_0)$ is given by $[\delta] = \mathbf{1}$; the inverse of the element $[\gamma]$ is given by $[\gamma^{-1}]$:$[\gamma]^{-1} = [\gamma^{-1}]$. If $\gamma \simeq 0$, then $\gamma \simeq \delta$, that is, $[\gamma] = \mathbf{1}$.

As an abstract group, the fundamental group $\pi_1(\Omega, c_0)$ does not depend on the choice of the point c_0, i.e., if $c \in \Omega$ is an arbitrary point, then $\pi_1(\Omega, c)$ and $\pi_1(\Omega, c_0)$ are isomorphic: $\pi_1(\Omega, c) \cong \pi_1(\Omega, c_0)$. We prove this now.

Let β be a curve connecting c_0 and c in Ω. If γ is an arbitrary closed curve with base point c_0 in Ω, then $\beta^{-1} \cdot \gamma \cdot \beta$ is a closed curve with base point c in Ω and $[\beta^{-1} \cdot \gamma \cdot \beta]$ is uniquely determined by $[\gamma]$ (by the corollary to Theorem 4.6). Conversely, $[\gamma]$ is uniquely determined by $[\beta^{-1} \cdot \gamma \cdot \beta]$. This is because $\beta \cdot \beta^{-1}$ is a closed curve with base point c_0 and $\beta\beta^{-1} \simeq 0$ in Ω by Theorem 4.5, hence $[\gamma] = [\beta \cdot \beta^{-1} \cdot \gamma \cdot \beta \cdot \beta^{-1}]$ by Theorem 4.6(3). We will prove that the one-to-one mapping $[\gamma] \to [\beta^{-1} \cdot \gamma \cdot \beta]$ from $\pi_1(\Omega, c_0)$ onto $\pi_1(\Omega, c)$ is an isomorphism. Since $[\gamma_1][\gamma_2]^{-1} = [\gamma_1 \gamma_2^{-1}]$ we have

$$[\beta^{-1}\gamma_1\beta][\beta^{-1}\gamma_2\beta]^{-1} = [\beta^{-1}\gamma_1\beta\beta^{-1}\gamma_2^{-1}\beta] = [\beta^{-1}\gamma_1\gamma_2^{-1}\beta]$$

proving that the mapping $[\gamma] \to [\beta^{-1} \cdot \gamma \cdot \beta]$ is a homomorphism. Finally, let $[\lambda] \in \pi_1(\Omega, c)$ be an arbitrary element; then λ is a closed curve with base point c in Ω. Putting $\gamma = \beta \cdot \lambda \cdot \beta^{-1}$, λ is a closed curve with base point c_0 and since $\beta^{-1} \cdot \beta \simeq 0$ we have

$$[\beta^{-1}\gamma\beta] = [\beta^{-1}\beta\lambda\beta^{-1}\beta] = [\lambda].$$

Theorem 4.7. Let Ω be a simply connected region and let $f_0(z)$ be a power series with center $c_0 \in \Omega$. If $f_0(z)$ is freely analytically continuable in Ω, then the collection of analytic continuations of $f_0(z)$ along curves with initial point c_0 in Ω defines a holomorphic, one-valued function.

Proof: Let c be an arbitrary point in Ω, let γ be a curve connecting c_0 and c in Ω, and let $f_c(z)$ denote the power series with center c which results from the analytic continuation of $f_0(z)$ along γ. Since Ω is simply connected, any pair of curves connecting c_0 and c in Ω is homotopic, hence $f_c(z)$ is uniquely determined by c by Theorem 4.3 and is independent of the choice of the curve γ.

Let $r(c)$ denote the radius of convergence of the power series $f_c(z)$, select $\varepsilon(c)$, $0 < \varepsilon(c) \leqq r(c)$, such that $U_{\varepsilon(c)}(c) \subset \Omega$ and put $U(c) = U_{\varepsilon(c)}(c)$. Of course $f_c(z)$ is a holomorphic function on $U(c)$. Since $\Omega = \bigcup_{c \in \Omega} U(c)$, in order to prove that the collection of all analytic continuations $f_c(z)$, $c \in \Omega$, of $f_0(z)$ forms a holomorphic, one-valued function on Ω it suffices to prove that if $U(c_1) \cap U(c_2) \neq \emptyset$ for two points c_1 and c_2 in Ω, then $f_{c_1}(z) = f_{c_2}(z)$ on $U(c_1) \cap U(c_2)$. Let $c \in U(c_1) \cap U(c_2)$. If we expand $f_{c_1}(z)$ into a power series with center c, we obtain the analytic continuation $f_c(z)$ of $f_0(z)$. Similarly, the power series expansion of $f_{c_2}(z)$ with center c is also $f_c(z)$. Hence $f_{c_1}(z)$ and $f_{c_2}(z)$ coincide on a neighborhood of c, and so they coincide on $U(c_1) \cap U(c_2)$ by Theorem 3.1.

Let $f_0(z)$ be a power series with center $c_0 \in \Omega$ which is freely continuable in the region Ω. Let c be an arbitrary point in Ω, γ a curve connecting c_0 and c in Ω, and let $f_c(z)$ be the power series with center c that is the result of the analytic continuation of $f_0(z)$ along γ. If Ω is not simply connected, in general $f_c(z)$ will be dependent not only on c, but also on the choice of γ:

$$f_c(z) = f_{c,\gamma}(z). \tag{4.19}$$

Theorem 4.8. Let c_0 and c be two fixed points of the region Ω. The collection of homotopy classes of curves connecting c_0 and c is finite or denumerable.

Proof: It suffices to show that there is a denumerable collection $\{\gamma_n : n = 1, 2, 3, \ldots\}$ of curves connecting c_0 and c in Ω, such that for each curve γ connecting c_0 and c in Ω we have $\gamma \simeq \gamma_n$ for some γ_n. The set Q of rational numbers is denumerable. Therefore, the set $Q(i)$ consisting of all complex numbers $r + is$ with $r, s \in Q$ is also denumerable. Obviously, $Q(i)$ is everywhere dense in $\mathbb{C}$. If $\sigma = \{c_1, c_2, \ldots, c_k, \ldots, c_m\}$ is a finite sequence of points in $Q(i)$, let $L(\sigma)$ denote the polygonal line consisting of the segments connecting the points $c_0, c_1, c_2, \ldots, c_k, c_{k+1}, \ldots, c_m, c$ in that order. Since $Q(i)$ is denumerable, the set of all finite sequences σ of points of $Q(i)$ is also denumerable, that is, the collection of all polygonal lines $L(\sigma)$ is denumerable. List all polygonal lines $L(\sigma) \subset \Omega$ as follows:

$$L(\sigma_1), L(\sigma_2), L(\sigma_3), \ldots, L(\sigma_n), \ldots$$

and put $\gamma_n = L(\sigma_n)$. By definition, all polygonal lines γ_n are curves connecting c_0 and c in Ω. It is easy to verify that an arbitrary curve γ connecting c_0 and c in Ω is homotopic with a polygonal line γ_n which is "sufficiently close to γ". Let $t \to \gamma(t)$, $0 \leqq t \leqq b$, be a parameter representa-

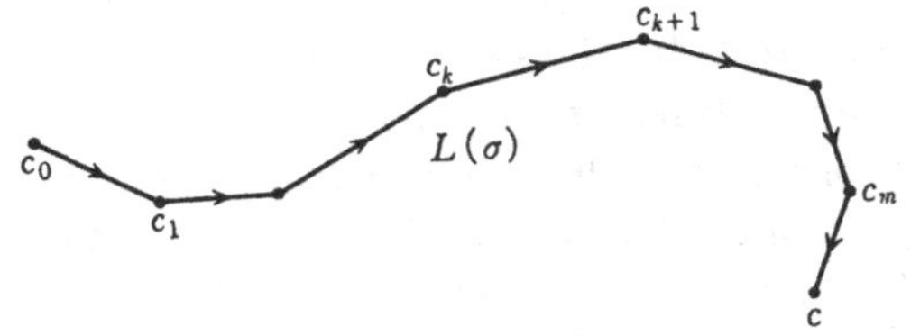

Fig. 4.12

tion of γ. Since $\{\gamma(t): 0 \leqq t \leqq b\}$ is a compact subset of Ω the distance ρ between it and $\mathbb{C} - \Omega$ is positive. Hence we have for all t.

$$\text{if } |z - \gamma(t)| < \rho, \qquad \text{then } z \in \Omega.$$

Put $\varepsilon = \rho/5$, then for some $\delta > 0$ we have

$$|\gamma(t) - \gamma(u)| < \varepsilon \qquad \text{if } |t - u| < \delta.$$

Select $t_1, t_2, \ldots, t_m$, such that $0 = t_0 < t_1 < t_2 < \cdots < t_m < t_{m+1} = b$ and $|t_{k+1} - t_k| < \delta$. Next, choose points $c_1, c_2, \ldots, c_m \in Q(i)$ such that

$$|c_k - \gamma(t_k)| < \varepsilon$$

and put $\sigma = \{c_1, c_2, \ldots, c_k, c_{k+1}, \ldots, c_m\}$.

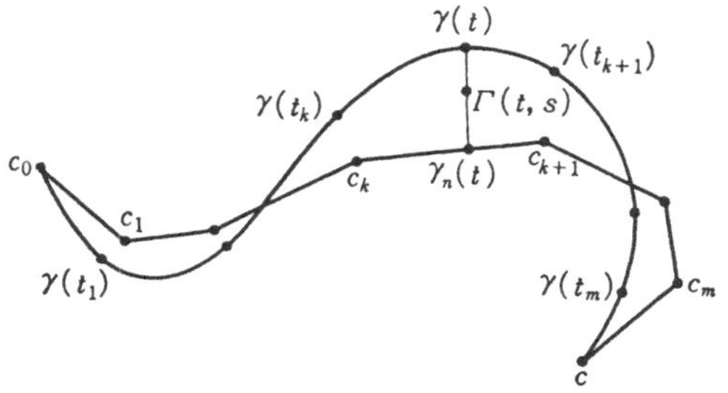

Fig. 4.13

Since $|\gamma(t_{k+1}) - \gamma(t_k)| < \varepsilon$, $|c_{k+1} - c_k| < 3\varepsilon$ and the segment connecting c_k and c_{k+1} is within the circle with center $\gamma(t_k)$ and radius 4ε. Since $\rho = 5\varepsilon$, we conclude $L(\sigma) \subset \Omega$, that is, $L(\sigma) = L(\sigma_n) = \gamma_n$ for some n. A parameter representation of $\gamma_n = L(\sigma)$ is given by $t \to \gamma_n(t)$, $0 \leqq t \leqq b$, with

$$\gamma_n(t) = \frac{(t_{k+1} - t)c_k + (t - t_k)c_{k+1}}{t_{k+1} - t_k}, \qquad t_k \leqq t \leqq t_{k+1}, k = 0, 1, 2, \ldots, m.$$

If $t_k \leqq t \leqq t_{k+1}$, then $|\gamma(t) - \gamma(t_k)| < \varepsilon$ and $|\gamma_n(t) - c_k| < 3\varepsilon$, hence $|\gamma(t) - \gamma_n(t)| < 5\varepsilon$, i.e.,

$$|\gamma(t) - \gamma_n(t)| < 5\varepsilon = \rho, \qquad 0 \leqq t \leqq b.$$

Hence the points $s\gamma(t) + (1 - s)\gamma_n(t)$, $0 \leqq s \leqq 1$, on the segment connecting $\gamma_n(t)$ and $\gamma(t)$ are in Ω. Therefore $\Gamma(t, s) = s\gamma(t) + (1 - s)\gamma_n(t)$ is a continuous function defined on the rectangle $K = \{(t, s): 0 \leqq t \leqq b, 0 \leqq s \leqq 1\}$. Since (i) $\Gamma(t, s) \in \Omega$ for all $(t, s) \in K$, (ii) $\Gamma(t, 0) = \gamma_n(t)$, $\Gamma(t, 1) = \gamma(t)$, and (iii) $\Gamma(0, s) = c_0$, $\Gamma(b, s) = c$ we conclude $\gamma \simeq \gamma_n$.

Putting $c = c_0$ we get the following corollary.

Corollary. The fundamental group $\pi_1(\Omega, c_0)$ of a region Ω is a finite group or a group with a denumerable number of elements.

According to Theorem 4.3, the power series, with center $c f_{c,\gamma}$, is uniquely determined by the homotopy class $[\gamma]$ of γ in Ω. Hence, by Theorem 4.8, there are among the $f_{c,\gamma}(z)$, only a finite or a denumerable number of different power series. Let $f_{c,k}(z)$, $k = 1, 2, 3, \ldots$, denote the different $f_{c,\gamma}(z)$, that is

$$f_c(z) = f_{c,k}(z), \qquad k = 1, 2, 3, \ldots$$

Let $r(c,k)$ denote the radius of convergence of the power series $f_{c,k}(z)$, choose $\varepsilon(c,k)$ such that $0 < \varepsilon(c,k) \leqq r(c,k)$ and $U_{\varepsilon(c,k)}(c) \subset \Omega$, and define $f(z)$ by $f(z) = f_{c,k}(z)$, $z \in U_{\varepsilon(c,k)}(c)$, $c \in \Omega$. The function $f(z)$ is an analytic function which is, in general, multivalued. Each power series $f_{c,k}(z)$ is called a *branch of the analytic function at c*. Each branch $f_{c,k}(z)$ is freely analytically continuable in Ω. Under these circumstances, we say that the analytic function $f(z)$ is *freely analytically continuable* in Ω. A power series obtained by the analytic continuation of an arbitrary branch of $f(z)$ along a curve in Ω is also a branch of $f(z)$. This fact is expressed by saying that the analytic function $f(z)$ is *complete* on Ω.

If $D \subset \Omega$ is a simply connected subregion of Ω and if $f_{c,k}(z)$ is a branch of $f(z)$ at c, then $f_{c,k}(z)$ is freely analytically continuable in Ω and the collection of all analytic continuations of $f_{c,k}(z)$ along curves in D starting from c is a holomorphic univalent function on D. This holomorphic function is denoted by $f_{D,k}(z)$ and is called the *branch of the analytic function $f(z)$ over D*. Obviously, $f_{c,k}(z)$ is the power series expansion of the holomorphic function $f_{D,k}(z)$ about c. The restriction $f_D(z)$ of the analytic function $f(z)$ to the simply connected region D consists of a (finite or denumerable) number of branches $f_{D,k}(z)$, $k = 1, 2, 3, \ldots$. In particular, if D is taken to be the largest disk $U_{r(c)}(c)$ (Section 1.3d) with center c contained in Ω, then, since $f_{D,k}(z)$ is holomorphic on $D = U_{r(c)}(c)$, its power series expansion, $f_{c,k}(z)$, about c is absolutely convergent on $U_{r(c)}(c)$. Hence, its radius of convergence $r(c,k)$ satisfies

$$r(c,k) \geqq r(c). \tag{4.20}$$

Next we need to consider the composition of a holomorphic and an analytic function. Let $f(z)$ be a freely analytically continuable, complete function on a region Ω and let $g(\zeta)$ be a holomorphic function defined on a region D of the ζ-plane, such that $g(D) \subset \Omega$. We consider the composite function $f(g(\zeta))$. Let α be an arbitrary point in D and put $c = g(\alpha)$. Let $f_{c,k}(z)$ denote a branch of $f(z)$ at c. Select $\varepsilon > 0$ such that $g(U_\varepsilon(\alpha)) \subset U_{r(c)}(c)$, then

$f_{c,k}(z)$ is a holomorphic function on $U_{r(c)}(c)$ by (4.20), hence $f_{c,k}(g(\zeta))$ is a holomorphic function of ζ on $U_\varepsilon(\alpha)$ which can be expanded in a power series about α. For the sake of simplicity, we denote this power series as $f_{c,k}(g(\zeta))$ also.

Let $0 \leqq t \leqq b$ and $\gamma: t \to \zeta = \gamma(t)$ be a curve in D with initial point $\alpha = \gamma(0) \in D$, then the holomorphic mapping g maps γ onto a curve $g(\gamma)$: $t \to z = g(\gamma(t))$ in Ω with initial point $c = g(\alpha)$. Under these circumstances, the following theorem is valid.

Theorem 4.9. Let $f_{c,k}(z)$ be a branch of $f(z)$ at $c = g(\alpha)$, and let $f = \{f(z,t)$: $0 \leqq t \leqq b\}$ be the analytic continuation of $f_{c,k}(z)$ along $g(\gamma)$, then the analytic continuation along γ of the power series $f_{c,k}(g(\zeta))$ with center α is given by $\{f(g(\zeta), t); 0 \leqq t \leqq b\}$.

Proof: Let $\mathscr{U}_t$ be the interior of the circle of convergence of the power series $f(z,t)$ with center $g(\gamma(t))$ for all considered t and let $U_t = U(\gamma(t))$ be a disk with center $\gamma(t)$ in the ζ-plane such that $g(U_t) \subset \mathscr{U}_t$. Now expand $f(g(\zeta), t)$, which is a holomorphic function of ζ on U_t, in a power series about $\gamma(t)$. In order to prove that the collection $\{f(g(\zeta), t): 0 \leqq t \leqq b\}$ is the analytic continuation of the power series $f_{c,k}(g(\zeta)) = f(g(\zeta), 0)$ along γ it suffices to show that if for each $s \in [0, b]$, we have picked an $\varepsilon(s) > 0$ such that

$$\gamma(t) \in U_s \qquad \text{if } |t - s| < \varepsilon(s),$$

then $f(g(\zeta), t) = f(g(\zeta), s)$ on $U_t \cap U_s$ if $|t - s| < \varepsilon(s)$. Since $\gamma(t) \in U_s$ implies $g(\gamma(t)) \subset \mathscr{U}_s$, $|t - s| < \varepsilon(s)$ implies $g(\gamma(t)) \in \mathscr{U}_s$, hence $f(z,t)$ is the power series expansion of $f(z,s)$ about $g(\gamma(t))$ by the corollary to Theorem 4.1. Therefore, $f(z,t) = f(z,s)$ on $\mathscr{U}_t \cap \mathscr{U}_s$ if $|t - s| < \varepsilon(s)$, hence $f(g(\zeta), t) = f(g(\zeta), s)$ on $U_t \cap U_s$.

We call the power series $f_{c,k}(g(\zeta))$ with center $\alpha \in D$, where $c = g(\alpha)$, a branch of the composite function $f(g(\zeta))$ at α, and can say by the above theorem that all branches $f_{c,k}(g(\zeta))$ of $f(g(\zeta))$ are freely analytically continuable in D. Putting $\gamma(b) = \beta$ in the above proof, $g(\gamma)$ is a curve connecting $c = g(\alpha)$ and $g(\beta)$, and the power series $f(z,b)$, with center $g(\beta)$, which is the result of the analytic continuation of $f_{c,k}(z) = f(z,0)$ along the curve $g(\gamma)$ in Ω, is one of the branches $f_{g(\beta),j}(z)$ at $g(\beta)$ of the complete analytic function $f(z)$. Hence the power series $f(g(\zeta), b) = f_{g(\beta),j}(g(\zeta))$ with center β obtained by analytic continuation of the branch $f_{c,k}(g(\zeta))$ of $f(g(\zeta))$ at α along the curve in D, is one of the branches of $f(g(z))$ at β. Hence the power series that is obtained by the analytic continuation of an arbitrary branch of $f(g(z))$ along a curve in D is again a branch of $f(g(z))$. Thus, we can say that the composite function $f(g(\zeta))$ is freely analytically continuable and complete on the region D.

If D is simply connected, then by Theorem 4.7 the collection of analytic continuations of a branch $f_{c,k}(g(\zeta))$ of $f(g(\zeta))$ at α, $c = g(\alpha)$, along a curve in Ω with initial point α constitutes a regular univalent function on D. This function is denoted by $f(g(\zeta))_k$ and is called a branch of the composite function $f(g(\zeta))$. Since $f_{ck}(g(\zeta))$ is the power series expansion of the holomorphic function $f(g(\zeta))_k$ about α, $f(g(\zeta))_k$ and $f(g(\zeta))_j$ are different holomorphic functions if $j \neq k$. We have proved

Theorem 4.10. If the region D is simply connected, then the composite function $f(g(\zeta))$ breaks into different branches $f(g(\zeta))_k$, $k = 1, 2, 3, \ldots$. Each branch is a holomorphic function defined on D.

If $f(z)$ is a freely analytically continuable and complete analytic function on a region Ω, then $f(z)$ is considered as the collection of all power series obtained by analytic continuation of a power series with center c_0 for some $c_0 \in \Omega$ along a curve in Ω. Hence, if $f_{c,k}(z)$ and $f_{d,j}(z)$ are arbitrary branches of $f(z)$ then $f_{d,j}(z)$ can be obtained by the analytic continuation of $f_{c,k}(z)$ along a suitable curve in Ω connecting c and d. In this sense, $f(z)$ is one analytic function. The composite function $f(g(\zeta))$, as considered above, is freely analytically continuable and complete, but does not necessarily constitute an analytic function in the above sense. In particular, if D is simply connected, the collection of all analytic continuations of a branch $f_{c,k}(g(\zeta))$ of $f(g(\zeta))$ at $\alpha \in D$, when $c = g(\alpha)$, along curves in D constitutes a holomorphic function $f(g(\zeta))_k$ defined on D; hence it is impossible to obtain by analytic continuation of $f_{c,k}(g(\zeta))$ along a curve in D, another branch $f_{c,j}(g(\zeta))$ at α. Also, if D is not simply connected, the composite function $f(g(\zeta))$ generally breaks into a number of freely analytically continuable and complete analytic functions on D.

4.3 Analytic continuation by integrals

Let Ω be a region in the complex plane, let $f(z)$ be a continuous function on Ω and let γ: $t \to \gamma(t)$, $a \leqq t \leqq b$, be a smooth curve in Ω. The integral of $f(z)$ along γ is given by Theorem 1.6.

$$\int_\gamma f(z)\,dz = \int_a^b f(\gamma(t))\gamma'(t)\,dt.$$

As explained in the previous section, the curve λ: $\tau \to \lambda(\tau) = \gamma(\phi(\tau))$, obtained from γ via the coordinate transformation $t = \phi(\tau)$, $\alpha \leqq \tau \leqq \beta$, where $\phi(\tau)$ is a continuously differentiable function such that $\phi'(\tau) > 0$ for all τ and $\phi(\alpha) = a$ and $\phi(\beta) = b$, is the same curve as γ. Therefore we have to verify

$$\int_\lambda f(z)\,dz = \int_\gamma f(z)\,dz.$$

Since $\lambda'(\tau)\,d\tau = \gamma'(t)\phi'(\tau)d\tau$, we have

$$\int_\alpha^\beta f(\lambda(\tau))\lambda'(\tau)\,d\tau = \int_a^b f(\gamma(t))\gamma'(t)dt,$$

proving our statement.

If $\gamma^{-1}: \tau \to \gamma^{-1}(\tau) = \gamma(a+b-\tau)$, $a \leqq \tau \leqq b$, is the curve obtained from γ by reversing the orientation, we have $d\gamma^{-1}(\tau)/d\tau = -\gamma'(b+a-\tau)$, hence

$$\int_{\gamma^{-1}} f(z)\,dz = -\int_a^b f(b+a-\tau))\gamma'(b+a-\tau)\,d\tau.$$

Putting $t = b+a-\tau$, we have

$$\int_{\gamma^{-1}} f(z)\,dz = \int_b^a f(\gamma(t))\gamma'(t)\,dt = -\int_a^b f(\gamma(t))\gamma'(t)\,dt,$$

that is,

$$\int_{\gamma^{-1}} f(z)\,dz = -\int_\gamma f(z)\,dz.$$

The curve $\gamma = \gamma_1 \cdot \gamma_2 \cdot \cdots \cdot \gamma_m$ obtained by piecing together smooth curves $\gamma_1, \gamma_2, \ldots, \gamma_m$ is called *piecewise smooth* and $\int_\gamma f(z)\,dz$ is defined by

$$\int_\gamma f(z)\,dz = \int_{\gamma_1} f(z)\,dz + \int_{\gamma_2} f(z)\,dz + \cdots + \int_{\gamma_m} f(z)\,dz.$$

Putting $\gamma_k: t \to \gamma_k(t)$, $a_{k-1} \leqq t \leqq a_k$, where $a = a_0 < a_1 < a_2 < \cdots < a_m = b$ then γ is given by $\gamma: t \to \gamma(t)$, $a \leqq t \leqq b$, with $\gamma(t) = \gamma_k(t)$, $t_{k-1} \leqq t \leqq t_k$. Therefore

$$\int_\gamma f(z)\,dz = \sum_{k=1}^m \int_{a_{k-1}}^{a_k} f(\gamma(t))\gamma_k'(t)\,dt.$$

$\gamma'(t)$ is continuous on $[a, b] - \{a_1, a_2, \ldots, a_{m-1}\}$ and $\gamma'(t) = \gamma_k'(t)$ if $a_{k-1} < t < a_k$. Hence

$$\int_\gamma f(z)\,dz = \int_a^b f(\gamma(t))\gamma'(t)\,dt.$$

Now let $f(z)$ be a holomorphic function defined on the region Ω. Let $c_0 \in \Omega$ be a fixed point and consider for arbitrary $z \in \Omega$,

$$F(z) = \int_\gamma f(z)\,dz,$$

where γ is a piecewise smooth curve in Ω connecting c_0 and z. For this purpose we write

$$F(z) = \int_{\gamma}^{z} f(\zeta)\,d\zeta, \tag{4.21}$$

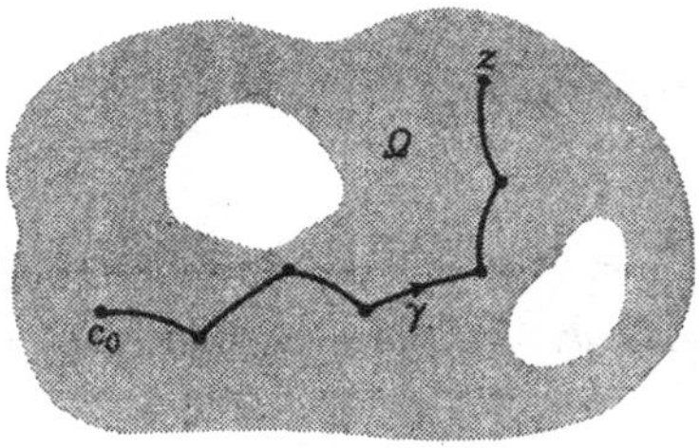

Fig. 4.14

where z represents the terminal point of the path of integration γ.

Let c be an arbitrary point in Ω and let $U(c) = U_{r(c)}(c)$ be the largest disk with center c contained in Ω. By Theorem 1.16, $f(z)$ can be expanded in a power series which absolutely converges in $U(c)$

$$f(z) = \sum_{n=0}^{\infty} a_n (z-c)^n.$$

Putting

$$F_c(z) = \sum_{n=0}^{\infty} \frac{a_n}{n+1}(z-c)^{n+1},$$

the power series $F_c(z)$ is also absolutely convergent on $U(c)$ and

$$\frac{d}{dz}F_c(z) = f(z), \qquad F_c(c) = 0,$$

that is, $F_c(z)$ is a primitive function of $f(z)$ on $U(c)$. Hence, if $z \in U(c)$ and if β is a piecewise smooth curve in $U(c)$ connecting c and z, then by (1.30),

$$\int_{\beta}^{z} f(z)\,dz = F_c(z).$$

Since the integral in the left-hand side of this equality does not depend on β, we can write

$$F_c(z) = \int_{c}^{z} f(z)\,dz, \qquad z \in U(c). \tag{4.22}$$

We put

$$F_0(z) = F_{c_0}(z) = \int_{c_0}^{z} f(z)\,dz, \qquad z \in U(c_0).$$

Next, for a piecewise smooth curve $\gamma: t \to \gamma(t)$, $0 \leqq t \leqq b$, starting from $c_0 = \gamma(0)$ in Ω we put

$$F(z, t) = \int_0^t f(\gamma(t))\gamma'(t)\,dt + F_{\gamma(t)}(z), \qquad z \in U(\gamma(t)). \tag{4.23}$$

$F(z, t)$ is a power series with center $\gamma(t)$ and absolutely convergent on the disk $U(\gamma(t))$. Since the radius $r(\gamma(t))$ of $U(\gamma(t))$ equals the distance between the point $\gamma(t)$ and $\mathbb{C} - \Omega$, we have

$$r(\gamma(t)) \geqq \rho > 0, \tag{4.24}$$

where ρ denotes the distance between $\{\gamma(t): 0 \leqq t \leqq b\}$ and $\mathbb{C} - \Omega$. $\{F(z,t): 0 \leqq t \leqq b\}$ is the analytic continuation of $F_0(z)$ along the curve γ. To prove this it suffices to show that for sufficiently small $\varepsilon > 0$, $|t - s| < \varepsilon$ implies that $F(z, t)$ is the direct analytic continuation of $F(z, s)$. Pick $\varepsilon > 0$ such that

$$|\gamma(u) - \gamma(s)| < \rho \qquad \text{if } |u - s| < \varepsilon$$

and s and t such that $|s - t| < \varepsilon$. For an arbitrary $z \in U(\gamma(t)) \cap U(\gamma(s))$, we have, by (4.23) and (4.22),

$$F(z, t) = \int_0^s f(\gamma(u))\gamma'(u)\,du + \int_s^t f(\gamma(u))\gamma'(u)\,du + \int_{\gamma(t)}^{z} f(z)\,dz.$$

Since $|\gamma(u) - \gamma(s)| < \rho$ if $s \leqq u \leqq t$, $\gamma(u) \in U(\gamma(s))$ by (4.24). Hence, letting $\int_{\gamma(t)}^{z} f(z)\,dz$ be the integral along the segment connecting $\gamma(t)$ and z, $\int_s^t f(\gamma(u))\gamma'(u)\,du + \int_{\gamma(t)}^{z} f(z)\,dz$ equals the integral of $f(z)$ along the segment in $U(\gamma(s))$ connecting $\gamma(s)$ and z. Therefore, by (4.22),

$$\int_s^t f(\gamma(u))\gamma'(u)\,du + \int_{\gamma(t)}^{z} f(z)\,dz = \int_{\gamma(s)}^{z} f(z)\,dz = F_{\gamma(s)}(z).$$

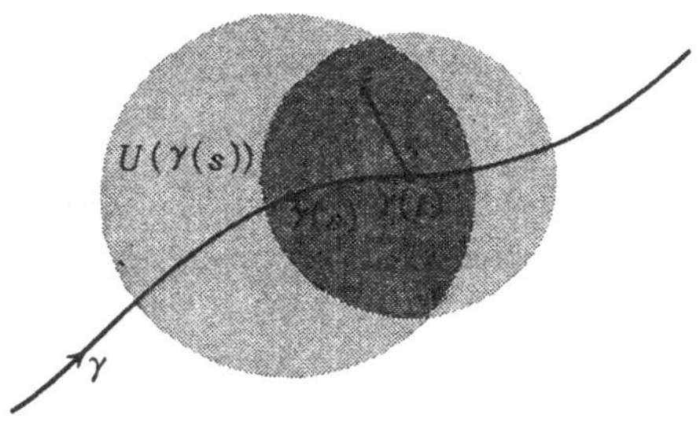

Fig. 4.15

Hence we have on $U(\gamma(t)) \cap U(\gamma(s))$

$$F(z, t) = \int_0^s f(\gamma(u))\gamma'(u)\,du + F_{\gamma(s)}(z) = F(z, s),$$

proving that $F(z, t)$ is the direct analytic continuation of $F(z, s)$.

Let γ: $t \to \gamma(t)$, $0 \leqq t \leqq \tau$, be a piecewise smooth curve in Ω connecting $c_0 = \gamma(0)$ and $z = \gamma(\tau)$. Select b, $0 < b < \tau$, such that $\gamma(t) \in U(\gamma(b))$ if $b \leqq t \leqq \tau$. The curve γ is divided by b into two curves; α: $t \to \gamma(t)$, $0 \leqq t \leqq b$, and β: $t \to \gamma(t)$, $b \leqq t \leqq \tau$. That is, $\gamma = \alpha \cdot \beta$. Under these circumstances

$$\int_\gamma f(\zeta)\,d\zeta = \int_\alpha f(\zeta)\,d\zeta + \int_\beta f(\zeta)\,d\zeta,$$

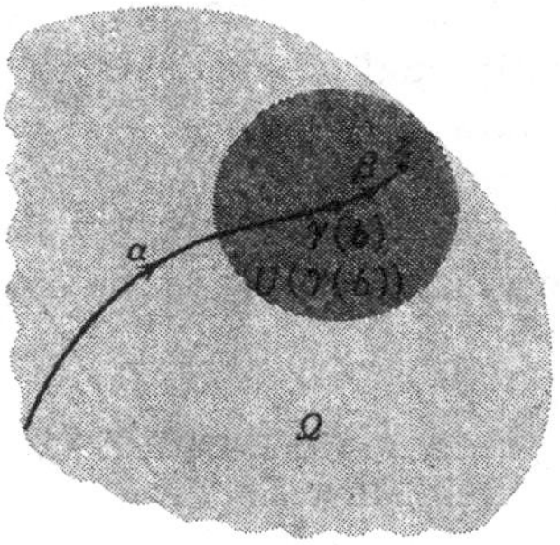

Fig. 4.16

hence, by (4.22),

$$\int_\beta f(\zeta)\,d\zeta = F_{\gamma(b)}(z).$$

Therefore

$$\int_\gamma^z f(\zeta)\,d\zeta = \int_0^b f(\gamma(t))\gamma'(t)\,dt + F_{\gamma(b)}(z). \tag{4.25}$$

Now we consider z as a variable varying over $U(\gamma(b))$ and allow β to vary with z in $U(\gamma(b))$ while α remains fixed. Then, by (4.23),

$$\int_\gamma^z f(\zeta)\,d\zeta = F(z, b), \qquad z \in U(\gamma(b)),$$

i.e., $\int_\gamma^z f(\zeta)\,d\zeta$ is the holomorphic function $F(z, b)$ which is the result of the analytic continuation of the power series $F_0(z)$ along the curve α: $t \to \gamma(t)$,

$0 \leqq t \leqq b$. Replacing $z = \gamma(\tau)$ by $w = \gamma(\tau)$ and representing the variable by z, $F(z, \tau)$ is the direct analytic continuation of $F(z, b)$, hence $F(z, \tau)$ is the power series expansion of $F(z, b)$ about $\gamma(\tau)$. Hence $\int_\gamma^w f(\zeta)\,d\zeta = F(w, b)$ equals the value $F(w, \tau)$ of the power series $F(z, \tau)$ at its center $w = \gamma(\tau)$. We have proved.

Theorem 4.11. Let $f(z)$ be a holomorphic function on Ω and put $F_0(z) = \int_{c_0}^z f(z)\,dz$, $z \in U(c_0) \subset \Omega$. If $z \in \Omega$ and γ: $t \to \gamma(t)$, $0 \leqq t \leqq \tau$, is a piecewise smooth curve connecting $c_0 = \gamma(0)$ and $z = \gamma(\tau)$ in Ω, then

$$F(z) = \int_\gamma^z f(z)\,dz$$

equals the value of the power series obtained by analytic continuation of $F_0(z)$ along γ at its center $z = \gamma(\tau)$. We express this fact by saying that $F(z)$ is the analytic continuation of $F_0(z)$ along γ.

$F(z)$ is a freely analytically continuable and complete analytic function on Ω. To prove this, we first show that $F_0(z)$ is analytically continuable along each curve λ: $t \to \lambda(t)$, $0 \leqq t \leqq b$, in Ω with initial point $c_0 = \lambda(0)$. This is clear, if λ is piecewise smooth, from Theorem 4.11. In order to reduce the case in which λ is not piecewise smooth to the case in which λ is piecewise smooth, let us denote for each $s \in [0, b]$ the curve λ: $t \to \lambda(t)$, $0 \leqq t \leqq s$, by λ^s, that is, λ^s: $t \to \lambda^s(t) = \lambda(t)$. In the proof of Theorem 4.8 it was shown that there exists a polygonal line γ_s such that $\gamma_s \simeq \lambda^s$. Since γ_s is a piecewise smooth curve in Ω connecting c_0 and $\lambda^s(s) = \lambda(s)$, $F_0(z)$ is analytically continuable along γ_s and the result of the analytic continuation is given by

$$F(z, s) = \int_{\gamma_s} f(z)\,dz + F_{\lambda(s)}(z)$$

by (4.23). By Theorem 4.3, $F(z, s)$ is uniquely determined by the homotopy class $[\gamma_s] = [\lambda^s]$ of γ_s in Ω. Hence, for a given curve λ, $F(z, s)$ is uniquely determined by s and independent of the choice of the polygonal line γ_s. Obviously, $F(z, 0) = F_0(z)$. In order to prove that $\{F(z, s): 0 \leqq s \leqq b\}$ is the analytic continuation of $F_0(z)$ along λ we pick an $\varepsilon > 0$ such that

$$|\lambda(t) - \lambda(s)| < \rho \qquad \text{if } |t - s| < \varepsilon,$$

where ρ was defined following (4.24). Let μ be the segment connecting $\lambda(s)$ and $\lambda(u)$ for $u \in (s, s+\varepsilon)$ and put $\gamma_u = \gamma_s \cdot \mu$. Since $\gamma_s \simeq \lambda^s$ in Ω and the segment μ and the curve λ_s^u: $t \to \lambda(t)$, $s \leqq t \leqq u$, are both in $U(\lambda(s)) \subset \Omega$, we have

$$\mu \simeq \lambda_s^u \qquad \text{in } \Omega,$$

hence $\gamma_u \simeq \lambda^u$ by Theorem 4.6(1). Therefore, the power series $F(z, u)$

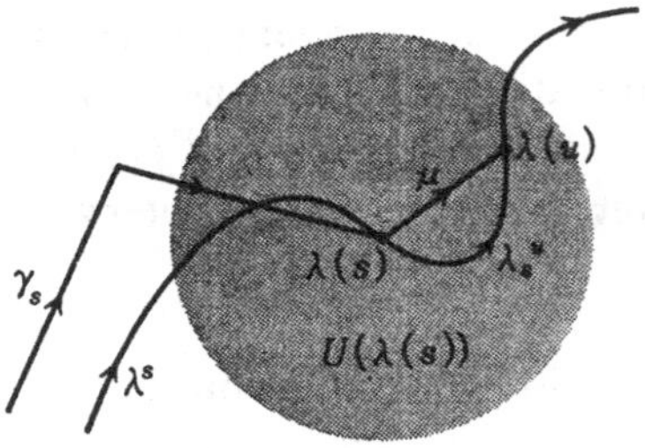

Fig. 4.17

obtained by analytic continuation of $F_0(z)$ along $\gamma_u = \gamma_s \cdot \mu$ can be obtained by analytic continuation of $F(z,s)$ along the segment μ. Since $F(z,s)$ is holomorphic on the disk $U(\lambda(s))$, $F(z,u)$ is the power series expansion of $F(z,s)$ about $\lambda(u)$. Therefore $F(z,u) = F(z,s)$ on $U(\lambda(u)) \cap U(\lambda(s))$ and $\{F(z,s): 0 \leqq s \leqq b\}$ is the analytic continuation of $F_0(z)$ along γ. The result of the analytic continuation can be represented as

$$F(z,b) = \int_{\gamma_b} f(z)\,dz + \int_{\lambda(b)}^{z} f(z)\,dz, \qquad z \in U(\lambda(b)),$$

where $\gamma_b \simeq \lambda$ is a piecewise smooth curve (actually a polygonal line). Since the terminal point of γ_b is $\lambda(b)$ we can piece together γ_b and the segment joining $\lambda(b)$ with z. Calling this curve γ we have

$$F(z,b) = \int_{\gamma}^{z} f(z)\,dz.$$

Hence $F(z) = \int_{\gamma}^{z} f(z)\,dz$ is freely analytically continuable and complete in Ω.

By (4.25)

$$\frac{d}{dz}\int_{\gamma}^{z} f(z)\,dz = f(z). \tag{4.26}$$

In general, $F(z)$ is a multivalued function and its value at z depends on z and on the curve γ connecting c_0 and z. However, if γ_1 is another curve connecting c_0 and z in Ω such that $\gamma_1 \simeq \gamma$ in Ω, then, by Theorem 4.3,

$$\int_{\gamma_1}^{z} f(z)\,dz = \int_{\gamma}^{z} f(z)\,dz. \tag{4.27}$$

Since for each curve γ there exists a smooth curve γ_1 such that $\gamma \simeq \gamma_1$, we may assume from the start that the path of integration γ is smooth. By (4.25), the branch $F_{c,k}(z)$ of the analytic function $F(z) = \int_{\gamma}^{z} f(z)\,dz$ at $c \in \Omega$ is given by

$$F_{c,k}(z) = \int_{\gamma_k} f(z)\,dz + F_c(z), \qquad k = 1, 2, 3, \ldots, \tag{4.28}$$

where γ_k is a piecewise smooth curve in Ω connecting c_0 and c and $F_c(z)$ is a power series with center c such that $F_c(c) = 0$. Putting $\beta = \gamma_k^{-1}\gamma_j$, β is a closed curve with base point c and we have

$$\int_\beta f(z)\,dz = \int_{\gamma_j} f(z)\,dz - \int_{\gamma_k} f(z)\,dz.$$

Hence

$$F_{c,j}(z) - F_{c,k}(z) = \int_\beta f(z)\,dz, \tag{4.29}$$

that is, the difference between two different branches of $F(z)$ at a point $c \in \Omega$ is constant.

If Ω is simply connected, then $F(z)$ is a single-valued holomorphic function on Ω by Theorem 4.7. In this case we write $F(z) = \int_{c_0}^{z} f(z)\,dz$. If Ω is not simply connected, then the branch of $F(z)$ over a simply connected subregion D of Ω is given by

$$F_{D,k}(z) = \int_c^z f(z)\,dz + \int_{\gamma_k} f(z)\,dz, \tag{4.30}$$

where c is a point in D and γ_k a curve in Ω connecting c_0 and c.

As an example of analytic continuation by integrals we want to consider the logarithmic function, as discussed in Section 4.1a. Let $\Omega = \mathbb{C}^* = \mathbb{C} - \{0\}$, $f(z) = 1/z$, $c_0 = 1$, and Ω be a smooth curve in $\mathbb{C}^*$ connecting 1 and z. We define the logarithmic function by

$$\log z = \int_\gamma^z \frac{dz}{z}. \tag{4.31}$$

We first prove that this definition is equivalent to the definition given in Section 4.1a. Putting $\gamma\colon t \to \gamma(t)$, $0 \leqq t \leqq \tau$, with $\gamma(0) = 1$ and $\gamma(\tau) = z$ we have

$$\int_\gamma^z \frac{dz}{z} = \int_0^\tau \frac{\gamma'(t)}{\gamma(t)}\,dt.$$

Writing $\gamma(t) = \rho(t)e^{i\theta(t)}$ with $\rho(t) = |\gamma(t)| > 0$ and $\theta(t)$ a continuously differentiable, real function of t with $\rho(0) = 1$ and $\theta(0) = 0$, we have

$$\gamma'(t) = \rho'(t)e^{i\theta(t)} + \rho(t)e^{i\theta(t)}\,i\theta'(t),$$

hence

$$\frac{\gamma'(t)}{\gamma(t)} = \frac{\rho'(t)}{\rho(t)} + i\theta'(t) = \frac{d}{dt}\log\rho(t) + i\theta'(t)$$

and therefore

$$\int_0^\tau \frac{\gamma'(t)}{\gamma(t)} dt = \log \rho(\tau) + i\theta(\tau).$$

Putting $z = |z|e^{i\theta}$, we conclude from $\rho(\tau)e^{i\theta(\tau)} = \gamma(\tau) = |z|e^{i\theta}$ that $\rho(\tau) = |z|$ and $\theta(\tau) = \theta + 2n\pi$, n an integer. Hence

$$\log z = \int_\gamma^z \frac{dz}{z} = \log|z| + i\theta + 2n\pi i, \qquad n \text{ an integer,}$$

which is just (4.1).

Defining $\log z$ by (4.31) we get at once from (4.26)

$$\frac{d}{dz} \log z = \frac{1}{z}.$$

Since $d(ze^{-\log z})/dz = 0$ we have $ze^{-\log z} = 1 \cdot e^{-\log 1} = 1$, that is, $w = \log z$ is the inverse function of $z = e^w$.

For an arbitrary $z = |z|e^{i\theta} \neq 0$, $-\pi < \theta \leqq \pi$, $\log|z| + i\theta$ is called the principal value of $\log z$ and denoted by $\operatorname{Log} z$. The power series expansion of $1/z$ about 1 is

$$\frac{1}{z} = \frac{1}{1 + (z-1)} = \sum_{n=0}^{\infty} (-1)^n (z-1)^n$$

and the radius of convergence of this power series equals 1. Hence

$$\operatorname{Log} z = \int_1^z \frac{dz}{z} = \sum_{n=1}^{\infty} \frac{(-1)^{n-1}}{n} (z-1)^n, \qquad |z-1| < 1.$$

Therefore, the branches of $\log z$ at the point 1 are given by

$$\log z = 2k\pi i + \sum_{n=1}^{\infty} \frac{(-1)^{n-1}}{n} (z-1)^n, \qquad k = 0, \pm 1, \pm 2, \pm 3, \ldots.$$

If $D = \{z: z = |z|e^{i\theta} \neq 0, -\pi < \theta < \pi\}$ is the region obtained from $\mathbb{C}^*$ by deleting the negative real axis, D is a simply connected subregion of $\mathbb{C}^*$ and the branches of $\log z$ over D are given by

$$\log z = \operatorname{Log} z + 2k\pi i, \qquad k = 0, \pm 1, \pm 2, \pm 3, \ldots.$$

Example 4.4. We want to determine

$$\int_0^{+\infty} \frac{x^{a-1}}{x+1} dx, \qquad 0 < a < 1,$$

using a (multivalued) analytic function.

Put $D = \{z: z = |z|e^{i\theta}, \varepsilon < |z| < R, 0 < \theta < 2\pi\}$, where $0 < \varepsilon < 1 < R$. Consider, over the simply connected region D, the branch $\log|z| + i\theta$,

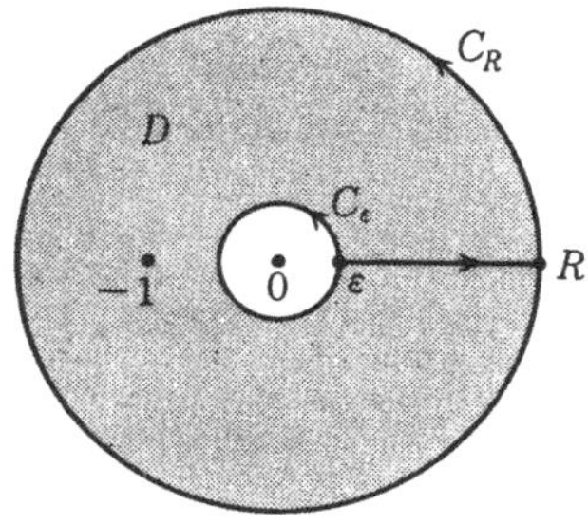

Fig. 4.18

$0 < \theta < 2\pi$, of $\log z$. Then

$$f(z) = \frac{z^{a-1}}{z+1}, \qquad z^{a-1} = \exp[(a-1)\log z],$$

is a single-valued holomorphic function on $D - \{-1\}$. The boundary of D is given by $[\varepsilon, R] \cup C_R \cup C_\varepsilon$ where C_ε and C_R are circles with center 0 and radius ε and R, respectively. The function $f(z)$ can be extended continuously to this boundary, but assumes two different values $f(x + i0)$ and $f(x - i0)$ at each $x \in [\varepsilon, R]$:

$$f(x+i0) = \lim_{\delta \to +0} f(x+i\delta) = \frac{x^{a-1}}{x+1},$$

$$f(x-i0) = \lim_{\delta \to +0} f(x-i\delta) = \frac{x^{a-1}\exp[(a-1)2\pi i]}{x+1}.$$

By dividing the closed region $[D]$ suitably into a number of cells, we see that the residue theorem (Theorem 2.5) takes the following form:

$$\int_\varepsilon^R f(x+i0)\,dx + \int_{C_R} f(z)\,dz - \int_\varepsilon^R f(x-i0)\,dx - \int_{C_\varepsilon} f(z)\,dz$$
$$= 2\pi i \,\mathrm{Res}_{z=-1}[f(z)].$$

By (2.63)

$$\mathrm{Res}_{z=-1}[f(z)] = \exp[(a-1)\pi i] = -e^{a\pi i},$$

further

$$\int_\varepsilon^R [f(x+i0) - f(x-i0)]\,dx = (1 - e^{2a\pi i}) \int_\varepsilon^R \frac{x^{a-1}}{x+1}\,dx.$$

From

$$\left|\int_{C_R} f(z)\,dz\right| < \frac{R^{a-1}}{R-1} 2\pi R = \frac{2\pi R^a}{R-1} \to 0, \qquad R \to +\infty,$$

$$\left|\int_{C_\varepsilon} f(z)\,dz\right| < \frac{\varepsilon^{a-1}}{1-\varepsilon} 2\pi\varepsilon = \frac{2\pi\varepsilon^a}{1-\varepsilon} \to 0,\ \varepsilon \to +0,$$

we conclude

$$(1 - e^{2a\pi i}) \int_0^{+\infty} \frac{x^{a-1}}{x+1} dx = -2\pi i e^{a\pi i}.$$

Hence

$$\int_0^{+\infty} \frac{x^{a-1}}{x+1} dx = \frac{\pi}{\sin a\pi}, \qquad 0 < a < 1.$$

4.4 Cauchy's Theorem (continued)

Let $f(z)$ be a holomorphic function of z defined on the region Ω, let $U(c_0)$ be a disk with center c_0 contained in Ω and put $F_0(z) = \int_{c_0}^{z} f(z)\,dz$ for $z \in U(c_0)$. If, for arbitrary $z \in \Omega$, γ is a piecewise smooth curve in Ω connecting c_0 and z, then $F(z) = \int_{\gamma}^{z} f(z)\,dz$ can be considered as the analytic continuation of $F_0(z)$ along γ by Theorem 4.11. We want to consider Cauchy's Theorem against this background.

Let γ be a closed curve with base point c_0 in Ω. The result $F_1(z)$ of the analytic continuation of $F_0(z)$ along γ is a power series with center c_0 and

$$\int_{\gamma} f(z)\,dz = F_1(c_0).$$

If $\gamma \simeq 0$ in Ω, then $F_1(z)$ and $F_0(z)$ coincide by the corollary to Theorem 4.3, hence $F_1(c_0) = F_0(c_0) = 0$. Therefore we have

Theorem 4.12. Let $f(z)$ be a holomorphic function on Ω and let γ be a piecewise smooth closed curve in Ω. If $\gamma \simeq 0$ in Ω, then

$$\int_{\gamma} f(z)\,dz = 0.$$

This theorem is called the *homotopy variant of Cauchy's Theorem.* If Ω is simply connected, then each closed curve is homotopic with 0 in Ω, hence $\int_{\gamma} f(z)\,dz = 0$ for each piecewise smooth closed curve. This is also clear from Theorem 4.7.

Next we want to introduce the concept of homology. If $\gamma = \gamma_1 \cdot \gamma_2 \cdot \cdots \cdot \gamma_m$ is the curve obtained by piecing together the piecewise smooth curves $\gamma_1, \gamma_2, \ldots, \gamma_m$ in the given order, then we have, for each continuous function $f(z)$ defined on Ω,

$$\int_{\gamma} f(z)\,dz = \int_{\gamma_1} f(z)\,dz + \int_{\gamma_2} f(z)\,dz + \cdots + \int_{\gamma_m} f(z)\,dz.$$

The right-hand side of this equality is independent of the order of the curves

$\gamma_1, \gamma_2, \ldots, \gamma_m$, hence $\gamma = \gamma_1 \cdot \gamma_2 \cdot \cdots \cdot \gamma_m$, as a path of integration, can be written as

$$\gamma = \gamma_1 + \gamma_2 + \cdots + \gamma_m. \tag{4.32}$$

(Compare Section 2.1a).

Extending this a bit further, let us consider linear combinations

$$\gamma = \sum_{k=1}^{m} n_k \gamma_k = n_1 \gamma_1 + n_2 \gamma_2 + \cdots + n_m \gamma_m, \qquad n_k \text{ an integer}, \tag{4.33}$$

of a finite number of curves $\gamma_1, \gamma_2, \ldots, \gamma_m$ with integer coefficients. Such a linear combination γ is called a *one-dimensional chain* or a 1-chain for short. Since (4.33) is a generalization of (4.32), we identify the curve $\gamma_1 \cdot \gamma_2 \cdot \cdots \cdot \gamma_m$ with the 1-chain $\gamma_1 + \gamma_2 + \ldots + \gamma_m$:

$$\gamma_1 \cdot \gamma_2 \cdot \cdots \cdot \gamma_m = \gamma_1 + \gamma_2 + \cdots + \gamma_m. \tag{4.34}$$

Further, we define for an arbitrary curve γ

$$\gamma^{-1} = -\gamma \tag{4.35}$$

and finally we denote the curve $\delta: t \to \delta(t) = c$, where c is a fixed point by 0:

$$\delta = 0. \tag{4.36}$$

It is easily verified that the collection of 1-chains in Ω can be made into a group, the *one-dimensional chain group* of Ω, denoted by $C_1(\Omega)$.

Similarly, a linear combination of points $c_1, c_2, \ldots, c_m \in \Omega$ with integer coefficients

$$\sum_{k=1}^{m} n_k c_k = n_1 c_1 + n_2 c_2 + \cdots + n_m c_m$$

is called a *0-chain* in Ω. The collection of all 0-chains can be made into a group, called the *zero-dimensional chain group* and denoted by $C_0(\Omega)$.

If γ is a curve with initial point c_0 and terminal point c_1, then the 0-chain $c_1 - c_0$ is called the *boundary* of γ. The boundary of γ is denoted by $\partial\gamma$:

$$\partial\gamma = c_1 - c_0.$$

If $\gamma = \sum_{k=1}^{m} n_k \gamma_k$ is an arbitrary 1-chain, its boundary is defined by

$$\partial\gamma = \sum_{k=1}^{m} n_k \partial\gamma_k. \tag{4.37}$$

We have to prove that this definition is consistent with (4.34), (4.35) and (4.36). For $\gamma = \gamma_1 \cdot \gamma_2 \cdot \cdots \cdot \gamma_m$ let c_0 be the initial point of γ_1 (and γ), let c_k be the terminal point of γ_k (and the initial point of γ_{k+1}) and let c_m be the terminal point of γ_m (and γ). Then

$$\sum_{k=1}^{m} \partial\gamma_k = \sum_{k=1}^{m} (c_k - c_{k-1}) = c_m - c_0 = \partial\gamma.$$

If γ is an arbitrary curve with initial point c_0 and terminal point c_1, then γ^{-1} has c_0 as its terminal point and c_1 as its initial point, hence

$$\partial\gamma^{-1} = c_0 - c_1 = -\partial\gamma.$$

For the curve δ of (4.36) we obviously have $\partial\delta = 0$. The map

$$\partial\colon \gamma \to \partial\gamma$$

from the chain group $C_1(\Omega)$ into the chain group $C_0(\Omega)$ is a homomorphism.

A 1-chain $\gamma \in C_1(\Omega)$ satisfying $\partial\gamma = 0$ is called a *1-cycle* in Ω. If γ is a curve, we have $\partial\gamma = c_1 - c_0$, where c_0 is the initial point and c_1 the terminal point of γ, hence $\partial\gamma = 0$ if and only if $c_0 = c_1$, that is, if and only if γ is a closed curve.

Theorem 4.13. An arbitrary 1-cycle in Ω can be written as a linear combination of a finite number of closed curves γ_k in Ω with integer coefficients:

$$\gamma = \sum_{k=1}^{m} n_k\gamma_k.$$

Proof: If, for some k, $n_k = 0$, then $n_k\gamma_k = 0$; if $n_k < 0$, then $n_k\gamma_k = -n_k\gamma_k^{-1} = |n_k|\gamma_k^{-1}$, hence we may assume that

$$\gamma = \sum_{k=1}^{m} n_k\gamma_k, \qquad n_k > 0. \tag{4.38}$$

We will prove the theorem by induction on the sum of coefficients $n = \sum_{k=1}^{m} n_k$.

(1) If $n = 1$, then $\gamma = \gamma_1$ and $\partial\gamma = 0$, hence γ is a closed curve.

(2) Let us assume that the theorem holds for all n with $n \leqq \nu - 1$. We will prove the truth of the theorem for $n = \nu$. If $\partial\gamma_1 = 0$, then γ_1 is a closed curve and

$$\gamma - n_1\gamma_1 = \sum_{k=2}^{m} n_k\gamma_k$$

is a 1-cycle. Since $\sum_{k=2}^{m} n_k = \nu - n_1 \leqq \nu - 1$, $\gamma - n_1\gamma_1$ is a linear combination of a finite number of closed curves in Ω. Therefore, γ is a linear combination of a finite number of closed curves in Ω by the induction hypothesis.

If $\partial\gamma_1 = c_1 - c_0 \neq 0$, we conclude from $\sum_{k=1}^{m} n_k\partial\gamma_k = 0$ that for at least one of the curves $\gamma_2, \gamma_3, \ldots, \gamma_m$, say γ_2, we have $\partial\gamma_2 = c_0 - c_2$, $c_2 \neq c_0$. Putting $\gamma_0 = \gamma_2 \cdot \gamma_1$ we have

$$\gamma = \gamma_0 + (n_1 - 1)\gamma_1 + (n_2 - 1)\gamma_2 + n_3\gamma_3 + \cdots + n_m\gamma_m$$

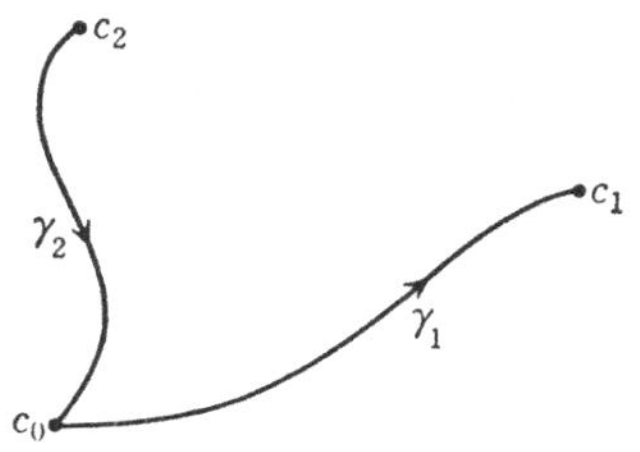

Fig. 4.19

and the sum of the coefficients on the right equals $\nu - 1$. Hence γ is a linear combination of a finite number of closed curves in Ω with integer coefficients by the induction hypothesis.

The above representation of a 1-cycle in Ω as a linear combination of closed curves in Ω is not unique.

Definition 4.5. If the 1-cycle γ in Ω can be represented as a linear combination

$$\gamma = \sum_{k=1}^{m} n_k \gamma_k, \qquad \gamma_k \simeq 0,\ n_k \text{ integers}, \tag{4.39}$$

of closed curves γ_k, homotopic to 0, then γ is said to be *homologous to* 0 *in* Ω, written $\gamma \sim 0$. Two 1-cycles β and γ in Ω are called *homologous in* Ω if $\beta - \gamma \sim 0$ in Ω, written $\beta \sim \gamma$.

If $\gamma = \sum_{k=1}^{m} n_k \gamma_k$ is a 1-chain such that all curves γ_k are piecewise smooth, then γ is called piecewise smooth. Now let $\gamma = \sum_{k=1}^{m} n_k \gamma_k$ be a piecewise smooth 1-chain and define the integral of the function $f(z)$ (continuous on Ω) along γ by

$$\int_\gamma f(z)\,dz = \sum_{k=1}^{m} n_k \int_{\gamma_k} f(z)\,dz, \qquad \gamma = \sum_{k=1}^{m} n_k \gamma_k. \tag{4.40}$$

This definition is consistent with (4.34), (4.35), and (4.36), that is, if $\gamma = \gamma_1 \cdot \gamma_2 \cdot \cdots \cdot \gamma_m$, where all γ_k are piecewise smooth curves, then

$$\int_\gamma f(z)\,dz = \sum_{k=1}^{m} \int_{\gamma_k} f(z)\,dz \tag{4.41}$$

and if γ is a piecewise smooth curve, then

$$\int_{\gamma^{-1}} f(z)\,dz = -\int_\gamma f(z)\,dz \tag{4.42}$$

while obviously $\int_\delta f(z)\,dz = 0$.

Theorem 4.14 (*Cauchy's Theorem*). Let $f(z)$ be a holomorphic function defined on the region Ω and let γ be a piecewise smooth 1-cycle in Ω. If $\gamma \sim 0$ in Ω, then

$$\int_\gamma f(z)\,dz = 0.$$

Proof. Since $\gamma \sim 0$, γ can be written as $\gamma = \sum_{k=1}^{q} n_k\gamma_k$, where γ_k is a closed curve such that $\gamma_k \simeq 0$ in Ω, $k = 1, \ldots, q$. Since γ is piecewise smooth, γ can also be represented as $\gamma = \sum_{h=1}^{p} m_h\lambda_h$, where λ_h is a smooth curve, $h = 1, \ldots, p$. Omitting one-point curves from the outset the meaning of the equality

$$\sum_{h=1}^{p} m_h\lambda_h = \sum_{k=1}^{q} n_k\gamma_k$$

is that $\sum_{h=1}^{p} m_h\lambda_h$ and $\sum_{k=1}^{q} n_k\gamma_k$ can be transformed into the same 1-cycle $\sum_{j=1}^{r} l_j\beta_j$ by using (4.34) and (4.35). To prove this in detail, first write γ_k and λ_k suitably as: $\gamma_k = \gamma_{k_1}\cdot\gamma_{k_2}\cdot\cdots\cdot\gamma_{k_\sigma}$ and $\lambda_h = \lambda_{h_1}\cdot\lambda_{h_2}\cdot\cdots\cdot\lambda_{h_\tau}$, $\sigma = \sigma(k)$ and $\tau = \tau(h)$, and let $\beta_1, \beta_2, \ldots, \beta_j$ represent all different curves among the curves $\gamma_{k_\nu}, \lambda_{h_\mu}$; $\nu = 1, 2, \ldots, \sigma(k)$, $\mu = 1, 2, \ldots, \tau(h)$, $k = 1, 2, \ldots, q$ and $h = 1, 2, \ldots, p$. If there is a pair β_i, β_j such that $\beta_i = \beta_j^{-1}$, then we omit either β_i or β_j from the sequence $\beta_1, \beta_2, \ldots, \beta_j, \ldots$. We write the resulting sequence as $\beta_1, \beta_2, \ldots, \beta_j, \ldots, \beta_\omega$. Each curve $\gamma_{k_\nu}, \lambda_{h_\mu}$ equals either β_j or β_j^{-1} for some j, hence γ_k and λ_h are linear combinations with integer coefficients of $\beta_1, \beta_2, \ldots, \beta_\omega$. Hence, $\sum_{h=1}^{p} m_h\lambda_h$ and $\sum_{k=1}^{q} n_k\gamma_k$ can be written as linear combinations with integer coefficients of $\beta_1, \beta_2, \ldots, \beta_\omega$ and both expressions are identical.

If all closed curves γ_k are piecewise smooth then

$$\int_\gamma f(z)\,dz = \sum_{k=1}^{q} n_k \int_{\gamma_k} f(z)\,dz$$

by (4.41) and (4.42) and the theorem is a direct consequence of Theorem 4.12.

If not all γ_k are piecewise smooth, it suffices to show that there exist piecewise smooth, closed curves $\hat\gamma_k$ such that $\hat\gamma_k \simeq \gamma_k$ in Ω and $\gamma = \sum_{k=1}^{q} n_k\hat\gamma_k$. Since not all γ_k are piecewise smooth, there are curves among the β_j which are not piecewise smooth. As in the proof of Theorem 4.8, for each curve β in Ω, there exists a polygonal line which is homotopic to β in Ω. Let $\hat\beta_j$ denote a polygonal line such that $\beta_j \simeq \hat\beta$ in Ω if β_j is not piecewise smooth and put $\hat\beta_j = \beta_j$ if β_j is piecewise smooth. We put $\hat\gamma_{k_\nu} = \hat\beta_j^{\pm 1}$ if $\gamma_{k_\nu} = \beta_j^{\pm 1}$ and $\hat\gamma_k = \hat\gamma_{k_1}\cdot\hat\gamma_{k_1}\cdot\cdots\cdot\hat\gamma_{k_\sigma}$, and similarly $\hat\lambda_{h_\mu} = \hat\beta_j^{\pm 1}$ if $\lambda_{h_\mu} = \beta_j^{\pm 1}$

and $\hat{\lambda}_h = \hat{\lambda}_{h_1} \cdot \hat{\lambda}_{h_2} \cdot \cdots \cdot \hat{\lambda}_{h_\tau}$. From

$$\sum_{h=1}^{p} m_h \lambda_h = \sum_{k=1}^{q} n_k \gamma_k$$

we get

$$\sum_{h=1}^{p} m_h \hat{\lambda}_h = \sum_{k=1}^{q} n_k \hat{\gamma}_k.$$

Since λ_h is a smooth curve, the λ_{h_μ} are smooth too, hence $\hat{\lambda}_{h_\mu} = \lambda_{h_\mu}$ and $\hat{\lambda}_h = \lambda_h$. Therefore

$$\gamma = \sum_{h=1}^{p} m_h \lambda_h = \sum_{k=1}^{q} n_k \hat{\gamma}_k.$$

Since $\gamma_k = \gamma_{k_1} \cdot \gamma_{k_2} \cdot \cdots \cdot \gamma_{k_\sigma}$ is a closed curve in Ω and $\hat{\gamma}_{k_\nu} \simeq \gamma_{k_\nu}$ in Ω, $\hat{\gamma}_k = \hat{\gamma}_{k_1} \cdot \hat{\gamma}_{k_2} \cdot \cdots \cdot \hat{\gamma}_{k_\sigma}$ is a closed curve in Ω, and $\hat{\gamma}_k \simeq \gamma_k \simeq 0$ by Theorem 4.6(1).

Cauchy's Theorem as stated in Theorem 2.2 can be considered as a special case of Theorem 4.14. This can be seen by rewriting the proof of Theorem 2.2 using homotopy. Remember that $[D]$, occurring in Theorem 2.2, was a bounded closed region, the boundary, C, of which consisted of a finite number of mutually disjoint piecewise smooth Jordan curves and that $f(z)$ was a holomorphic function defined on a region $\Omega \supset [D]$. Denoting the cellular decomposition of $[D]$, as obtained in the proof of Theorem 2.1, by $\{\Gamma_\lambda(K)\colon \lambda = 1, 2, \ldots, \mu\}$, we have $\partial\Gamma_\lambda(K) = \sum_\nu C_{\lambda\nu}$ or $\partial\Gamma_\lambda(K) = C_\lambda + \sum_\nu C_{\lambda\nu}$, $C_{\lambda\nu} = -C_{\nu\lambda}$ and

$$\sum_{\lambda=1}^{\mu} \partial\Gamma_\lambda(K) = \sum_\lambda C_\lambda = C$$

(cf. Section 2.3a).

The boundary $\partial\Gamma_\lambda(K)$ of each cell is a piecewise smooth Jordan curve and $\partial\Gamma_\lambda(K) \simeq 0$ in Ω. Hence the 1-cycle C is homologous to 0 in Ω and so $\int_C f(z)\,dz = 0$ by Theorem 4.14.

If $f(z)$ is a holomorphic function defined on the region Ω, c_0 a fixed point in Ω and γ a curve in Ω connecting c_0 and an arbitrary point $z \in \Omega$, then $F(z) = \int_\gamma^z f(z)\,dz$ is the analytic continuation along γ of the power series $F_0(z) = \int_{c_0}^z f(z)\,dz$ about c_0, $z \in U(c_0)$, by Theorem 4.11. This result can be extended to multivalued functions $f(z)$ as discussed below.

Let $f(z)$ be a freely continuable, complete analytic function on the region Ω and let $f_{c,k}(z)$ represent the branches of $f(z)$ at $c \in \Omega$, $k = 1, 2, 3, \ldots$. Each branch $f_{c,k}(z)$ is holomorphic on the greatest disk $U(c)$ with center c and contained in Ω. In order to define the integral

$$\int_{\gamma} f(z)\,dz = \int_a^b f(\gamma(t))\gamma'(t)\,dt,$$

where γ: $t \to \gamma(t)$, $a \leqq t \leqq b$, is a piecewise smooth curve in Ω, we have to determine which of the values $f_{\gamma(t),k}(\gamma(t))$, $k = 1, 2, 3, \ldots$, assumed by $f(z)$ at $\gamma(t)$ we want to consider. For this purpose, select a branch $f_{\gamma(a),k}(z)$ at the initial point $\gamma(a)$ of γ and let $f = \{f(z, t)\colon a \leqq t \leqq b\}$ be the analytic continuation along γ of $f_{\gamma(a),k}(z)$. All $f(z, t)$ are branches of $f(z)$ at $\gamma(t)$ and holomorphic on the disks $U(\gamma(t))$. Pick $\varepsilon > 0$ such that

$$\gamma(t) \in U(\gamma(s)) \qquad \text{if } |t - s| < \varepsilon. \tag{4.43}$$

Then, if $|t - s| < \varepsilon$, we have $f(z, t) = f(z, s)$ on $U(\gamma(t)) \cap U(\gamma(s))$, hence $f(\gamma(t), t) = f(\gamma(t), s)$ and $f(\gamma(t), t)$ is a continuous function of t. Therefore, we define

$$\int_{\gamma} f(z)\,dz = \int_a^b f(\gamma(t), t)\gamma'(t)\,dt.$$

The value of this integral of course depends on the choice of the branch $f(z, a) = f_{\gamma(a),k}(z)$ at the initial point $\gamma(a)$ of γ.

Now let γ: $t \to \gamma(t)$, $0 \leqq t \leqq \tau$, be a piecewise smooth curve in Ω connecting c_0 with an arbitrary point $z \in \Omega$ and consider the integral

$$F(z) = \int_{\gamma}^{z} f(z)\,dz = \int_0^{\tau} f(\gamma(t), t)\gamma'(t)\,dt$$

where $\{f(z, t)\colon 0 \leqq t \leqq \tau\}$ is the analytic continuation along γ of the branch $f(z, 0) = f_{c_0,k}(z)$ of $f(z)$ at c_0. Put

$$F_0(z) = \int_{c_0}^{z} f_{c_0,k}(z)\,dz, \qquad z \in U(c_0). \tag{4.44}$$

Theorem 4.15. $F(z) = \int_{\gamma}^{z} f(z)\,dz$ is the analytic continuation along γ of $F_0(z)$.

Proof: We put, just as in the proof of Theorem 4.11,

$$F(z, t) = \int_0^t f(\gamma(t), t)\gamma'(t)\,dt + \int_{\gamma(t)}^{z} f(z, t)\,dz, \qquad z \in U(\gamma(t)).$$

It suffices to show that $\{F(z, t)\colon 0 \leqq t \leqq \tau\}$ is the analytic continuation along γ of $F_0(z)$. To do this, it is sufficient to show that $F(z, t) = F(z, s)$ on $U(\gamma(t)) \cap U(\gamma(s))$ if $0 \leqq t - s < \varepsilon$, where ε is determined as in (4.43).

If $0 \leqq t - s < \varepsilon$, then $z \in U(\gamma(t)) \cap U(\gamma(s))$ implies $f(z, t) = f(z, s)$ and $s \leqq u \leqq t$ implies $f(\gamma(u), u) = f(\gamma(u), s)$. Hence we have for an arbitrary $z \in U(\gamma(t)) \cap U(\gamma(s))$

$$\begin{aligned} F(z, t) &= \int_0^s f(\gamma(u), u)\gamma'(u)\,du + \int_s^t f(\gamma(u), u)\gamma'(u)\,du + \int_{\gamma(t)}^z f(z, t)\,dz \\ &= \int_0^s f(\gamma(u), u)\gamma'(u)\,du + \int_s^t f(\gamma(u), s)\gamma'(u)\,du + \int_{\gamma(t)}^z f(z, s)\,dz \\ &= \int_0^s f(\gamma(u), u)\gamma'(u)\,du + \int_{\gamma(s)}^z f(z, s)\,dz = F(z, s). \end{aligned}$$

The fact that $F(z)$ is a freely continuable, complete analytic function on Ω is proved in the same way as when $f(z)$ is a single-valued holomorphic function. Since the homotopy variant of Cauchy's Theorem (Theorem 4.12) is a direct consequence of the fact that $F(z)$ is the analytic continuation along γ of $F_0(z)$, it is also valid if $f(z)$ is a multivalued function. That is, if $f(z)$ is a freely continuable, complete analytic function on Ω and if γ is a piecewise smooth curve in Ω connecting c_0 and c_1, then $\int_\gamma f(z)\,dz$ equals the value of the power series $F_1(z)$ obtained by analytic continuation along γ of $F_0(z)$ at its center c_1:

$$\int_\gamma f(z)\,dz = F_1(c_1),$$

by Theorem 4.15.

If β is a piecewise smooth curve homotopic with γ in Ω, then the result of the analytic continuation of $F_0(z)$ along β is the same power series $F_1(z)$, hence

$$\int_\beta f(z)\,dz = \int_\gamma f(z)\,dz, \qquad \beta \simeq \gamma. \tag{4.45}$$

Of course, $F_1(c_1)$ depends in general on the choice of the branch $f_{c_0,k}(z)$ appearing in $F_0(z)$ (4.44). Equality (4.45) holds under the assumption that both integrands are analytic continuations along the path of integration of the same branch $f_{c_0,k}(z)$ at the initial point c_0. If $c_1 = c_0$, that is, if γ is a closed curve with base point c_0 and if $\gamma \simeq 0$ in Ω, then $\gamma \simeq \delta$ (δ is the one-point curve c_0). Hence

$$\int_\gamma f(z)\,dz = \int_\delta f(z)\,dz = 0.$$

Therefore we have

Theorem 4.16. Let $f(z)$ be a freely continuable, complete analytic function on the region Ω and let γ be a piecewise closed curve in Ω. If $\gamma \simeq 0$ then

$$\int_\gamma f(z)\,dz = 0. \tag{4.46}$$

However, if γ is homologous to 0, then (4.46) is not necessarily true.

Example 4.5. The function $f(z) = \sqrt{z}/(z-2)$ is a holomorphic, two-valued function on the region $\Omega = \mathbb{C} - \{0, 2\}$. Let α and β be circles with base point 1 and center 2, respectively 0, then $\gamma = \alpha \cdot \beta \cdot \alpha^{-1} \cdot \beta^{-1}$ is a piecewise smooth closed curve with base point 1 in Ω. Let us evaluate $\int_\gamma f(z)\,dz$. Fixing a branch for $\sqrt{z}$ by $\sqrt{z} = e^{\mathrm{Log}\,z/2}$, we have

$$f(z) = f_\pm(z), \qquad f_+(z) = \frac{\sqrt{z}}{z-2}, \qquad f_-(z) = \frac{-\sqrt{z}}{z-2}.$$

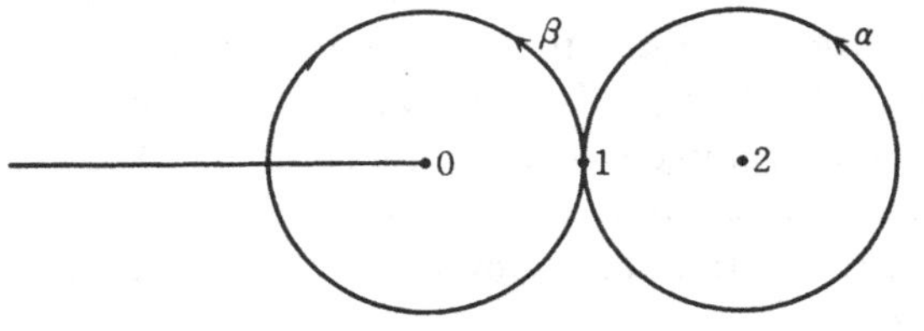

Fig. 4.20

Writing $z = |z|e^{i\theta}$, $-\pi < \theta \leqq \pi$, we have $\sqrt{z} = \sqrt{(|z|)}e^{i\theta/2}$. Let D be the region obtained by deleting the negative real axis $\mathbb{R}^-$ from $\mathbb{C}$. D is simply connected and the branches of $f(z)$ over D are $f_+(z)$ and $f_-(z)$. Since $f_+(z)$ is holomorphic on D with the exception of the point 2, which is a pole of the first order, we have by the residue theorem (Theorem 2.5)

$$\int_\alpha f(z)\,dz = \int_\alpha f_+(z)\,dz = 2\pi i\,\mathrm{Res}_{z=2}[f_+(z)] = 2\pi i\sqrt{2}.$$

For sufficiently small ε, let $\lambda = [\varepsilon, 1]$ be the segment connecting ε and 1 and let C_ε be the circle with center 0 and radius ε, then $\beta \simeq \lambda^{-1} \cdot C_\varepsilon \cdot \lambda$ in Ω. By analytic continuation along the circle C_ε with initial point ε, $f_+(z)$ becomes $f_-(z)$, hence, by (4.45)

$$\int_\beta f(z)\,dz = \int_1^\varepsilon f_+(x)\,dx + \int_{C_\varepsilon} f(z)\,dz + \int_\varepsilon^1 f_-(x)\,dx$$
$$= -2\int_\varepsilon^1 \frac{\sqrt{x}}{x-2}\,dx + \int_{C_\varepsilon} f(z)\,dz.$$

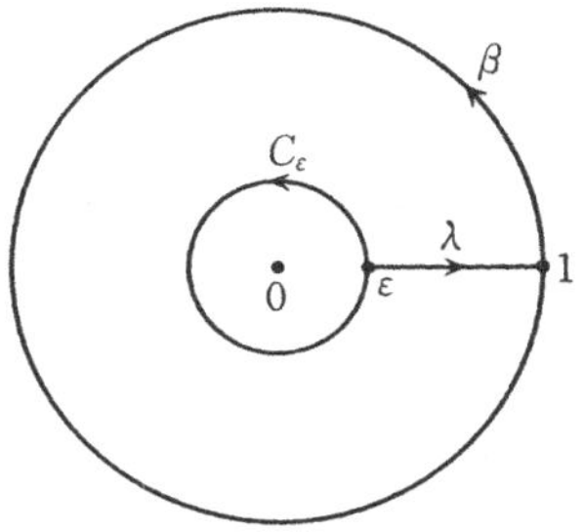

Fig. 4.21

Letting ε tend to 0 from the right, we have

$$\int_{C_\varepsilon} f(z)\,dz = \int_0^{2\pi} \frac{\sqrt{\varepsilon}\,e^{i\theta/2}}{\varepsilon e^{i\theta}-2}\,\varepsilon i e^{i\theta}\,d\theta \to 0, \qquad \varepsilon \to +0,$$

hence

$$\int_\beta f(z)\,dz = -2\int_0^1 \frac{\sqrt{x}}{x-2}\,dx.$$

Since by analytic continuation along $\alpha\cdot\beta$, $f_+(z)$ becomes $f_-(z)$, we have

$$\int_{\alpha^{-1}} f(z)\,dz = -\int_\alpha f_-(z)\,dz = \int_\alpha f_+(z)\,dz = 2\pi i\sqrt{2}.$$

Since at the initial point 1 of β^{-1}, $f(z) = f_-(z)$ also, we have

$$\int_{\beta^{-1}} f(z)\,dz = \int_1^\varepsilon f_-(x)\,dx + \int_{C_\varepsilon^{-1}} f(z)\,dz + \int_\varepsilon^1 f_+(x)\,dx,$$

hence

$$\int_{\beta^{-1}} f(z)\,dz = 2\int_0^1 \frac{\sqrt{x}}{x-2}\,dx.$$

Therefore

$$\int_\gamma f(z)\,dz = \left(\int_\alpha + \int_\beta + \int_{\alpha^{-1}} + \int_{\beta^{-1}}\right) f(z)\,dz = 4\pi i\sqrt{2}.$$

Since $\gamma = \alpha\cdot\beta\cdot\alpha^{-1}\cdot\beta^{-1} = \alpha+\beta-\alpha-\beta = 0$ by (4.34) and (4.35), obviously $\gamma \sim 0$ in Ω. Since $\int_\gamma f(z)\,dz \neq 0$, we conclude from Theorem 4.16 that γ is not homotopic with 0.

In general, if γ is a closed curve in a region Ω, then $\gamma \simeq 0$ in Ω implies $\gamma \sim 0$, but the converse is not necessarily true, as shown by the above example.

5

Riemann's Mapping Theorem

5.1 Riemann's Mapping Theorem

We gave in Section 3.3c some examples of conformal mappings $f: z \to w = f(z)$ mapping a region D "of simple shape" in the z-plane into the unit disk in the w-plane. Given an arbitrary region D, in general it will be impossible to find a conformal mapping f which maps D onto the unit disk U by the composition of suitably chosen known functions. However, we have the following theorem concerning the existence of conformal mappings which map D onto U.

Riemann's Mapping Theorem. Let D be a region in the complex plane $\mathbb{C}$, z_0 a point in D, and $U = \{w: |w| < 1\}$ the unit disk in the w-plane. If D is simply connected and $D \neq \mathbb{C}$, then there exists exactly one conformal mapping $f: z \to w = f(z)$ from D onto U that satisfies $f(z_0) = 0$ and $f'(z_0) > 0$.

Obviously for a conformal mapping f between D and U to exist, it is necessary that D be simply connected and $D \neq \mathbb{C}$. Since, by Liouville's Theorem (Theorem 1.24), a function that is holomorphic and bounded on $\mathbb{C}$ is a constant, there cannot exist a conformal mapping from $\mathbb{C}$ onto U. If a conformal mapping f from D onto U exists, then f is a one-to-one continuous mapping from D onto U and its inverse mapping f^{-1} is continuous. Therefore, D has to be simply connected by the simple connectedness of U and the definition of simple connectedness (Definition 4.4). This section is devoted to proving Riemann's Theorem. We first give an outline of the proof.

(a) We first show that there exists at least one conformal mapping $f: z \to w = f(z)$ from D onto a subregion of U such that $f(z_0) = 0$ and $f'(z_0) > 0$. By Theorem 3.5, this is equivalent to saying that there exists at least one univalent, holomorphic function defined on D satisfying $|f(z)| < 1$ for all $z \in D$.

(b) Let $\mathscr{F}$ denote the collection of all univalent, holomorphic functions

$f(z)$ defined on D and satisfying $|f(z)| < 1$ for all $z \in D$, $f(z_0) = 0$ and $f'(z_0) > 0$. By (a), $\mathscr{F}$ is nonempty.

If a conformal mapping $f: z \to w = f(z)$ from D onto U that satisfies $f(z_0) = 0$ and $f'(z_0) > 0$ exists, then of course $f(z)$ belongs to $\mathscr{F}$. Now, picking an arbitrary $g(z) \in \mathscr{F}$, we put

$$\phi = g \circ f^{-1}: w \to \phi(w) = g(f^{-1}(w)).$$

By Theorem 3.4, ϕ is a conformal mapping from $U = f(D)$ onto $g(D) \subset U$, that is, $\phi(w)$ is holomorphic for $|w| < 1$ and satisfies $|\phi(w)| < 1$, and obviously $\phi(0) = 0$. By Schwarz's Lemma, we have $|\phi(0)| \leqq 1$ and the case $|\phi'(0)| = 1$ occurs only if $\phi(z) = cz$ for some c with $|c| = 1$. Since $\phi(f(z)) = g(z)$, we have

$$\phi'(0)f'(z_0) = g'(z_0),$$

and, by our assumption, $f'(z_0) > 0$ and $g'(z_0) > 0$. Therefore, $g'(z_0) \leqq f'(z_0)$ and $g'(z_0) = f'(z_0)$ if and only if $g(z) = f(z)$ identically. Let s be defined by

$$s = \sup_{g(z) \in \mathscr{F}} g'(z_0).$$

If a conformal mapping $f: z \to w = f(z)$ from D onto U exists, then $f(z) \in \mathscr{F}$ and $f'(z_0) \leqq s$. We prove that there exists a function $f(z) \in \mathscr{F}$ satisfying $f'(z_0) = s$. To this end, we show that if $\{f_n(z)\}$ is an arbitrary sequence of functions $f_n(z) \in \mathscr{F}$ with $\lim_{n\to\infty} f_n'(z_0) = s$, then there exists a subsequence $\{f_{n_j}(z)\}$, $n_1 < n_2 < \cdots < n_j < \cdots$ that is uniformly convergent on all compact subsets of D. By Theorem 1.19, the limit function $f(z) = \lim_{j\to\infty} f_{n_j}(z)$ is holomorphic on D and $f'(z) = \lim_{j\to\infty} f'_{n_j}(z)$. Hence, $f(z_0) = 0$ and $f'(z_0) = s > 0$.

(c) We show that this limit function $f(z)$ is univalent on D and hence a holomorphic function belonging to $\mathscr{F}$.

(d) Finally, we prove that the range of this function $f(z)$ is U, that is, $f: z \to w = f(z)$ is a conformal mapping, from D onto U.

(**a**) Since $D \subsetneqq \mathbb{C}$ by assumption, there is a point σ in $\mathbb{C}$ such that $\sigma \notin D$. Putting $\Omega = \mathbb{C} - \{\sigma\}$, $h(z) = \sqrt{(z-\sigma)}$ is a freely analytically continuable and complete two-valued analytic function on Ω. Hence, as explained in Section 4.2, $h(z)$ has two branches $h_{D,1}(z)$ and $h_{D,2}(z)$ on the simply connected subregion D of Ω. Putting $h(z) = h_{D,1}(z)$ we have $h_{D,2}(z) = -h(z)$, where $h(z)$ is a holomorphic function defined on D, satisfying

$$h(z)^2 = z - \sigma.$$

The function $h(z)$ is univalent on D: if $h(z_1) = h(z_2)$, then $z_1 = \sigma + h(z_1)^2$

$= \sigma + h(z_2)^2 = z_2$. The ranges $h(D)$ and $-h(D) = \{-h(z): z \in D\}$ of $h(z)$, respectively $-h(z)$, have no common points. (If $h(z_1) = -h(z_2)$, then $z_1 = \sigma + h(z_1)^2 = \sigma + h(z_2)^2 = z_2$, hence $2h(z_1) = h(z_1) - h(z_1) = 0$, that is, $z_1 = \sigma$, contradicting the assumption $\sigma \notin D$.)

Since

$$h'(z_0) = \tfrac{1}{2}\sqrt{(z_0 - \sigma)} \neq 0$$

we conclude from Theorem 3.2 that $-h(D)$ contains a sufficiently small neighborhood $U_\varepsilon(-h(z_0))$, $\varepsilon > 0$. Therefore, $h(D) \cap U_\varepsilon(-h(z_0)) = \emptyset$. Hence, if $z \in D$, then $|h(z) + h(z_0)| > \varepsilon$. Therefore, putting

$$g(z) = \varepsilon/(h(z) + h(z_0)),$$

the function $g(z)$ is also a univalent, holomorphic function defined on D satisfying $|g(z)| < 1$ for all z, that is, $g: z \to w = g(z)$ is a conformal mapping from D onto a subregion of U, but $g(z_0) \neq 0$. Putting $\alpha = g(z_0)$, we consider the mapping

$$\phi: w \to \phi(w) = e^{i\theta} \cdot \frac{w - \alpha}{1 - \bar{\alpha}w'}, \qquad \theta \text{ a real number,}$$

from the unit disk U onto itself (Theorem 3.6), and the composition, $f = \phi \circ g$, of g and ϕ. Clearly, $f: z \to w = f(z)$ is a conformal mapping from D onto a subregion of U satisfying

$$f(z_0) = \phi(g(z_0)) = \phi(\alpha) = 0,$$

and, since $\phi'(w) = e^{i\theta}(1 - |\alpha|^2)/(1 - \bar{\alpha}w)^2$,

$$f'(z_0) = e^{i\theta}(1 - |\alpha|^2)^{-1} g'(z_0).$$

Since g is conformal, we have $g'(z_0) \neq 0$. Taking for θ the opposite of the argument of $g'(z_0)$, we arrive at $f'(z_0) > 0$. So we have established the existence of a conformal mapping $f: z \to w = f(z)$ from D onto a subregion of U satisfying $f(z_0) = 0$ and $f'(z_0) > 0$.

(b) A sequence of functions $\{f_n(z)\}$ defined on D is called *uniformly bounded* on D if there exists a constant M such that $|f_n(z)| < M$ for all $z \in D$.

Theorem 5.1 (*Montel's Theorem*). If the sequence $\{f_n(z)\}$ of holomorphic functions defined on the region D is uniformly bounded on D, then there exists a subsequence $\{f_{n_j}(z)\}$ that converges uniformly on each compact subset of D.

Proof: (1) Let $Q(i)$ denote the set of all complex numbers of the form $r + is$, where r and s are rational numbers and let the countable set $D \cap Q(i)$ be represented by

$$D \cap Q(i) = \{c_1, c_2, c_3, \ldots, c_k, \ldots\}.$$

$D \cap Q(i)$ is everywhere dense in D. We first prove that it is possible to select a subsequence $\{f_{n_j}(z)\}$ of $\{f_n(z)\}$ such that $\{f_{n_j}(c_k)\}$ converges for all points c_k. For this purpose, we let

$$\{n_{jm}\}_< = \{n_{j1}, n_{j2}, \ldots, n_{jm}, \ldots\}_<$$

represent monotonic increasing sequences of natural numbers, that is, $n_{j1} < n_{j2} < \cdots < n_{jm} < \cdots$. The sequence $\{n_{jm}\}_<$ is a subsequence of $\{n_{km}\}_<$ if and only if the set of natural numbers $\{n_{jm}\}$ is a subset of $\{n_{km}\}$, so we denote the fact that $\{n_{jm}\}_<$ is a subsequence of $\{n_{km}\}_<$ by $\{n_{km}\}_< \supset \{n_{jm}\}_<$.

Obviously,

$$\{n_{km}\}_< \supset \{n_{jm}\}_< \text{ implies } n_{km} \leqq n_{jm}. \tag{5.1}$$

For the sake of legibility, we put $f(z, n) = f_n(z)$. Substituting $z = c_1$ in the function sequence $\{f(z, n)\}$, we obtain the bounded number sequence $\{f(c_1, n)\}$, which has a convergent subsequence $\{f(c_1, n_{1m})\}$, $(n_{11} < n_{12} < \ldots < n_{1m} < \ldots)$. Substituting $z = c_2$ in $\{f(z, n_{im})\}$ we obtain the bounded sequence $\{f(c_2, n_{1m})\}$, which has a convergent subsequence $\{f(c_2, n_{2m})\}$, where $\{n_{1m}\}_< \supset \{n_{2m}\}_<$. Continuing this way, we arrive at a sequence of subsequences of a sequence of natural numbers

$$\{n_{1m}\}_< \supset \{n_{2m}\}_< \supset \{n_{3m}\}_< \supset \cdots \supset \{n_{km}\}_< \supset \cdots$$

such that for each k, the number sequence $\{f(c_k, n_{km})\}$ converges. Now we put

$$n_j = n_{jj}.$$

Then we have by (5.1)

$$n_k = n_{kk} < n_{kj} \leqq n_{jj} = n_j \qquad \text{if } k < j,$$

hence $\{n_j\} = \{n_1, n_2, n_3, \ldots, n_j, \ldots\}$ is a monotone increasing sequence of natural numbers. Since $k < j$ implies $n_j = n_{jj} \in \{n_{jm}\} \supset \{n_{km}\}$, the sequence $\{n_k, n_{k+1}, n_{k+2}, \ldots, n_j, \ldots\}$ is a subsequence of $\{n_{km}\}_<$. Since the sequence $\{f(c_k, n_{km})\}$ converges, so does the sequence $\{f(c_k, n_j)\}$. Therefore, the function sequence $\{f_{n_j}(z)\} = \{f(z, n_j)\}$ converges at all points c_k, $k = 1, 2, 3, \ldots$.

(2) By assumption, the $f_n(z)$ are holomorphic on the region D and for all $z \in D$ we have

$$|f_n(z)| < M, \qquad M \text{ a constant.}$$

For $c \in D$, let $\rho > 0$ be such that $U_{3\rho}(c) \subset D$, where $U_{3\rho}(c)$ is the disk with center c and radius 3ρ. Then

$$|f_n(z)-f_n(c)| < \frac{M}{\rho}|z-c| \qquad \text{if } |z-c|<\rho. \tag{5.2}$$

For, if C is a circle with center c and radius 2ρ, we have by Cauchy's Theorem

$$f_n(z)-f_n(c) = \frac{1}{2\pi i}\int_C\left(\frac{1}{\zeta-z}-\frac{1}{\zeta-c}\right)f_n(\zeta)\,d\zeta.$$

For $\zeta\in C$ we have

$$\left|\frac{1}{\zeta-z}-\frac{1}{\zeta-c}\right| \leqq \frac{|z-c|}{|\zeta-z|\,|\zeta-c|} \leqq \frac{|z-c|}{\rho\cdot 2\rho}$$

while $|f_n(\zeta)|<M$, hence

$$\left|\frac{1}{2\pi i}\int_C\left(\frac{1}{\zeta-z}-\frac{1}{\zeta-c}\right)f_n(\zeta)\,d\zeta\right| \leqq \frac{|z-c|}{2\rho^2}M\cdot 2\rho = \frac{M}{\rho}|z-c|.$$

(3) We prove that the subsequence $\{f_{n_j}(z)\}$ of $\{f_n(z)\}$, which converges at all points c_k of $D\cap Q(i)$, converges uniformly on all compact subsets $K\subset D$. Let r denote the distance between K and $\mathbb{C}-D$. Since K is compact we have $r>0$. We have to prove that for arbitrary $\varepsilon>0$ there exists a natural number $j_0(\varepsilon)$ such that if $z\in K$,

$$|f_{nj}(z)-f_{nh}(z)|<\varepsilon \qquad \text{if } h>j>j_c(\varepsilon).$$

Put $\rho=r/4$ and select $\delta>0$ such that $M\delta<\varepsilon\rho/3$ and $\delta<\rho$. Since for each $z\in K$ there exists a point $c_k\in D\cap Q(i)$ such that $|c_k-z|<\delta$, K is covered by the disks $U_\delta(c_k)$, $k=1,2,3,\ldots$. Since K is compact, K is already covered by a finite number of the disks $U_\delta(c_k)$. Hence we have

$$K\subset\bigcup_{k=1}^{m}U_\delta(c_k), \qquad U_\delta(c_k)\cap K\neq\varnothing$$

for some m (possibly after rearranging the point sequence $\{c_k\}$).

Since the number sequence $\{f_{nj}(c_k)\}$ converges for all points c_k, $k=1,2,\ldots,m$, we have for a sufficiently large $j_0(\varepsilon)$:

$$\text{if } h>j>j_0(\varepsilon) \text{ then } |f_{nj}(c_k)-f_{nh}(c_k)|<\varepsilon/3, \qquad k=1,2,\ldots,m.$$

An arbitrary $z\in K$ belongs to at least one of the $U_\delta(c_k)$, $k=1,2,\ldots,m$. Suppose $z\in U_\delta(c_k)$, then $|c_k-z|<\delta<\rho$, hence

$$U_{3\rho}(c_k)\subset U_{4\rho}(z)=U_r(z)\subset D.$$

Therefore, by (5.2),

$$|f_n(z)-f_n(c_k)|<\frac{M}{\rho}|z-c_k|<\frac{M\delta}{\rho}<\frac{\varepsilon}{3}$$

and hence

$$|f_n(z)-f_i(z)| < \frac{2\varepsilon}{3} + |f_n(c_k)-f_i(c_k)|.$$

We conclude that

$$|f_{nj}(z)-f_{nh}(z)| < \frac{2\varepsilon}{3} + |f_{nj}(c_k)-f_{nh}(c_k)| < \varepsilon$$

for $h > j > j_0(\varepsilon)$.

Let $\mathscr{F}$ be the collection of all univalent, holomorphic functions on D, such that $|f(z)| < 1$ for all $z \in D$, $f(z_0) = 0$ and $f'(z_0) > 0$ for some fixed $z_0 \in D$. By (a), $\mathscr{F}$ is nonempty. The set of all $f'(z_0)$ for $f(z) \in \mathscr{F}$ is bounded. To see this, select $\rho > 0$ such that $U_\rho(z_0) \subset D$ and put $h(z) = f(z_0 + \rho z)$. Since $h(z)$ is holomorphic for $|z| < 1$, $|h(z)| \leqq 1$ and $h(0) = 0$, we have, by Schwarz's Lemma, $|h'(0)| \leqq 1$ hence $|f'(z_0)| = |h'(0)|/\rho \leqq 1/\rho$.

Put

$$s = \sup_{f(z)\in\mathscr{F}} f'(z_0) \tag{5.3}$$

and select a function sequence $\{f_n(z)\}$, $f_n(z) \in \mathscr{F}$, such that $\lim_{n\to\infty} f'_n(z_0) = s$. By Theorem 5.1, there exists a subsequence $\{f_{nj}(z)\}$ of $\{f_n(z)\}$ which converges uniformly on each compact subset of D.

Defining $f(z)$ by

$$f(z) = \lim_{j\to\infty} f_{nj}(z)$$

we know, by Theorem 1.19, that $f(z)$ is holomorphic on D, while $f'(z) = \lim_{j\to\infty} f'_{nj}(z)$. Since $f_{nj}(z_0) = 0$, we have $f(z_0) = 0$ and

$$f'(z_0) = \lim_{f\to\infty} f'_{nj}(z_0) = s.$$

(c) We prove that this function $f(z) = \lim_{j\to\infty} f_{nj}(z)$ is univalent on D. Let us assume the contrary, that is, there are two points $z_1, z_2 \in D$ such that $z_1 \neq z_2$ and $f(z_1) = f(z_2)$. Since $f'(z_0) = s > 0$, $f(z)$ is not a constant. Hence, putting $a = f(z_1)$, we know by Theorem 3.1 and (1.60), that there exists an $\varepsilon > 0$ such that

$$\text{if } 0 < |z - z_1| \leqq \varepsilon \text{ then } f(z) - a \neq 0.$$

Now, for sufficiently large j, $f_{nj}(z)$ has at least one a-point on the neighborhood $U_\varepsilon(z_1)$ of z_1. We prove this now.

Let μ be the minimal value assumed by the continuous function $|f(z) - a|$ on the circle $C = \partial[U_\varepsilon(z_1)]$ with center z_1 and radius ε. Then

$$|f(z) - a| \geqq \mu > 0, \qquad \text{if } z \in C,$$

hence, since the function sequence $\{f_{nj}(z)\}$ converges uniformly on the compact subset C of D, we have for sufficiently large j

$$|f_{nj}(z) - a| \geqq \frac{\mu}{2} > 0, \qquad \text{if } z \in C. \tag{5.4}$$

Now, assume that $f_{nj}(z)$ does not possess any a-points in $U_\varepsilon(z_1)$, then $1/(f_{nj}(z) - a)$ is a holomorphic function of z on $[U_\varepsilon(z_1)]$. Hence, by the maximum principle (corollary to Theorem 1.21), $1/(|f_{nj}(z) - a|)$ assumes its maximum on C. Therefore, by (5.4)

$$\frac{1}{|f_{nj}(z) - a|} \leqq \frac{2}{\mu}, \qquad \text{if } z \in U_\varepsilon(z_1),$$

that is,

$$|f_{nj}(z) - a| \geqq \frac{\mu}{2} > 0, \qquad \text{if } z \in U_\varepsilon(z_1).$$

This contradicts $\lim_{j\to\infty} f_{nj}(z_1) = f(z_1) = a$.

In a similar way it may be proved that $f_{nj}(z)$ has at least one a-point in $U_\varepsilon(z_2)$. We may assume that $U_\varepsilon(z_1)$ and $U_\varepsilon(z_2)$ are disjoint, hence it follows that $f_{nj}(z)$ has at least two a-points in D contradicting the fact that $f_{nj}(z)$ is univalent on D. Since $|f(z)| = \lim_{j\to\infty} |f_{nj}(z)| \leqq 1$ for all $z \in D$, we conclude from the maximum principle that $|f(z)| < 1$. Since, as observed above, $f(z_0) = 0$ and $f'(z_0) = s > 0$, we conclude $f(z) \in \mathscr{F}$.

(d) In order to prove that the range $f(D)$ of $f(z)$ coincides with the unit disk U, we assume $f(D) \subsetneqq U$. From this assumption we will derive the existence of a function $g(z) \in \mathscr{F}$ with $g'(z_0) > s$, which contradicts the definition, (5.3), of s. If $f(D) \subsetneqq U$, there is a point $\tau \in U$ such that $\tau \notin f(D)$. Since $f(z_0) = 0$, we have $0 < |\tau| < 1$. The mapping $w \to (w - \tau)/(1 - \bar{\tau}w)$ is a conformal mapping from U onto itself mapping τ onto 0; hence

$$\psi(z) = \frac{f(z) - \tau}{1 - \bar{\tau} f(z)} \tag{5.5}$$

is a univalent holomorphic function of z defined on D and satisfying $0 < |\psi(z)| < 1$ for all z. $\sqrt{w}$ is a freely analytically continuable, complete, two-valued analytic function defined on the region $\mathbb{C}^* = \mathbb{C} - \{0\}$. Hence, since D is simply connected, we have by Theorem 4.10, that $\sqrt{[\psi(z)]}$ breaks into two branches $h(z)$ and $-h(z)$. The function $h(z)$ is a holomorphic function defined on D such that

$$h(z)^2 = \psi(z), \qquad 0 < |h(z)| < 1. \tag{5.6}$$

Therefore, since $\psi(z)$ is univalent on D, so is $h(z)$. Observing that $h(z_0) \neq 0$, we put

$$g(z) = e^{i\theta} \cdot \frac{h(z) - \beta}{1 - \bar{\beta} h(z)}, \qquad \beta = h(z_0),\ \theta = \arg \beta.$$

Since $w \to e^{i\theta}(w - \beta)/(1 - \bar{\beta} w)$ is a conformal mapping which maps U onto U and β onto 0, $g(z)$ is a univalent, holomorphic function defined on D such that $|g(z)| < 1$ for all $z \in D$ and $g(z_0) = 0$. Since $h(z_0) = \beta$, we have

$$g'(z_0) = e^{i\theta} \cdot \frac{h'(z_0)}{1 - |\beta|^2}.$$

Since $f(z_0) = 0$, we have from (5.6) and (5.5)

$$2h(z_0)h'(z_0) = \psi'(z_0) = (1 - |\tau|^2) f'(z_0)$$

hence

$$2|\beta| e^{i\theta} h'(z_0) = (1 - |\beta|^4) f'(z_0)$$

since $h(z_0) = \beta = |\beta| e^{i\theta}$ and $-\tau = \psi(z_0) = h(z_0)^2 = \beta^2$. Therefore

$$g'(z_0) = \frac{1 + |\beta|^2}{2|\beta|} f'(z_0).$$

Since $0 < |\beta| < 1$, we have $1 + |\beta|^2 > 2|\beta|$, hence

$$g'(z_0) > f'(z_0) = s.$$

We have proved the existence of a conformal mapping f: $z \to w = f(z)$ mapping D onto U and satisfying $f(z_0) = 0$ and $f'(z_0) > 0$.

To prove the uniqueness, let g: $z \to w = g(z)$ be another conformal mapping from D onto U satisfying $g(z_0) = 0$ and $g'(z_0) > 0$.

The mapping

$$\phi = g \circ f^{-1}: w \to \phi(w) = g(f^{-1}(w))$$

is a conformal mapping from U onto itself, such that $\phi(f(z)) = g(z)$. Since $g(z_0) = 0$, we have by the corollary to Schwarz's Lemma that $\phi(w) = e^{i\theta} w$, $\theta \in \mathbb{R}$. Therefore $e^{i\theta} f(z) = g(z)$. Since $g'(z_0) > 0$, we have $e^{i\theta} f'(z_0) = g'(z_0) > 0$, i.e., $e^{i\theta} = 1$ and $g(z) = f(z)$ identically.

This completes the proof of Riemann's Mapping Theorem.

(e) If D is a region in the complex plane $\mathbb{C}$, then the complement $F = S - D$, of D with respect to the Riemann sphere $S = \mathbb{C} \cup \{\infty\}$ is a closed set. F is called *connected* if it is impossible to find two nonempty, closed sets F_0 and F_1 satisfying $F = F_0 \cup F_1$ and $F_0 \cap F_1 = \varnothing$.

Definition 5.1. If the complement $F = S - D$ of D is connected, then D is called a *region without holes.*

Figure 5.1 shows an example of a region with holes.

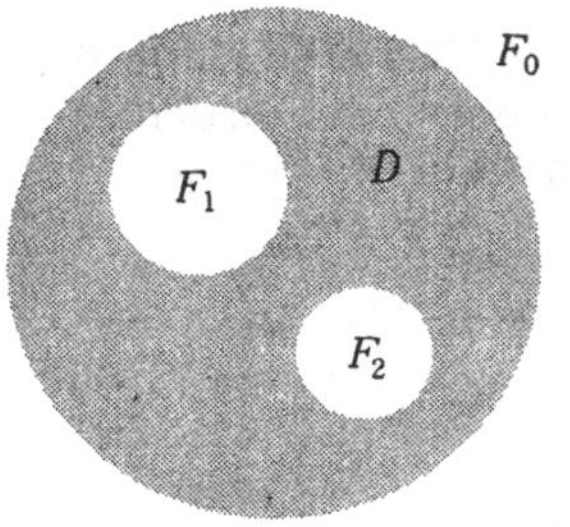

Fig. 5.1

Theorem 5.2. (See L. Ahlfors, 1966, Chapter 4, Theorem 16.) If $D \subset \mathbb{C}$ is a region without holes, $f(z)$ a function holomorphic on D, and γ a piecewise smooth closed curve in D, then

$$\int_\gamma f(z)\,dz = 0.$$

Proof: Let γ be represented by $\gamma\colon t \to \gamma(t)$, $a \leqq t \leqq b$, and let $|\gamma| = \{\gamma(t)\colon a \leqq t \leqq b\}$. Let Ω denote the connected component of the open set $S - |\gamma|$ containing ∞ (i.e., the largest connected open subset of $S - |\gamma|$ containing ∞) and Ω_1 the open set defined by

$$S - |\gamma| = \Omega \cup \Omega_1, \qquad \Omega \cap \Omega_1 = \varnothing.$$

Since $|\gamma| \subset D$, we have $F = S - D \subset \Omega \cup \Omega_1$. Since F is connected by assumption, we have $F \subset \Omega$ ($F_0 = F \cap \Omega = F - \Omega_1$ and $F_1 = F \cap \Omega_1 = F - \Omega$ are closed sets with $F = F_0 \cup F_1$ and $F_0 \cap F_1 = \varnothing$. Since both F and Ω contain ∞, we conclude $F_0 \neq \varnothing$ and hence $F_1 = \varnothing$.) Since the region Ω contains a neighborhood $\{z\colon |z| > R\} \cup \{\infty\}$ of ∞, $\Gamma = |\gamma| \cup \Omega_1 = S - \Omega$ is a bounded subset of $\mathbb{C}$ and since $F \subset \Omega$ we have $\Gamma \subset S - F = D$. Since Γ is compact, the distance between Γ and F in $\mathbb{C}$ is positive. Pick a $\delta > 0$ such that $2\delta < r$ and divide the complex plane into an infinite number of squares with side-length δ

$$Q_{hk} = \{x + iy\colon h\delta \leqq x \leqq h\delta + \delta, k\delta \leqq y \leqq k\delta + \delta\}, \quad h, k = 0, \pm 1, \pm 2, \ldots,$$

(compare Section 2.2b).

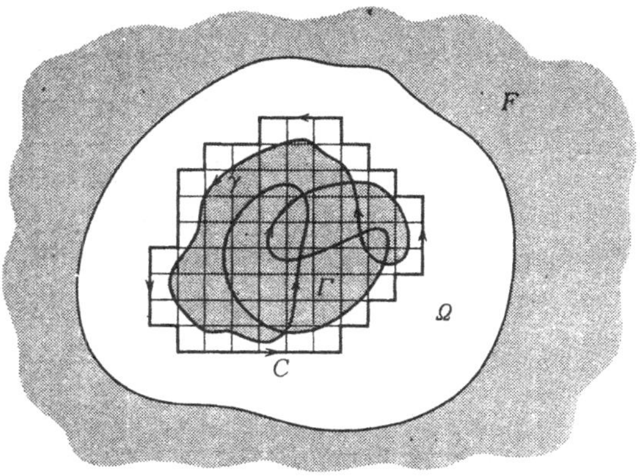

Fig. 5.2

Arrange all squares Q_{hk} that have a nonempty intersection with Γ in a sequence:

$$Q_1, Q_2, \ldots, Q_\lambda, \ldots, Q_\nu, \qquad Q_\lambda = Q_{hk}, \qquad Q_{hk} \cap \Gamma \neq \emptyset,$$

and denote the union of those squares by K

$$K = \bigcup_{\lambda=1}^{\nu} Q_\lambda.$$

Since the diameter of each square Q_λ equals $\sqrt{2}\delta$, we have $K \cap F = \emptyset$, hence $\Gamma \subset K \subset D$. The boundary ∂Q_λ of a square Q_λ is a 1-cycle made up by the four sides. Put $\partial Q_1 = C_1 + C_2 + C_3 + C_4$, $\partial Q_2 = C_5 + C_6 + C_7 + C_8, \ldots$. Then

$$\partial Q_\lambda = C_{4\lambda-3} + C_{4\lambda-2} + C_{4\lambda-1} + C_{4\lambda}$$

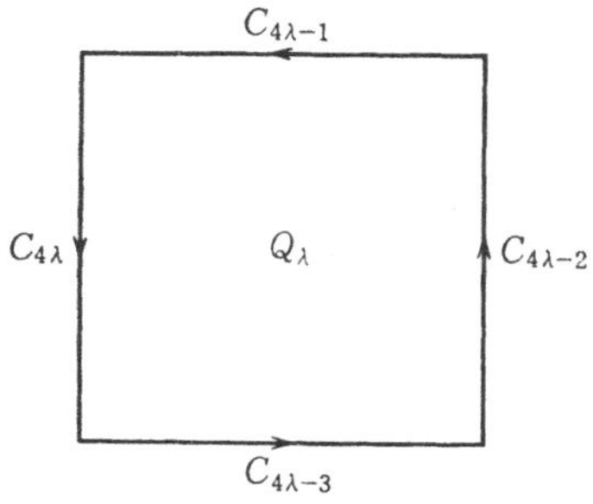

Fig. 5.3

Ignoring the orientation, the four sides of Q_λ are $|C_j|$, $j = 4\lambda - 3$, $4\lambda - 2$, $4\lambda - 1$, 4λ. If the side $|C_j|$ of $Q_\lambda = Q_{hk}$ is also a side of $Q_{h'k'}$, where $(h', k') = (h-1, k)$, $(h+1, k)$, $(h, k-1)$ or $(h, k+1)$, then we write C_j as C'_j if $Q_{h'k'} \subset K$ and as C''_j if $Q_{h'k'} \not\subset K$. The sides $|C'_j|$ of Q_λ coincide with a side $|C'_l|$ of some other square $Q_\mu = Q_{h'k'}$. Taking orientations into account, we have $C'_j = -C'_l$. The sides $|C''_j|$ of Q_λ are not contained in any square $Q_\mu \subset K$ with $\mu \neq \lambda$. Hence, taking the sum of all 1-cycles ∂Q_λ, $\lambda = 1, \ldots, \nu$, all C'_j are cancelled:

$$C = \sum_{\lambda=1}^{\nu} \partial Q_\lambda = \sum C''_j. \tag{5.7}$$

Obviously, the boundary of the point set K is $|C| = \bigcup |C''_j|$. If $z \in \Gamma$, then $z \in Q_{hk}$ for some square. Since $Q_{hk} \subset K$, we have $z \notin |C''_j|$, hence $z \notin |C|$. Therefore Γ is contained in the interior of K:

$$\Gamma \subset (K) = K - |C|.$$

We will prove that for $z \in (K)$ the equality

$$f(z) = \frac{1}{2\pi i} \int_C \frac{f(\zeta)}{\zeta - z} d\zeta, \qquad z \in (K) \tag{5.8}$$

holds. The function $f(z)$ *is* holomorphic on D by assumption and $K \subset D$. If $z \in (K)$, then $z \in Q_\lambda$ for some square $Q_\lambda \subset K$. If z belongs to the interior of Q_λ we have by Cauchy's Integral formula (Theorem 2.4)

$$f(z) = \frac{1}{2\pi i} \int_{\partial Q_\lambda} \frac{f(\zeta)}{\zeta - z} d\zeta$$

and by Cauchy's Theorem (Theorem 2.2)

$$0 = \frac{1}{2\pi i} \int_{\partial Q_\mu} \frac{f(\zeta)}{\zeta - z} d\zeta, \qquad \mu \neq \lambda.$$

Therefore, by (5.7) and (4.40),

$$f(z) = \sum_{\mu=1}^{\nu} \frac{1}{2\pi i} \int_{\partial Q_\mu} \frac{f(\zeta)}{\zeta - z} d\zeta = \frac{1}{2\pi i} \int_C \frac{f(\zeta)}{\zeta - z} d\zeta,$$

that is, equality (5.8) holds if z belongs to the interior of Q_λ. The integral on the right-hand side of (5.8) is a holomorphic function of z on (K) by Theorem 1.17, and therefore a continuous function of z. Therefore (5.8) is also valid if z is a point on the boundary of the square Q_λ.

Since $|\gamma| \subset \Gamma \subset (K)$, we obtain from (5.8)

$$\int_\gamma f(z)\, dz = \frac{1}{2\pi i} \int_\gamma dz \int_C \frac{f(\zeta)}{\zeta - z} d\zeta.$$

Therefore

$$\int_\gamma f(z)\,dz = \frac{1}{2\pi i}\int_C f(\zeta)\,d\zeta \int_\gamma \frac{dz}{\zeta - z}.$$

Putting $\psi(\zeta) = \int_\gamma (dz/(\zeta - z))$, the function $\psi(\zeta)$ is a holomorphic function of ζ on the region $\Omega - \{\infty\} \subset \mathbb{C} - |\gamma|$ by Theorem 1.17 and $\psi'(\zeta) = -\int_\gamma (dz/(\zeta - z)^2)$. Since γ is closed, we have by (1.30)

$$\psi'(\zeta) = -\int_\gamma \frac{d}{dz}\left(\frac{1}{\zeta - z}\right)dz = 0.$$

Therefore $\psi(\zeta)$ is constant on the region $\Omega - \{\infty\}$, while $\lim_{\zeta \to \infty} \psi(\zeta) = 0$. We conclude that $\psi(\zeta) = 0$ identically on $\Omega - \{\infty\}$. Hence $\int_\gamma f(z)\,dz = 0$.

Now let $D \subset \mathbb{C}$ be a region without holes, $z_0 \in D$ and $f(z)$ a holomorphic function on D without zeros. If γ is a piecewise smooth curve connecting z_0 with an arbitrary point $z \in D$, the value of the integral $\int_\gamma^z (f'(z)/f(z))\,dz$ is uniquely determined by z and is independent of the choice of γ. For, if β is another curve in D connecting z_0 and z, then $\gamma^{-1}\beta$ is a closed curve in D. Since, moreover, $f'(z)/f(z)$ is holomorphic on D, we have, by Theorem 5.2,

$$\int_\beta \frac{f'(z)}{f(z)}dz - \int_\gamma \frac{f'(z)}{f(z)}dz = \int_{\gamma^{-1}\cdot\beta} \frac{f'(z)}{f(z)}dz = 0.$$

Writing $\int_\gamma^z (f'(z)/f(z))\,dz$ as $\int_{z_0}^z (f'(z)/f(z))\,dz$, the function

$$l(z) = \int_{z_0}^z \frac{f'(z)}{f(z)}dz$$

is a holomorphic function defined on D. By (4.26), $dl(z)/dz = f'(z)/f(z)$, hence $d(e^{l(z)}/f(z))/dz = 0$. Therefore $e^{l(z)}/f(z)$ is constant and since $l(z_0) = 0$ we conclude

$$e^{l(z)} = \frac{f(z)}{f(z_0)}.$$

Selecting an arbitrary value of $\log f(z_0)$ we have

$$\exp[l(z) + \log f(z_0)] = f(z),$$

that is, the holomorphic function $l(z) + \log f(z_0)$ is a branch over D of the multiple-valued analytic function $\log f(z)$. Hence

$$\log f(z) = \int_{z_0}^z \frac{f'(z)}{f(z)}dz + \log f(z_0). \tag{5.9}$$

In the above proof of Riemann's Mapping Theorem we assumed D is simply connected only to show that $\sqrt{(z - \sigma)}$ and $\sqrt{[\psi(z)]}$ have two single-valued holomorphic branches over D each. Therefore, this assumption can be replaced by the assumption that for each function $f(z)$ that is holomorphic and without zeros on D, $\sqrt{[f(z)]}$ has two single-valued, holomorphic branches $h(z)$ and $-h(z)$ over D. If $D \subset \mathbb{C}$ is a region without holes, $\log f(z)$, as occurring in (5.9), is a single-valued, holomorphic function on D and $\exp[\log f(z)] = f(z)$. Hence $\sqrt{[f(z)]}$ has two single-valued holomorphic branches $h(z) = \exp[\log f(z)]/2$ and $-h(z)$ over D. Therefore the following theorem holds.

Theorem 5.3. If $D \subsetneq \mathbb{C}$ is a region without holes and z_0 a point in D, then there exists exactly one conformal mapping $f: z \to w = f(z)$, from D onto the unit disk and satisfies $f(z_0) = 0$ and $f'(z_0) > 0$.

As stated above, if there exists a conformal mapping from D onto the unit disk U, then D is simply connected. Hence,

Corollary. A region without holes in the complex plane is simply connected.

This fact follows directly from equality (5.8) in the proof of Theorem 5.2. For, suppose that the complement with respect to $S = \mathbb{C} \cup \{\infty\}$ of the simply connected region $E \subset \mathbb{C}$ is the disjoint union of two nonempty closed sets F_0 and F_1, i.e., $S - E = F_0 \cup F_1$ and $F_0 \cap F_1 = \emptyset$. If $\infty \in F_0$, then F_1 is a bounded closed subset of $\mathbb{C}$. Letting F_0, F_1, and $E \cup F_1$ play the role of F, Γ, and D, respectively, occurring in the proof of Theorem 5.2, we find that there exists a 1-cycle C such that equality (5.8) holds for an arbitrary function that is holomorphic on $S - F_0$ and an arbitrary point $z \in F_1$.

Putting $f(z) = 1$, we get

$$1 = \frac{1}{2\pi i}\int_C \frac{1}{\zeta - z}\, d\zeta. \tag{5.10}$$

By Theorem 4.13 and the definition of homology (Definition 4.5) all 1-cycles are homologous to 0 on a simply connected region E. Therefore (5.10) contradicts Cauchy's Theorem (Theorem 4.14).

Ahlfors has defined a region without holes in $\mathbb{C}$ as a simply connected region (see L. Ahlfors, 1966, Section 4.2, Definition 1). By that definition, Theorem 5.3 becomes Riemann's Mapping Theorem. In order to prove that $\sqrt{(z-\sigma)}$ and $\sqrt{[\psi(z)]}$ have two single-valued holomorphic branches over D in the proof of Riemann's Mapping Theorem we used the Monodromy Theorem (Theorem 4.3). Note that the Monodromy Theorem was not used for the proof of Theorem 5.3.

If the boundary C of a bounded region $D \subset \mathbb{C}$ is a connected closed set, D is a region without holes.

For, suppose $S - D = \mathbb{C} \cup \{\infty\} - D = F_0 \cup F_1$ for some closed sets F_0 and F_1 with $F_0 \cap F_1 = \varnothing$. It suffices to prove that $F_0 = \varnothing$ or $F_1 = \varnothing$. Since $C = [D] - D \subset F_0 \cup F_1$ is a connected closed set, we have $C \subset F_0$ or $C \subset F_1$. Let us assume $C \subset F_0$; the case $C \subset F_1$ is treated similarly. Putting $F = D \cup F_0$, $F = D \cup C \cup F_0 = [D] \cup F_0$ is a closed set and $S = F \cup F_1$ and $F \cap F_1 = \varnothing$. Since S is of course a connected closed set, we conclude $F_1 = \varnothing$.

Hence, by the above corollary, if the boundary C of a bounded region $D \subset \mathbb{C}$ is a connected closed set, then D is simply connected. In particular, if the boundary C of the closed region D is a Jordan curve, then D is simply connected.

Theorem 5.4. If $D \subsetneqq \mathbb{C}$ and $\Delta \subsetneqq \mathbb{C}$ are simply connected regions and $z_0 \in D$, $\zeta \in \Delta$, and the angle θ are given, then there exists exactly one conformal mapping $f\colon z \to \zeta = f(z)$ from D onto Δ that satisfies $f(z_0) = \zeta_0$ and $\arg f'(z_0) = \theta$.

Proof: By Riemann's Theorem there exists one conformal mapping $g\colon z \to w = g(z)$ from D onto the unit disk U that satisfies $g(z_0) = 0$ and $g'(z_0) > 0$ and one conformal mapping $h\colon \zeta \to w = h(\zeta)$ from Δ onto U that satisfies $h(\zeta_0) = 0$ and $h'(\zeta_0) > 0$. Putting $\phi(w) = e^{i\theta}w$, $\phi\colon w \to \phi(w)$ is a conformal mapping from U onto itself that satisfies $\phi(0) = 0$. Hence

$$f = h^{-1} \circ \phi \circ g\colon z \to \zeta = f(z) = h^{-1}(e^{i\theta} g(z))$$

is a conformal mapping from D onto Δ and satisfies $f(z_0) = h^{-1}(0) = \zeta_0$:

$$\begin{array}{ccc} D & \xrightarrow{\ f\ } & \Delta \\ {\scriptstyle g}\downarrow & & \downarrow{\scriptstyle h} \\ U & \xrightarrow[\ \varphi\ ]{} & U \end{array}$$

Differentiating both sides of the equality $h(f(z)) = e^{i\theta} g(z)$ with respect to z and putting $z = z_0$ yields

$$h'(\zeta_0) f'(z_0) = e^{i\theta} g'(z_0).$$

Therefore

$$f'(z_0) = |f'(z_0)| e^{i\theta}, \qquad |f'(z_0)| = \frac{g'(z_0)}{h'(\zeta_0)},$$

that is, $\arg f'(z_0) = \theta$.

If $f: z \to \zeta = f(z)$ is a conformal mapping from D onto Δ and z_0 onto ζ_0, then $\phi = h \circ f \circ g^{-1}$ is a conformal mapping from U onto U that maps 0 onto 0. Hence, by the corollary to Schwarz's Lemma, we have $\phi: w \to \phi(w) = e^{i\theta}w$, so $\theta = \arg f'(z_0)$. We conclude that there exists only one conformal mapping from D onto Δ and z_0 onto ζ_0 that satisfies $\arg f'(z_0) = \theta$.

Regions D and Δ that can be conformally mapped onto each other are called *conformally equivalent*. Conformally equivalent regions can be identified with each other as far as the analytic functions on these regions are concerned. By Theorem 5.4, all simply connected regions $D \subsetneq \mathbb{C}$ are conformally equivalent.

5.2 Correspondence of boundaries

Let D and Δ be simply connected regions in $\mathbb{C}$ and let C and Γ be their respective boundaries. There are several results known pertaining to the question of what kind of correspondence between C and Γ is induced by a conformal mapping between D and Δ. The degree of difficulty of this problem depends on the restrictions imposed on C and Γ. In this paragraph we assume that both C and Γ are piecewise smooth Jordan curves.

If the boundary C of a bounded region D is a piecewise smooth Jordan curve, then D is the interior of its closure $[D] = D \cup C$.

To see this, let Ω represent the interior of $[D]$; then Ω is a region. Since $[D] \supset \Omega \supset D$, the assumption $D \subsetneq \Omega$ implies $C \cap \Omega \neq \emptyset$. Let $\gamma: t \to \gamma(t)$, $0 \leqq t \leqq 1$ with $\gamma(0) = \gamma(1)$, be a parameter representation of C such that $\gamma(0) \in \Omega$. There is an a, $0 < a < 1$, such that $\gamma(t) \in \Omega$ for $0 \leqq t < a$ and $\gamma(a) \notin \Omega$. We divide C into two Jordan curves, $C_1: t \to \gamma(t)$, $0 \leqq t \leqq a$, and $C_2: t \to \gamma(t)$, $a \leqq t \leqq 1$. Since C is a piecewise smooth by assumption, there exists a sufficiently small $\delta > 0$ such that for all r, $0 < r \leqq \delta$, the circle λ_r with center a and radius r intersects C_1 as well as C_2 in exactly one point

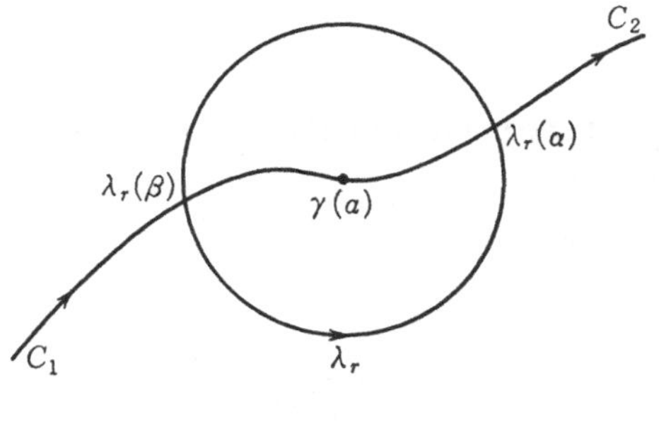

Fig. 5.4

$\lambda_r(\beta)$, respectively $\lambda_r(\alpha)$ (see Section 2.1b). Since $[\Omega] = [D]$ and $\Omega \supset D$, the boundary $[\Omega] - \Omega$ of Ω is a subset of C and $C_1 - \{\gamma(a)\} \subset \Omega$. Therefore, the boundary of Ω is a subset of C_2. Since Ω is a region and $\lambda_r(\beta) \in \Omega$, the circle λ_r minus its point of intersection, $\lambda_r(\alpha)$, with C_2, is contained in Ω (see Section 1.3a); hence $U_\delta(\gamma(a)) - C_2 \subset \Omega$. Therefore

$$[U_\delta(\gamma(a))] = [U_\delta(\gamma(a)) - C_2] \subset [\Omega].$$

Hence $\gamma(a) \notin \Omega$ turns out to be in an interior point of $[\Omega] = [D]$, contradicting the fact that Ω is the interior of $[D]$. Therefore $D = \Omega$, i.e., D is the interior of $[D]$.

So, if the boundary C of the bounded region D is a piecewise smooth Jordan curve, then D is the interior of $[D]$, hence $C = [D] - D$ is the boundary of the bounded closed region $[D]$. Therefore, orienting C and putting $C = \partial[D]$, we have $[D] - D = |C|$ (see Definition 2.2).

A mapping f from a point set in $\mathbb{C}$ one-to-one onto another point set in $\mathbb{C}$ such that both f and its inverse mapping are continuous is called a *homeomorphism*.

Theorem 5.5 (see Hurwitz, 1929, Chapter 6, Section 4). Let D and Δ be bounded regions in $\mathbb{C}$ and let $f: z \to \zeta = f(z)$ be a conformal mapping from D onto Δ. If the boundaries of D and Δ are piecewise smooth Jordan curves, the conformal mapping f can be extended to a homeomorphism $\tilde{f}$ from $[D]$ onto $[\Delta]$. The mapping $\tilde{f}$ maps the boundary $C = \partial[D]$ of D onto the boundary $\Gamma = \partial[\Delta]$ of Δ preserving orientations: $\tilde{f}(C) = \Gamma$.

Proof: (1) By way of preparation let us first consider what happens in the neighborhood of a point $q \in C$. Let $\gamma: t \to \gamma(t), 0 \leqq t \leqq 1$ with $\gamma(0) = \gamma(1)$ be a parameter representation of the piecewise smooth Jordan curve C and put $q = \gamma(a)$ for some a with $0 < a < 1$. Let

$$\lambda_r: \theta \to \lambda_r(\theta) = q + re^{i\theta}$$

represent the circle with center q and radius r. We saw in Section 2.1b that for a sufficiently small $\delta > 0$ all circles λ_r with $0 < r \leqq \delta$ intersect the curve C in exactly two points $\lambda_r(\alpha) = \gamma(u)$, $u > a$, and $\lambda_r(\beta) = \gamma(v)$, $v < a$. (We proved this for a point $\gamma(a) = \gamma(a_k)$, where the curve C is not smooth; obviously the same result holds also for points $\gamma(a)$ where C is smooth.) Of course, $\alpha = \alpha(r)$, $\beta = \beta(r)$, $u = u(r)$ and $v = v(r)$ are all functions of r. We will examine the properties of these functions below.

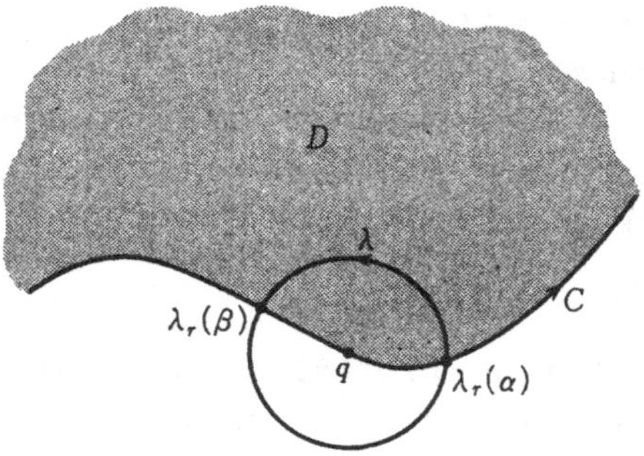

Fig. 5.5

As stated earlier, $|\gamma(t)-q|$ is a continuously differentiable, monotone increasing function of t on the closed interval $[a, a+\kappa]$ for some $\kappa > 0$ and

$$\frac{d}{dt}|\gamma(t)-q| > 0, \qquad |\gamma(\kappa)-q| > \delta.$$

The function $u = u(r)$ is determined by the equation $|\gamma(u)-q| = r$, i.e., $t = u(r)$ is the inverse function of the function $r = |\gamma(t)-q|$. Therefore, $u(r)$ is a continuously differentiable, monotone increasing function of r on the closed interval $[0, \delta]$ and $u'(r) > 0$. The function $\alpha(r)$ is determined by

$$q + re^{i\alpha(r)} = \lambda_r(\alpha(r)) = \gamma(u(r)),$$

hence $\alpha(r)$ is a continuously differentiable function of r. Since $q = \gamma(a) = \gamma(u(0))$, we have

$$\lim_{r\to+0} e^{i\alpha(r)} = \lim_{r\to+0} \frac{\gamma(u(r))-q}{r} = \gamma'(a)u'(0).$$

Since $((d|\gamma(t)-q|)/dt)_{t=+0} = \gamma'(a)$ by (2.16), we have $u'(0) = 1/|\gamma'(a)|$. Hence

$$\lim_{r\to+0} e^{i\alpha(r)} = \frac{\gamma'(a)}{|\gamma'(a)|}.$$

This proves that $\alpha(0) = \lim_{r\to+0}\alpha(r)$ exists and equals the argument of $\gamma'(a)$. Hence $\alpha(r)$ is a continuous function of r on $[0, \delta]$. Here, $\gamma'(a)$ denotes the right derivative $D^+\gamma(a)$ of $\gamma(t)$ at $t = a$.

Similarly, $\beta(r)$ is a continuous function of r on the closed interval $[-\delta, 0]$ and $\beta(0)-\pi$ equals the left derivative $D^-\gamma(a)$. Therefore, $\beta(0) = \alpha(0)+\pi$ if C is smooth at $\gamma(a)$ and $\beta(0) \neq \alpha(0)+\pi$ if C is not smooth at $\gamma(a)$. Since

$$\lambda_r \cap D = \{\lambda_r(\theta): \alpha(r) < \theta < \beta(r)\} \tag{5.11}$$

we have

$$U_\delta(q) \cap D = \{q + re^{i\theta}: 0 < r < \delta,\ \alpha(r) < \theta < \beta(r)\}, \tag{5.12}$$

where $U_\delta(q)$ is the disk with center q and radius δ. Hence $U_\delta(q) \cap D$ is a region and therefore a connected open set.

(2) We want to prove that the holomorphic function $f(z)$ is uniformly continuous on D, that is for each $\varepsilon > 0$ there exists a $\delta(\varepsilon) > 0$ such that if

$$|z - w| < \delta(\varepsilon), \qquad \text{then } |f(z) - f(w)| < \varepsilon \tag{5.13}$$

on D. Let us assume that $f(z)$ is not uniformly continuous on D and deduce a contradiction. Since by assumption, $f(z)$ is not uniformly continuous on D, there exists an $\varepsilon_0 > 0$ such that for each natural number m there are z_m and $w_m \in D$ such that

$$|z_m - w_m| < 1/m, \qquad |f(z_m) - f(w_m)| \geqq \varepsilon_0.$$

Since D and Δ are bounded, $\{z_m\}$, $\{w_m\}$, $\{f(z_m)\}$, and $\{f(w_m)\}$ are bounded sequences. It is well-known that every bounded sequence possesses a convergent subsequence. Therefore, it is possible to find a sequence $\{m_n\}$, $m_1 < m_2 < \cdots < m_n < \cdots$, of natural numbers such that all subsequences $\{z_{m_n}\}$, $\{w_{m_n}\}$, $\{f(z_{m_n})\}$, and $\{f(w_{m_n})\}$ are convergent. Since $m_n \geqq n$ we have

$$|z_{m_n} - w_{m_n}| < \frac{1}{m_n} \leqq \frac{1}{n}.$$

Hence, rewriting z_{m_n} and w_{m_n} as z_n, respectively w_n, we have

$$|z_n - w_n| < \frac{1}{n}, \qquad |f(z_n) - f(w_n)| \geqq \varepsilon_0, \tag{5.14}$$

and the sequences $\{z_n\}$, $\{w_n\}$, $\{f(z_n)\}$, and $\{f(w_n)\}$ are convergent. Since $\lim_{n\to\infty} z_n = \lim_{n\to\infty} w_n$ by (5.14), we put

$$q = \lim_{n\to\infty} z_n = \lim_{n\to\infty} w_n.$$

Now, $q \in [D]$. Suppose $q \in D$; then $\lim_{n\to\infty} f(z_n) = \lim_{n\to\infty} f(w_n) = f(q)$, contradicting (5.14). Hence $q \in \partial[D] = C$. We put $\zeta_n = f(z_n)$, $\omega_n = f(w_n)$, and

$$P = \lim_{n\to\infty} \zeta_n = \lim_{n\to\infty} f(z_n), \qquad Q = \lim_{n\to\infty} \omega_n = \lim_{n\to\infty} f(w_n).$$

Now $P \in [\Delta]$. Suppose $P \in \Delta$, then $q = \lim_{n\to\infty} z_n = \lim_{n\to\infty} f^{-1}(\zeta_n) = f^{-1}(P) \in D$ since the inverse map f^{-1} of f maps Δ conformally onto D, contradicting $q \in C$. We conclude that $P \in \Gamma$ and similarly $Q \in \Gamma$. By (5.14)

$$|P - Q| \geqq \varepsilon_0 > 0.$$

For sufficiently small ρ, $0 < \rho < \varepsilon_0/3$, $U_\rho(P) \cap \Delta$ and $U_\rho(Q) \cap \Delta$ are regions by (1). Since $\lim_{n\to\infty} \zeta_n = P$ and $\lim_{n\to\infty} \omega_n = Q$, we have for a

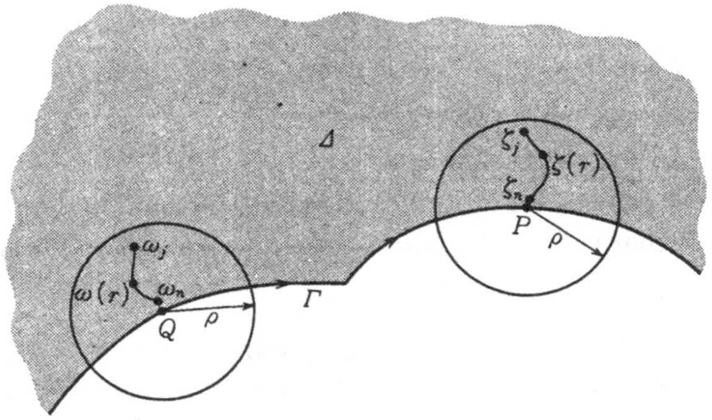

Fig. 5.6

sufficiently large natural number j:

$$\text{if} \quad n \geqq j, \qquad \text{then } \zeta_n \in U_\rho(P),\ \omega_n \in U_\rho(Q). \tag{5.15}$$

If we choose δ, as defined in (1), sufficiently small, we have for this j:

$$z_j \notin U_\delta(q), \qquad w_j \notin U_\delta(q).$$

Since $\lim_{n\to\infty} z_n = \lim_{n\to\infty} w_n = q$, for arbitrary ε, $0 < \varepsilon < \delta$, we have

$$z_n \in U_\varepsilon(q), \qquad w_n \in U_\varepsilon(q)$$

where $n > j$ is sufficiently large.

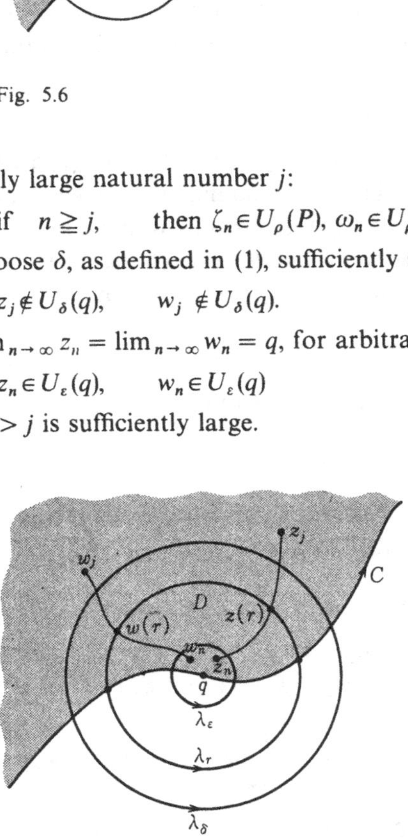

Fig. 5.7

Since $\zeta_n = f(z_n) \in U_\rho(P)$ by (5.15), it is possible to connect ζ_j and ζ_n by a curve σ: $s \to \sigma(s)$, $0 \leqq s \leqq 1$, in the region $U_\rho(P)$. The curve $f^{-1}(\sigma)$: $s \to f^{-1}(\sigma(s))$, $0 \leqq s \leqq 1$, connects z_j and z_n in D. Since $z_j \notin U_\delta(q)$ and

$z_n \in U_\varepsilon(q)$, the curve $f^{-1}(\sigma)$ intersects the circle λ_r in at least one point $z(r)$ for all r with $\varepsilon < r < \delta$. Since $z(r) \in \lambda_r \cap D$, we have

$$z(r) = \lambda_r(\phi), \qquad \alpha(r) < \phi < \beta(r)$$

by (5.11), where the angle ϕ depends on r. Similarly, connecting ω_j and ω_n by a curve τ in $U_\rho(Q)$, the curve $f^{-1}(\tau)$ connects w_j and w_n in D and intersects the circle λ_r in at least one point:

$$w(r) = \lambda_r(\psi), \qquad \alpha(r) < \psi < \beta(r).$$

Since $\zeta_r = f(z(r))$ is a point on the curve σ, we have $f(z(r)) \in U_\rho(P)$ and similarly, $\omega(r) = f(w(r)) \in U_\rho(Q)$. Since $|P - Q| \geqq \varepsilon_0 > 3\rho$, we have

$$|f(\lambda_r(\psi)) - f(\lambda_r(\phi))| = |f(w(r)) - f(z(r))| > \rho.$$

The two cases $\phi < \psi$ and $\psi < \phi$ are treated in exactly the same way, so let us assume $\phi < \psi$. Then

$$f(\lambda_r(\psi)) - f(\lambda_r(\phi)) = \int_\phi^\psi \frac{d}{d\theta} f(\lambda_r(\theta))\, d\theta = \int_\phi^\psi f'(\lambda_r(\theta))\lambda_r'(\theta)\, d\theta.$$

Since $\lambda_r(\theta) = q + re^{i\theta}$, we have $\lambda_r'(\theta) = ire^{i\theta}$, hence

$$\rho < |f(\lambda_r(\psi)) - f(\lambda_r(\phi))| \leqq \int_\phi^\psi |f'(\lambda_r(\theta))|\, r d\theta.$$

Writing $|f'(\lambda_r(\theta))|\, r$ as the product of $\sqrt{r}$ and $|f'(\lambda_r(\theta))|\sqrt{r}$ we get, by Schwarz's inequality,

$$\rho^2 < \left(\int_\phi^\psi |f'(\lambda_r(\theta))| r d\theta \right)^2 \leqq \int_\phi^\psi r d\theta \int_\phi^\psi |f'(\lambda_r(\theta))|^2 r d\theta.$$

Hence, since $\alpha(r) < \phi < \psi < \beta(r) < \alpha(r) + 2\pi$, we obtain

$$\rho^2 < 2\pi r \int_{\alpha(r)}^{\beta(r)} |f'(q + re^{i\theta})|^2 r\, d\theta.$$

Dividing both sides of the above inequality by $2\pi r$ and integrating from ε to δ, we arrive at the inequality

$$\frac{\rho^2}{2\pi} \log \frac{\delta}{\varepsilon} < \int_\varepsilon^\delta \int_{\alpha(r)}^{\beta(r)} |f'(q + re^{i\theta})|^2 r\, dr\, d\theta. \tag{5.16}$$

The integral on the right is an integral over the subregion

$$W_\varepsilon = (U_\delta(q) - [U_\varepsilon(q)]) \cap D = \{q + re^{i\theta} : \varepsilon < r < \delta, \alpha(r) < \theta < \beta(r)\}$$

of D representing the area of the subregion $f(W_\varepsilon)$ of Δ. For, writing $z = q + re^{i\theta} = x + iy$ and considering (r, θ) as polar coordinates with center q in the (x, y) plane, we have by the formula for the change of variables:

$$\int_\varepsilon^\delta \int_{\alpha(r)}^{\beta(r)} |f'(z)|^2 r\, dr\, d\theta = \int_{W_\varepsilon} |f'(z)|^2\, dx\, dy.$$

Now putting $f(z) = u + iv$, we have $|f'(z)|^2 = \partial(u, v)/\partial(x, y)$ by (3.5), hence

$$\int_{W_\varepsilon} |f'(z)|^2 \, dx \, dy = \int_{W_\varepsilon} \frac{\partial(u, v)}{\partial(x, y)} \, dx \, dy = \text{area of } f(W_\varepsilon).$$

Putting A = area of Δ we get from (5.16)

$$\frac{\rho^2}{2\pi} \log \frac{\delta}{\varepsilon} < A$$

contradicting $\lim_{\varepsilon \to +0} \log \delta/\varepsilon = +\infty$. Hence $f(z)$ is uniformly continuous on D.

(3) In order to extend a uniformly continuous function $f(z)$ on D to a uniformly continuous function $\tilde{f}(z)$ on $[D] = D \cup |C|$, it suffices to define $\tilde{f}(z) = f(z)$ if $z \in D$ and $\tilde{f}(z) = \lim_{n \to \infty} f(z_n)$ if $z \in C$, where $\{z_n\}$ is a sequence of points $z_n \in D$ converging to z. This is easily verified as follows: Select $c \in C$ and points $z_n \in D$ such that $\{z_n\}$ converges to c. For each $\delta > 0$ there exists an $n_0(\delta)$ satisfying

$$|z_n - z_m| < \delta \qquad \text{if } m > n > n_0(\delta).$$

Since $f(z)$ is uniformly continuous on D for each $\varepsilon > 0$ we can find a $\delta(\varepsilon) > 0$ satisfying

$$|f(z) - f(w)| < \varepsilon \qquad \text{if } |z - w| < \delta(\varepsilon), \qquad z \in D, w \in D$$

and therefore

$$|f(z_n) - f(z_m)| < \varepsilon \qquad \text{if } m > n > n_0(\delta(\varepsilon)),$$

that is, the sequence $\{f(z_n)\}$ converges. Put $\zeta = \lim_{n \to \infty} f(z_n)$. If $|w - c| < \delta(\varepsilon)$ and $w \in D$, then $|w - z_n| < \delta(\varepsilon)$ for sufficiently large n, since $\{z_n\}$ converges to c. Hence $|f(w) - f(z_n)| < \varepsilon$. Letting n tend to ∞, we conclude

$$|f(w) - \zeta| \leqq \varepsilon \qquad \text{if } |w - c| < \delta(\varepsilon), \qquad w \in D. \tag{5.17}$$

Hence, if $\{w_n\}$, $w_n \in D$, is a sequence converging to c, the sequence $\{f(w_n)\}$ converges to ζ, that is, $\zeta = \lim_{n \to \infty} f(z_n)$ is uniquely determined by c and independent of the choice of the sequence $\{z_n\}$. Therefore we write

$$\tilde{f}(c) = \lim_{n \to \infty} f(z_n), \qquad \lim_{n \to \infty} z_n = c, \qquad z_n \in D.$$

In this way the function $f(z)$ defined on D is extended to a function $\tilde{f}(z)$ defined on $[D] = D \cup |C|$. In order to prove that $\tilde{f}(z)$ is uniformly continuous on $[D]$, we replace c by z in (5.17)

$$|f(w) - \tilde{f}(z)| \leqq \varepsilon \qquad \text{if } |w - z| < \delta(\varepsilon), \qquad w \in D.$$

For $w \in C$ such that $|w - z| < \delta(\varepsilon)$ let $\{w_n\}$, $w_n \in D$, be a sequence converging to w. Since $|w_n - z| < \delta(\varepsilon)$ for sufficiently large n, we have $|f(w_n) - \tilde{f}(z)| \leqq \varepsilon$, hence $|\tilde{f}(w)-\tilde{f}(z)| \leqq \varepsilon$. Therefore

$$|\tilde{f}(w)-\tilde{f}(z)| \leqq \varepsilon, \qquad \text{if } |w-z| < \delta(\varepsilon)$$

on $[D]$, that is, $\tilde{f}(z)$ is uniformly continuous on $[D]$. $\tilde{f}$: $z \to \zeta = \tilde{f}(z)$ is a continuous mapping from $[D]$ onto $[\Delta]$. The inverse mapping f^{-1} of f maps Δ conformally onto D. Therefore it is possible to extend $g = f^{-1}$ to a continuous mapping $\tilde{g}$: $\zeta \to z = \tilde{g}(\zeta)$ which maps $[\Delta]$ onto $[D]$ in the same way as above. $\tilde{g}$ is the inverse of $\tilde{f}$: $\tilde{g} = \tilde{f}^{-1}$. (For, if $z \in C$ and if $\{z_n\}$, $z_n \in D$, is a sequence converging to z, then

$$\tilde{g}(\tilde{f}(z)) = \tilde{g}\left(\lim_{n\to\infty} f(z_n)\right) = \lim_{n\to\infty} g(f(z_n)) = \lim_{n\to\infty} z_n = z.$$

Similarly, if $\zeta \in \Gamma$, then $\tilde{f}(\tilde{g}(\zeta)) = \zeta$.) Therefore $\tilde{f}$ is a homeomorphism mapping $[D]$ onto $[\Delta]$. Since $\tilde{f}(D) = f(D) = \Delta$, $\tilde{f}(|C|) = |\Gamma|$, that is, the restriction of $\tilde{f}$ to $|C|$ is a homeomorphism mapping $|C|$ onto Γ.

(4) We prove $\tilde{f}(C) = \Gamma$, that is, $\tilde{f}$ preserves orientations. Let g be a conformal mapping which maps D onto the unit disk U and put $h = f \circ g^{-1}$. Then h is a conformal mapping which maps U onto Δ and $f = h \circ g$. Therefore it suffices to consider the case in which D is the unit disk. $D = \{z: |z| < 1\}$. Then $C = \partial[D]$ is the unit circle. From the parameter representation $\theta \to z = e^{i\theta}$, $0 \leqq \theta \leqq 2\pi$, of C, we obtain the parameter representation

$$\tilde{f}(C)\colon \theta \to \zeta = \tilde{f}(re^{i\theta}), \qquad 0 \leqq \theta \leqq 2\pi$$

for the Jordan curve $\tilde{f}(C)$. Since $|\tilde{f}(C)| = \tilde{f}(|C|) = |\Gamma|$, we have $\tilde{f}(C) = \Gamma$ or $\tilde{f}(C) = -\Gamma$ (see Section 2.1a).

Putting $U_r = \{z: |z| < r\}$, $C_r = \partial[U_r]$, and $\Delta_r = f(U_r)$, Δ_r is a subdomain of Δ and $f(C_r)$ is a smooth curve in Δ, represented by $\theta \to \zeta = f(re^{i\theta})$, $0 \leqq \theta \leqq 2\pi$. $\tilde{f}$ maps the radius $[0,1]$ of C onto the Jordan curve λ: $s \to \lambda(s) = \tilde{f}(s)$, $0 \leqq s \leqq 1$, in $[\Delta]$. The curve λ is smooth except at its end point $\lambda(1)$. Since f is a conformal mapping λ crosses $f(C_r)$ at $\lambda(r) = f(r)$ from left to right. Hence $\lambda(s) \in \Delta_r$ if $s < r$ and $\lambda(s) \notin [\Delta_r]$ if $s > r$, i.e., Δ_r is to the left of $f(C_r)$. Therefore $f(C_r) = \partial[\Delta_r]$, where $\partial[\Delta_r]$ is defined by Definition 2.2. Since the boundary of the closed region $[\Delta] - \Delta_r$ is given by

$$\partial([\Delta] - \Delta_r) = \Gamma - f(C_r)$$

we have $\Gamma - f(C_r) \sim 0$ in an arbitrary region containing $[\Delta] - \Delta_r$, for example, $\Omega = \mathbb{C} - \{f(0)\}$, as explained in connection with Theorem 4.14 (Cauchy's Theorem).

In order to prove that $\tilde{f}(C) - f(C_r) \sim 0$ in Ω, we consider the continuous mapping

$$M\colon (s, \theta) \to M(s, \theta) = \tilde{f}(se^{i\theta})$$

from the rectangle $K = \{(s, \theta): r \leqq s \leqq 1, 0 \leqq \theta \leqq 2\pi\}$ into $[\Delta] - \Delta_r$. Putting $\mu(s) = M(s, 0)$, $\mu: s \to \mu(s)$, $r \leqq s \leqq 1$, is a curve connecting $f(r)$ and $\tilde{f}(1)$ in $[\Delta] - \Delta_r \subset \Omega$. The mapping M maps the boundary ∂K of K onto the closed curve with base point $f(r)$:

$$\mu \cdot \tilde{f}(C) \cdot \mu^{-1} \cdot f(C_r)^{-1}.$$

Hence, $\mu \cdot \tilde{f}(C) \cdot \mu^{-1} \cdot f(C_r)^{-1} \simeq 0$ in Ω, hence

$$\tilde{f}(C) - f(C_r) \sim 0$$

in Ω by (4.34), (4.35), and the definition of homology (Definition 4.5). Also $\Gamma - f(C_r) \sim 0$ in Ω. Hence

$$\Gamma - \tilde{f}(C) \sim 0$$

in Ω.

Now suppose $\tilde{f}(C) = -\Gamma$, then $2\Gamma \sim 0$ in Ω. Since the function $1/(\zeta - f(0))$ of ζ is holomorphic on $\Omega = \mathbb{C} - f(0)$, we conclude from Cauchy's Theorem (Theorem 4.14)

$$2 \int_\Gamma \frac{d\zeta}{\zeta - f(0)} = 0.$$

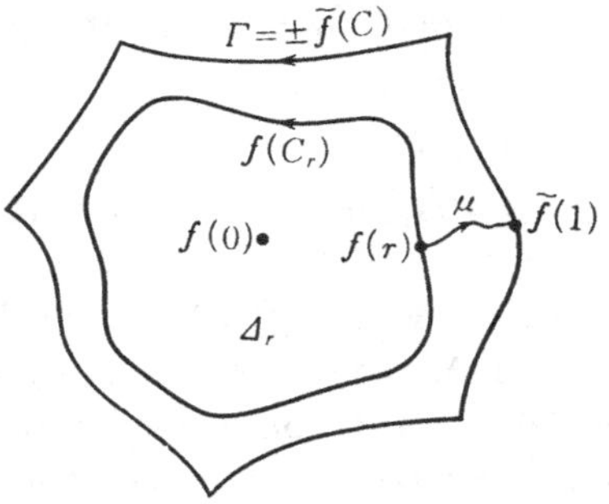

Fig. 5.8

Since $f(0) \in \Delta$, $\Gamma = \partial[\Delta]$ we have by Cauchy's integral formula (Theorem 2.4)

$$1 = \frac{1}{2\pi i} \int_\Gamma \frac{d\zeta}{\zeta - f(0)}.$$

So we have arrived at a contradiction and hence $\tilde{f}(C) = \Gamma$.

Now let us assume that both D and Δ are simply connected bounded regions and that $C = \partial[D]$ and $\Gamma = \partial[\Delta]$ are piecewise smooth Jordan curves. Using the orientation defined on C, we have for three distinct points a_1, a_2, and a_3 on C either $a_1 \prec a_2 \prec a_3$ or $a_3 \prec a_2 \prec a_1$ (see Section 3.3b, where the case of the circle was treated). Let $f: z \to \zeta = f(z)$ be a conformal mapping which maps D onto Δ and let $\tilde{f}: z \to \zeta = \tilde{f}(z)$ be the homeomorphism of Theorem 5.5, from $[D]$ onto $[\Delta]$.

Theorem 5.6. Given three different points a_1, a_2, and a_3 on C such that $a_1 \prec a_2 \prec a_3$ and three different points α_1, α_2, and α_3 on Γ such that $\alpha_1 \prec \alpha_2 \prec \alpha_3$, there exists exactly one $\tilde{f}$ satisfying $\tilde{f}(a_1) = \alpha_1, \tilde{f}(a_2) = \alpha_2$, and $\tilde{f}(a_3) = \alpha_3$.

Proof: Let h be a conformal mapping from Δ onto the unit disk $U = \{w: |w| < 1\}$. If g is a conformal mapping from D onto U, then $f = h^{-1} \circ g$ is a conformal mapping from D onto Δ.

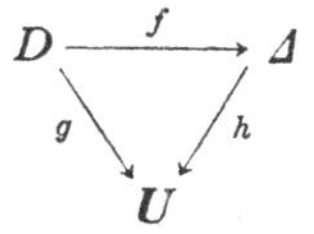

Conversely, if f is a conformal mapping from D onto Δ, $g = h \circ f$ is a conformal mapping from D onto U. Since $\tilde{h}(\Gamma) = +\partial[U]$ by Theorem 5.5, $w_k = \tilde{h}(\alpha_k)$, where here and below $k = 1, 2, 3$, are three points on the unit circle $\partial[U]$ satisfying $w_1 \prec w_2 \prec w_3$. Since $\tilde{f}(a_k) = \alpha_k$ if and only if $\tilde{g}(a_k) = w_k$, it suffices to prove that there exists exactly one conformal mapping $g: z \to w = g(z)$ from D onto U that satisfies $\tilde{g}(a_k) = w_k$, that is, it suffices to prove the theorem for the case that Δ is the unit disk. Let g be a conformal mapping from D onto $\Delta = \{\zeta: |\zeta| < 1\}$. For an arbitrary conformal mapping f from D onto Δ, let ϕ be the conformal mapping from Δ onto itself determined by $f = \phi \circ g$. By Theorem 3.6, $\phi: \zeta \to \phi(\zeta)$ is given by a linear fractional function $\phi(\zeta)$. The linear fractional transformation ϕ is a homeomorphism from the Riemann sphere S onto itself (Section 3.3a), hence $\tilde{f} = \phi \circ \tilde{g}$. Since $\tilde{g}(C) = +\Gamma$ by Theorem 5.5, $\zeta_k = \tilde{g}(a_k)$ are three points on the unit circle Γ, such that $\zeta_1 \prec \zeta_2 \prec \zeta_3$. By the corollary to Theorem 3.8 there exists exactly one linear fractional transformation ϕ such that $\phi(\zeta_k) = \alpha_k$. Since $\alpha_1 \prec \alpha_2 \prec \alpha_3$ by assumption, ϕ maps Δ onto itself by Section 3.3b. Hence there exists exactly one conformal mapping from D onto Δ that satisfies $\tilde{f}(a_k) = \alpha_k$.

By Theorem 3.13, the mapping

$$h: \zeta \to w = h(\zeta) = \frac{\zeta - i}{\zeta + i}$$

is a conformal mapping from the upper half-plane H^+ of the ζ-plane onto the unit disk U. Replacing Δ by H^+ in the first half of the above proof, all conformal mappings f from D onto H^+ can be written as $f = h^{-1} \circ g$, where g is a conformal mapping from D onto U. Since the linear fractional transformation h is a homeomorphism from the Riemann sphere onto itself, $h^{-1}([U])$ is the closure $[H^+] = H^+ \cup \mathbb{R} \cup \{\infty\}$ of H^+ in S and $\tilde{f} = h^{-1} \circ \tilde{g}$ is a homeomorphism from $[D]$ onto $[H^+]$. Since $h(0) = -1$, $h(1) = -i$ and $h(\infty) = 1$, h^{-1} maps the unit circle $\partial[U]$ onto the positively oriented line $\mathbb{R} \cup \{\infty\}$. Therefore $\tilde{f}$ maps $C = \partial[D]$ onto $\mathbb{R} \cup \{\infty\}$. If c_0, c_1, and c_∞ are three points on C such that $c_0 \prec c_1 \prec c_\infty$, then $\tilde{f}(c_0) = 0, \tilde{f}(c_1) = 1$, and $\tilde{f}(c_\infty) = \infty$ if and only if $\tilde{g}(c_0) = -1, \tilde{g}(c_1) = -i$, and $\tilde{g}(c_\infty) = 1$. Therefore we get the following result.

Theorem 5.7. (1) The conformal mapping f from D onto H^+ can be extended to a homeomorphism $\tilde{f}$ from $[D]$ onto $[H^+] = H^+ \cup \mathbb{R} \cup \{\infty\}$ that $C = \partial[D]$ onto the positively oriented line $\mathbb{R} \cup \{\infty\}$.

(2) If c_0, c_1, and c_∞ are three different points on $C = \partial[D]$ such that $c_0 \prec c_1 \prec c_\infty$, then there exists exactly one conformal mapping $f: z \to \zeta = f(z)$ from D onto H^+ satisfying $\tilde{f}(c_0) = 0, \tilde{f}(c_1) = 1$ and $\tilde{f}(c_\infty) = \infty$.

5.3 The principle of reflection

a. The principle of reflection

For an arbitrary region D in the complex plane let $\bar{D}$ be defined by $\bar{D} = \{\bar{z}: z \in D\}$. If $f(z)$ is a holomorphic function of z on D, then $\overline{f(\bar{z})}$ is a holomorphic function of z on $\bar{D}$.

Proof: Expand $f(z)$ into a power series around an arbitrary point $c \in D$:

$$f(z) = \sum_{n=0}^{\infty} a_n (z - c)^n.$$

Then

$$\overline{f(\bar{z})} = \sum_{n=0}^{\infty} \bar{a}_n (z - \bar{c})^n$$

is the power series expansion of $\overline{f(\bar{z})}$ around $\bar{c} \in \bar{D}$. Hence $\overline{f(\bar{z})}$ is a holomorphic function of z on $\bar{D}$.

Let D be a region in the upper half-plane of the z-plane such that the open interval (a, b), of the real axis is part of the boundary of D, that is $(a, b) \subset [D] - D$ and $D \cup (a, b) \cup \bar{D}$ is a region in the z-plane. Similarly, let Δ

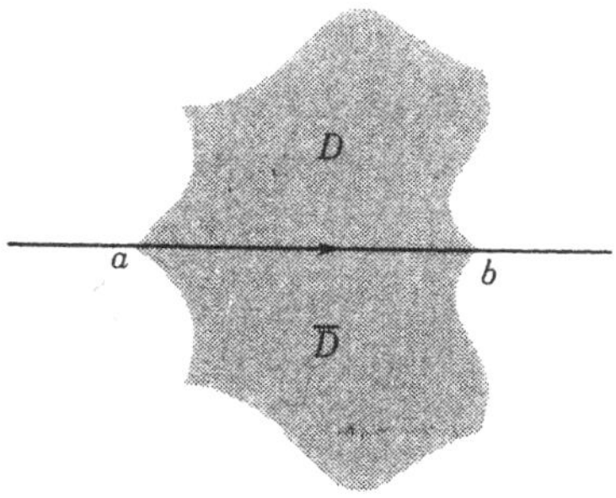

Fig. 5.9

be a region in the upper half-plane of the ζ-plane such that the open interval (α, β) of the real axis is part of the boundary of Δ, that is $(\alpha, \beta) \subset [\Delta] - \Delta$ and such that $\Delta \cup (\alpha, \beta) \cup \bar{\Delta}$ is a region in the ζ-plane. If D and Δ are conformally equivalent, the following theorem holds.

Theorem 5.8. If it is possible to extend a conformal mapping $f: z \to \zeta = f(z)$ from D onto Δ to a homeomorphism $\tilde{f}: z \to \zeta = \tilde{f}(z)$ from $D \cup (a, b)$ onto $\Delta \cup (\alpha, \beta)$, then it is possible to extend f to a conformal mapping g: $z \to \zeta = g(z)$ from $D \cup (a, b) \cup \bar{D}$ onto $\Delta \cup (\alpha, \beta) \cup \bar{\Delta}$. The mapping $g(z)$ is given by

$$g(z) = \begin{cases} f(z), & z \in D \\ \tilde{f}(z), & z \in (a,b), \\ \overline{f(\bar{z})}, & z \in \bar{D}. \end{cases} \tag{5.18}$$

Since $g(z)$ is an analytic continuation of $f(z)$, $f(z)$ can be extended in only one way.

Proof: Since $\overline{f(\bar{z})}$ is a holomorphic function of z on D, $\overline{\tilde{f}(\bar{z})}$ is holomorphic on $\bar{D}$ and continuous on $\bar{D} \cup (a, b)$. Since $\tilde{f}$ is a homeomorphism mapping $D \cup (a,b)$ onto $\Delta \cup (\alpha, \beta)$ such that $\tilde{f}(D) = \tilde{f}(\Delta)$, we have $\tilde{f}((a, b)) = (\alpha, \beta)$, that is, if $z \in (a, b)$, then $\zeta = \tilde{f}(z)$ is a real number and so $\overline{\tilde{f}(\bar{z})} = \tilde{f}(z)$. Therefore the continuous functions $\overline{\tilde{f}(\bar{z})}$ and $\tilde{f}(z)$ coincide on (a, b). This proves that the function $g(z)$ defined in (5.18) is continuous on $D \cup (a, b) \cup \bar{D}$ and holomorphic on $D \cup \bar{D}$. In order to prove that $g(z)$ is holomorphic on $D \cup (a, b) \cup \bar{D}$ it suffices to prove that it is holomorphic on some sufficiently small neighborhood $U_\varepsilon(c)$ of each point $c \in (a,b)$. Since $[U_\varepsilon(c)] \subset D \cup$

$(a, b) \cup \overline{D}$ for sufficiently small ε, the function

$$h(z) = \frac{1}{2\pi i}\int_{C_\varepsilon} \frac{g(w)}{w-z}\,dw, \qquad C_\varepsilon = \partial[U_\varepsilon(c)]$$

is holomorphic on $U_\varepsilon(c)$ by Theorem 1.17.

In order to prove that $g(z) = h(z)$ for $z \in U_\varepsilon(c)$, we put $V = U_\varepsilon(c) \cap D$ and $\overline{V} = U_\varepsilon(c) \cap \overline{D}$, whereby $[U_\varepsilon(c)]$ is divided into two cells $[V]$ and $[\overline{V}]$. Since $C_\varepsilon = \partial[V] + \partial[\overline{V}]$, we have, for $z \in V$ or $z \in \overline{V}$,

$$h(z) = \frac{1}{2\pi i}\int_{\partial[V]} \frac{g(w)}{w-z}\,dw + \frac{1}{2\pi i}\int_{\partial[\overline{V}]} \frac{g(w)}{w-z}\,dw.$$

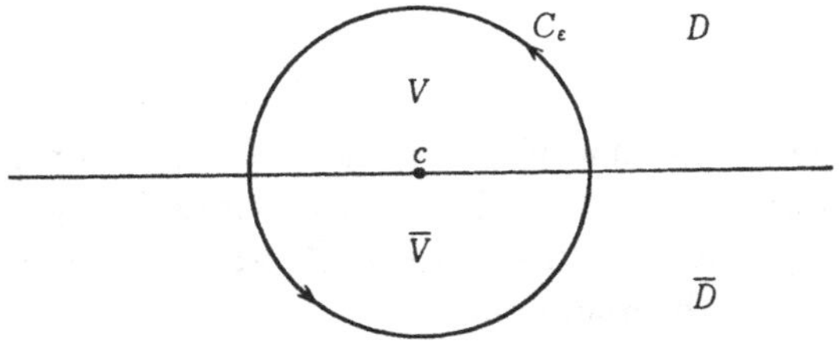

Fig. 5.10

If $z \in \overline{V}$, then the function $g(w)$ of w is continuous on $[V]$ and holomorphic on V, hence $g(w)/(w-z)$ is also continuous on $[V]$ and holomorphic on V. Hence

$$\int_{\partial[V]} \frac{g(w)}{w-z}\,dw = 0$$

by the strong version of Cauchy's Theorem (Theorem 2.3). Since $g(w)$ is continuous on $[\overline{V}]$ and holomorphic on $\overline{V}$, we have by Cauchy's integral formula (2.57),

$$g(z) = \frac{1}{2\pi i}\int_{\partial[\overline{V}]} \frac{g(w)}{w-z}\,dw.$$

Therefore, $h(z) = g(z)$. If $z \in V$, then it is proved in the same way that $h(z) = g(z)$. Hence, $h(z) = g(z)$ if $z \in U_\varepsilon(c)$, $\operatorname{Im} z \neq 0$. Since $h(z)$ and $g(z)$ are continuous on $U_\varepsilon(c)$, we conclude that $h(z) = g(z)$ holds identically on $U_\varepsilon(z)$. Hence $g(z)$ is holomorphic on $U_\varepsilon(c)$. We have proved that $g(z)$ is a holomorphic function on the region $D \cup (a, b) \cup \overline{D}$.

That $g(z)$ is a one-to-one mapping from $D \cup (a, b) \cup \overline{D}$ onto $\Delta \cup (\alpha, \beta) \cup \overline{\Delta}$ follows from the fact that $\tilde{f}\colon z \to \zeta = \tilde{f}(z)$ is a homeomorphism mapping

$D \cup (a, b)$ onto $\Delta \cup (\alpha, \beta)$ and the definition of $g(z)$. Therefore, by Theorem 3.5, g: $z \to \zeta = g(z)$ is a conformal mapping from $D \cup (a, b) \cup \bar{D}$ onto $\Delta \cup (\alpha, \beta) \cup \bar{\Delta}$.

Let l be a line in the complex plane. The correspondence $z \to z^*$ which assigns to each point z the point z^*, such that z and z^* are symmetric with respect to l, is called the *reflection with respect to l*. The point z^* is called the *reflection of z with respect to l*. The reflection with respect to the real axis $\mathbb{R}$ is, of course, given by $z \to \bar{z}$. A reflection with respect to an arbitrary line l can be reduced to a reflection with respect to $\mathbb{R}$ with the help of a linear transformation. To see this, select two different points c_0 and c_1 on l and put

$$\lambda(w) = (c_1 - c_0)w + c_0. \tag{5.19}$$

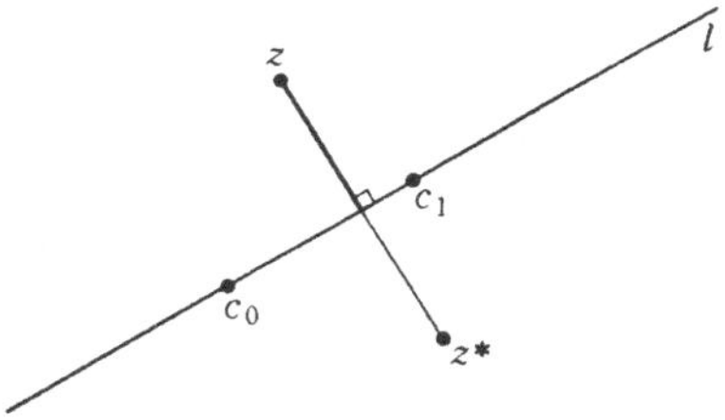

Fig. 5.11

The linear transformation λ: $w \to z = \lambda(w)$ maps the points $0, 1$, and ∞ of the extended plane onto the points c_0, c_1, and ∞, respectively, and therefore the real axis $\mathbb{R}$ onto l. Hence, $u \to z = \lambda(u)$, $-\infty < u < +\infty$, is a parameter representation of l. If $z = \lambda(w)$, then $z^* = \lambda(\bar{w})$. (For, putting $w = u + iv$, $u, v \in \mathbb{R}$, $V \neq 0$, we have

$$\tfrac{1}{2}(\lambda(w) + \lambda(\bar{w})) = \lambda(u), \qquad \lambda(w) - \lambda(\bar{w}) = 2(c_1 - c_0)iv.$$

Hence, the segment connecting $\lambda(\bar{w})$ and $\lambda(w)$ is perpendicular to the line l and its midpoint $\lambda(u)$ is on l. Hence the linear transformation λ transforms the reflection $w \to \bar{w}$ with respect to $\mathbb{R}$ into the reflection $z \to z^*$ with respect to l.)

The above λ is a special example of a linear transformation mapping the line $\mathbb{R} \cup \{\infty\}$ of the extended complex plane onto $l \cup \{\infty\}$. If μ: $w \to z = \mu(w)$ is another linear fractional transformation mapping $\mathbb{R} \cup \{\infty\}$ onto $l \cup \{\infty\}$, then $z = \mu(w)$ implies $z^* = \mu(\bar{w})$.

To see this, put $\phi = \lambda^{-1} \circ \mu$; then ϕ: $w \to \phi(w)$ is a linear fractional transformation mapping $\mathbb{R} \cup \{\infty\}$ onto itself. Putting $a_0 = \phi^{-1}(0)$, $a_1 = \phi^{-1}(1)$, and $a_\infty = \phi^{-1}(\infty)$, a_0, a_1, and a_∞ are real numbers and

$$\phi(w) = \frac{(w - a_0)(a_1 - a_\infty)}{(w - a_\infty)(a_1 - a_0)}.$$

Hence $\phi(\bar{w}) = \overline{\phi(w)}$, therefore using $z = \mu(w) = \lambda(\phi(w))$, we have $\mu(\bar{w}) = \lambda(\phi(\bar{w})) = \lambda\overline{(\phi(w))} = z^*$.

If for example $a_0 = \infty$, $a_1 = 1$, and $a_\infty = 0$, then $\phi(w) = 1/w$. In this case, ϕ maps the upper half-plane onto the lower half-plane and vice versa. Now we want to define reflection with respect to a circle. Let C be a circle in the z-plane, let c_0, c_1, and c_∞ be three different points on C and put

$$\lambda(w) = \frac{(c_0 - c_\infty)(c_\infty - c_1)}{(c_1 - c_0)w + (c_\infty - c_1)} + c_\infty. \tag{5.20}$$

Since $\lambda(0) = c_0$, $\lambda(1) = c_1$, and $\lambda(\infty) = c_\infty$, the linear fractional transformation λ: $w \to z = \lambda(w)$ maps the line $\mathbb{R} \cup \{\infty\}$ onto C. We define the reflection of $z = \lambda(w)$ with respect to C as $z^* = \lambda(\bar{w})$. Just as above in the case of reflection with respect to a line, it is proved that the reflection $z = \lambda(w) \to z^* = \lambda(\bar{w})$ with respect to C is independent of the choice of λ.

Let c be the center of C, and R its radius. If $c_0 \prec c_1 \prec c_\infty$ on C, λ maps the upper half-plane conformally onto the interior $U_R(c)$ of C (Section 3.3b). Putting $w_0 = \lambda^{-1}(c)$ we have by Theorem 3.13

$$\lambda(w) = c + Re^{i\psi} \cdot \frac{w - w_0}{w - \bar{w}_0}, \qquad \psi \text{ real}. \tag{5.21}$$

If $c_\infty \prec c_1 \prec c_0$, then λ maps the lower half-plane onto $U_R(c)$. Again putting $w_0 = \lambda^{-1}(c)$, (5.21) is valid again. Hence

$$z - c = Re^{i\psi} \cdot \frac{w - w_0}{w - \bar{w}_0}, \qquad z^* - c = Re^{i\psi} \cdot \frac{\bar{w} - w_0}{\bar{w} - \bar{w}_0},$$

where $z = \lambda(w)$ and $z^* = \lambda(\bar{w})$. Thus

$$(\bar{z} - \bar{c})(z^* - c) = R^2.$$

Therefore, writing $z - c = re^{i\theta}$, $r > 0$, $\theta \in \mathbb{R}$, we have $z^* - c = (R^2/r)e^{i\theta}$. Hence, for $z \neq c$ and $z \neq \infty$, z^* is on the ray starting at c through z and its distance from c is given by $|z^* - c| = R^2/|z - c|$. This also shows that the reflection $z \to z^*$ with respect to C is independent of the choice of the linear fractional transformation from $\mathbb{R} \cup \{\infty\}$ onto C. The reflections of c and ∞ with respect to C are given by $c^* = \infty$ and $\infty^* = c$. Reflections with respect to a circle or a line are invariant under linear fractional transformations, that is, if C is a circle or a line in $\mathbb{C}$ and ψ a linear fractional transformation of z,

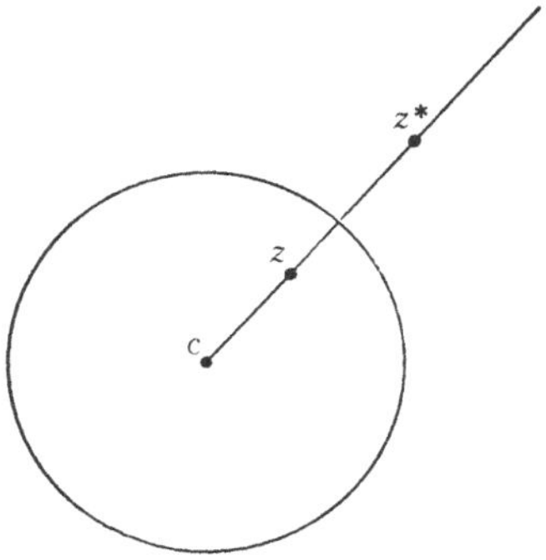

Fig. 5.12

then $\psi(z^*)$ is the reflection of $\psi(z)$ with respect to the circle or line $\psi(C)$: $\psi(z^*) = \psi(z)^*$.

To see this, let $\lambda: w \to z = \lambda(w)$ be a linear fractional transformation from $\mathbb{R} \cup \{\infty\}$ onto C, then $\psi \circ \lambda: w \to \psi(\lambda(w))$ is a linear fractional transformation from $\mathbb{R} \cup \{\infty\}$ onto $\psi(C)$. Therefore the reflection of $\psi(z)$, $z = \lambda(w)$, with respect to $\psi(C)$ is given by $\psi(z)^* = (\psi \circ \lambda)(\bar{w}) = \psi(\lambda(\bar{w})) = \psi(z^*)$.

Theorem 5.9. Let C be a line or a circle in the z-plane and let c be the center of C in case C is a circle.

(1) Let γ_1 and γ_2 be smooth curves in the z-plane intersecting at the point $a \neq c$ and let θ be the angle between γ_1 and γ_2 at a. If γ_1^*, γ_2^*, and a^* are the reflections of γ_1, γ_2, and a, respectively, with respect to C, then the angle between γ_1^* and γ_2^* at a^* equals $-\theta$.

(2) The reflection $z \to z^*$ with respect to C maps circles or lines into circles or lines.

Proof: If $\lambda: w \to z = \lambda(w)$ is a linear fractional transformation mapping the line $\mathbb{R} \cup \{\infty\}$ onto C, then the reflection $z \to z^*$ with respect to C is obtained from the reflection $w \to \bar{w}$ with respect to $\mathbb{R}$ via the transformation λ. If $z = \lambda(w)$, then $z^* = \lambda(\bar{w})$. For the reflection $w \to \bar{w}$ the theorem is obviously true and λ is a conformal mapping, from which the theorem follows.

We now want to derive from Theorem 5.8 the principle of reflection.

Let C be a circle in the z-plane with center c and radius R and let $z \to z^*$ be the reflection with respect to C. Further, let D be a region contained in the

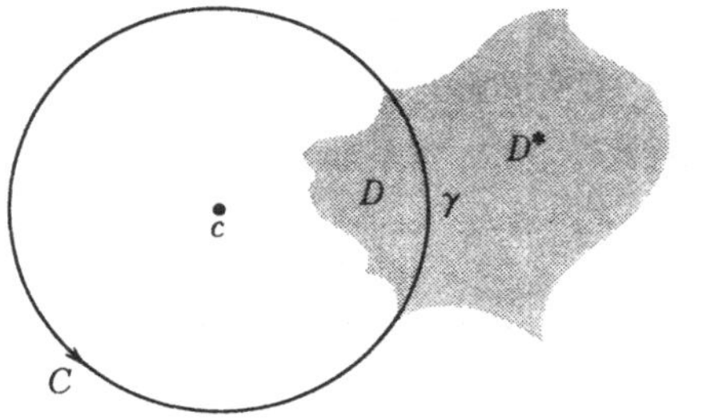

Fig. 5.13

interior or the exterior of C, such that $c \notin D$ and let the arc $\gamma \subsetneq C$ be part of the boundary of D: $\gamma \subset [D] - D$. If (γ) represents the arc γ from which the two end points have been removed, we further assume that $D \cup (\gamma) \cup D^*$ is one region in the z-plane. Here, $D^* = \{z^*: z \in D\}$. If $z \in C$, then $z^* = z$, hence $\gamma \subset [D^*] - D^*$ and each point $z \in (\gamma)$ is an interior point of $D \cup (\gamma) \cup D^*$ by assumption. Let Δ be a region in the upper half-plane of the ζ-plane that is conformally equivalent to D, let the open interval (α, β) of the real axis of the ζ-plane be a part of the boundary of Δ and assume that $\Delta \cup (\alpha, \beta) \cup \bar{\Delta}$ is a region in the ζ-plane.

Theorem 5.10 (principle of reflection). If the conformal mapping $f: z \to \zeta = f(z)$ for D onto Δ can be extended to a homeomorphism $\tilde{f}$: $z \to \zeta = \tilde{f}(z)$ mapping $D \cup (\gamma)$ onto $\Delta \cup (\alpha, \beta)$, then f can be extended to a conformal mapping $g: z \to \zeta = g(z)$ from $D \cup (\gamma) \cup D^*$ onto $\Delta \cup (\alpha, \beta) \cup \Delta^*$ and g is given by

$$g(z) = \begin{cases} f(z), & z \in D, \\ \tilde{f}(z), & z \in (\gamma), \\ \overline{f(z^*)}, & z \in D^*. \end{cases} \tag{5.22}$$

Proof: Let w_0 be a point in the w-plane such that $\operatorname{Im} w_0 > 0$ if D is contained in the interior of C and $\operatorname{Im} w_0 < 0$ if D is contained in the exterior of C, and put

$$\lambda(w) = c + Re^{i\psi} \frac{w - w_0}{w - \bar{w}_0}, \qquad \psi \in \mathbb{R}.$$

$\lambda: w \to z = \lambda(w)$ is a linear fractional transformation which maps the real axis $\mathbb{R} \cup \{\infty\}$ of the extended w-plane onto C and $\lambda(w_0) = c$ and $\lambda(\bar{w}_0) = \infty$. $\lambda(\infty) = c + Re^{i\psi}$ is a point on C. Since $\gamma \subsetneq C$, we assume from the outset that

ψ has been chosen such that $\lambda(\infty) \notin \gamma$. If we restrict the domain of definition of the inverse transformation λ^{-1}: $z \to w = \lambda^{-1}(z)$ to $\mathbb{C} - \{\lambda(\infty)\}$, λ^{-1} becomes a conformal mapping from $\mathbb{C} - \{\lambda(\infty)\}$ onto $\mathbb{C} - \{\bar{w}_0\}$ (Section 3.3a). If D is contained in the interior of C, then $c \notin D$ by assumption, hence $\infty \notin D^*$, hence $D \cup (\gamma) \cup D^*$ is contained in $\mathbb{C} - \{\lambda(\infty)\}$. Thus λ^{-1} maps $D \cup (\gamma) \cup D^*$ conformally onto the region $\lambda^{-1}(D) \cup \lambda^{-1}((\gamma)) \cup \lambda^{-1}(D^*)$ in the w-plane. If $\operatorname{Im} w_0 > 0$ then λ^{-1} maps the interior of C onto the upper half-plane of the w-plane and the exterior of C onto the lower half-plane of the w-plane. If $\operatorname{Im} w_0 < 0$, then λ^{-1} maps the exterior of C onto the upper half-plane and the interior onto the lower half-plane of the w-plane. If D is contained in the interior of C, then $\operatorname{Im} w_0 > 0$ and if D is contained in the exterior of C, then $\operatorname{Im} w_0 < 0$, hence in both cases, $\lambda^{-1}(D)$ is contained in the upper half-plane of the w-plane and $\lambda^{-1}(D^*)$ is contained in the lower half-plane of the w-plane. Further, if $z = \lambda(w)$ then $z^* = \lambda(\bar{w})$, hence $w = \lambda^{-1}(z)$ implies $\bar{w} = \lambda^{-1}(z^*)$; therefore $\lambda^{-1}(D^*) = \bar{E}$, where $E = \lambda^{-1}(D)$. Since $\lambda(\infty) \notin \gamma$, $\lambda^{-1}((\gamma))$ is an open interval (a, b) on the real axis of the w-plane. Therefore

$$\lambda^{-1}(D) \cup \lambda^{-1}((\gamma)) \cup \lambda^{-1}(D^*) = E \cup (a, b) \cup \bar{E}.$$

Since by assumption $f: z \to \zeta = f(z)$ maps D conformally onto Δ and f can be extended to a homeomorphism $\tilde{f}$ mapping $D \cup (\gamma)$ onto $\Delta \cup (\alpha, \beta)$, $f \circ \lambda$: $w \to \zeta = f(\lambda(w))$ maps $E = \lambda^{-1}(D)$ conformally onto Δ and can be extended to a homeomorphism $\tilde{f} \circ \lambda$ which maps $E \cup (a, b) = \lambda^{-1}$ $(D \cup (\gamma))$ onto $\Delta \cup (\alpha, \beta)$. Therefore, by Theorem 5.8, $f \circ \lambda$ can be extended to a conformal mapping G: $w \to \zeta = G(w)$ from $E \cup (a, b) \cup \bar{E}$ onto $\Delta \cup (\alpha, \beta) \cup \bar{\Delta}$. Hence, the conformal mapping f can be extended to a conformal mapping $g = G \circ \lambda^{-1}$: $z \to \zeta = G(\lambda^{-1}(z))$ from $D \cup (\gamma) \cup D^* = \lambda(E \cup (a, b) \cup \bar{E})$ onto $\Delta \cup (\alpha, \beta) \cup \bar{\Delta}$ and by (5.18):

$$G(w) = \begin{cases} f(\lambda(w)), & w \in E, \\ \tilde{f}(\lambda(w)), & w \in (a, b), \\ \overline{f(\lambda(\bar{w}))}, & w \in \bar{E}. \end{cases}$$

Putting $w = \lambda^{-1}(z)$, we have $z = \lambda(w)$ and $z^* = \lambda(\bar{w})$, hence $g(z) = G(\lambda^{-1}(z))$ is given by (5.22).

We assumed C to be a circle in the z-plane, but the principle of reflection is also valid if C is a line. In that case, we have to assume that D is a region on one side of C, that $\gamma \subset C$ is a segment such that $\gamma \subset [D] - D$.

It is important to note that by the principle of reflection, a function $\tilde{f}(z)$ that is continuous on (γ) can be extended to a function that is holomorphic in each point of (γ).

Let C be a circle or a line in the z-plane and let $z \to z^*$ be the reflection with respect to C. Further, let D be a simply connected and bounded region in the z-plane, such that $D \cap C = \varnothing$ and the boundary of D is a piecewise smooth Jordan curve and let the arc or segment, $\gamma \subsetneqq C$, be a part of the boundary $\partial[D]$ of D, orientations considered; in other words, $\gamma \subset \partial[D]$. Since $D \cap C = \varnothing$, D is either in the interior or the exterior of C, if C is a circle, and on one side of C if C is a line. If $f: z \to \zeta = f(z)$ is a conformal mapping from D onto the upper half-plane H^+ of the ζ-plane, then f can be extended to a homeomorphism $\tilde{f}: z \to \zeta = \tilde{f}(z)$ mapping $[D]$ onto $[H^+] = H^+ \cup \mathbb{R} \cup \{\infty\}$. We assume $\tilde{f}^{-1}(\infty) \notin (\gamma)$. Since $\tilde{f}$ maps $\partial[D]$ onto the positively oriented line $\mathbb{R} \cup \{\infty\}$, $\tilde{f}((\gamma))$ is an open segment on $\mathbb{R}$, say, $\tilde{f}((\gamma)) = (\alpha, \beta)$, $-\infty \leqq \alpha < \beta \leqq +\infty$. Since $\gamma \subsetneqq C$, $(\alpha, \beta) \subsetneqq \mathbb{R}$. Applying Theorem 5.10 to $\Delta = H^+$ yields:

Theorem 5.11. A conformal mapping $f: z \to \zeta = f(z)$ from D onto the upper half-plane H^+ can be extended to a conformal mapping $g: z \to \zeta = g(z)$ from $D \cup (\gamma) \cup D^*$ onto $H^+ \cup (\alpha, \beta) \cup H$. The function $g(z)$ is given by (5.22).

We have derived the reflection principle (5.10) from the reflection principle for the real axis (Theorem 5.8) by using the linear fractional transformation λ: $w \to z = \lambda(w)$, (5.20). By replacing this transformation by an arbitrary conformal mapping, a generalization of the principle of reflection is obtained as follows.

A smooth Jordan arc γ: $t \to \gamma(t)$, $a \leqq t \leqq b$, in the complex plane with the property that both Re $\gamma(t)$ and Im $\gamma(t)$ are real analytic functions is called an *analytic Jordan arc*. Let $W \subset \mathbb{C}$ be a region which is symmetric with respect to the real axis (i.e., $\bar{W} = W$), $[a, b] \subset \mathbb{R} \cap W$, $a < b$, and λ: $w \to z = \lambda(w)$ a conformal mapping from W onto the region $\lambda(W)$ in the z-plane. Under these circumstances, γ: $t \to \lambda(t)$, $a \leqq t \leqq b$, is an analytic Jordan arc in the z-plane. We define, for points $z \in \lambda(W)$, the reflection z^* of z with respect to the curve γ by

$$z = \lambda(w) \to z^* = \lambda(\bar{w}), \qquad w \in W.$$

Similarly, let Ω be a region in $\mathbb{C}$ which is symmetric with respect to $\mathbb{R}$, $[\alpha, \beta] \subset \mathbb{R} \cap \Omega$, $\alpha < \beta$, and μ: $\omega \to \zeta = \mu(\omega)$ a conformal mapping which maps Ω onto $\mu(\Omega)$ and define δ: $\tau \to \zeta = \mu(\tau)$, $\alpha \leqq \tau \leqq \beta$. The reflection of a point $\zeta \in \mu(\Omega)$ with respect to the analytic Jordan arc δ is given by

$$\zeta = \mu(\omega) \to \zeta^* = \mu(\bar{\omega}), \qquad \omega \in \Omega.$$

Let D be a region in the z-plane, such that $D \subset \lambda(W^+)$, where

$W^+ = \{w \in W\colon \operatorname{Im} w > 0\}$, $\gamma \subset [D] - D$ and $D \cup (\gamma) \cup D^*$ is a region in the z-plane. Here (γ) represents $\{\gamma(t)\colon a < t < b\}$. Similarly, let $\Delta \subset \mu(\Omega^+)$ be a region in the ζ-plane such that $\delta \subset [\Delta] - \Delta$ and $\Delta \cup (\delta) \cup \Delta^*$ is a region in the ζ-plane. If D and Δ are conformally equivalent, we have the following theorem.

Theorem 5.12 (principle of reflection). If the conformal mapping $f\colon z \to \zeta = f(z)$ from D onto Δ can be extended to a homeomorphism $\tilde{f}\colon z \to \zeta = \tilde{f}(\zeta)$ mapping $D \cup (\gamma)$ onto $\Delta \cup \delta$, then f can be extended to a conformal mapping $g\colon z \to \zeta = g(z)$ from $D \cup (\gamma) \cup D^*$ onto $\Delta \cup (\delta) \cup \Delta^*$. The function $g(z)$ is given by:

$$g(z) = \begin{cases} f(z), & z \in D, \\ \tilde{f}(z), & z \in (\gamma), \\ f(z^*)^*, & z \in D^*. \end{cases} \tag{5.23}$$

b. Modular functions

As an application of the principle of reflection we now want to construct an example of a so-called modular function.

Let $U = \{z\colon |z| < 1\}$ be the unit disk in the z-plane and $C = \{z\colon |z| = 1\}$. Let c_1, c_2, and c_3 be three points on C, l_1, l_2, and l_3 be the tangents to C at c_1, c_2, and c_3, respectively, and let P_1 be the point of intersection of l_2 and l_3, P_2 that of l_1 and l_3 and P_3 that of l_1 and l_2. The circle C_1 with center P_1 and passing through c_2 and c_3 intersects the unit circle C at right angles, as do the circle C_2 with center P_2 and passing through c_3 and c_1 and the circle C_3 with center P_3 and passing through c_1 and c_2. The intersections of C_1, C_2, and C_3 with the closed disk $[U] = U \cup C$ are arcs γ_1, γ_2, and γ_3 and the end points of γ_1 are c_2 and c_3, the end points of γ_2 are c_3 and c_1, and the end points of γ_3 are c_1 and c_2. Let D be the region enclosed by the three arcs γ_1, γ_2, and γ_3. The closure $[D]$ of D is a "three-side" with sides γ_1, γ_2, and γ_3 and vertices c_1, c_2, and c_3. Since c_1, c_2, and c_3 are on the unit circle C, we will call $[D]$ a three-side of arcs inscribed in C.

Considering the unit disk U as a model of the non-Euclidean plane, the lines of the non-Euclidean plane U are the arcs in U of circles that intersect C at right angles. The open arcs (γ_1), (γ_2), and (γ_3), obtained by omitting the end points from γ_1, γ_2, and γ_3 are therefore lines in the non-Euclidean plane and $[D] \cap U$ is a three-side with the lines (γ_1), (γ_2), and (γ_3) as sides. If w is an arbitrary point in the interior of that three-side, a line (γ) passing through w will intersect at least one of the sides (γ_1), (γ_2), and (γ_3).

For $\nu = 1, 2, 3$ we will call the reflection with respect to C_ν the reflection

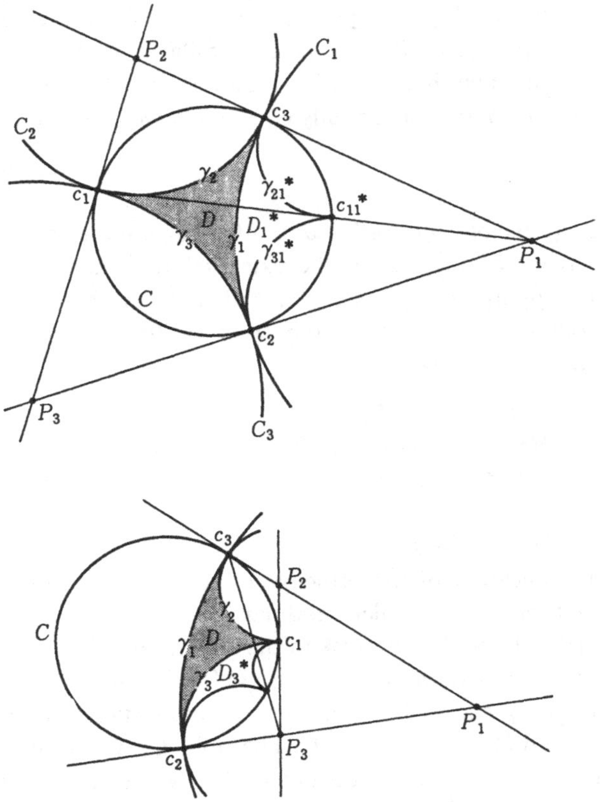

Fig. 5.14

with respect to the arc γ_ν and denote this reflection by $z \to z_\nu^*$. By Theorem 5.9 the reflection $z \to z_\nu^*$ with respect to γ_ν maps the circle C, which intersects γ_ν at right angles, into a circle which maps C at right angles again, while the points of intersection of C and γ_ν are invariant. Hence the reflection $z \to z_\nu^*$ maps C onto itself. Since it also maps ∞ onto P_ν, the exterior of C is mapped onto itself. Thus the refection $z \to z_\nu^*$ maps the interior U of C onto itself. Now let us consider the reflections with respect to the three sides γ_1, γ_2, and γ_3 of the three-side of arcs $[D]$ inscribed in C. Let D_1^* denote the image of D under the reflection with respect to γ_1 and let $c_{\lambda 1}^*, \gamma_{\lambda 1}^*$ denote the images of c_λ and γ_λ, $\lambda = 1, 2, 3$, under the same reflection. Since the image of C under the reflection with respect to γ_1 is C, c_{11}^* is the point of intersection of C and the line through P_1 and c_1, while $c_{21}^* = c_2$ and $c_{31}^* = c_3$. Obviously, $\gamma_{11}^* = \gamma_1$.

By using Theorem 5.9, the image γ_{21}^* of the arc γ_2, which intersects C in c_3 and c_1 is the arc intersecting C at right angles in $c_{31}^* = c_3$ and c_{11}^*. Similarly, γ_{31}^* is the arc intersecting C at right angles in c_2 and c_{11}^*. Therefore, the image $[D_1^*]$ of $[D]$ under the reflection with respect to γ_1 is a three-side of arcs inscribed in C with sides γ_1, γ_{21}^*, and γ_{31}^* and vertices c_{11}^*, c_2, and c_3.

The image of $[D]$ under the reflections with respect to γ_2 and γ_3 is found in a similar way. To state the result, let the image of D, c_ν, γ_λ, and γ_μ (ν, λ, $\mu = 1, 2, 3$, and λ, μ and ν all different) under the reflection with respect to γ_ν, be denoted by D_ν^*, $c_{\nu\nu}^*$, $\gamma_{\lambda\nu}^*$, and $\gamma_{\mu\nu}^*$, respectively. Then $[D_\nu^*]$ is a three-side of arcs inscribed in C with, as sides, the arcs γ_ν, $\gamma_{\lambda\nu}^*$, and $\gamma_{\mu\nu}^*$ that intersect C at right angles and as vertices the points $c_{\nu\nu}^*$, c_λ, and c_μ. We see that the image of a three-side of arcs inscribed in C under reflection with respect to any one of its sides is again a three-side of arcs inscribed in C. Repeating this process, starting from a three-side of arcs $[D]$ inscribed in C we arrive at an infinite number of three-sides of arcs inscribed in C. In order to prove this let us write the images of D, c_ν, γ_λ, and γ_μ, under the reflection with respect to γ_ν, as D_ν, $c_{\nu\nu}$, $\gamma_{\lambda\nu}$, and $\gamma_{\mu\nu}$, suppressing the *. The image of $[D_1]$ under reflection with respect to $\gamma_{11} = \gamma_1$ is the original $[D]$; the image of D_1 and $\gamma_{\lambda 1}$ under reflection with respect to one of the other sides, say, γ_{21}, we denote by D_{12}, and $\gamma_{\lambda 12}$, respectively. $[D_{12}]$ is again a three-side of arcs inscribed in C and its sides are the arcs γ_{112}, $\gamma_{212} = \gamma_{21}$, and γ_{312}, all of which

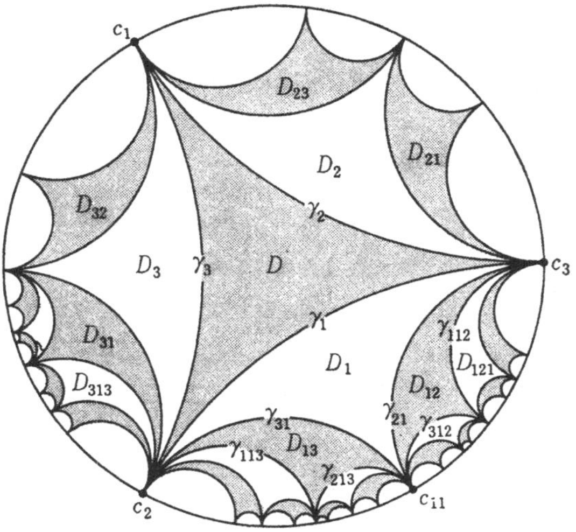

Fig. 5.15

intersect C at right angles. Denoting the images of D_{12} and $\gamma_{\lambda 12}$ under reflection with respect to γ_{112} by D_{121}, respectively $\gamma_{\lambda 121}$, $[D_{121}]$ is a three-side of arcs inscribed in C with sides $\gamma_{1121} = \gamma_{112}$, γ_{2121}, and γ_{3121}, all of which intersect C at right angles. Denoting the images of D_{121} and $\gamma_{\lambda 121}$, under reflection with respect to γ_{3121}, by D_{1213}, respectively $\gamma_{\lambda 1213}$, $[D_{1213}]$ is a three-side of arcs inscribed in C with sides γ_{11213}, γ_{21213}, and $\gamma_{31213} = \gamma_{3121}, \ldots$.

For an arbitrary sequence $\{\nu, \mu, \rho, \sigma, \ldots\}$ consisting of 1's, 2's, or 3's, such that consecutive terms are different, a sequence $[D_\nu]$, $[D_{\nu\mu}]$, $[D_{\nu\mu\rho}]$, $[D_{\nu\mu\rho\sigma}], \ldots$ of three-sides of arcs inscribed in C is defined similarly. In other words, D_ν, $\gamma_{\lambda\nu}$ are the images of D and γ_λ under reflection with respect to γ_ν; $D_{\nu\mu}$ and $\gamma_{\lambda\nu\mu}$ are images of D_ν and $\gamma_{\lambda\nu}$ under reflection with respect to $\gamma_{\mu\nu}$; $D_{\mu\nu\rho}$ and $\gamma_{\lambda\nu\mu\rho}$ are the images of $D_{\nu\mu}$ and $\gamma_{\lambda\nu\mu}$ under reflection with respect to $\gamma_{\rho\nu\mu}$; and $D_{\nu\mu\rho\sigma}$ and $\gamma_{\lambda\nu\rho\sigma}$ are the images of $D_{\nu\mu\rho}$ and $\gamma_{\lambda\nu\mu\rho}$ under reflection with respect to $\gamma_{\sigma\nu\mu\rho}, \ldots$.

Let us denote the reflection with respect to the arc γ_ν by $z \to z_\nu$, suppressing the *, and similarly, let $z_\nu \to z_{\nu\mu}$ represent the reflection with respect to the arc $\gamma_{\mu\nu}$; $z_{\nu\mu} \to z_{\nu\mu\rho}$ the reflection with respect to $\gamma_{\rho\nu\mu}$; $z_{\nu\mu\rho} \to z_{\nu\mu\rho\sigma}$ the reflection with respect to $\gamma_{\sigma\nu\mu\rho}, \ldots$. By Theorem 5.9, the composition of an even number of reflections is a conformal mapping. Since all reflections $z \to z_\nu$, $z_\nu \to z_{\nu\mu}$, $z_{\nu\mu} \to z_{\nu\mu\rho}, \ldots$ map U onto itself, the compositions of reflections $z \to z_{\nu\mu}$, $z \to z_{\nu\mu\rho\sigma}, \ldots$ are conformal mappings which map U onto itself. For brevity's sake let us write $\{\nu_j\} = \{\nu_1, \nu_2, \nu_3, \ldots, \nu_j, \nu_{j+1}, \ldots\}$ instead of $\{\nu, \mu, \rho, \sigma, \ldots\}$, where $\nu_j = 1, 2, 3$, $j = 1, 2, 3, \ldots$, and $\nu_j \neq \nu_{j+1}$. For each such sequence $\{\nu_j\}$ a sequence $[D_{\nu_1}]$, $[D_{\nu_1\nu_2}], \ldots$ of three-sides of arcs inscribed in C is defined. $[D_{\nu_1\nu_2\ldots\nu_j}]$ is obtained from $[D]$ by repeated reflections and the side γ_λ of $[D]$ corresponds with the side $\gamma_{\lambda\nu_1\nu_2\ldots\nu_j}$ of $[D_{\nu_1\nu_2\ldots\nu_j}]$, $\lambda = 1, 2, 3$. The side $\gamma_{\nu_j\nu_1\ldots\nu_{j-1}\nu_j} = \gamma_{\nu_j\nu_1\ldots\nu_{j-1}}$ is a common side of $[D_{\nu_1\nu_2\ldots\nu_{j-1}\nu_j}]$ and $[D_{\nu_1\nu_2\ldots\nu_{j-1}}]$ is the image of $[D_{\nu_1\nu_2\ldots\nu_{j-1}}]$ under the reflection with respect to $\gamma_{\nu_j\nu_1\ldots\nu_{j-1}}$.

The reflection with respect to $\gamma_{\nu_j\nu_1\nu_2\ldots\nu_{j-1}}$ is represented by

$$z_{\nu_1\nu_2\ldots\nu_{j-1}} \to z_{\nu_1\nu_2\ldots\nu_j}. \tag{5.24}$$

The composition of the reflections $z \to z_{\nu_1}$, $z_{\nu_1} \to z_{\nu_1\nu_2}, \ldots, z_{\nu_1\nu_2\ldots\nu_{j-1}} \to z_{\nu_1\nu_2\ldots\nu_{j-1}\nu_j}$ is represented by

$$z \to z_{\nu_1\nu_2\ldots\nu_{j-1}\nu_j}. \tag{5.25}$$

The mapping $z \to z_{\nu_1\nu_2\ldots\nu_j}$ maps the unit disk onto itself and is a conformal mapping for even j by Theorem 5.9. We denote the intersection of $[D_{\nu_1\nu_2\ldots\nu_j}]$ and U by $[D_{\nu_1\nu_2\ldots\nu_j}]'''$, that is, $[D_{\nu_1\nu_2\ldots\nu_j}]'''$ is obtained from

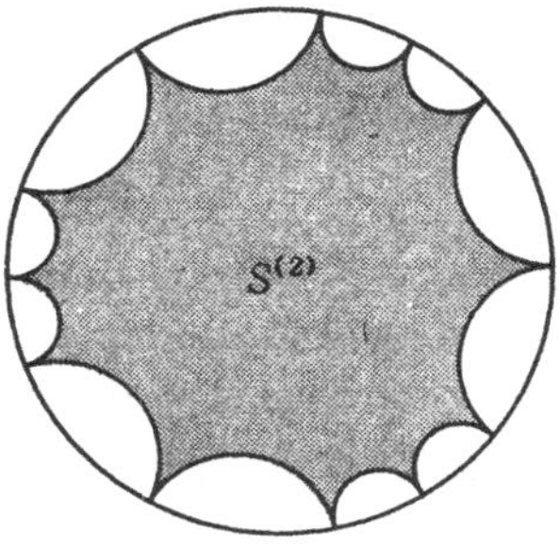

Fig. 5.17

also.) $S^{(2)}$ can be represented as

$$S^{(2)} = \bigcup_{\mu \neq \nu} D_{\nu\mu} \cup ([S^{(1)}] \cap U).$$

For $m = 3, 4, 5, \ldots$ we define the region $S^{(m)}$ by

$$S^{(m)} = \bigcup D_{\nu_1\nu_2\ldots\nu_m} \cup ([S^{(m-1)}] \cap U). \tag{5.30}$$

A holomorphic function $g(z)$ on $S^{(2)}$ can be analytically continued to yield holomorphic functions $g(z)$ defined on $S^{(m)}$, $m = 3, 4, 5, 6, \ldots$. The union $\bigcup D_{\nu_1\nu_2\ldots\nu_m}$ occurring on the right-hand side of (5.30) denotes the union of $3 \times 2^{m-1}$ regions $D_{\nu_1\nu_2\ldots\nu_m}$, where $\nu_1, \nu_2, \ldots, \nu_{m-1}, \nu_m$ take one of the values 1, 2, or 3 and $\nu_1 \neq \nu_2 \neq \nu_3 \neq \cdots \neq \nu_{m-1} \neq \nu_m$. $[S^{(m-1)}]$ is a polygon with $3 \times 2^{m-1}$ arcs $\gamma_{\nu_m\nu_1\nu_2\ldots\nu_{m-1}}$ that intersect the unit circle C at right angles. The side $\gamma_{\nu_m\nu_1\nu_2\ldots\nu_{m-1}}$ is common to the three-side $[D_{\nu_1\nu_2\ldots\nu_m}]$ and the polygon $[S^{(m-1)}]$. Letting $z \to z_{\nu_1\nu_2\ldots\nu_m}$ be defined as in (5.25), the analytic continuation $g(z)$ is given by

$$g(z_{\nu_1\nu_2\ldots\nu_m}) = \overline{f(z)}, \qquad z_{\nu_1\nu_2\ldots\nu_m} \in (\gamma_{\nu_m\nu_1\ldots\nu_{m-1}}) \cup D_{\nu_1\nu_2\ldots\nu_m}, \tag{5.31}$$

if m is odd and by

$$g(z_{\nu_1\nu_2\ldots\nu_m}) = \tilde{f}(z), \qquad z_{\nu_1\nu_2\ldots\nu_m} \in (\gamma_{\nu_m\nu_1\ldots\nu_{m-1}}) \cup D_{\nu_1\nu_2\ldots\nu_m}, \tag{5.32}$$

if m is even, just as (5.26) and (5.29).

In this way the holomorphic function $f(z)$ defined on D is analytically extended to yield holomorphic functions $g(z)$ defined on $S^{(1)}, S^{(2)}, S^{(3)}, \ldots, S^{(m)}$. Since

$$D \subset S^{(1)} \subset S^{(2)} \subset \cdots \subset S^{(m)} \subset \cdots, \qquad \bigcup_{m=1}^{\infty} S^{(m)} = U,$$

$f(z)$ can be analytically continued to yield a holomorphic function $g(z)$

defined on U. Putting

$$E_\nu = D \cup (\gamma_\nu) \cup D_\nu, \qquad E_{\nu\mu} = D_\nu \cup (\gamma_{\mu\nu}) \cup D_{\nu\mu},$$

and generally

$$E_{\nu_1\nu_2\ldots\nu_{m-1}\nu_m} = D_{\nu_1\nu_2\ldots\nu_{m-1}} \cup (\gamma_{\nu_m\nu_1\ldots\nu_{m-1}}) \cup D_{\nu_1\nu_2\ldots\nu_{m-1}\nu_m}.$$

$E_{\nu_1\nu_2\ldots\nu_{m-1}\nu_m}$ is a subregion of U and the collection $\{E_{\nu_1\nu_2\ldots\nu_m}\}$ of all $E_{\nu_1\nu_2\ldots\nu_m}$ is an open covering of U. On each $E_{\nu_1\nu_2\ldots\nu_m}$ the holomorphic function $g(z)$ is univalent and $z \to w = g(z)$ is a conformal mapping from $E_{\nu_1\nu_2\ldots\nu_m}$ onto $H^+ \cup (I_{\nu_m}) \cup H^-$. Therefore $g'(z) \neq 0$ for all $z \in U$. The function $g(z)$ is an example of a so-called *modular function*. Let $z = f^{-1}(w)$ be the inverse function of the holomorphic function $w = f(z)$ on D; then f^{-1}: $w \to z = f^{-1}(w)$ is a conformal mapping that maps the upper half-plane H^+ of the w-plane onto D. By repeated application of the principle of reflection, $f(z)$ is analytically continued to yield the holomorphic function $g(z)$ on U. From this we see that the inverse $f^{-1}(w)$ is freely analytically continuable in the region $\mathbb{C} - \{0, 1\}$ and that its analytic continuation $g^{-1}(w)$ is a holomorphic multivalued analytic function on $\mathbb{C} - \{0, 1\}$. Since the identity $f(f^{-1}(w)) = w$ is invariant under analytic continuation (Section 4.1a), we have $g(g^{-1}(w)) = w$, that is, $z = g^{-1}(w)$ is the inverse of $w = g(z)$.

In order to verify directly, using the analytic continuation $g(z)$ of $f(z)$, that $f^{-1}(w)$ is freely analytically continuable in $\mathbb{C} - \{0,1\}$, we denote the restriction of $g(z)$ to $E_{\nu_1\nu_2\ldots\nu_m}$ by $g_{\nu_1\nu_2\ldots\nu_m}(z)$ and we put $W_\lambda = H^+ \cup (I_\lambda) \cup H^-$, $\lambda = 1, 2, 3$. Obviously

$$W_1 \cup W_2 \cup W_3 = \mathbb{C} - \{0, 1\}.$$

Since $g_{\nu_1\nu_2\ldots\nu_m}$: $z \to w = g_{\nu_1\nu_2\ldots\nu_m}(z)$ is a conformal mapping from $E_{\nu_1\nu_2\ldots\nu_m}$ onto W_{ν_m}, $g^{-1}_{\nu_1\nu_2\ldots\nu_m}$: $w \to z = g^{-1}_{\nu_1\nu_2\ldots\nu_m}(w)$ is a conformal mapping from W_{ν_m} onto $E_{\nu_1\nu_2\ldots\nu_m}$. Let β: $t \to w = \beta(t)$, $0 \leqq t \leqq b$, with $\beta(0) = c_0 \in H^+$ be a curve in $\mathbb{C} - \{0, 1\}$ and let us consider the analytic continuation of the holomorphic function $f^{-1}(w)$ defined on H^+ along β. Divide β into n curves β_h: $t \to w = \beta(t)$, $t_{h-1} \leqq t \leqq t_h$, $0 = t_0 < t_1 < \cdots < t_{h-1} < t_h < \cdots < t_n = b$, such that each β_h is contained in one region $W_\lambda(h)$ of the regions W_1, W_2, and W_3; that is, $|\beta_h| \subset W_{\lambda(h)}$. Since $f^{-1}(w)$ can be analytically continued to yield the holomorphic function $g^{-1}_{\lambda(1)}(w)$ defined on $W_{\lambda(1)}$, the analytic continuation of $f^{-1}(w)$ along β_1 is given by $g^{-1}_{\lambda(1)}(w)$. Using complete induction with respect to h, we assume that $f^{-1}(w)$ has been analytically continued along β to $\beta(t_{h-1})$ and we denote this analytic continuation by $\{g^{-1}(w, t): 0 \leqq t \leqq t_{h-1}\}$. (See Definition 4.1 and notice that the power series $g^{-1}(w, t)$ of $w - \beta(t)$ are all inverse functions of w

$= g(z)$.) Since $\beta(t_{h-1})$ is the initial point of β_h and $|\beta_h| \subset W_{\lambda(h)}$, $g^{-1}(w, t_{h-1})$ coincides with some $g^{-1}_{\nu_1\nu_2\ldots\nu_{m-1}\lambda(h)}(w)$ on a neighborhood of $\beta(t_{h-1})$. Since $g^{-1}_{\nu_1\nu_2\ldots\nu_{m-1}\lambda(h)}(w)$ is a holomorphic single-valued function on $W_{\lambda(h)}$, it is the analytic continuation of $g^{-1}(w, t_{h-1})$ along β_h. We conclude that $f^{-1}(w)$ is freely analytically continuable along β. Hence $f^{-1}(w)$ is freely analytically continuable on the region $\mathbb{C} - \{0,1\}$ and this analytic continuation is just $g^{-1}(w)$. $g^{-1}(w)$ is freely analytically continuable and complete on $\mathbb{C} - \{0, 1\}$.

c. *Picard's Theorem*

Let $F(z)$ be an entire function of z, i.e., $F(z)$ is a holomorphic function of z defined on the whole z-plane $\mathbb{C}$.

Theorem 5.13 (*Picard's Theorem*). A nonconstant entire function $F(z)$ assumes all complex values with at most one exception. In other words, the range of an entire function $F(z)$ is either $\mathbb{C}$ or $\mathbb{C} - \{w_0\}$ for some $w_0 \in \mathbb{C}$.

Proof: Assume $F(\mathbb{C}) \subset \mathbb{C} - \{w_0, w_1\}$ for some $w_0, w_1 \in \mathbb{C}$ $(w_0 \neq w_1)$. Since the linear transformation $w \to (w - w_0)/(w_1 - w_0)$ maps w_0 onto 0 and w_1 onto 1 we see right away that $F(\mathbb{C}) \subset \mathbb{C} - \{0, 1\}$. Substituting ζ for z in the modular function $g(z)$ discussed above, $\zeta = g^{-1}(w)$ is the inverse of $w = g(\zeta)$. The function $g^{-1}(w)$ is a multivalued analytic function, holomorphic on $\mathbb{C} - \{0, 1\}$ and moreover freely analytically continuable and complete on $\mathbb{C} - \{0, 1\}$. Therefore, the composite function $g^{-1}(f(z)$ is freely analytically continuable on $\mathbb{C}$ by Theorem 4.9. Therefore, $g^{-1}(F(z))$ is a single-valued holomorphic function on $\mathbb{C}$, that is, $g^{-1}(F(z))$ is an entire function, by Theorem 4.7. On the other hand, $\zeta = g^{-1}(w) \in U$, hence $|g^{-1}(F(z))| < 1$. Hence, $g^{-1}(F(z))$ is a constant by Liouville's Theorem (Theorem 1.24) and $F(z)$ reduces to a constant, in contradiction with our assumption.

If $F(z)$ is a polynomial in z, then $F(\mathbb{C}) = \mathbb{C}$ (by the Fundamental Theorem of Algebra). If $F(\mathbb{C}) = \mathbb{C} - \{w_0\}$, the value w_0 is called the *exceptional value* of the transcendental entire function $F(z)$. For example, the exceptional value of the exponential function e^z is 0 (see Example 1.6), while the function $\cos z$ has no exceptional value.

d. *The Schwarz–Christoffel formula*

If D is a simply connected and bounded region in the complex plane bounded by a Jordan curve consisting of n segments, then the closed region

$[D]$ is called an n-gon. Let $[D]$ be an n-gon and let its boundary be given by

$$\partial[D] = \gamma_1 \cdot \gamma_2 \cdot \cdots \cdot \gamma_\lambda \cdot \gamma_{\lambda+1} \cdot \cdots \cdot \gamma_n,$$

where each γ_λ connects the vertices $c_{\lambda-1}$ and c_λ. Let θ_λ, $-\pi < \theta_\lambda < \pi$, be the angle between $\gamma_{\lambda+1}$ and γ_λ at c_λ (the so-called exterior angle). If $f: z \to w = f(z)$ is a conformal mapping from D onto the upper half-plane H^+ of the w-plane, then f can be extended to a homeomorphism $\tilde{f}$ from $[D]$ onto $[H^+] = H^+ \cup \mathbb{R} \cup \{\infty\}$ by Theorem 5.7 (1) and $\tilde{f}$ maps $\partial[D]$ onto the positively oriented line $\mathbb{R} \cup \{\infty\}$. Hence, putting $a_\lambda = \tilde{f}(c_\lambda)$, we may assume

$$-\infty < a_1 < a_2 < \cdots < a_{\lambda-1} < a_\lambda < \cdots < a_n < +\infty.$$

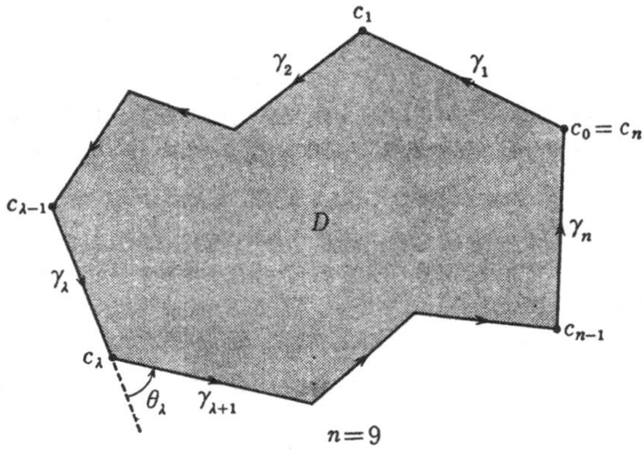

Fig. 5.18

Let ϕ denote the inverse of f: $\phi = f^{-1}$. Then $\phi: w \to z = \phi(w)$ is a conformal mapping from H^+ onto D. We will now establish a formula for $\phi(w)$ using only elementary functions of w. The reflection with respect to (the line determined by) the segment γ_λ is given by

$$z \to z_\lambda^* = \frac{c_\lambda - c_{\lambda-1}}{\bar{c}_\lambda - \bar{c}_{\lambda-1}}(\bar{z} - \bar{c}_{\lambda-1}) + c_{\lambda-1}. \tag{5.33}$$

We denote the image of D under the reflection with respect to γ_λ by D_λ^*. $[D_\lambda^*]$ is an n-gon that is the symmetric image of $[D]$ with respect to γ_λ. Since $\tilde{f}((\gamma_\lambda)) = (a_{\lambda-1}, a_\lambda)$, in case $[D]$ is a convex n-gon, the conformal mapping f from D onto H^+ can be extended to a conformal mapping g_λ: $z \to w = g_\lambda(z)$ from $D \cup (\gamma_\lambda) \cup D_\lambda^*$ onto $H^+ \cup (a_{\lambda-1}, a_\lambda) \cup H^+$ by Theorem 5.11. (If $\lambda = 1$, then we let (a_0, a_1) represent $(a_n, +\infty) \cup \{\infty\} \cup (-\infty, a_1)$.) The function

$g_\lambda(z)$ is given by

$$\begin{cases} g_\lambda(z) = f(z), & z \in D, \\ g_\lambda(z_\lambda^*) = \overline{f}(z), & z_\lambda^* \in (\gamma_\lambda) \cup D_\lambda^*. \end{cases} \tag{5.34}$$

We denote the inverse g_λ^{-1} of g_λ by ψ_λ. The conformal mapping ϕ: $w \to z = \phi(w)$ from H^+ onto D can be extended to a homeomorphism $\tilde{\phi}$ from $[H^+]$ onto $[D]$. The mapping ψ_λ: $w \to z = \psi_\lambda(w)$ is a conformal mapping from $H^+ \cup (a_{\lambda-1}, a_\lambda) \cup H^-$ onto $D \cup (\gamma_\lambda) \cup D_\lambda^*$ and an extension of ϕ. Since $\tilde{\phi} = \tilde{f}^{-1}$, $\psi_\lambda(w)$ is given by

$$\begin{cases} \psi_\lambda(w) = \phi(w), & w \in H^+ \\ \psi_\lambda(\bar{w}) = \tilde{\phi}(w)_\lambda^*, & \bar{w} \in (a_{\lambda-1}, a_\lambda) \cup H^- \end{cases} \tag{5.35}$$

by (5.34). Since all ψ_λ are conformal mappings, $\psi'_\lambda(w) \neq 0$. Also if $[D]$ is not convex, ψ_λ: $w \to z = \psi_\lambda(w)$ as defined in (5.35) is a holomorphic mapping from $H^+ \cup (a_{\lambda-1}, a_\lambda) \cup H^-$ onto $D \cup (\gamma_\lambda) \cup D_\lambda^*$ and $\psi'_\lambda(w) \neq 0$. Hence, the holomorphic function $\phi(w)$ on H^+ can be analytically continued to yield an analytic function $\psi(w) = \psi_\lambda(w)$, $\lambda = 1, 2, \ldots, n$, which is holomorphic on $\mathbb{R} \cup \{\infty\} - \{a_1, a_2, \ldots, a_n\}$. For $w \in H^-$ $\psi_\lambda(w) = \tilde{\phi}(\bar{w})_\lambda^*$ and $\psi_\mu(w) = \tilde{\phi}(\bar{w})_\mu^*$ do not coincide if $\lambda \neq \mu$, hence $\psi(w)$ is a multivalued function. There exists a very simple relationship between its branches $\psi_\lambda(w)$ and $\psi_\mu(w)$: Since the composition of the reflections $z_\lambda^* \to z$ and $z \to z_\mu^*$ takes the form

$$z_\lambda^* \to z_\mu^* = \alpha_{\mu\lambda} z_\lambda^* + \beta_{\mu\lambda}, \qquad \alpha_{\mu\lambda} \neq 0$$

by (5.33), we have

$$\psi_\mu(w) = \alpha_{\mu\lambda} \psi_\lambda(w) + \beta_{\mu\lambda}, \qquad w \in H^-.$$

Hence

$$\frac{\psi''_\mu(w)}{\psi'_\mu(w)} = \frac{\psi''_\lambda(w)}{\psi'_\lambda(w)}, \qquad w \in H^-.$$

Therefore the function

$$h(w) = \frac{\psi''(w)}{\psi'(w)}$$

is a holomorphic single-valued function on the Riemann sphere $\mathbb{R} \cup \{\infty\}$, from which the points $a_0, a_1, \ldots, a_{n-1}$ have been deleted. In particular, since $\psi_1(w)$ is a holomorphic univalent function on a neighborhood of $w = \infty$, we have on a neighborhood of ∞

$$\psi_1(w) = \psi_1(\infty) + \frac{1}{w}\beta\left(\frac{1}{w}\right),$$

where $\beta(1/w)$ is a power series in $1/w$ such that $\beta(0) = 0$. Hence

$$h(\infty) = 0. \tag{5.36}$$

In order to study the behavior of $h(w)$ in a neighborhood of each point a_λ, we consider $\tilde{\phi}(w)$ on a neighborhood of a_λ. For sufficiently small $\varepsilon > 0$ and a suitably chosen angle κ_λ the mapping $q_\lambda: z \to \zeta = q_\lambda(z)$, where $q_\lambda(z)$ is defined by

$$q_\lambda(z) = e^{i\kappa_\lambda}(z - c_\lambda)^{\pi/\omega_\lambda}, \qquad \omega_\lambda = \pi - \theta_\lambda,$$

maps the fan-shaped region $S_\lambda = \{z \in D: |z - c_\lambda| < \varepsilon\}$ on the upper half disk $U_\rho^+(0) = \{\zeta: \operatorname{Im}\zeta > 0 \text{ and } |\zeta| < \rho\}$ with center 0 and radius $\rho = \varepsilon^{\pi/\omega_\lambda} W_\lambda = f(S_\lambda)$ is a subregion of the upper half-plane H^+ of the w-plane and ϕ maps W_λ conformally onto S_λ. Hence the mapping $\Phi_\lambda: w \to \zeta = \Phi_\lambda(w)$, where $\Phi_\lambda(w)$ is defined by

$$\Phi_\lambda(w) = e^{i\kappa_\lambda}(\phi(w) - c_\lambda)^{\pi/\omega_\lambda}, \qquad w \in W_\lambda \tag{5.37}$$

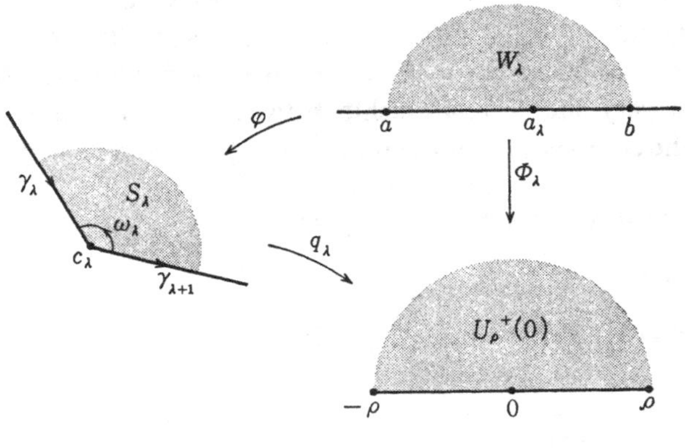

Fig. 5.19

is a conformal mapping from W_λ onto $U_\rho^+(0)$. $\tilde{f}$ maps that part of $\gamma_\lambda \cdot \gamma_{\lambda+1}$ contained within a circle with center c_λ and radius ε onto the open interval (a, b) of the real axis of the w-plane. Obviously, $a_\lambda = \tilde{f}(c_\lambda) \in (a, b)$. The open interval (a, b) is part of the boundary of W_λ. It follows from (5.37) that the conformal mapping Φ_λ from W_λ onto $U_\rho^+(0)$ can be extended to a homomorphism

$$w \to \zeta = e^{i\kappa_\lambda}(\tilde{\phi}(w) - c_\lambda)^{\pi/\omega_\lambda}, \qquad w \in W_\lambda \cup (a,b),$$

from $W_\lambda \cup (a, b)$ onto $U_\rho^+(0) \cup (-\rho, \rho)$. Thus Φ_λ can be extended to a conformal mapping $\Psi_\lambda: w \to \zeta = \Psi_\lambda(w)$ from $W_\lambda \cup (a, b) \cup \overline{W}_\lambda$ onto $U_\rho^+(0) \cup (-\rho, \rho) \cup \overline{U_\rho^+(0)} = U_\rho(0)$. Since $a_\lambda \in (a,b)$ and $\tilde{\phi}(a_\lambda) = c_\lambda$, we have $\Psi_\lambda(a_\lambda) = 0$. Hence

$$\Psi_\lambda(w) = (w - a_\lambda) G_\lambda(w),$$

where $G_\lambda(w)$ is holomorphic on $W_\lambda \cup (a, b) \cup \overline{W}_\lambda$ and $G_\lambda(w) \neq 0$ for all w. Since $\Psi_\lambda(w) = \Phi_\lambda(w)$ for $w \in W_\lambda$ we have

$$\phi(w) - c_\lambda = (w - a_\lambda)^{\omega_\lambda/\pi} F_\lambda(w), \qquad F_\lambda(w) = (e^{-i\kappa_\lambda} G_\lambda(w))^{\omega_\lambda/\pi}$$

by (5.37). Here, $F_\lambda(w)$ is holomorphic on a neighborhood of a_λ and $F_\lambda(a_\lambda) \neq 0$. Since $\omega_\lambda/\pi - 1 = -\theta_\lambda/\pi$, we have

$$\phi'(w) = \frac{\omega_\lambda}{\pi}(w - a_\lambda)^{-\theta_\lambda/\pi}\left(F_\lambda(w) + \frac{\pi}{\omega_\lambda}(w - a_\lambda) F'_\lambda(w)\right),$$

and hence

$$h(w) = \frac{\phi''(w)}{\phi'(w)} = \frac{d}{dw}\log \phi'(w) = \frac{-\theta_\lambda/\pi}{w - a_\lambda} + H_\lambda(w),$$

where

$$H_\lambda(w) = \frac{d}{dw}\log\left(F_\lambda(w) + \frac{\pi}{\omega_\lambda}(w - a_\lambda) F'_\lambda(w)\right).$$

Since $F_\lambda(a_\lambda) \neq 0$, $H_\lambda(w)$ is holomorphic on a neighborhood of a_λ and $h(w)$ is holomorphic on a neighborhood of a_λ from which a_λ has been deleted. Therefore, the expression for $h(w)$, which was obtained under the assumption that $w \in W_\lambda$ is valid on a neighborhood of a_λ. Since $h(w)$ is holomorphic on the w-plane from which $a_1, a_2, \ldots, a_n$ have been removed,

$$h(w) + \sum_{\lambda=1}^{n} \frac{\theta_\lambda/\pi}{w - a_\lambda}$$

is an entire function of w and $\lim_{w \to \infty} h(w) = 0$ by (5.36). We conclude from Liouville's Theorem

$$h(w) = -\sum_{\lambda=1}^{n} \frac{\theta_\lambda/\pi}{w - a_\lambda}.$$

Therefore

$$\frac{d}{dw}\log \phi'(w) = h(w) = -\sum_{\lambda=1}^{n} \frac{\theta_\lambda/\pi}{w - a_\lambda}$$

for $w \in H^+$ and hence

$$\log \phi'(w) = -\sum_{\lambda=1}^{n} (\theta_\lambda/\pi) \log(w - a_\lambda) + \text{constant}$$

from which we conclude

$$\phi'(w) = C_0 \prod_{\lambda=1}^{n} (w - a_\lambda)^{-\theta_\lambda/\pi}, \qquad C_0 \text{ a constant.}$$

Fixing an arbitrary $w_0 \in H^+$, we have

$$\phi(w) = C_0 \int_{w_0}^{w} \prod_{\lambda=1}^{n} (w - a_\lambda)^{-\theta_\lambda/\pi}\, dw + C_1, \tag{5.38}$$

where C_0 and C_1 are constants, $C_0 \neq 0$. We have proved:

Theorem 5.14. A conformal mapping that maps the upper half-plane H^+ of the w-plane onto the interior D of an n-gon can be represented as

$$w \to z = \phi(w) = C_0 \int_{w_0}^{w} \prod_{\lambda=1}^{n} (w - a_\lambda)^{-\theta_\lambda/\pi}\, dw + C_1,$$

where C_0 and C_1 are constants, $C_0 \neq 0, w_0$ a fixed point in H^+ and $\theta_1, \theta_2, \ldots, \theta_\lambda, \ldots, \theta_n$ the exterior angles at the vertices of the n-gon. (5.38) is called the *Schwarz–Christoffel formula.*

6

Riemann surfaces

This chapter deals with analytic functions on Riemann surfaces following Weyl's famous book *The Idea of Riemannian Surface.*

6.1 Differential forms

To lay the groundwork for what follows we study differential forms.

a. Differential forms

We say that the real-valued function $u(z) = u(x + iy)$ of the complex variable $z = x + iy$, defined on a region D in the complex plane is continuously differentiable, of class C^∞, real analytic, ... if the function $u(x, y) = u(x + iy)$ of two real variables is continuously differentiable, of class C^∞, real analytic .... If $u = u(z) = u(x, y)$ is continuously differentiable, the total differential (or differential for short) of u is given by

$$du = u_x\, dx + u_y\, dy, \qquad u_x = u_x(x, y), \qquad u_y = u_y(x, y).$$

If $\varphi_1(x, y)$ and $\varphi_2(x, y)$ are arbitrary functions of x and y, the expression

$$\varphi = \varphi_1\, dx + \varphi_2\, dy, \qquad \varphi_1 = \varphi_1(x, y), \qquad \varphi_2 = \varphi_2(x, y),$$

which generalizes the previous expression for du, is called a *differential form of degree 1* or a *1-form*. Generally, the coefficients $\varphi_1 = \varphi_1(x, y)$ and $\varphi_2 = \varphi_2(x, y)$ can be complex-valued functions, but in this section we assume that φ_1 and φ_2 are real-valued functions, unless the contrary is indicated. When it is necessary to make clear that the coefficients φ_1 and φ_2 are real-valued, $\varphi = \varphi_1\, dx + \varphi_2\, dy$ is called a *real 1-form*.

If the coefficients φ_1 and φ_2 are continuous functions of x and y on the domain D in the complex plane (or on the point set S), then the 1-form $\varphi = \varphi_1\, dx + \varphi_2\, dy$ is said to be continuous on D (or on S). If φ_1 and φ_2 are continuously differentiable, of class C^n, of class C^∞, ... on D, then the 1-form φ is continuously differentiable, of class C^n, of class C^∞, ...

on D. Using the complex variable $z = x + iy$, the 1-form φ is also written as

$$\varphi(z) = \varphi_1(z)dx + \varphi_2(z)dy.$$

If the function $u = u(x, y)$ is continuously differentiable, of class C^n, of class C^∞, ..., on the region D then the 1-form du is continuous, of class C^{n-1}, of class C^∞, ..., on D. The linear combination $a\varphi + b\psi$ of two 1-forms $\varphi = \varphi_1\, dx + \varphi_2\, dy$ and $\psi = \psi_1\, dx + \psi_2\, dy$ is defined by

$$a\varphi + b\psi = (a\varphi_1 + b\psi_1)dx + (a\varphi_2 + b\psi_2)dy, \qquad a, b \in \mathbb{R}$$

and the product of the function u and the 1-form $\varphi = \varphi_1\, dx + \varphi_2\, dy$ is defined by

$$u\varphi = (u\varphi_1)dx + (u\varphi_2)dy.$$

The following two rules are obvious:

$$d(au + bv) = a\, du + b\, dv, \qquad a, b \in \mathbb{R}$$

$$d(uv) = v\, du + u\, dv.$$

We now want to introduce a new operation $\wedge$ for 1-forms called the *exterior product*. The exterior product is anti-symmetric, i.e.

$$\psi \wedge \varphi = -\varphi \wedge \psi$$

for 1-forms φ and ψ and moreover satisfies

$$(a\varphi + b\chi) \wedge \psi = a(\varphi \wedge \psi) + b(\chi \wedge \psi),$$

$$\varphi \wedge (a\psi + b\chi) = a(\varphi \wedge \psi) + b(\varphi \wedge \chi).$$

Putting $\varphi = \varphi_1\, dx + \varphi_2\, dy$ and $\psi = \psi_1\, dx + \psi_2\, dy$ and observing that $dx \wedge dx = 0$, $dx \wedge dy = -dy \wedge dx$ and $dy \wedge dy = 0$, we have

$$\varphi \wedge \psi = (\varphi_1\psi_2 - \varphi_2\psi_1)dx \wedge dy.$$

An expression of the form

$$\omega = \omega_{12}\, dx \wedge dy, \qquad \omega_{12} = \omega_{12}(x, y),$$

which generalizes the previous expression for $\varphi \wedge \psi$, is called a *differential form of degree 2* or a *2-form*. In this section we will assume that the coefficient $\omega_{12}(x, y)$ of the 2-form ω is real-valued unless the contrary is indicated. If $\omega_{12} = \omega_{12}(x, y)$ is continuous, continuously differentiable, of class C^n, ..., then the 2-form $\omega = \omega_{12}\, dx \wedge dy$ is called continuous, continuously differentiable, of class C^n, ... The product of a function u and a 2-form ω is defined by

$$u\omega = (u\omega_{12})dx \wedge dy,$$

and the linear combination $a\omega + b\psi$ of two 2-forms ω and ψ by

$$a\omega + b\psi = (a\omega_{12} + b\psi_{12})dx \wedge dy, \qquad a, b \in \mathbb{R}.$$

Putting $\omega_{12} = -\omega_{21}$, the 2-form ω can be written as

$$\omega = \frac{1}{2}(\omega_{12}\, dx \wedge dy + \omega_{21}\, dy \wedge dx).$$

This formula is a special case of the result for 2-forms in n variables:

$$\omega = \frac{1}{2}\sum_{j,k=1}^{n} \omega_{jk}\, dx_j \wedge dx_k,$$

where $\omega_{jk} = -\omega_{kj}$, by putting $n = 2$, $x_1 = x$ and $x_2 = y$. Since $\omega_{21}\, dx \wedge dy = \omega_{12}\, dy \wedge dx$, we arrive at $\omega = \omega_{12}\, dx \wedge dy$. In the case of two variables the only independent coefficient is ω_{12}, hence the subscript is really not necessary, but is used to distinguish the 2-form ω from the coefficient ω_{12}.

Obviously,

$$u\varphi \wedge \psi = \varphi \wedge u\psi = u(\varphi \wedge \psi).$$

If φ is a continuously differentiable 1-form defined on a domain, than the *exterior derivative* of φ is defined by

$$d\varphi = d\varphi_1 \wedge dx + d\varphi_2 \wedge dy, \qquad \varphi = \varphi_1\, dx + \varphi_2\, dy.$$

This can be reduced to

$$d\varphi = \left(\frac{\partial \varphi_2}{\partial x} - \frac{\partial \varphi_1}{\partial y}\right) dx \wedge dy. \tag{6.1}$$

We have

$$d(a\varphi + b\psi) = ad\varphi + bd\psi, \qquad a, b \in \mathbb{R}$$

and, since for a continuously differential function u, $d(u\varphi_1) \wedge dx = du \wedge \varphi_1\, dx + ud\varphi_1 \wedge dx$, we have

$$d(u\varphi) = du \wedge \varphi + ud\varphi. \tag{6.2}$$

If u is twice continuously differentiable, then $du = u_x\, dx + u_y\, dy$ and $u_{yx} = u_{xy}$, hence by (6.1)

$$d\, du = 0. \tag{6.3}$$

b. Line integrals

Let $u(z) = u(x, y)$ be a continuously differentiable function of $z = x + iy$ defined on a region Ω in the complex plane and let $\gamma : t \to z = \gamma(t)$, $a \leqslant t \leqslant b$, be a smooth curve in Ω. Writing $\gamma(t) = x(t) + iy(t)$, the function $u(\gamma(t)) = u(x(t), y(t))$ is continuously differentiable and

$$\frac{d}{dt}u(\gamma(t)) = u_x(\gamma(t))x'(t) + u_y(\gamma(t))y'(t).$$

Hence

$$u(\gamma(b)) - u(\gamma(a)) = \int_a^b (u_x(\gamma(t))x'(t) + u_y(\gamma(t))y'(t))dt. \tag{6.4}$$

For a 1-form $\varphi = \varphi_1(z)dx + \varphi_2(z)dy$, which is continuous on Ω, we define the integral of φ along γ by

$$\int_\gamma \varphi = \int_a^b (\varphi_1(\gamma(t))x'(t) + \varphi_2(\gamma(t))y'(t))dt \tag{6.5}$$

generalizing the integral occurring in (6.4). It is easy to see that the above definition of $\int_\gamma \varphi$ also can be used if the curve γ is piecewise smooth. We call $\int_\gamma \varphi$ a *line integral*. Using the notation of line integrals we have

$$u(\gamma(b)) - u(\gamma(a)) = \int_\gamma du. \tag{6.6}$$

For two continuous 1-forms φ and ψ we obviously have

$$\int_\gamma (c_1\varphi + c_2\psi) = c_1\int_\gamma \varphi + c_2\int_\gamma \psi, \qquad c_1, c_2 \in \mathbb{R}.$$

Let $f(z)$ be a complex-valued continuous function on Ω. Writing $f(z) = u + iv$, $u = u(z)$ and $v = v(z)$, we have

$$f(z)dz = (u + iv)(dx + i\,dy) = u\,dx - v\,dy + i(v\,dx + u\,dy).$$

Generally, the dual form $*\varphi$ of a 1-form φ is defined by

$$*\varphi = -\varphi_2\,dx + \varphi_1\,dy, \qquad \varphi = \varphi_1\,dx + \varphi_2\,dy.$$

Putting $\varphi = u\,dx - v\,dy$ we obtain

$$f(z)dz = \varphi + i * \varphi. \tag{6.7}$$

Now, according to definition (6.6), the integral of $f(z)$ along γ is given by

$$\int_\gamma f(z)dz = \int_a^b f(\gamma(t))\gamma'(t)dt.$$

From

$$f(\gamma(t))\gamma'(t) = ux'(t) - vy'(t) + i(vx'(t) + uy'(t))$$

we conclude

$$\int_\gamma f(z)dz = \int_\gamma \varphi + i\int_\gamma * \varphi. \tag{6.8}$$

For a real 1-form $\varphi = \varphi_1(z)dx + \varphi_2(z)dy$, which is continuous on Ω, we put $f(z) = \varphi_1(z) - i\varphi_2(z)$. Then $f(z)dz = \varphi + i * \varphi$, hence

$$\int_\gamma \varphi = \mathrm{Re}\int_\gamma f(z)dz, \qquad f(z) = \varphi_1(z) - i\varphi_2(z). \tag{6.9}$$

If $t(\tau)$ is a continuously differentiable function of τ, $\alpha \leqslant \tau \leqslant \beta$, satisfying $t'(\tau) > 0$ for all τ, $t(\alpha) = a$, $t(\beta) = b$, then the curve $\lambda : \tau \to \lambda(\tau)$

$= \gamma(t(\tau))$, $\alpha \leqslant \tau \leqslant \beta$, can be identified with the curve $\gamma : t \to \gamma(t)$, $a \leqslant t \leqslant b$ (see Section 4.2). The following equalities,

$$\int_{\lambda} \varphi = \int_{\gamma} \varphi$$

and

$$\int_{\gamma^{-1}} \varphi = -\int_{\gamma} \varphi,$$

can be proved by using (6.9). If $\gamma = \gamma_1 \cdot \gamma_2 \cdot \gamma_3 \cdot \dots \cdot \gamma_m$ is the curve resulting from pasting together the piecewise smooth curves γ_1, $\gamma_2, \dots, \gamma_m$ in that order, then, obviously

$$\int_{\gamma} \varphi = \int_{\gamma_1} \varphi + \int_{\gamma_2} \varphi + \dots + \int_{\gamma_m} \varphi.$$

Let $\gamma_1, \dots, \gamma_m$ be piecewise smooth curves in Ω, let $n_1, \dots, n_m$ be integers and let $\gamma = \sum_{k=1}^{m} n_k \gamma_k$ be a 1-chain. The integral $\int_{\gamma} \varphi$, where φ is a 1-form which is continuous on Ω, is defined by

$$\int_{\gamma} \varphi = \sum_{k=1}^{m} n_k \int_{\gamma_k} \varphi, \qquad \gamma = \sum_{k=1}^{m} n_k \gamma_k. \tag{6.10}$$

By the above results, this definition is compatible with the standard definitions, for example (4.34), (4.35) and (4.36).

If the 2-form $\omega = \omega_{12}(x, y) dx \wedge dy$ is continuous on D and if

$$\int_D |\omega_{12}(x, y)| dx\, dy < +\infty$$

then we define

$$\int_D \omega = \int_D \omega_{12}(x, y) dx\, dy.$$

The integral $\int_D \omega$ is said to *converge absolutely*. Further, if ω is continuous on the closed region $[D]$, then we define

$$\int_{[D]} \omega = \int_{[D]} \omega_{12}(x, y) dx\, dy.$$

It is clear that $\int_D \omega$ as well as $\int_{[D]} \omega$ is linear in ω. If the closed region $[D]$ is decomposed into the closed regions $[D_1], [D_2], \dots, [D_\mu]$, then

$$\int_{[D]} \omega = \sum_{\lambda=1}^{\mu} \int_{[D_\lambda]} \omega. \tag{6.11}$$

Theorem 6.1 (Green's Theorem). Let $[D]$ be a bounded closed region such that its boundary $\partial[D]$ consists of a finite number of mutually disjoint piecewise smooth Jordan curves. If the 1-form φ is continuously differentiable on a region $\Omega \supset [D]$, then

$$\int_{\partial[D]} \varphi = \int_{[D]} d\varphi. \tag{6.12}$$

Proof: First we note that $[D]$ can be decomposed into cells (see Theorem 2.1). Let

$$[D] = \Gamma_1(K) \cup \Gamma_2(K) \cup \cdots \cup \Gamma_\lambda(K) \cup \cdots \cup \Gamma_\mu(K) \tag{6.13}$$

represent the cellular decompositon of $[D]$ obtained in the proof of Theorem 2.1. Since $\partial\Gamma_\lambda(K) = \sum_\nu C_{\lambda\nu}$ or $\partial\Gamma_\lambda(K) = C_\lambda = \sum_\nu C_{\lambda\nu}$, $C_{\lambda\nu} = -C_{\nu\lambda}$ and $\partial[D] = \sum_\lambda C_\lambda$ we have

$$\partial[D] = \sum_{\lambda=1}^{\mu} \partial\Gamma_\lambda(K) \tag{6.14}$$

(see Section 2.3a). Hence

$$\int_{\partial[D]} \varphi = \sum_{\lambda=1}^{\mu} \int_{\partial\Gamma_\lambda(K)} \varphi.$$

By (6.11) and (6.13) we have

$$\int_{[D]} d\varphi = \sum_{\lambda=1}^{\mu} \int_{\Gamma_\lambda(K)} d\varphi$$

therefore it suffices to prove that

$$\int_{\partial\Gamma_\lambda(K)} \varphi = \int_{\Gamma_\lambda(K)} d\varphi$$

holds for all cells $\Gamma_\lambda(K)$. As is clear from the proof of Theorem 2.1, the cells $\Gamma_\lambda(K)$ are all of the form of the cells of Example 2.3, i.e.

$$\Gamma_\lambda(K) = \{z : z = x + iy,\ a \leqslant x \leqslant b,\ \gamma(x) \leqslant y \leqslant \delta(x)\},$$

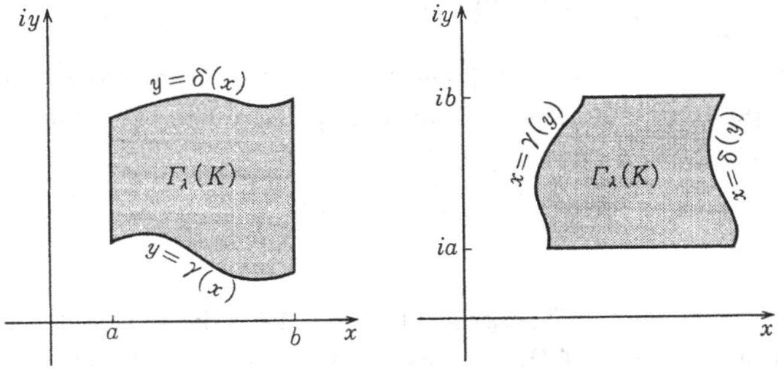

or

$$\Gamma_\lambda(K) = \{z : z = x + iy,\ a \leqslant y \leqslant b,\ \gamma(y) \leqslant x \leqslant \delta(y)\},$$

where $\gamma(z)$ and $\delta(x)$ are continuously differentiable functions of x defined on the closed interval $[a, b]$ such that $\gamma(x) < \delta(x)$ if $a < z < b$ and similarly, $\gamma(y)$ and $\delta(y)$ are continuously differentiable functions defined on $[a, b]$ such that $\gamma(y) < \delta(y)$ if $a < y < b$. We will prove the equality

$$\int_{\partial\Gamma(K)} \varphi = \int_{\Gamma(K)} d\varphi \tag{6.15}$$

for a cell $\Gamma(K)$ given by

$$\Gamma(K) = \{z : z = x + iy,\ a \leqslant x \leqslant b,\ \gamma(x) \leqslant y \leqslant \delta(x)\},$$

the other case being treated similarly.

Writing $\varphi = u\,dx + v\,dy$, $u = u(x, y)$ and $v(x, y)$ are, by assumption, continuously differentiable functions on Ω. The boundary of the cell $\Gamma(K) \subset \Omega$ consists of two line segments:

$$\alpha : y \to z = a + iy, \qquad \gamma(a) \leqslant y \leqslant \delta(a),$$
$$\beta : y \to z = b + iy, \qquad \gamma(b) \leqslant y \leqslant \delta(b),$$

and two smooth curves:

$$\gamma : x \to z = x + i\gamma(x), \qquad a \leqslant x \leqslant b,$$
$$\delta : x \to z = x + i\delta(x), \qquad a \leqslant x \leqslant b.$$

i.e.

$$\partial\Gamma(K) = \gamma + \beta - \delta - \alpha.$$

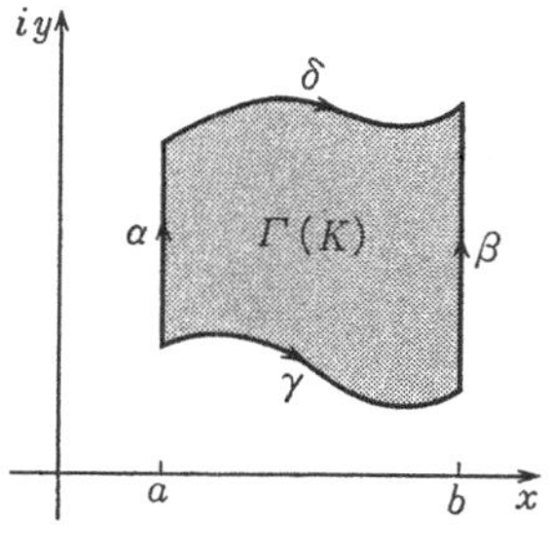

Since $d\varphi = (v_x - u_y)dx \wedge dy$, we have

$$\int_{\Gamma(K)} d\varphi = \int_{\Gamma(K)} (v_x - u_y)dx\,dy = \int_a^b dx \int_{\gamma(x)}^{\delta(x)} (v_x - u_y)dy.$$

We first observe that

$$\int_a^b dx \int_{\gamma(x)}^{\delta(x)} u_y(x, y)dy = \int_a^b [u(x, \delta(x)) - u(x, \gamma(x))]\, dx$$
$$= \int_a^b u(x, \delta(x))dx - \int_a^b u(x, \gamma(x))dx$$
$$= \int_\delta u\, dx - \int_\gamma u\, dx.$$

Next we put

$$\Phi(x, t, s) = \int_s^t v(x, y)dy$$

and we notice that $\Phi(x, t, s)$ is a continuously differentiable function of (x, t) and s such that

$$\Phi_x(x, t, s) = \int_s^t v_x(x, y)dy;$$
$$\Phi_t(x, t, s) = v(x, t);$$
$$\Phi_s(x, t, s) = -v(x, s).$$

Hence

$$\frac{d}{dx}\int_{\gamma(x)}^{\delta(x)} v(x, y)dy = \int_{\gamma(x)}^{\delta(x)} v_x(x, y)dy + v(x, \delta(x))\delta'(x) - v(x, \gamma(x))\gamma'(x).$$

Integration of both sides with respect to x from a to b yields the equality

$$\int_{\gamma(b)}^{\delta(b)} v(b, y)dy - \int_{\gamma(a)}^{\delta(a)} v(a, y)dy$$
$$= \int_a^b dx \int_{\gamma(x)}^{\delta(x)} v_x(x, y)dy$$
$$+ \int_a^b v(x, \delta(x))\delta'(x)dx - \int_a^b v(x, \gamma(x))\gamma'(x)dx.$$

Hence

$$\int_a^b ds \int_{\gamma(x)}^{\delta(x)} v_x(x, y)dy = \int_\beta v\, dy - \int_\alpha v\, dy - \int_\delta v\, dy + \int_\gamma v\, dy.$$

Since obviously $\int_\alpha u\, dx = \int_\beta u\, dx = 0$, we conclude

$$\int_{\Gamma(K)} d\varphi = \int_\gamma \varphi + \int_\beta \varphi - \int_\delta \varphi - \int_\alpha \varphi = \int_{\partial\Gamma(K)} \varphi.$$

Formula (6.15) is the counterpart for 1-forms of formula (6.6).

A continuous function $u(x, y)$ defined on a closed region $[D]$ such that $u(x, y)$ is continuously differentiable on the interior of $[D]$ and $u_x(x, y)$ and $u_y(x, y)$ can be extended to continuous functions $\tilde{u}_x(x, y)$ and $\tilde{u}_y(x, y)$ on $[D]$ is said to be *continuously differentiable* on $[D]$ and the partial derivatives $u_x(x, y)$ and $u_y(x, y)$ on $[D]$ are defined by $u_x(x, y) = \tilde{u}_x(x, y)$ and $\tilde{u}_y(x, y) = u_y(x, y)$. A continuous 1-form $\varphi = \varphi_1\, dx + \varphi_2\, dy$ defined on $[D]$ is said to be *continuously differentiable* on $[D]$ if φ_1 and φ_2 are continuously differentiable on $[D]$. Green's Theorem is still valid for 1-forms φ which are continuously differentiable on $[D]$ in the following sense:

Theorem 6.2. Let $[D]$ be a bounded closed region such that its boundary consists of a finite number of mutually disjoint piecewise smooth Jordan curves. If the 1-form φ is continuously differentiable on $[D]$, then

$$\int_{\partial[D]} \varphi = \int_{[D]} d\varphi.$$

Proof: It suffices to prove equality (6.15):

$$\int_{\partial\Gamma(K)} \varphi = \int_{\Gamma(K)} d\varphi$$

assuming that φ is continuously differentiable on the cell

$$\Gamma(K) = \{z + iy : a \leqslant x \leqslant b,\ \gamma(x) \leqslant y \leqslant \delta(x)\}.$$

To this end, put

$$\gamma^{\varepsilon}(x) = (1 - \varepsilon)\gamma(x) + \varepsilon\delta(x), \qquad \delta^{\varepsilon}(x) = \varepsilon\gamma(x) + (1 - \varepsilon)\delta(x)$$

for a sufficiently small $\varepsilon > 0$ and consider the cell:

$$\Gamma(K^{\varepsilon}) = \{x + iy : a + \varepsilon \leqslant x \leqslant b - \varepsilon,\ \gamma^{\varepsilon}(x) \leqslant y \leqslant \delta^{\varepsilon}(x)\}$$

(see the proof of Lemma 2.3). φ is continuously differentiable on

$$\Gamma(E) = \{x + iy : a < x < b,\ \gamma(x) < y < \delta(x)\}$$

and $\Gamma(K^{\varepsilon}) \subset \Gamma(E)$, hence

$$\int_{\partial\Gamma(K)} \varphi = \int_{\Gamma(K^{\varepsilon})} d\varphi$$

by Theorem 6.1. Letting ε tend to $+0$ we obtain:

$$\int_{\partial\Gamma(K)} \varphi = \int_{\Gamma(K)} d\varphi.$$

Theorem 1.14 of Section 1.3 can be derived directly from Theorem 6.2. To see this we first observe that equation (1.42),

$$\frac{\partial}{\partial s}(f(\Gamma(t, s))\Gamma_t(t, s)) = \frac{\partial}{\partial t}(f(\Gamma(t, s))\Gamma_s(t, s)),$$

can be written as:

$$d(f(\Gamma(t, s))\Gamma_t(t, s)dt + f(\Gamma(t, s))\Gamma_s(t, s)ds) = 0,$$

i.e.

$$d(f(\Gamma(t, s))d\Gamma(t, s)) = 0.$$

From

$$d(f(\Gamma(t, s))d\Gamma(t, s)) = f'(\Gamma(t, s))d\Gamma(t, s) \wedge d\Gamma(t, s) = 0$$

it is obvious that a holomorphic function $f(z)$ of z satisfies this equation. So we have established (1.42) for holomorphic functions. Letting K be the rectangle

$$k = \{t + is : a \leqslant t \leqslant b, 0 \leqslant s \leqslant 1\}$$

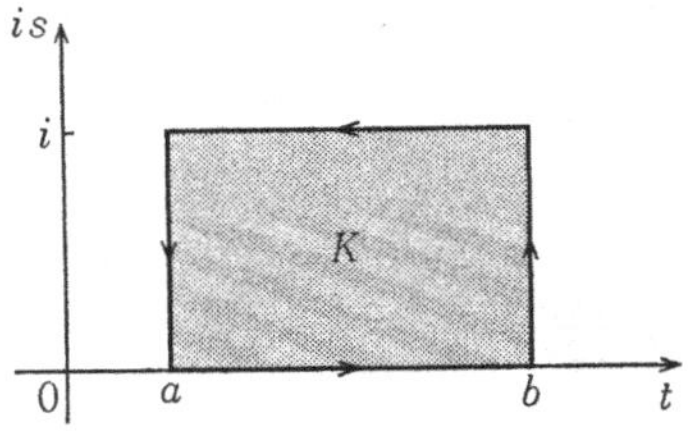

Theorem 6.2 yields

$$\int_{\partial K} f(\Gamma(t, s))d\Gamma(t, s) = \int_K d(f(\Gamma(t, s))d\Gamma(t, s)) = 0$$

and the left-hand member of this equality equals

$$\int_a^b f(\Gamma(t, 0))\Gamma_t(t, 0)dt + \int_0^1 f(\Gamma(b, s))\Gamma_s(b, s)ds$$
$$- \int_a^b f(\Gamma(t, 1))\Gamma_t(t, 1)dt - \int_0^1 f(\Gamma(a, s))\Gamma_s(a, s)ds$$
$$= \int_{\gamma_0} f(x)dz + \int_{{}_b\gamma} f(z)dz - \int_{\gamma_1} f(z)dz - \int_{\gamma_a} f(z)dz.$$

This yields (1.44).

A 1-form φ that can be represented as $\varphi = du$ with u a continuously differentiable function is called an *exact differential form*. A continuously differentiable 1-form φ satisfying $d\varphi = 0$ is called a *closed differential form*. Since $ddu = 0$ by (6.3), every continuously differentiable exact form φ is closed. Locally, the converse of this statement is valid too.

Theorem 6.3. Let the 1-form φ be continuously differentiable on the region Ω such that $d\varphi = 0$. Let c be an arbitrary point of Ω and choose $r > 0$ such that $U_r(c) \subset \Omega$. Then there exists a twice continuously differentiable function u such that $\varphi = du$ on the disk $U_r(c)$.

Proof: Put $\varphi = \varphi_1\, dx + \varphi_2\, dy$, where $\varphi_1 = \varphi_1(x, y)$ and $\varphi_2 = \varphi_2(x, y)$ are continuously differentiable functions on Ω. Putting $c = a + ib$, we define a function u on $U_r(c)$ by

$$u = u(x, y) = \int_a^x \varphi_1(\xi, b)d\xi + \int_b^y \varphi_2(x, \eta)d\eta.$$

Now $u(x, y)$ is differentiable with respect to y and

$$u_y(x, y) = \varphi_2(x, y).$$

$u(x, y)$ is also differentiable with respect to x and

$$u_x(x, y) = \varphi_1(x, b) + \int_b^y \frac{\partial}{\partial x}\varphi_2(x, \eta)d\eta.$$

Since, by assumption

$$\left(\frac{\partial \varphi_2}{\partial x} - \frac{\partial \varphi_1}{\partial y}\right) dx \wedge dy = d\varphi = 0,$$

we have

$$\int_b^y \frac{\partial}{\partial x}\varphi_2(x, \eta)d\eta = \int_b^y \frac{\partial}{\partial \eta}\varphi_1(x, \eta)d\eta = \varphi_1(x, y) - \varphi_1(x, b).$$

Hence:

$$u_x(x, y) = \varphi_1(x, y).$$

Therefore $u = u(x, y)$ is twice continuously differentiable on $U_r(c)$ and

$$du = \varphi_1(x, y)dx + \varphi_2(x, y)dy = \varphi.$$

c. *Harmonic forms*

If $f(z)$ is a holomorphic function on the region Ω, the 1-form $f(z)dz$ is called a holomorphic 1-form. Writing $f(z) = u + iv$, we have $f'(z) = p + iq$ with $p = u_x + v_y$ and $q = -u_y + v_x$. Hence

$$df(z) = du + i\,dv = (p + iq)(dx + i\,dy) = f'(z)dz.$$

Since $f'(z)$ is a holomorphic function, the differential $f'(z)dz$ is a holomorphic 1-form. From this fact it follows that the exterior derivative of a holomorphic 1-form is always 0:

$$d(f(z)dz) = 0. \tag{6.16}$$

(This is because $d(f(z)dz) = df(z) \wedge dz = f'(z)dz \wedge dz = 0$.) By (6.7) the holomorphic 1-form $f(z)dz$ can be split into real and imaginary parts as follows:

$$f(z)dz = \varphi + i * \varphi, \qquad \varphi = u\,dx - v\,dy,$$

where the real part φ and the imaginary part $*\varphi$ are both closed forms, so that $d\phi = d(*\varphi) = 0$.

Definition 6.1. A 1-form φ, which is continuously differentiable on the region Ω and satisfies

$$d\varphi = d(*\varphi) = 0 \tag{6.17}$$

is called *harmonic on* Ω.

By the above result, the real part φ of a holomorphic 1-form $f(z)dz$ is harmonic. Conversely:

Theorem 6.4. If the 1-form φ is harmonic on some region then $\varphi + i * \varphi$ is a holomorphic 1-form.

Proof: Writing $\varphi = u\,dx - v\,dy$ with $u = u(x, y)$ and $v = (x, y)$ continuously differentiable functions we have

$$0 = d\varphi = d(u\,dx - v\,dy) = (-v_x - u_y)dx \wedge dy,$$
$$0 = d(*\varphi) = d(v\,dx + u\,dy) = (u_x - v_y)dx \wedge dy,$$

i.e. u and v satisfy the Cauchy–Riemann equations

$$u_x = v_y, \qquad u_y = -v_x.$$

Therefore, by Theorem 1.4, $f(z) = u + iv$ is a holomorphic function of $z = x + iy$ and

$$\varphi + i * \varphi = f(z)dz.$$

For $\varphi = \varphi_1\,dx + \varphi_2\,dy$ we have $**\varphi = *(-\varphi_2\,dx + \varphi_1\,dy) = -\varphi_1\,dx - \varphi_2\,dy$, i.e.

$$**\varphi = -\varphi. \tag{6.18}$$

We conclude that the dual form $*\varphi$ of a harmonic form is again harmonic. Hence, the imaginary part of a holomorphic 1-form $f(z)dz = \varphi + i * \varphi$ is also harmonic.

d. Harmonic functions

Let $u = u(z) = u(x, y)$ and $v = v(z) = v(x, y)$ be continuously differentiable functions defined on a region in the complex plane. Since $*du = -u_y\,dx + u_x\,dy$ the Cauchy–Riemann equations $u_x = v_y$ and $u_y = -v_x$ can be written as

$$*du = dv. \tag{6.19}$$

Hence for a holomorphic function $f(z) = u + iv$ we have by (6.3) $d(*du) = ddv = 0$. Generally, let $u = u(z) = u(x, y)$ be a twice continuously differentiable function, then,

$$d(*du) = d(-u_y\, dx + u_x\, dy) = (u_{xx} + u_{yy})dx \wedge dy.$$

Putting

$$\Delta u = \frac{\partial^2 u}{\partial x^2} + \frac{\partial^2 u}{\partial y^2}$$

we have

$$d(*du) = \Delta u\, dx \wedge dy. \tag{6.20}$$

Obviously Δu is linear in u. The linear differential operator $\Delta = \partial^2/\partial x^2 + \partial^2/\partial y^2$ is called the *Laplacian*.

Definition 6.2. If a twice continuously differentiable function $u = u(z)$ defined on a region Ω of the complex plane satisfies the Laplace equation $\Delta u = 0$, then u is called *a harmonic function.*

As we have seen, the real part $u = \operatorname{Re} f(z)$ of a holomorphic function $f(z) = u + iv$ is a harmonic function. Hence its imaginary part $v = \operatorname{Im} f(z) = \operatorname{Re}(-if(z))$ is also a harmonic function.

Theorem 6.5. A harmonic function $u = u(z)$ defined on a region Ω can be written as the real part of a holomorphic function $f(z)$ on a sufficiently small neighborhood of any point of Ω.

Proof: Since by assumption $*du$ is a continuously differentiable 1-form on Ω satisfying $d(*du) = 0$, there exists a twice continuously differentiable function $v = v(z)$ on a sufficiently small neighborhood of any point of Ω such that $*du = dv$. Since $*du = dv$ is just another way of writing the Cauchy–Riemann equations, we conclude that $f(z) = u + iv$ is holomorphic.

Corollary. Harmonic functions are real analytic functions.

For a harmonic function u defined on a region Ω, the function v satisfying $dv = *dy$ is uniquely determined on each neighborhood apart from a real constant (for if $dv_1 = *du$, then $d(v_1 - v) = 0$ hence $v_1 - v$ is a constant). This means that the holomorphic function $f(z)$ satisfying $u = \operatorname{Re} f(z)$ is, apart from a constant ib, $b \in \mathbb{R}$, uniquely determined on each neighborhood. Therefore, the derivative $f'(z)$ of $f(z)$ is uniquely determined by u and a holomorphic function defined on Ω. Since $ddu = d(*du) = 0$, du is a harmonic 1-form on Ω and

$$du + i * du = f'(z)dz \tag{6.21}$$

is a holomorphic 1-form. Fixing a point $c_0 \in \Omega$ and connecting an arbitrary point $z \in \Omega$ with c_0 by means of a piecewise smooth curve γ we put

$$f(z) = \int_{\gamma}^{z} f'(z)dz + u(c_0).$$

Now, $f(z)$ is a freely analytically continuable and complete analytic function on Ω, which, in general, is multi-valued. By (6.6) we have

$$f(z) = \int_{\gamma}^{z} du + i\int_{\gamma}^{z} * du + u(c_0) = u(z) + i\int_{\gamma}^{z} * du.$$

Therefore we have proved:

Theorem 6.6. If $u = u(z)$ is a harmonic function on a region Ω, then

$$f(z) = u(z) + i\int_{\gamma}^{z} * du \tag{6.22}$$

is a freely analytically continuable and complete analytic function on Ω.

Hence a harmonic function $u(z)$ is the real part of a holomorphic multi-valued function $f(z)$: $u(z) = \operatorname{Re} f(z)$.

6.2 Riemann surfaces

a. Hausdorff spaces

As an example let us first consider the plane $\mathbb{R}^2$ with its open sets. Let $\mathfrak{O}$ denote the collection of open sets of $\mathbb{R}^2$, then

(i) $\varnothing \in \mathfrak{O}$ and $\mathbb{R}^2 \in \mathfrak{O}$,
(ii) $\cup_{u \in \mathfrak{S}} U \in \mathfrak{O}$ for arbitrary subsets $\mathfrak{S} \in \mathfrak{O}$, and
(iii) If $U \in \mathfrak{O}$ and $v \in \mathfrak{O}$ then $U \cap V \in \mathfrak{O}$.

In general, if Σ is a set and if it is agreed which sets are called open (such that (i), (ii) and (iii) above are satisfied), then Σ is called a *topological space*.

Definition 6.3. A set Σ together with a collection $\mathfrak{O}$ consisting of subsets of Σ satisfying

(i) $\varnothing \in \mathfrak{q}$, $\Sigma \in \mathfrak{O}$,
(ii) $\cup_{u \in \mathfrak{S}} U \in \mathfrak{O}$ for each subset $\mathfrak{S}$ of $\mathfrak{O}$;

(iii) if $U \in \mathfrak{D}$ and $V \in \mathfrak{D}$, then $U \cap V \in \mathfrak{D}$

is called a *topological space* and $\mathfrak{D}$ is called a *system of open sets*.

We say that the system of open sets *defines* a topology on Σ. If $\mathfrak{D}$ also satisfies the following condition (iv):

(iv) for arbitrary $P \in \Sigma$ and $Q \in \Sigma$, $(P \neq Q)$ there exists $U \in \mathfrak{D}$ and $V \in \mathfrak{D}$ such that $P \in U$, $Q \in V$ and $U \cap V = \emptyset$,

then the topological space Σ is called a *Hausdorff space*.

Let Σ be a Hausdorff space and let $\mathfrak{D}$ be its system of open sets. For $P \in \Sigma$, a subset U of Σ such that $P \in U$ and $U \in \mathfrak{D}$ is called an open *neighborhood* of P. We use the notation $U(P)$ to denote an arbitrary open neighborhood of the point P.

Just as in $\mathbb{R}^2$ it is possible to base the definitions of interior point, boundary point, closure, open set, closed set and continuous function on the idea of neighborhoods $U_\varepsilon(P)$, so in a Hausdorff space it is possible to define these concepts using neighborhoods $U(P)$.

Let $S \in \Sigma$ be an arbitrary subset of Σ. Then $P \in \Sigma$ is called an *interior point* of S if there exists a neighborhood $U(P) \subset \Sigma$ of P. The collection of all interior points of S is called the *interior* of S and denoted by (S). The set of all points P, such that each neighborhood $U(P)$ has a nonempty intersection with S is called the *closure* of S and denoted by $[S]$. Obviously, $(S) \subset S \subset [S]$. The set $[S] - (S)$ is called the *boundary* of S and a point belonging to the boundary is called a *boundary point* of S. If $S = [S]$ then S is called a *closed set*. It is clear that all sets $U \in \mathfrak{D}$ are open sets and conversely all open sets S belong to $\mathfrak{D}$ because as $S = (S)$, there exists for all $P \in S$ an open neighborhood $U(P) \in \mathfrak{D}$ such that $U(P) \subset S$. Hence by (ii) of Definition 6.3 we have:

$$S = \bigcup_{P \in S} U(P) \in \mathfrak{D}.$$

Therefore $\mathfrak{D}$ is precisely the set of all open sets of the Hausdorff space Σ.

Let $S \subset \Sigma$ be a subset of the Hausdorff space Σ and put $\mathfrak{D}|S = \{U \cap S : U \in \mathfrak{D}\}$, where $\mathfrak{D}$ denotes the system of open sets of Σ. Then S becomes a Hausdorff space with $\mathfrak{D}|S$ a system of open sets, called a subspace of Σ. Unless the contrary is stated, subsets S of a Hausdorff space Σ will always be considered a subspaces in this sense. Since the real line $\mathbb{R}$, the plane $\mathbb{R}^2$ and so on are all Hausdorff spaces, subsets of $\mathbb{R}$, $\mathbb{R}^2$ and so on can be considered as Hausdorff spaces.

A subset $U \subset \Sigma$ of the Hausdorff space Σ is called *connected* if it is not possible to write U as the union of two disjoint, nonempty open sets (compare Section 1.1b). A connected open subset is called a *region*.

Next, we want to consider mappings between two Hausdorff spaces Σ and T with systems of open subsets $\mathfrak{D}_\Sigma$ and $\mathfrak{D}_T$ respectively. Let $f : P \to f(P)$ be a mapping from Σ onto T. If it is possible for each point $P \in \Sigma$ and for each neighborhood $V(f(P)) \in \mathfrak{D}_T$ of $f(P) \in T$ to find a neighborhood $U(P) \in \mathfrak{D}_\Sigma$ such that

$$f(U(P)) \supset V(f(P)) \tag{6.23}$$

then the mapping $f : P \to f(P)$ is called *continuous*. This means that if $f(P) \in V \in \mathfrak{D}_T$, i.e. $P \in f^{-1}(V)$ and $V \in \mathfrak{D}_T$, then there exists a neighborhood $U(P) \in \mathfrak{D}$ such that $U(P) \geqslant f^{-1}(V)$. Therefore, a mapping f from Σ into T is continuous if and only if the inverse image $f^{-1}(V)$ of all open subsets $V \in \mathfrak{D}_T$ are open in $\Sigma : f^{-1}(V) \in \mathfrak{D}_\Sigma$.

Now let f be a one-to-one continuous map sending the Hausdorff space Σ onto the Hausdorff space T. If the inverse map f^{-1} is also continuous, then f is called a *homeomorphism*. If there exists a homeomorphism mapping Σ onto T, then Σ and T are called *homeomorphic*. A homeomorphism mapping Σ onto T induces a one-to-one correspondence between $\mathfrak{D}_\Sigma$ and $\mathfrak{D}_T$. Therefore, homeomorphic Hausdorff spaces can be identified as topological spaces.

Let $f(P)$ be a real-valued or complex-valued function defined on a subset $S \subset \Sigma$ of a Hausdorff space Σ and let $Q \in S$. If for each $\varepsilon > 0$ there exists a neighborhood $U(Q)$ such that

$$P \in U(Q) \cap S, \qquad |f(P) - f(Q)| < \varepsilon \tag{6.24}$$

then the function $f(P)$ is said to be *continuous* at Q. If $f(P)$ is continuous at all points of its domain S, then $f(P)$ is called a continuous function. Since (6.24) can be written as

$$f(U(Q) \cap S) \subset U_\varepsilon(f(Q))$$

$f(P)$ is a continuous function if and only if $f : P \to f(P)$ is a continuous function from S, as a Hausdorff space with $\mathfrak{D}|S$ as a system of open sets, into $\mathbb{R}$ or $\mathbb{C}$.

Let S be a subset of the Hausdorff space Σ. A collection $\mathfrak{U} \subset \mathfrak{D}_\Sigma$ of open sets of Σ such that $S = \bigcup_{U \in \mathfrak{U}} U$ is called an *open covering* of S. If $\mathfrak{U}$ contains a finite number of sets, the covering is called *finite*. If $\mathfrak{V}$ is another covering of S such that $\mathfrak{V} \subset \mathfrak{U}$, then $\mathfrak{V}$ is called a *subcovering* of $\mathfrak{U}$.

Definition 6.4. S is called *compact* if each open covering of S contains a finite subcovering.

Proposition. A compact subset $S \subset \Sigma$ is closed.

Proof: It suffices to show that $Q \notin S$ implies $Q \notin [S]$. According to Definition 6.3 (iv) there exists for each point $P \in S$ open sets U and V such that $P \in U$, $Q \in V$ and $U \cap V = \emptyset$. Denoting this U by $U(P)$ we have $S \subset \bigcup_{P \in S} U(P)$. Since S is compact, S can be covered by a finite number $U(P_k)$, $k = 1, 2, \ldots, m$ from among the $U(P) : S \subset \bigcup_{k=1}^{m} U(P_k)$. Denoting the set V corresponding $U = U(P_k)$ by V_k and putting $V(Q) = \cap_{k=1}^{m} V_k$ we have $Q \in V(Q)$ and $S \bigcap V(Q) = \emptyset$ since $Q \in V_k$ and $U(P_k) \bigcap V_k = \emptyset$. Since $V(Q)$ is open by Definition 6.3 (iii) we conclude that $Q \notin [S]$.

Theorem 6.7. If f is a continuous map from the subset S of the Hausdorff space Σ into the Hausdorff space T and if S is compact, then its image $f(S)$ is also compact.

Proof: Let $\mathfrak{B} \subset \mathfrak{D}_T$ be an open subcovering of $f(S)$. Since f is continuous, the inverse image $U = f^{-1}(V)$ of each $V \in \mathfrak{B}$ is open, hence $\mathfrak{U} = \{U : U = f^{-1}(V),\ V \in \mathfrak{B}\}$ is an open covering of S. Since S is compact, S can be covered by a finite number $U_k \in \mathfrak{U}$, $k = 1, 2, \ldots, m : S \subset \bigcup_{k=1}^{m} U_k$. Letting V_k denote the sets from $\mathfrak{B}$ corresponding with the U_k, we have $f(S) = \bigcup_{k=1}^{m} V_k$, i.e. S is compact.

b. *Definition of Riemann surfaces*

On the Riemann sphere $\mathbf{S} = \{\xi : \xi = (\xi_1, \xi_2, \xi_3),\ \xi_1^2 + \xi_2^2 + \xi_3^2 = 1\}$ there are defined two complex coordinates

$$\xi \to z = z(\xi) = \frac{\xi_1 + i\xi_2}{1 - \xi_3}, \qquad \xi \to \hat{z} = \hat{z}(\xi) = \frac{\xi_1 - i\xi_2}{1 + \xi_3}$$

(Section 3.2c). The domain of definition of $z(\xi)$ is $\mathbf{S} - \{N\} = \{\xi \in \mathbf{S} : \xi_3 < 1\}$. Denoting the restriction of z to, for example, $U_1 = \{\xi \in \mathbf{S} : \xi_3 < 1/2\}$ by $z_1(\xi) = z(\xi)$, $\xi \to z_1 = z_1(\xi)$ determines a local complex coordinate on U_1. Denoting in the same way the restriction of $\hat{z}$ to $U_2 = \{\xi \in \mathbf{S} : \xi_3 > -1/2\}$ by $z_2(\xi) = \hat{z}(\xi)$, $\xi \to z_2 = z_2(\xi)$ determines a local complex coordinate on U_2. Since $\mathbf{S} = U_1 \cup U_2$, each point ξ of $\mathbf{S}$ has at least one local complex coordinate: $z_1(\xi)$ or $z_2(\xi)$. If $\xi \in U_1 \cap U_2$, then ξ has two local coordinates, $z_1(\xi)$ and $z_2(\xi)$, while $z_2(\xi)$ can be obtained from $z_1(\xi)$ through the coordinate transformation

$$z_2(\xi) = \frac{1}{z_2(\xi)}. \tag{6.25}$$

Apart from z_1 and z_2, other local complex coordinates can be defined on S. For example, identifying the (ξ_1, ξ_2) plane with the complex plane A, the stereographic projection from (1, 0, 0) onto the (ξ_1, ξ_2) plane

$$\xi \to z_3 = z_3(\xi) = \frac{\xi_2 + \xi_3}{1 - \xi_1}, \qquad \xi_1 < 0,$$

yields a local complex coordinate on the "Western hemisphere" $U_3 = \{\xi \in \mathbf{S} : \xi_1 < 0\}$. According to (3.17) and (3.16) we have

$$\xi_1 = \frac{z + \overline{z}}{|z|^2 + 1}, \qquad \xi_2 = i\frac{\overline{z} - z}{|z|^2 + 1}, \qquad \xi_3 = \frac{|z|^2 - 1}{|z|^2 + 1}$$

hence

$$z_3 = i\frac{\overline{z} - z + |z|^2 - 1}{|z|^2 + 1 - z - \overline{z}} = i\frac{z + 1}{z - 1}.$$

Therefore, if $\xi = U_1 \cap U_3$, then

$$z_3 = i\frac{z_1 + 1}{z_1 - 1}, \qquad z_1 = z_1(\xi), \qquad z_3 = z_3(\xi),$$

hence, by (6.25), for $\xi = U_2 \cap U_3$,

$$z_3 = i\frac{1 + z_2}{1 - z_2}, \qquad z_2 = z_2(\xi), \qquad z_3 = z_3(\xi).$$

We see that the coordinate transformations between different local complex coordinates are biholomorphic. The property of being connected by biholomorphic coordinate transformations is the most fundamental property of local complex coordinates.

The Riemann sphere is an example of a Riemann surface. In order to give a general definition of a Riemann surface, let Σ be a connected Hausdorff space, the points of which we denote by $p, q \dots$. Let Σ be covered by a finite or countably infinite number of regions $U_1, U_2, \dots, U_j, \dots$, and let there be given complex-valued continuous functions $z_j(p)$ on each U_j, such that the maps $z_j : p \to z_j(p)$ are homeomorphisms mapping U_j onto regions $\mathcal{U}_j \subset \mathbb{C}$. If $U_j \cap U_k \neq \emptyset$, then the sets $\mathcal{U}_{jk} = \{z_j(p) : p \in U_j \cap U_k\} \subset \mathcal{U}_j$ and $\mathcal{U}_{k_j} = \{z_k(p) : p \in U_k \cap U_j\} \subset \mathcal{U}_k$ are both open sets in $\mathbb{C}$ and

$$\tau_{jk} : z_k(p) \to z_j(p), \qquad p \in U_j \cap U_k, \tag{6.26}$$

is a homeomorphism from $\mathcal{U}_{kj}$ onto $\mathcal{U}_{jk}$. If all homeomorphisms $\tau_{jk} : \mathcal{U}_{kj} \to \mathcal{U}_{jk}$ for j and k with $U_j \cap U_k \neq \emptyset$ are biholomorphic then the maps $z_j : p \to z_j(p)$ are called *local complex coordinates* on U_j and the

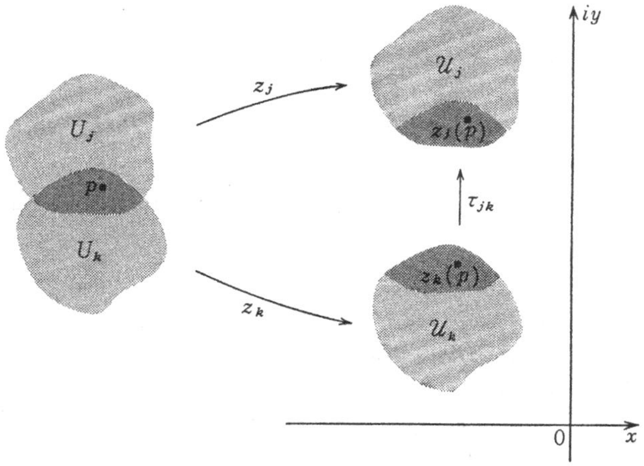

collection $\{z_1, z_2, z_3, \ldots, z_j, \ldots\}$ of local complex coordinates is called a *system* of local complex coordinates on Σ.

Definition 6.5. A connected Hausdorff space Σ on which a system of local complex coordinates $\{z_1, z_2, \ldots, z_j, \ldots\}$ is defined is called a *Riemann surface*, and denoted by $\mathcal{R}$, and $\{z_1, z_2, \ldots, z_j \ldots\}$ is called the system of local complex coordinates for $\mathcal{R}$.

Let $\mathcal{R}$ be a Riemann surface, let $\{z_1, z_2, \ldots, z_j, \ldots\}$ be its system of local complex coordinates and let U_j be the domain of the local complex coordinate (or simply complex coordinate) $z_j : p \to z_j(p)$. Of course $\mathcal{R}$ is covered by the $U_j : \mathcal{R} \subset \bigcup_j U_j$. The complex number $z_j(p)$ is called the local (complex) coordinate of the point p. For a point $p \in \mathcal{R}$, its local coordinate $z_j = z_j(p)$ is uniquely determined once we have selected a region U_j containing p. Of course, $z_j = z_j(p)$ is a complex number belonging to the range $\mathcal{U}_j = z_j(U_j)$ of the function z_j. If $p \in U_j \cap U_k$ then the coordinate transformation

$$\tau_{jk} : z_k \to z_j = \tau_{jk}(z_k), \tag{6.27}$$

which maps the local coordinate $z_k = z_k(p)$ onto $z_j = z_j(p)$, is biholomorphic by definition.

Since a point $p \in \mathcal{R}$ is determined by its local coordinate $z_j = z_j(p)$, we can identify the complex number $z_j = z_j(p)$ with the point $p \in U_j$. If we identify $z_j \in \mathcal{U}_j$ with the point p it corresponds with, then $\mathcal{U}_j$ becomes identified with U_j. Hence, the Riemann surface $\mathcal{R}$ can be considered as the surface $\mathcal{R} = \bigcup_j \mathcal{U}_j$ obtained by piecing together the regions $\mathcal{U}_1$, $\mathcal{U}_2$, $\ldots \mathcal{U}_j, \ldots$ of the complex plane in such a way that the points $z_k \in \mathcal{U}_{kj} \subset \mathcal{U}_k$ and $z_j = \tau_{jk}(z_k) \in \mathcal{U}_{jk} \subset \mathcal{U}_j$ coincide. Considered this way, the points $z_j \in \mathcal{U}_j$ and $z_k \in \mathcal{U}_k$ represent the same point of $\mathcal{R}$ if $z_j = \tau_{jk}(z_k)$.

Example 6.1. Regions $\mathcal{U}$ in the complex plane are Riemann surfaces with only one local coordinate $z \to z$.

Example 6.2. The Riemann sphere **S** is a Riemann surface. We have seen that **S** can be covered by two coordinate neighborhoods U_1 and U_2. Since, by (3.16) $|z_1^2| = (1 + \xi_3)/(1 - \xi_3) < 3$ if $\xi_3 < 1/2$, $\mathcal{U}_1 = z_1(U_1) = \{z_1 : |z_1| < \sqrt{3}\}$. Similarly, $\mathcal{U}_2 = \{z_2 : |z_2| < \sqrt{3}\}$. According to (6.25) the coordinate transformation is given by

$$z_2 = \tau_{21}(z_1) = \frac{1}{z_1}.$$

Hence **S** is the Riemann surface obtained by pasting together two disks $\mathcal{U}_1$ and $\mathcal{U}_2$ with radius $\sqrt{3}$ in the complex plane in such a way that $z_1 \in \mathcal{U}_1$, $1/\sqrt{3} < |z_1| < \sqrt{3}$ and $z_2 = 1/z_1 \in \mathcal{U}_2$ coincide.

Next we want to consider functions defined on a Riemann surface $\mathcal{R}$. Let $u = u(p)$ be a real-valued function defined on a region $D \subset \mathcal{R}$. On $D \cap U_j$, p is represented by the local coordinate $z_j(p)$ hence putting

$$u(p) = u_j(z_j(p)), \qquad p \in D \cap U_j,$$

$u_j(z_j)$ is a function of the complex variable z_j defined on the open set $\mathcal{D}_j = z_j(D \cap U_j) \subset \mathcal{U}_j$. Since $z_j(p) = \tau_{jk}(z_k(p))$ by (6.27), $u_j(z_j) = u_k(z_k)$ if $z_j = \tau_{jk}(z_k)$. Writing $z_j = x_j + iy_j$ and $u_j(x_j, y_j) = u_j(z_j)(x, j)$, we have

$$u(p) = u_j(x_j, y_j), \qquad x_j + iy_j = z_j(p), \qquad p \in D \cap U_j.$$

The pair (x_j, y_j) is also called a local coordinate of p. Since $z_j : p \to z_j(p)$ is a homeomorphism, $u(p)$ is continuous on D if and only if for all j satisfying $D \cap U_j \neq \emptyset$, we have $u_j(x_j, y_j)$ is a continuous function of x_j and y_j. The situation is similar with respect to complex-valued functions $f(p)$ defined on a region $D \subset \mathcal{R}$. Putting

$$f(p) = f_j(z_j(p)), \qquad p \in D \cap U_j,$$

$f_j(z_j)$ is a function of z_j defined on $\mathcal{D}_j$ and $f_j(z_j) = f_k(z_k)$ if $z_j = \tau_{jk}(z_k)$. The function $f(p)$ is continuous on D if and only if for all j with $D \cap U_j \neq \emptyset$, we have $f_j(z_j)$ is a continuous function of z_j.

Definition 6.6. If for all j satisfying $D \cap U_j \neq \emptyset$, the functions $u_j(x_j, y_j)$

of x_j and y_j are continuously differentiable, of class C^n, of class C^∞, ..., then the function $u = u(p)$ is continuously differentiable, of class C^n, of class C^∞, ... on the region $D \subset \mathcal{R}$.

If for all j satisfying $D \cap U_j \neq \emptyset$, $f_j(z_j)$ is a holomorphic function of the complex variable z_j, then $f(p)$ is a holomorphic function defined on the region $D \subset \mathcal{R}$.

If we consider the Riemann surface $\mathcal{R}$ as a surface obtained by pasting together the regions $\mathcal{U}_j$, $j = 1, 2, 3, \ldots$, of the complex plane, then the point $z_j \in \mathcal{U}_j$ is identified with the point $p \in \mathcal{R}$ and the functions $u(p)$ and $f(p)$ can be written as $u(z_j)$ and $f(z_j)$, i.e. the subscript j of u_j and f_j occurring in $u_j(z_j)$ and $f_j(z_j)$ becomes superfluous. Since $z_j \in \mathcal{U}_j$ and $z_k \in \mathcal{U}_k$ are the same points on $\mathcal{R}$ if $z_j = \tau_{jk}(z_k)$, we have $u(z_j) = u(z_k)$ and $f(z_j) = f(z_k)$ if $z_j = \tau_{jk}(z_k)$, i.e. $u(z_j)$ and $f(z_j)$ are different from the expressions obtained by substituting z_j for z_k in $u(z_k)$ and $f(z_k)$ respectively. As usual, we write $u(z_j) = u(x_j + iy_j)$ as $u(x_j, y_j)$. With this notation, a function $u(p) = u(x_j, y_j)$ defined on a region $D \subset \mathcal{R}$ is continuous, continuously differentiable, of class C^∞, ..., if and only if it is a continuous, continuously differentiable, C^∞, ..., function of the local coordinates x_j and y_j and $f(p) = f(z_j)$ is holomorphic if and only if it is a holomorphic function of the local coordinate z_j. Here, it is essential that the coordinate transformations $\tau_{jk} : z_k \to \tau_{jk}(z_k)$ are biholomorphic because, as τ_{jk} is biholomorphic, saying that $f(z_j) = f(z_k)$ is holomorphic with respect to z_j on $\mathcal{U}_j \cap \mathcal{U}_k$ is the same as saying that it is holomorphic with respect to z_k on $\mathcal{U}_j \cap \mathcal{U}_k$.

If $f(p)$ is a holomorphic function defined on the region $D \subset \mathcal{R}$, then its range $\mathcal{D} = f(D)$ is a region in the complex plane, since holomorphic mappings are open mappings. We say that $f : p \to f(p)$ maps D *conformally* onto $\mathcal{D} = f(D)$ if $f(p)$ is a single-valued holomorphic function on the region $D \subset \mathcal{R}$. (The fact that $f(p)$ is single-valued, of course, means that $p \neq q$ implies $f(p) \neq f(q)$.) If $w(p)$ is a single-valued, holomorphic function defined on the region $W \subset \mathcal{R}$, $w : p \to w(p)$ can be used as a local complex coordinate on W. If $\mathcal{R}$ is covered by a finite or countably infinite number of regions $W_1, W_2, \ldots, W_n, \ldots$ and if on each W_n a single-valued, holomorphic function $w_n(p)$ is defined, then the collection $\{w_1, w_2, \ldots, w_n, \ldots\}$ of local complex coordinates $w_n : p \to w_n(p)$ constitutes a system of local complex coordinates of $\mathcal{R}$. There are infinitely many different ways of choosing a system of local complex coordinates for $\mathcal{R}$.

6.3 Differential forms on a Riemann surface

a. Differential forms

Let $\mathcal{R}$ be a Riemann surface, $\{z_1, z_2, \ldots, z_j, \ldots\}$ its system of local coordinates, U_j the domain of the local coordinate $z_j : p \to z_j(p)$ and $\mathcal{U}_j = z_j(U_j)$ its range.

If we identify the point p with its local coordinate $z_j = z_j(p)$ then, as we have seen in the previous section, $\mathcal{U}_j$ becomes identified with U_j and $\mathcal{R}$ becomes obtained by pasting together the regions $\mathcal{U}_j \subset A : \mathcal{R} = \bigcup_j \mathcal{U}_j$. In this section too we identify $\mathcal{U}_j$ with $U_j : U_j = \mathcal{U}_j$. The coordinate transformations $\tau_{jk} : z_k \to z_k = \tau_{jk}(z_j)$ are biholomorphic and $z_j \in U_k$ and $z_k \in U_k$ are identical if $z_j = \tau_{jk}(z_k)$. U_j and U_k are both regions of the complex plane, but we assume that $U_j \cap U_k$ denotes the common part of U_j and U_k considered as subsets of $\mathcal{R}$. Thus, in Example 6.2, the Riemann sphere $\mathbf{S} = U_1 \cup U_2$, both U_1 and U_2 are disks with center 0 and radius $\sqrt{3}$ in the complex plane, but $U_1 \cap U_2 = \{z_1 : 1/\sqrt{3} < |z_1| < \sqrt{3}\}$. If $u = u(p)$ is a continuous function defined on the region $D \subset \mathcal{R}$ and if $D \cap U_j \neq \emptyset$, then $u_j(z_j) = u(z_j)$ is a continuous function of z_j defined on $D \cap U_j$ and $u_j(z_j) = u_k(z_k)$ if $z_j = \tau_{jk}(z_k)$. Conversely, if continuous functions $u_j(z_j)$ are defined on $D \cap U_j \neq \emptyset$ and if $u_j(z_j) = u_k(z_k)$ for $z_j = \tau_{jk}(z_k)$ on $D \cap U_j \cap U_k \neq \emptyset$ then a continuous function $u(p)$ defined on D can be obtained by "pasting together in the right way" continuous functions $u_j(z_j)$ defined on each $D \cap U_j \neq \emptyset$.

Now 1-forms on D are defined in a similar way. If a 1-form,

$$\varphi_j(z_j) = \varphi_{j1}(z_j)dx_j + \varphi_{j2}(z_j)dy, \qquad z_j = x_j + iy_j, \tag{6.28}$$

is given on each $D \cap U_j \neq \emptyset$ and if on $D \cap U_j \cap U_k \neq \emptyset$

$$\varphi_j(z_j) = \varphi_k(z_k), \qquad z_j = \tau_{jk}(z_k), \tag{6.29}$$

then on D a 1-form φ is defined by putting $\varphi(z_j) = \varphi_j(z_j)$. Equality (6.29) means that $\varphi_j(z_j)$ is transformed into $\varphi_k(z_k)$ by the coordinate transformation $z_k \to z_j = \tau_{jk}(z_k)$. That is to say, writing

$$x_j + iy_j = z_j = \tau_{jk}(z_k) = \tau_{jk}(x_k + iy_k)$$

and

$$\tau'_{jk}(z_k) = \alpha_{jk}(z_k) + i\beta_{jk}(z_k),$$

we conclude from $dx_j + i\,dy_j = \tau'_{jk}(z_k)(dx_k + i\,dy_k)$ that

$$\left.\begin{aligned} dx_j &= \alpha_{jk}\,dx_k - \beta_{jk}\,dy_k, \qquad \alpha_{jk} = \alpha_{jk}(z_k), \\ dy_j &= \beta_{jk}\,dx_k + \alpha_{jk}\,dy_k, \qquad \beta_{jk} = \beta_{jk}(z_k). \end{aligned}\right\} \tag{6.30}$$

hence by (6.28)

$$\varphi_j(z_j) = (\alpha_{jk}\varphi_{j1} + \beta_{jk}\varphi_{j2})dx_k + (-\beta_{jk}\varphi_{j1} + \alpha_{jk}\varphi_{j2})dy_k.$$

Equality (6.29) implies that the 1-form in the right-hand side of the above equality equals $\varphi_k(z_k)$, i.e.

$$\left.\begin{aligned}\varphi_{k1}(z_k) &= \alpha_{jk}\varphi_{j1}(z_j) + \beta_{jk}\varphi_{j2}(z_j),\\ \varphi_{k2}(z_k) &= -\beta_{jk}\varphi_{j1}(z_j) + \alpha_{jk}\varphi_{j2}(z_j).\end{aligned}\right\} \tag{6.31}$$

The 1-form φ is called continuous, continuously differentiable, of class C^∞, ... if all $\varphi_j(z_j)$ are continuous, continuously differentiable, of class C^∞, ...

Similarly, if on each $D \cap U_j \neq \varnothing$ a 2-form

$$\omega_j(Z_j) = \omega_{j12}(z_j)dx_j \wedge dx_j$$

is given and if on $D \cap U_j \cap U_k \neq \varnothing$

$$\omega_j(z_j) = \omega_k(z_k), \qquad z_j = \tau_{jk}(z_k), \tag{6.32}$$

then a 2-form ω is defined on D by putting $\omega(z_j) = \omega_j(z_j)$ on each $D \cap U_j \neq \varnothing$. Since $dx_j \wedge dy_j = (\alpha_{jk}^2 + \beta_{jk}^2)dx_k \wedge dy_k$, i.e.

$$dx_j \wedge dy_j = |\tau'_{jk}(z_k)|^2 dx_k \wedge dy_k \tag{6.33}$$

equality (6.32) becomes

$$\omega_{k12}(z_k) = |\tau'_{jk}(z_k)|^2 \omega_{j12}(z_j). \tag{6.34}$$

The 2-form ω is called continuous, continuously differentiable, of class C^∞, ... if all $\omega_j(z_j)$ are continuous, continuously differentiable, of class C^∞, ...

In order to define the differential of a continuously differentiable function u defined on a region $D \subset \mathcal{R}$, we put $u_j(z_j) = u(z_j)$ on $D \cap U_j \neq \varnothing$, where $u_j(z_j) = u_j(x_j + iy_j)$ is a continuously differentiable function of x_j and y_i, satisfying

$$u_j(z_j) = u_k(z_k), \qquad z_j = \tau_{jk}(z_k),$$

on $D \cap U_j \cap U_k \neq \varnothing$.

We will prove:

$$du_j(z_j) = du_k(z_k),\ z_j = \tau_{jk}(z_k). \tag{6.35}$$

We have

$$du_j(z_j) = \frac{\partial u_j}{\partial x_j}\,dx_j + \frac{\partial u_j}{\partial y_j}\,dy_j, \qquad du_k(z_k) = \frac{\partial u_k}{\partial x_k}\,dx_k + \frac{\partial u_k}{\partial y_k}\,dy_k$$

and $u_j(z_j) = u_k(z_k)$, hence

$$\frac{\partial u_k}{\partial x_k} = \frac{\partial x_j}{\partial x_k}\frac{\partial u_j}{\partial x_j} + \frac{\partial y_j}{\partial x_k}\frac{\partial u_j}{\partial y_j},$$
$$\frac{\partial u_k}{\partial y_k} = \frac{\partial x_j}{\partial y_k}\frac{\partial u_j}{\partial x_j} + \frac{\partial y_j}{\partial y_k}\frac{\partial u_j}{\partial y_j}.$$

Since $\partial x_j/\partial x_k = \partial y_j/\partial y_k = \alpha_{jk}$ and $\partial y_k/\partial x_k = \partial x_j/\partial y_k = \beta_{jk}$, by (6.30), we have

$$\frac{\partial u_k}{\partial x_k} = \alpha_{jk}\frac{\partial u_j}{\partial x_j} + \beta_{jk}\frac{\partial u_j}{\partial y_j},$$
$$\frac{\partial u_k}{\partial y_k} = -\beta_{jk}\frac{\partial u_j}{\partial x_j} + \alpha_{jk}\frac{\partial u_j}{\partial y_j}.$$

Hence $du_j(z_j) = du_k(z_k)$ by (6.31).

(6.35) shows that we obtain a continuous 1-form φ on D by putting $\varphi(z_j) = du_j(z_j)$ on each $D \cap U_j \neq \emptyset$. This 1-form φ is called the *differential* of u and denoted by du. Equality (6.35) shows that the differential of a continuously differentiable function is invariant under the coordinate transformation $z_k \to z_k = \tau_{jk}(z_k)$.

Next we want to define the exterior differential of a continuously differentiable 1-form defined on D. Let the continuously differentiable 1-form φ be given by $\varphi(z_j) = \varphi_j(z_j)$ on each $D \cap U_j$, while

$$\varphi_j(z_j) = \varphi_k(z_k), \qquad z_j = \tau_{jk}(z_k), \tag{6.36}$$

i.e.

$$\varphi_{j1}(z_j)dx_j + \varphi_{j2}(z_j)dy_j = \varphi_{k1}(z_k)dx_k + \varphi_{k2}(z_k)dy_k.$$

This means that if we substitute $\alpha_{jk}\,dx_k - \alpha_{jk}\,dy_k$ and $\beta_{jk}\,dx_k + \alpha_{jk}\,dy_k$ for dx_j and dy_j respectively on the left-hand side we get the right-hand expression

$$\varphi_{j1}(z_j)(\alpha_{jk}\,dx_k - \beta_{jk}\,dy_k) + \varphi_{j2}(z_j)(\beta_{jk}\,dx_k + \alpha_{jk}\,dy_k) = \varphi_k(z_k).$$

Considering x_j and y_j as functions of x_k and y_k their differentials are given by

$$dx_j = \alpha_{jk}\,dx_k - \beta_{jk}\,dy_k, \qquad dy_j = \beta_{jk}\,dx_k + \alpha_{jk}\,dy_k$$

hence, by (6.3)

$$d(\alpha_{jk}\,dx_k - \beta_{jk}\,dy_k) = ddx_j = 0,$$
$$d(\beta_{jk}\,dx_k + \alpha_{jk}\,dy_k) = ddy_j = 0.$$

Therefore by (6.2)

$$\begin{aligned} d\varphi_k(z_k) &= d\varphi_{j1} \wedge (\alpha_{jk}\,dx_k - \beta_{jk}\,dy_k) + d\varphi_{j2} \wedge (\beta_{jk}\,dx_k + \alpha_{jk}\,dy_k) \\ &= d\varphi_{j1}(z_j) \wedge dx_j + d\varphi_{j2}(z_j) \wedge dy_j = d\varphi_j(z_j), \end{aligned}$$

i.e.

$$d\varphi_j(z_j) = d\varphi_k(z_k), \qquad z_j = \tau_{jk}(z_k) \tag{6.37}$$

on $D \cap U_j \cap U_k \neq \emptyset$. So if we put $\omega(x_j) = d\varphi_j(z_j)$ on each $D \cap U_j \neq \emptyset$ we obtain a continuous 2-form. This 2-form ω is called the *exterior derivative* of φ and is denoted by $d\varphi$. (6.37) shows that the exterior derivative of a continuously differentiable 1-form is invariant under coordinate transformations $z_k \to z_j = \tau_{jk}(z_k)$.

We want to define the dual form $*\varphi$ of a 1-form φ defined on the region D in a similar way. We only have to verify that if $\varphi_j(z_j) = \varphi_k(z_k)$ and $z_j = \tau_{jk}(z_k)$ then $*\varphi_j(z_j) = *\varphi_k(z_k)$. Using (6.30) and (6.31) we have

$$\begin{aligned}
&*(\varphi_{j1}\, dx_j + \varphi_{j2}\, dy_j)\\
&\quad = -\varphi_{j2}\, dx_j + \varphi_{j1}\, dy_j\\
&\quad = -\varphi_{j2}(\alpha_{jk}\, dx_k - \beta_{jk}\, dy_k) + \varphi_{j1}(\beta_{jk}\, dx_k + \alpha_{jk}\, dy_k)\\
&\quad = (\beta_{jk}\varphi_{j1} - \alpha_{jk}\varphi_{j2})dx_k + (\alpha_{jk}\varphi_{j1} + \beta_{jk}\varphi_{j2})dy_k\\
&\quad = -\varphi_{k2}\, dx_k + \varphi_{k1}\, dy_k\\
&\quad = *(\varphi_{k1}\, dx_k + \varphi_{k2}\, dy_k).
\end{aligned}$$

Hence $*\varphi_j(x_j) = *\varphi_k(z_k)$. So, if we put $\psi(z_j) = *\varphi_j(z_j)$ on each $D \cap U_j \neq \varnothing$, then the dual form $*\varphi = \psi$ of φ is defined on D.

In this way all local properties, operations and so on, pertaining to functions and differential forms defined on a region in the complex plane that are invariant under biholomorphic transformations can be considered as properties, operations and so on pertaining to functions and differential forms defined on a region of a Riemann surface. In what follows we will apply concepts, operations and so on defined for functions or differential forms on the complex plane to functions or differential forms defined on a Riemann surface, whenever the transition is trivial. For example, if φ is a real, continuously differentiable 1-form on the region D of the Riemann surface $\mathcal{R}$ satisfying $d\varphi = d(*\varphi) = 0$, then φ is called *harmonic* on D (Definition 6.1). A real-valued, twice continuously differentiable function u, defined on D and satisfying $d(*du) = 0$ is called a *harmonic function* (Definition 6.2). According to Theorem 6.4, a 1-form φ is harmonic on a region D of $\mathcal{R}$ if and only if there are holomorphic functions $f_j(z_j)$ on all $U_j \cap D \neq \varnothing$ such that $\varphi + i * \varphi$ can be written as

$$\varphi + i * \varphi = f_j(z_j)dz_j. \tag{6.38}$$

If the real-valued function u is twice continuously differentiable on the region $D \subset \mathcal{R}$ then, by (6.20)

$$d(*du) = \Delta_j u(z_j)dx_j \wedge dy_j, \qquad \Delta_j = \frac{\partial^2}{\partial x_j^2} + \frac{\partial^2}{\partial y_j^2} \tag{6.39}$$

on each $D \cap U_j \neq \varnothing$. Hence, by (6.34)

$$\Delta_k u(z_k) = |\tau'_{jk}(z_k)|^2 \Delta_j u(z_j), \qquad z_j = \tau_{jk}(z_k) \tag{6.40}$$

on $D \cap U_j \cap U_k \neq \varnothing$. This is the formula for the coordinate transformation for the Laplacian.

b. Line integrals

A continuous map $\gamma : t \to \gamma(t)$ which maps the closed interval $[a, b] \subset \mathbb{R}$ into the Riemann surface $\mathcal{R}$ is called a *curve*. Since $|\gamma| = \{\gamma(t) : a \leqslant t \leqslant b\}$ is compact by Theorem 6.7, it is possible to find a finite number of coordinate neighborhoods, U_j covering $|\gamma|$. If $\gamma(t) \in U_j$, then the local coordinate of $\gamma(t)$ can be given as

$$\gamma_j(t) = z_j(\gamma(t)) = z_j(t) + iy_j(t).$$

If the local coordinate $\gamma_j(t)$ is continuously differentiable and if $\gamma_j'(t) \neq 0$ for all t, then γ is called a *smooth* curve. If γ is a smooth curve such that all $x_j(t)$ and $y_j(t)$ are real analytic functions of t, then γ is called an *analytic curve* (cf. Section 5.3a). The definition of piecewise smooth and piecewise analytic curves should be clear.

Since the concept of a "line segment" is not invariant under biholomorphic transformations, it makes no sense to consider "line segments" on a general Riemann surface, but we can use analytic Jordan arcs instead of line segments. Two line segments in the plane are either disjoint or their intersection consists of one point or a line segment. The situation for Jordan arcs on a Riemann surface is similar: if the intersection $|\gamma| \cap |\delta|$ of two analytic Jordan arcs $\gamma : t \to \gamma(t)$, $a \leqslant t \leqslant b$, and $\delta : s \to \delta(s)$, $c \leqslant s \leqslant d$, is non-empty, then it consists of a finite number of points and at most two analytic Jordan arcs. To see this note we may assume that $|\gamma| \cap |\delta|$ is an infinite set. A compact, infinite set has an accumulation point. Let $p_0 = \gamma(t_0) = \delta(s_0)$, $a \leqslant t_0 \leqslant b$ and $c \leqslant s_0 \leqslant d$, be such an accumulation point, then there exists a point sequence $\{p_m\}$ with $p_m = \gamma(t_m) = \delta(s_m) \neq p_0$, $a \leqslant t_m \leqslant b$ and $c \leqslant s_m \leqslant d$, which converges to p_0. Since γ and δ are both Jordan arcs, we have $t_m \to t_0$ and $s_m \to s_0$ if $m \to \infty$. Let U_j be a neighborhood containing $p_0 = \gamma(t_0) = \delta(s_0)$ and expand $\gamma_j(t)$ and $\delta_j(s)$ in power series around t_0 and s_0 respectively.

$$\gamma_j(t) = \sum_{n=0}^{\infty} \gamma_{jn}(t - t_0)^n, \qquad \delta_j(t) = \sum_{n=0}^{\infty} \delta_{jn}(t - t_0)^n.$$

Hence $\gamma_j(t)$ and $\delta_j(s)$ are extended to holomorphic functions of t and s respectively, defined on a neighborhood of t_0 and s_0 respectively. Since $\gamma_j(t_0) = \delta_j(s_0)$ and $\gamma_j'(t_0) \neq 0$ and $\delta_j'(s_0) \neq 0$ by assumption, it is possible according to Theorem 3.2 to solve the equation $\delta_j(s) = \gamma_j(t)$ with respect to s on a neighborhood of t_0 and to write $s = \lambda(t)$ with λ a holomorphic function of $t : \delta_j(\lambda(t)) = \gamma_j(t)$. Hence $\delta_j'(s_0)\lambda'(t_0) = \gamma_j'(t_0)$, so $\lambda'(t_0) \neq 0$. Since $\gamma_j(t_m) = \delta_j(s_m)$, we have $\lambda(t_m) = s_m$. The function $\overline{\lambda(\bar{t})}$ is a holomorphic function of t (see Section 5.3a). Hence $\mu(t) = \lambda(t) = \overline{\lambda(\bar{t})}$ is also a holomorphic function of t on a neighborhood of t_0. Since both t_m and

$s_m = \lambda(t_m)$ are real, $\mu(\bar{t}) = 0$, hence $t_0 = \lim_{m\to\infty} t_m$ is an accumulation point of the set of zeros of $\mu(t)$. Hence $\mu(t) = 0$ identically, by Theorem 3.1, i.e. $x(\bar{t}) = x(t)$ identically. Hence, $\lambda(t)$ is real for real values of t. Therefore, on a neighborhood of t_0 to the curve $t \to \gamma_j(t) = \delta_j(\lambda(t))$ is obtained from the curve $s \to \delta_j(s)$ by the coordinate transformation $s = \lambda(t)$. Hence, if $a < t_0 < b$ and $c < s_0 < d$, then for a sufficiently small $\varepsilon > 0$ we have

$$\delta(t) = \gamma(\lambda(t)) \in |\gamma| \cap |\delta| \text{ if } t_0 - \varepsilon \leqslant t \leqslant t_0 + \varepsilon. \tag{6.41}$$

Notice that $s = \lambda(t)$ is a monotone function of t, since $\lambda(t)$ is a real analytic function of t on $[t_0 - \varepsilon,\ t_0 + \varepsilon]$ with $\lambda'(t) \neq 0$.

Let A be the set of accumulation points of $|\gamma| \cap |\delta|$. Since $|\gamma| \cap |\delta|$ is compact, it consists of A and a finite number of isolated points. Putting $T = \{t : \gamma(t) \in A\}$ and $S = \{s : \delta(s) \in A\}$, then of course $T \subset [a, b]$ and $S \subset [c, d]$. Since both γ and δ are Jordan arcs, there exists a one-to-one correspondence between $t \in T$ and $s \in S$ given by $\gamma(t) = \delta(s)$. Since A is a closed set, so are T and S. If $t \in T$, $a < t < b$ and $c < s = \lambda(t) < d$ then t is an interior point of T and s an interior point of S by (6.41). Hence, if t, $a < t < b$, is a boundary point of T, then $\lambda(t) = c$ or $\lambda(t) = d$. If $\lambda(t) = c$, then we write $t = t_c$; if $\lambda(t) = d$, we write $t = t_d$. Therefore, T has at most two boundary points in (a, b).

(i) The case that T has two boundary points t_c and t_d in (a, b). We assume $t_c < t_d$, the case $t_d < t_c$ being treated similarly. Then either $T = [t_c, t_d]$ or $T = [a, t_c] \cup [t_d, b]$. Hence $|\gamma| \cap |\delta|$ consists either of a finite number of isolated points and an analytic Jordan arc $\{\gamma(t) : t_c \leqslant t \leqslant t_d\}$ or of two analytic Jordan arcs $\gamma(t) : a \leqslant t \leqslant t_c\}$ and $\{\gamma(t) : t_d \leqslant t \leqslant b\}$.

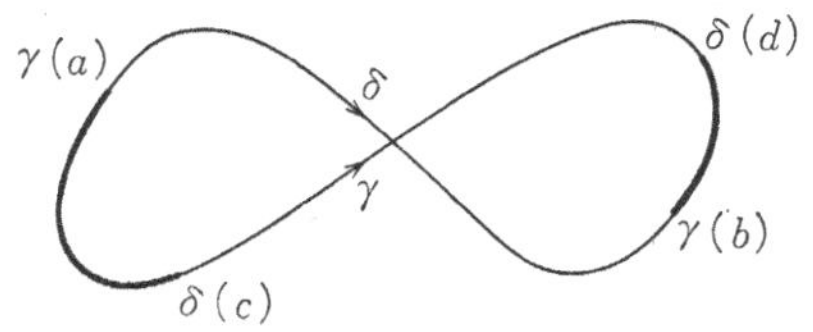

(i)

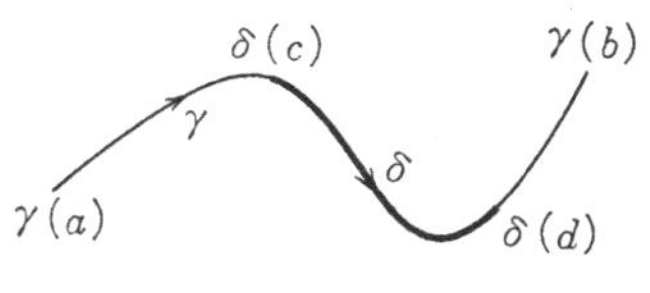

(i)

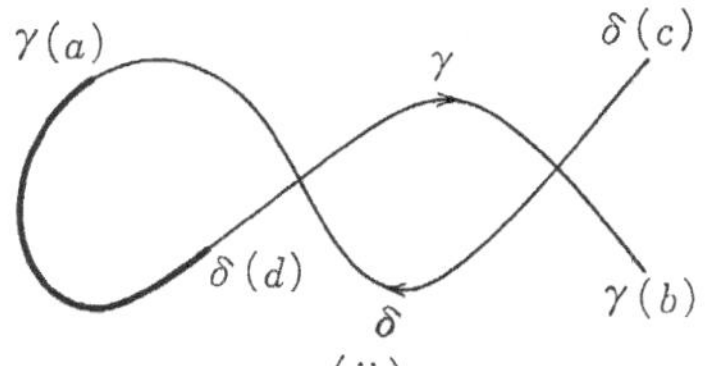

(ii)

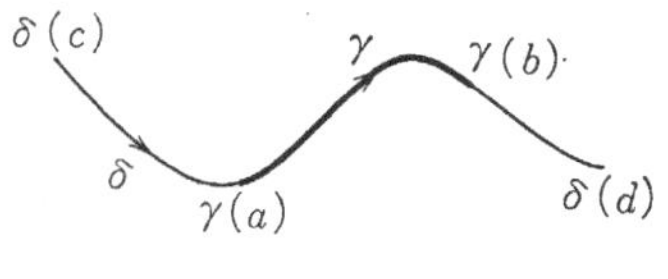

(iii)

(ii) The case that T has only one boundary point t_c or t_d in (a, b). Denoting that boundary point by t_*, we have $T = [a, t_*]$ or $T = [t_*, b]$. Hence $|\gamma| \cap |\delta|$ consists of a finite number of isolated points and an analytic Jordan arc.

(iii) The case that T has no boundary points on (a, b). Then $T = [a, b]$, hance $|\gamma| \cap |\delta| = |\gamma|$.

If both γ and δ are analytic Jordan curves, then it is clear from the above considerations that either $|\gamma| = |\delta|$ or $|\gamma| \cap |\delta|$ consists of a finite number of points.

Just as one proves that an open set $U \in A$ is connected if and only if for each pair of points in U there exists a polygonal line in U connecting these points, it is proved that an open set U of a Riemann surface $\mathcal{R}$ is connected if and only if for each pair of points in U there exists a piecewise analytic curve in U connecting those points.

Now, let $\gamma : t \to \gamma(t)$, $a \leqslant t \leqslant b$, be a smooth curve on the Riemann surface $\mathcal{R}$. Letting

$$\gamma_j(t) = z_j(\gamma(t)) = x_j(t) + iy_j(t)$$

denote the local coordinate of $\gamma(t) \in U_j$, we have

$$\gamma_j(t) = \tau_{jk}(\gamma_k(t))$$

if $\gamma(t) \in U_j \cap U_k$, where $\tau_{jk} : z_k \to z_j = \tau_{jk}(z_k)$ denotes the coordinate transformation. Differentiating both sides with respect to t yields

$$\gamma_j'(t) = \tau_{jk}'(\gamma_k(t))\gamma_k'(t).$$

Since $\tau_{jk}'(z_k) = \alpha_{jk}(z_k) + i\beta_{jk}(z_k)$, we have

$$\left.\begin{aligned} x_j'(t) &= \alpha_{jk}x_k'(t) - \beta_{jk}y_k'(t), \qquad \alpha_{jk} = \alpha_{jk}(\gamma_k(t)),\\ y_j'(t) &= \beta_{jk}x_k'(t) + \alpha_{jk}y_k'(t), \qquad \beta_{jk} = \beta_{jk}(\gamma_k(t)).\end{aligned}\right\} \tag{6.42}$$

Next we want to define the integral of a continuous 1-form φ defined on a region $D \in \mathcal{R}$ along a curve γ satisfying $|\gamma| \in D$. Put

$$\varphi_{j1}(z_j)dx_j + \varphi_{j2}\,dy_j = \varphi_j(z_j) = \varphi(z_j)$$

on each $D \cap U_j \neq \emptyset$. We have $\varphi_j(z_j) = \varphi_k(z_k)$ on $D \cap U_j \cap U_k \neq \emptyset$, which means that substitution of (6.30) in the expression for $\varphi_j(z_j)$ yields $\varphi_k(z_k)$. Hence we obtain

$$\varphi_{j1}(\gamma_j(t))x_j'(t) + \varphi_{j2}(\gamma_j(t))y_j'(t) = \varphi_{k1}(\gamma_k(t))x_k'(t) + \varphi_{k2}(\gamma_k(t))y_k'(t)$$

by comparing (6.42) and (6.30). Putting

$$\varphi(\gamma(t)) = \varphi_{j1}(\gamma_j(t))x_j'(t) + \varphi_{j2}(\gamma_k(t))y_j'(t) \tag{6.43}$$

for $\gamma(t) \in U_j$, the above equation shows that $\varphi(\gamma(t))$ is a continuous function of t defined on $[a, b]$. Therefore we define the integral of φ along γ by

$$\int_\tau \varphi = \int_a^b \varphi(\gamma(t))dt. \tag{6.44}$$

The right-hand side of (6.43) is invariant under a coordinate transformation $z_k \to z_j = \tau_{jk}(z_k)$. Therefore we can omit the subscript j and write (6.43) as

$$\varphi(\gamma(t)) = \varphi_1(\gamma(t))x'(t) + \varphi_2(\gamma(t))y'(t)$$

which by (6.44) becomes

$$\int_\tau \varphi = \int_a^b (\varphi_1(\gamma(t))x'(t) + \varphi_2(\gamma(t))y'(t))dt. \tag{6.45}$$

Since the differential forms $\varphi_{j1}\, dx_j + \varphi_{j2}\, dy_j$ and $\omega_{j12}\, dx_j \wedge dy_j$ are invariant under coordinate transformations we sometimes suppress the subscript j and write $\varphi_1\, dx + \varphi_2\, dy$ and $\omega_{12}\, dx \wedge dy$. If we do so it should be kept in mind that φ_1, φ_2 and ω_{12} depend on the choice of the local coordinate z_j. For example, the coefficient $f(z) = f_j(z)$ occurring in the holomorphic 1-form $f(z)dz$ is a holomorphic function which is transformed into $f_k(z) = \tau'_{jk}(z)f_j(z)$ under the coordinate transformation $z_k \to z_j = \tau_{jk}(z_k)$ and not simply a holomorphic function.

Let u by a continuously differentiable function defined on the domain $D \subset \mathcal{R}$. On each $D \cap U_j \neq \varnothing$ we have

$$du(z_j) = \frac{\partial u_j}{\partial x_j}\, dx_j + \frac{\partial u_j}{\partial y_j}\, dy_j, \qquad u_j = u(z_j).$$

Hence

$$du(\gamma(t)) = \frac{\partial u_j}{\partial x_j} x'_j(t) + \frac{\partial u_j}{\partial y_j} y'_j(t) = \frac{d}{dt} u(\gamma(t)),$$

and therefore

$$\int_\tau du = u(\gamma(b)) - u(\gamma(a)). \tag{6.46}$$

This is an extension of formula (6.6) that is valid for the complex plane to an arbitrary Riemann surface.

c. *Locally finite open coverings*

If $z(p)$ is a single-valued holomorphic function defined on a region U of a Riemann surface $\mathcal{R}$, then, as explained at the end of Section 6.2b, $z : p \to z(p)$ can be used as a local coordinate on U. If $z_q(p)$ is a single-valued holomorphic function defined on a region $U(q)$ containing q and if $z_q(p) = 0$, then $z_q : p \to z_q(p)$ is called a local coordinate with center q. In this case $z_q(U_q)$ is a region in the complex plane containing $0 = z_q(q)$. If for some $r > 0$ the closed disk $\{z \in \mathbb{C} : |z| \leqslant r\}$ is contained in $z_q(U(q))$, then

$$U_r(q)n = \{p : |z_q(p)| < r\}$$

is called a *coordinate disk* with center q and radius r on $\mathcal{R}$. Obviously, $[U_r(q)]$ is a compact subset of $\mathcal{R}$. It hardly needs to be said that the coordinate disk $U_r(q)$ depends on the choice of the local coordinate z_q. If $\{z_1, z_2, \ldots, z_j, \ldots\}$ is a system of local coordinates on $\mathcal{R}$, U_j is the domain of z_j, $q \in \mathcal{R}$ and U_j a coordinate neighborhood such that $q \in U_j$, then

$$z_q(p) = z_j(p) - z_j(q)$$

is a local coordinate with center q.

In general, if Σ is a Hausdorff space and $\mathcal{U}$ an open covering of Σ such that for each $P \in \Sigma$ it is possible to find a neighborhood $U(P)$ of P such that $\mathcal{U} \cap U(P) \neq \varnothing$ for only finitely many $\mathcal{U} \in \mathfrak{U}$ then $\mathfrak{U}$ is called a *locally finite open covering*. It is obvious that if $\mathfrak{U}$ is a locally finite open covering of Σ and K is a compact subset of Σ, then there are only finitely many $\mathcal{U} \in \mathfrak{U}$ such that $\mathcal{U} \cap K \neq \varnothing$.

Theorem 6.8. Let each $q \in \mathcal{R}$ be given a local coordinate z_q with center q and a coordinate disk $U_{r(q)}(q)$. It is possible to select a finite or countably infinite number of coordinate disks:

$$U_j = U_{r(j)}(q_j), \qquad 0 < r(j) \leqslant r(q_j), \qquad j = 1, 2, 3, \ldots,$$

such that $\mathfrak{U} = \{U_j : j = 1, 2, 3, \ldots\}$ is a locally finite open covering of $\mathcal{R}$.

Proof: If $\mathcal{R}$ is compact, then $\mathcal{R}$ can be covered by a finite number of coordinate disks $U_{R(q)}(q)$ and the theorem is obviously true. So we may assume that $\mathcal{R}$ is not compact.

(1) $\mathcal{R}$ is the union of a countably infinite number of open sets B_1, $B_2, \ldots, B_n, \ldots$ satisfying the following two conditions:

(i) each $[B_n]$ is compact,
(ii) $B_1 \subset [B_1] \subset B_2 \subset [B_2] \subset \cdots \subset B_n \subset [B_n] \subset B_{n+1} \subset \cdots$.

In order to prove this, let $\{w_1, w_2, \ldots, w_j, \ldots\}$ be a system of local coordinates on $\mathcal{R}$ and let W_j be the domain of w_j. As explained at the beginning of Section 6.3a, we can consider each W_j as a region $\mathcal{W}_j = w_j(W_j)$ in $\mathcal{R}$. Therefore, there exists for each W_j a sequence

$$A_{j1} \subset A_{j2} \subset \cdots \subset A_{jm} \subset \cdots \subset W_j, \qquad \bigcup_{m=1}^{\infty}(A_{jm}) = W_j,$$

of compact sets $A_{j1}, A_{j2}, \ldots$ (here (A_{jm}) denotes the interior of A_{jm}). Not only it is true that $A_{jm} \subset (A_{jm+1})$ but for each m, $A_{jm} \subset (A_{jn})$ holds if

$n > m$ is large enough. Hence we can pick natural numbers $m_1 < m_2 < \cdots < m_n < \cdots$ such that

$$A_{jm_1} \subset (A_{jm_2}) \subset A_{jm_2} \subset \cdots \subset A_{jm_n} \subset (A_{jm_{n+1}}) \subset \cdots$$

Denoting A_{jm_n} again by A_{jn}, we have $W_j = \bigcup_{n=1}^{\infty}(A_{jn})$ and

$$A_{j1} \subset (A_{j2}) \subset A_{j2} \subset (A_{j3}) \subset \cdots \subset A_{jn} \subset (A_{jn+1}) \subset \cdots.$$

Considering the A_{jn} as subsets of $\mathcal{R}$ (i.e. if $z_j \in A_{jn}$ and $z_k \in A_{kn}$ and $z_j = \tau_{jk}(z_k)$ then z_j and z_k represent the same point of $\mathcal{R}$) and putting

$$B_n = \bigcup_{j \leqslant n}(A_{jn})$$

we have

$$[B_n] = \bigcup_{j \leqslant n}[(A_{jn})] = \bigcup_{j \leqslant n} A_{jn} \subset \bigcup_{j \leqslant n}(A_{jn+1}) \subset B_{n+1}.$$

Hence the sequence of open sets of $\{B_n\}$ satisfies the above conditions (i) and (ii). Since $W_j = \bigcup_{n=1}^{\infty} A_{jn} \subset \bigcup_{n=1}^{\infty} B_n$ we have $\bigcup_{n=1}^{\infty} B_n = \mathcal{R}$.

(2) Put $K_1 = [B_2]$, $K_2 = [B_3] - B_2, \ldots, K_n = [B_{n+1}] - B_n, \ldots$. Each K_n is compact. $K_1 \cup K_2 \cup \ldots \cup K_n = [B_{n+1}]$, hence

$$\mathcal{R} = \bigcup_{n=1}^{\infty} K_n. \tag{6.47}$$

Since $B_n \supset [B_{n-1}]$ by (ii), we have

$$K_n \subset \mathcal{R} - [B_{n-1}], \qquad n = 2, 3, 4.$$

K_1 can be covered by a finite number $U_j = U_{r(j)}(q_j)$, $r(j) = R(q_j)$, $j = 1, 2, \ldots, j(1)$ of coordinate disks $U_{R(q)}(q)$ with $q \in K_1$:

$$K_1 \subset U_1 \cup U_2 \cup \cdots \cup U_{j(1)}.$$

Since $K_2 \subset \mathcal{R} - [B_1]$, K_2 can be covered by a finite number $U_j = U_{r(j)}(q_j)$, $r(j) = r(w_j)$, $j = j(1)+1$, $j(1)+2, \ldots, j(z)$ of coordinate disks $U_{r(q)}(q)$ with $q \in K_2$ and $r(q)$, $0 < r(q) \leqslant R(q)$, for each q chosen in such a way that $U_{r(q)}(q) \subset \mathcal{R} - [B_1]$.

Similarly, it is possible to cover K_n with a finite number of coordinate disks $U_j = U_{r(j)}(q_j) \subset \mathcal{R} - [B_{n-1}]$, $0 < r(j) \leqslant R(q_j)$, $j = j(n-1)+1$, $j(n-1)+2, \ldots, j(n)$:

$$K_n \subset U_{j(n-1)+1} \cup U_{j(n-1)+2} \cup \cdots \cup U_{j(n)} \subset \mathcal{R} - [B_{n-1}]. \tag{6.48}$$

Since $\mathcal{R} = \bigcup_{n=1}^{\infty} K_n$, the collection $\mathcal{U} = \{U_j : j = 1, 2, 3, \ldots\}$ of coordinate disks determined this way is an open covering of $\mathcal{R} : \mathcal{R} = \bigcup_j U_j$. Since all $[U_j]$ are compact and $\mathcal{R}$ is not compact, $\mathcal{U}$ consists of infinitely many coordinate disks. Hence $j(n) \to \infty$ if $n \to \infty$. Replacing n by $n+1$ in

(6.48) we see that $U_j \cap [B_n] = \varnothing$ for $j(n) \leqslant j \leqslant j(n+1)$. If $j(n+1) < j$, picking $m > n$ such that $j(m) \leqslant j < j(m+1)$ we have $U_j \cap [B_m] = \varnothing$ and $[B_n] \subset [B_m]$, hence $U_j \cap [B_n] = \varnothing$. Therefore, $U_j \cap [B_n] = \varnothing$ for $j > j(n)$. Since $\mathcal{R} = \bigcup_{n=1}^{\infty} B_n$, each $q \in \mathcal{R}$ belongs to some B_n. This B_n is an open neighborhood of q and there are at most $j(n)$ sets $U_j \in \mathcal{U}$ such that $U_j \cap B_n \neq \varnothing$. Hence $\mathcal{U}$ is a locally finite open covering of $\mathcal{R}$.

d. Partition of unity

If $\mathcal{U}$ is a locally finite open covering of the Riemann surface $\mathcal{R}$ and K a compact subset of $\mathcal{R}$, then there are only finitely many open sets $U \in \mathcal{U}$ such that $U \cap K \neq \varnothing$. Hence, if $\mathcal{R}$ is compact, $\mathcal{U}$ consists of a finite number of open sets. If R is not compact, then $R = \bigcup_{n=1}^{\infty} K_n$ by (6.47), where all K_n are compact, hence $\mathcal{U}$ consists of at most a countably infinite number of open sets. In both cases we can write

$$\mathcal{U} = \{U_j : j = 1, 2, 3, \ldots\}.$$

If $u = u(p)$ is a function on $\mathcal{R}$, the closure of the set of points p such that $u(p) \neq 0$ is called the *support* of u:

$$\operatorname{supp} u = [\{p \in \mathcal{R} : u(p) \neq 0\}].$$

The support of a 1-form $\varphi = \varphi(p)$ or a 2-form $\omega = \omega(p)$ on $\mathcal{R}$ is defined similarly:

$$\operatorname{supp} \varphi = [\{p \in \mathcal{R} : \varphi(p) \neq 0\}],$$
$$\operatorname{supp} \omega = [\{p \in \mathcal{R} : \omega(p) \neq 0\}].$$

Definition 6.7. Let $\mathcal{U} = \{U_j : j = 1, 2, 3, \ldots\}$ be a locally finite open covering of $\mathcal{R}$. A collection $\{p_j : j = 1, 2, 3, \ldots\}$ of C^∞ functions $\rho_j = \rho_j(p)$ defined on $\mathcal{R}$ is called a *partition of unity* subordinated to $\mathcal{U}$ if the following conditions are satisfied:

(i) $\rho_i(p) \geqslant 0$ for all $p \in \mathcal{R}$ and $\operatorname{supp} \rho_j \subset U_j$;

(ii) the equality

$$1 = \sum_j \rho_j(p) \tag{6.49}$$

is valid.

If $\rho_j(p) \neq 0$, then $p \in U_j$. Since $\mathcal{U} = \{U_j\}$ is locally finite, $\rho_j(p) = 0$ for all j except finitely many at each point $p \in \mathcal{R}$. Hence the infinite sum on the right-hand side of (6.49) reduces to a finite sum at each point $p \in \mathcal{R}$.

Theorem 6.9. For each locally finite open covering $\mathcal{U} = \{U_j\}$ of $\mathcal{R}$ there exists a partition of unity $1 = \sum_j \rho_j(p)$ subordinated to it.

Proof: (1) We first consider the case where $\mathcal{U} = \{U_j\}$ consists of coordinate disks, as in Theorem 6.8. Putting $z_j(p) = z_{q_j}(p)$ in order to simplify the notation, we have

$$U_j = U_{r(j)}(q_j) = \{p : |z_j(p)| < r(j)\}.$$

We define the disk U_j^ε, $\varepsilon > 0$, by:

$$U_j^\varepsilon = \{p : |z_j(p)|^2 < r(j)^2 - \varepsilon\}.$$

If $\varepsilon \geqslant r(j)^2$, then we put $U_j^\varepsilon = \emptyset$. Since $U_j = \bigcup_{\varepsilon>0} U_j^\varepsilon$ we have, by (6.48),

$$K_n \subset \bigcup_{\varepsilon>0} \bigcup_{j(n-1)<j\leqslant j(n)} U_j^\varepsilon.$$

Hence, since K_n is compact, we have for some $\varepsilon > 0$,

$$K_n \subset \bigcup_{j(n-1)<j\leqslant j(n)} U_j^\varepsilon.$$

Hence, for each j it is possible to find a sufficiently small $\varepsilon(j) > 0$ such that

$$K_n \subset \bigcup_{j(n-1)<j\leqslant j(n)} U_j^{\varepsilon(j)},$$

so, by (6.47),

$$\mathcal{R} = \bigcup_{n=1}^{\infty} K_n = \bigcup_j U_J^{\varepsilon(j)}.$$

We define a C^∞ function $h_j(s)$ of the real variable s as follows:

$$\begin{cases} h_j(s) = 1 & \text{for } s \leqslant r(j)^2 - \varepsilon(j), \\ 0 < h_j(s) < 1 & \text{for } r(j)^2 - \varepsilon(j) < s < r(j)^2 - \varepsilon(j)/2, \\ h_j(s) = 0 & \text{for } s \geqslant r(j)^2 - \varepsilon(j)/2. \end{cases} \tag{6.50}$$

Now, $h_j(x^2 + y^2)$ is a C^∞ function of x and y, hence

$$\eta_j(p) = \begin{cases} h_j(|z_j(p)|^2), & p \in U_j, \\ 0, & p \in \mathcal{R} - U_j, \end{cases}$$

is a C^∞ function on $\mathcal{R}$ with $\operatorname{supp} \eta_j = [U_j^{\varepsilon(j)/2}] \subset U_j$. if $p \in U_j^{\varepsilon(j)}$, then $\eta_j(p) = 1$, hence

$$\sum_j \eta_j(p) \geqslant 1.$$

by (6.50). Since $\sum_j \eta_j(p)$ is actually a finite sum, it is a C^∞ function of p. Hence

$$\rho_j(p) = \frac{\eta_j(p)}{\sum_j \eta_j(p)}$$

defines a C^∞ function on $\mathcal{R}$ such that $\operatorname{supp} \rho_j \subset U_j$ and $\sum_j \rho_j(p) = 1$. Therefore, $\{\rho_j : j = 1, 2, 3, \ldots\}$ constitutes a partition of unity subordinated to $\mathfrak{U} = \{U_j\}$.

(2) For the general case, let $\mathfrak{W} = \{W_m : m = 1, 2, 3, \ldots\}$ be an arbitrary locally finite open covering of $\mathcal{R}$. If $q \in \mathcal{R}$, then there exists only finitely many W_m with $q \in W_m$, hence it is possible to select a coordinate disk $U_{R(q)}(q)$ such that if $q \in W_m$ then $U_{R(q)}(q) \subset W_m$. Applying Theorem 6.8 to the collection of coordinate disks $U_{R(q)}(q)$ obtained in this way, we obtain a locally finite open covering $\mathfrak{U} = \{U_j\}$ of $\mathcal{R}$ with $U_j = U_{r(j)}(q_j)$ and $0 < r(j) \leqslant R(q_j)$. Let $m(j)$ represent a choice of m such that $q_j \in W_m$. Since $U_j \subset U_{r(q_j)}(q_j)$ we have $U_j \subset W_{m(j)}$. Now let $\{\rho_j\}$ be a partition of unity subordinated to $\mathfrak{U} = \{U_j\}$ as obtained above and put

$$\sigma_m(p) = \sum_{m(j)=m} \rho_j(p).$$

$\sigma_m = \sigma_m(p)$ is a C^∞ function on $\mathcal{R}$ and $\sigma_m(p) \geqslant 0$ for all $p \in \mathcal{R}$. The set $\operatorname{supp} \rho_j$ is closed, $\operatorname{supp} \rho_j \subset U_j$ and $\mathfrak{U} = \{U_j\}$ is locally finite, hence $\bigcup_{m(j)=m} \operatorname{supp} \rho_j$ is also closed. Hence

$$\operatorname{supp} \sigma_m = \bigcup_{m(j)=m} \operatorname{supp} \rho_j \subset \bigcup_{m(j)=m} U_j \subset W_m.$$

Further:

$$\sum_m \sigma_m(p) = \sum_m \sum_{m(j)=m} \rho_j(p) = \sum_j \rho_j(p) = 1$$

so $\{\sigma_m\}$ is a partition of unity subordinated to $\mathfrak{W} = \{W_m\}$.

Next we want to define the integral of a continuous 2-form ω over $\mathcal{R}$, using the idea of a partition of unity. Let $\{z_1, z_2, \ldots, z_j, \ldots\}$ be a system of local coordinates of $\mathcal{R}$, U_j a region, such that $[U_j]$ is contained in the domain of z_j, while $\mathcal{R} = \bigcup_j U_j$. If $U_j \cap U_k \neq \emptyset$, then $z_j = \tau_{jk}(z_k)$. We further assume that all $[U_j]$ are compact and that the open covering $\mathfrak{U} = \{U_j\}$ of $\mathcal{R}$ is locally finite. Writing

$$\omega(z_j) = \omega_{j12}(z_j) dx_j \wedge dy_j, \qquad z_j = x_j + iy_j,$$

on each U_j, we have on each $U_j \cap U_k \neq \emptyset$,

$$\omega_{k12}(z_k) = |\tau'_{jk}(z_k)|^2 \omega_{j12}(z_j)$$

by (6.34). Hence

$$|\omega_{k12}(z_k)| = |\tau'_{jk}(z_k)|^2 |\omega_{j12}(z_j)|,$$

so

$$|\omega|(z_j) = |\omega_{j12}(z_j)| dx_j \wedge dy_j$$

defines a continuous 2-form on $\mathcal{R}$. Let $\{\rho_j\}$ be a partition of unity subordinated to $\mathfrak{U} = \{U_j\}$. Since $\sum_j \rho_j = 1$, we have

$$\omega = \sum_j \rho_j \omega.$$

Each $\rho_j \omega$ is a continuous 2-form on $\mathcal{R}$ which is identically equal to zero on $\mathcal{R} - U_j$. Considering $\rho_j \omega$ as a 2-form defined on the region U_j in the z_j plane, we define

$$\int_{\mathcal{R}} \omega = \sum_j \int_{U_j} \rho_j \omega_{j12}(z_j) dx_j \, dy_j \tag{6.51}$$

where we have put $\rho_j \omega_{j12}(z_j) = \rho_j(z_j)\omega_{j12}(z_j)$, $\rho_j(z_j) = \rho_j(p)$ and $z_j = z_j(p)$. If $\mathcal{R}$ is compact, the sum on the right-hand side of (6.51) is finite. If $\mathcal{R}$ is not compact, this sum is infinite and we have to ascertain its convergence. To do this we consider the integral of $|\omega|$:

$$\int_{\mathcal{R}} |\omega| = \sum_j \int_{U_j} \rho_j |\omega_{j12}(z_j)| dx_j \, dy_j. \tag{6.52}$$

Since all terms on the right-hand side are nonnegative, the series either converges or diverges to $+\infty$.

Since

$$\left| \int_{U_j} \rho_j \omega_{j12}(z_j) dx_j \, dy_j \right| \leqslant \int_{U_j} \rho_j |\omega_{j12}(z_j)| dx_j \, dy_j$$

the series on the right-hand side of (6.51) converges absolutely if $\int_{\mathcal{R}} |\omega| < +\infty$ and

$$\left| \int_{\mathcal{R}} \omega \right| \leqslant \int_{\mathcal{R}} |\omega|. \tag{6.53}$$

If $\int_{\mathcal{R}} |\omega| = +\infty$ we shall say that the value of $\int_{\mathcal{R}} \omega$ is not determined. Obviously $\int_{\mathcal{R}} \omega$ is linear in ω.

We have to show that the value of $\int_{\mathcal{R}} \omega$ as defined above is independent of the choice of the system of local coordinates, the open covering $\{U_j\}$ and the partition of unity $\{\rho_j\}$. To see this, let $\{w_1, w_2, \ldots, w_n, \ldots\}$ be another system of local coordinates, W_n be regions such that $[W_n]$ is compact and is contained in the domain of w_n and $\mathfrak{W} = \{W_n\}$ is a locally finite open covering of $\mathcal{R}$ and let $\{\sigma_n\}$ be a partition of unity subordinated to $\mathfrak{W}$.

On W_n we put

$$\omega(w_n) = \omega_{n12}(w_n)du_n \wedge dv_n, \qquad w_n = u_n + iv_n.$$

We have to verify that $\int_{\mathcal{R}}\omega$ as defined in (6.51) is also given by

$$\int_{\mathcal{R}} \omega = \sum_n \int_{\mathcal{R}} \sigma_n \omega, \tag{6.54}$$

$$\int_{\mathcal{R}} \sigma_n \omega = \int_{W_n} \sigma_n \omega_{n12}(w_n)du_n\, dv_n. \tag{6.55}$$

We first prove (6.55). By (6.51) we have

$$\int_{\mathcal{R}} \sigma_n \omega = \sum_j \int_{U_j} \rho_j \sigma_n \omega_{j12}(z_j)dx_j\, dy_j.$$

If $\rho_j\sigma_n$ is not identically equal to 0, then $\operatorname{supp} \rho_j\sigma_n \subset U_j \cap W_n$ is compact, hence $\operatorname{supp} \rho_j\sigma_n$ is covered by finitely many connected components of $U_j \cap W_n$. The coordinate transformation $\tau_{jn} : w_n \to z_j = \tau_{jn}(w_n)$ is biholomorphic on each component and

$$\omega_{n12}(w_n) = |\tau'_{jn}(w_n)|^2 \omega_{j12}(z_j)$$

by (6.34). Since $|\tau'_{jn}(w_n)|^2 = \partial(x_j, y_j)/\partial(u_n, v_n)$, we get

$$\begin{aligned}\int_{U_j} \rho_j\sigma_n\omega_{j12}(z_j)dx_j\, dy_j &= \int_{W_n} \rho_j\sigma_n\omega_{j12}(z_j)\frac{\partial(x_j, y_j)}{\partial(u_n, v_n)}\, du_n\, dv_n \\ &= \int_{W_n} \rho_j\sigma_n\omega_{j12}(z_j)|\tau'_{jn}(w_n)|^2 du_n\, dv_n \\ &= \int_{W_n} \rho_j\sigma_n\omega_{n12}(w_n)du_n\, dv_n.\end{aligned}$$

Since $[W_n]$ is compact, there are only a finite number of sets U_j such that $U_j \cap [W_n] \neq \varnothing$. Hence, for some h, we have $U_j \cap W_n \neq \varnothing$ if $j > h$. Therefore, $\sum_{j=1}^{h} \rho_j(p) = 1$ on W_n and $\rho_j(p) = 0$ if $j > h$. Hence

$$\int_{\mathcal{R}} \sigma_n\omega = \sum_{j=1}^{h} \int_{W_n} \rho_j\sigma_n\omega_{n12}(w_n)du_n\, dv_n = \int_{W_n} \sigma_n\omega_{n12}(w_n)du_n\, dv_n,$$

proving (6.55).

If $\mathcal{R}$ is compact, then $\mathfrak{W}$ is a finite covering. $\mathfrak{W} = \{W_n : n = 1, 2, \ldots, m\}$ and obviously $\sum_{n=1}^{m} \sigma_n = 1$ by (6.54). If $\mathcal{R}$ is not compact, we shall prove (6.54) assuming that $\int_{\mathcal{R}}|\omega| < +\infty$. Since

$$\sum_{j=1}^{\infty} \int_{\mathcal{R}} \rho_j|\omega| = \int_{\mathcal{R}} |\omega| < +\infty$$

by (6.52), there exists for each $\varepsilon > 0$ an h such that

$$\sum_{j=h+1}^{\infty} \int_{\mathcal{R}} \rho_j |\omega| < \varepsilon.$$

Hence, by (6.51) and (6.53):

$$\left| \int_{\mathcal{R}} \omega - \sum_{j=1}^{h} \int_{U_j} \rho_j \omega \right| = \left| \sum_{j=h+1}^{\infty} \int_{U_j} \rho_j \omega \right| \leqslant \sum_{j=h+1}^{\infty} \int_{U_j} \rho_j |\omega| < \varepsilon.$$

Replacing w by $\sum_{n=1}^{m} \sigma_n \omega$ and observing that

$$\int_{U_j} \rho_j \left| \sum_{n=1}^{m} \sigma_n \omega \right| \leqslant \int_{U_j} \rho_j |\omega|,$$

we get

$$\left| \int_{\mathcal{R}} \sum_{n=1}^{m} \sigma_n \omega - \sum_{j=1}^{h} \int_{U_j} \rho_j \sum_{n=1}^{m} \sigma_n \omega \right| < \varepsilon.$$

Since $\bigcup_{j=1}^{h} [U_j]$ is compact, there is an m such that $W_n \cap \bigcup_{j=1}^{h} [U_j] \neq \emptyset$ if $n > m$. Therefore $\rho_j(p) \sum_{n=1}^{m} \sigma_n(p) = \rho_j(p) \sum_{n=1}^{\infty} \sigma_n(p) = \rho_j(p)$ if $1 \leqslant j \leqslant h$. Therefore

$$\left| \sum_{n=1}^{m} \int_{\mathcal{R}} \sigma_n \omega - \sum_{j=1}^{h} \int_{U_j} \rho_j \omega \right| < \varepsilon,$$

and hence

$$\left| \int_{\mathcal{R}} \omega - \sum_{n=1}^{m} \int_{\mathcal{R}} \sigma_n \omega \right| < 2\varepsilon,$$

proving (6.54).

In Section 6.3b we already defined the integral of a continuous 2-form $\omega = \omega_{12}(z) dx \wedge dy$, $z = x + iy$, over a region $D \subset A$ as $\int_D \omega = \int_D \omega_{12}\, dx\, dy$. We have to verify that this definition is consistent with definition (6.51). Let $\mathcal{U} = \{U_j : j = 1, 2, 3, \ldots\}$ be a locally finite open covering of D consisting of disks $U_j \subset D$, and $\{\rho_j\}$ be a partition of unity subordinated to $\mathcal{U} = \{U_j\}$. Then the integral of ω over D according to definition (6.51) is given by

$$\sum_{j=1}^{\infty} \int_D \rho_j \omega.$$

Therefore it suffices to show that

$$\int_D \omega = \sum_{j=1}^{\infty} \int_D \rho_j \omega \tag{6.56}$$

assuming $\int_D |\omega| < +\infty$. Selecting a sequence $\{A_m\}$ of sets A_m which are unions of rectangles and satisfy $A_1 \subset A_2 \subset \cdots \subset A_m \subset \ldots$, $A_m \subset D$ and $\bigcup_{m=1}^{\infty} (A_m) = D$, we have by the original definition of the integral,

$$\left|\int_D \omega - \int_{A_m} \omega\right| \leqslant \int_D |\omega| - \int_{A_m} |\omega| \to 0 \quad m \to \infty. \tag{6.57}$$

Replacing ω by $\sum_{j=1}^{j} \rho_j \omega$ and observing that

$$\int_D \left|\sum_{j=1}^{h} \rho_j \omega\right| \leqslant \int_D \sum_{j=1}^{h} \rho_j |\omega| \leqslant \int_D |\omega|,$$

we get

$$\left|\int_D \sum_{j=1}^{h} \rho_j \omega - \int_{A_m} \sum_{j=1}^{h} \rho_j \omega\right| \leqslant \int_D |\omega| - \int_{A_m} \left|\sum_{j=1}^{h} \rho_j \omega\right|.$$

Since $\mathcal{U} = \{U_j\}$ is locally finite, for each m we have $U_j \cap A_m = \varnothing$ if $j > h(m)$ for some sufficiently large $h(m)$. Hence $\sum_{j=1}^{h(m)} \rho_j \omega = \omega$ on A_m. Therefore

$$\left|\sum_{j=1}^{h(m)} \int_D \rho_j \omega - \int_{A_m} \omega\right| \leqslant \int_D |\omega| - \int_{A_m} |\omega|,$$

and hence by (6.57)

$$\left|\int_D \omega - \sum_{j=1}^{h(m)} \int_D \rho_j \omega\right| \to 0 \quad m \to \infty,$$

proving (6.56).

e. *Green's Theorem*

Let C be a piecewise smooth Jordan curve on the Riemann surface $\mathcal{R}$, $\gamma : t \to \gamma(t)$, $0 \leqslant t \leqslant 1$, $\gamma(0) = \gamma(1)$, a parameter representation of C and $q = \gamma(a)$ with $0 < a < 1$. let $z_q : p \to z_q(p)$ be a local coordinate with center q and consider the coordinate disk $U_{R(q)}(q)$ as a disk with center 0 and radius $R(q)$ in the z_q plane. The map

$$\lambda_r : \theta \to z_q = \lambda_r(\theta) = re^{i\theta} 0 < \theta < 2\pi$$

with $0 < r \leqslant R(q)$ defines a circle with center 0 and radius r contained in $U_{R(q)}(q)$. Assuming from the beginning that $R(q) > 0$ has been taken sufficiently small, we know from the first part of the proof of Theorem 5.5 that the circle λ_r intersects C in two points $\lambda_r(\alpha) = \gamma(u)$ and $\lambda_r(\beta) = \gamma(v)$, $v < a < u$, $\alpha < \beta < \alpha + 2\pi$, and that $\alpha = \alpha(r)$ and $\beta = \beta(r)$ are continuous functions of r, $0 \leqslant r \leqslant R(q)$. hence C divides the disk $U_{R(q)}(q)$ into two regions:

$$U^+ = \{re^{i\theta} : 0 < r < R(q), \alpha(r) < \theta < \beta(r)\},$$

$$U^- = \{re^{i\theta} : 0 < r < R(q), \beta(r) < \theta < \alpha(r) + 2\pi\},$$

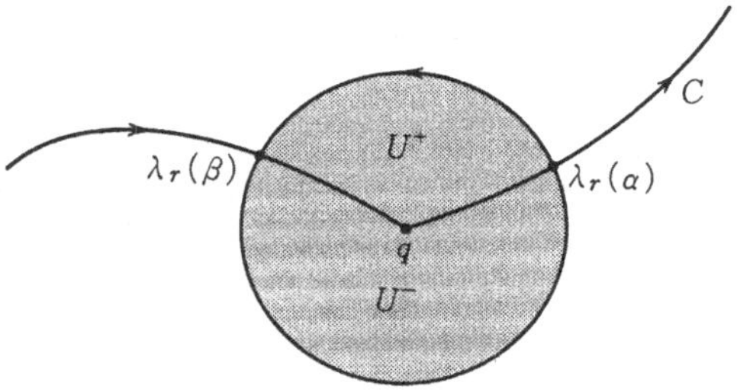

i.e.

$$U_{r(q)}(q) - C = U^+ \cup U^-, \qquad U^+ \cap U^- = \emptyset. \tag{6.58}$$

Since λ_r intersects C in $\lambda_r(\alpha)$ from left to right and in $\lambda_r(b)$ from right to left, U^+ is the left of C and U^- is to the right of C.

For each $q \in C$ we pick a disk $U_{R(q)}$ as defined above and for each $q \notin C$ a disk $U_{R(q)}$ such that $[U_{R(q)}(q)] \cap C \neq \emptyset$. By Theorem 6.8, there exists a locally finite open covering of $\mathcal{R}$ of the form

$$\begin{aligned} &\mathcal{U} = \{U_j : j = 1, 2, 3, \ldots\}, \qquad U_j = U_{r(j)}, \\ &0 < r(j) \leqslant R(q_j). \end{aligned} \tag{6.59}$$

If $U_j \cap C \neq \emptyset$, then $q_j \in C$ and by (6.58) C divides U_j into U_j^+ (on the left) and U_j^- (on the right):

$$U_j - C = U_j^+ \cup U_j^-, \qquad U_j^+ \cap U_j^- = \emptyset.$$

Since the coordinate transformation $z_k \to z_j = \tau_{jk}(z_k)$ is a conformal mapping, right and left with respect to C do not change under this transformation. Hence, if $U_j \cap U_k \cap C \neq \emptyset$ then $U_j^+ \cap U_k^+ \neq \emptyset$, $U_j^- \cap U_k^- \neq \emptyset$, $U_j^+ \cap U_k^- = \emptyset$ and $U_j^- \cap U_k^+ = \emptyset$.

Putting

$$U(C) = \bigcup_{q_j \in C} U_j, \qquad U^+(C) = \bigcup_{q_j \in C} U_j^+, \qquad U^-(C) = \bigcup_{q_j \in C} U_j^-, \tag{6.60}$$

$U(C)$ is a region containing C and C divides $U(C)$ into two regions $U^+(C)$ and $U^-(C)$:

$$U(C) - C = U^+(C) \cup U^-(C), \qquad U^+(C) \cap U^-(C) = \emptyset. \tag{6.61}$$

The set $U^+(C)$ is to the left of C and $U^-(C)$ is to the right of C. We assume that C is the boundary of a compact closed region $[D]$ where $D = ([D])$. If $D \cap U^+(C) \neq \emptyset$, then $U^+(C) \subset D$, for since C does not pass through $U^+(C)$, $V = U^+(C) - [D]$ is an open set, hence the connected open set $U^+(C)$ is the union of the two disjoint open sets V and

$U^+(C) \cap D$, hence $V = \varnothing$. Similarly we can prove that $U^-(C) \cap D \neq \varnothing$ implies $U^-(C) \subset D$. But, $U^+(C) \subset D$ and $U^-(C) \subset D$ cannot occur simultaneously since, assuming that $U(C) - C \subset D$, we have $C \subset U(C) \subset [D]$, contradicting the fact that C is the boundary of $[D]$. If $U^+(C) \subset D$, i.e. if D is to the left of C, C is called the *boundary* of the closed region $[D] : C = \partial[D]$. If D is to the right of C, then we replace C by C^{-1} so that again $C = \partial[D]$.

Since $U_j = U_{r(j)}(q_j) \subset U_{R(q_j)}$, $q_j \notin C$ implies $[U_j] \cap C = \varnothing$, hence $[U_j] \subset D$ or $[U_j] \subset \mathcal{R} - [D]$. Therefore, $[U_j] \subset D$ if $q_j \in D$ and $[U_j] \subset \mathcal{R} - [D]$ if $q_j \in \mathcal{R} - [D]$. Since $[D]$ is compact and the open covering $\mathcal{U} = \{U_j\}$ is locally finite, there are only finitely many U_j with $U_j \cap D \neq \varnothing$. We further observe that $U_j \cap [D] \neq \varnothing$ if and only if $q_j \in [D] = C \cup D$. Therefore, we may assume that $q_j \in C$ if $1 \leqslant j \leqslant h$ and $q_j \in D$ is $h+1 \leqslant j \leqslant m$ and $[U_j] \cap [D] = \varnothing$ if $j \geqslant m+1$. Under these circumstances we have $U_j \cap D = U_j^+$ if $q_j \in C$, hence:

$$D = U_1^+ \cup U_2^+ \cup \cdots \cup U_h^+ \cup U_{h+1} \cup \cdots \cup U_m. \tag{6.62}$$

Let ω be a continuous 2-form on the closed region $[D]$ and let $\{\rho_j : j = 1, 2, 3, \ldots\}$ be a partition of unity subordinated to the open covering $\mathcal{U} = \{U_j\}$ of $\mathcal{R}$. Then $U_j \cap [D] = \varnothing$ for $j > m$, hence $\rho_j(p) = 0$ identically on $[D]$ for $j > m$. Hence $1 = \sum_{j=1}^m \rho_j(p)$ on $[D]$ and so

$$\omega = \sum_{j=1}^{m} \rho_j \omega.$$

Now think of D as a Riemann surface itself and consider

$$\int_D \omega = \sum_{j=1}^{m} \int_D \rho_j \omega,$$

as defined in Section 6.3d above. Since $\operatorname{supp} \rho_j \subset U_j$ we have

$$\int_D \omega = \sum_{j=1} h \int_{u_j^+} \rho_j \omega + \sum_{j=h+1}^{m} \int_{U_j} \rho_j \omega. \tag{6.63}$$

As explained at the end of Section 6.3d, we can consider $\int_{U_j}^+ \rho_j \omega$ and $\int_{U_j} \rho_j \omega$ as integrals over the regions U_j^+ and U_j of the z_j-plane. The boundary $\lambda_j = \partial[U_j]$ of $[U_j]$ is the circle with center 0 and radius $r(j)$ in the z_j-plane. The boundary of $[U_j^+]$ is a Jordan curve consisting of

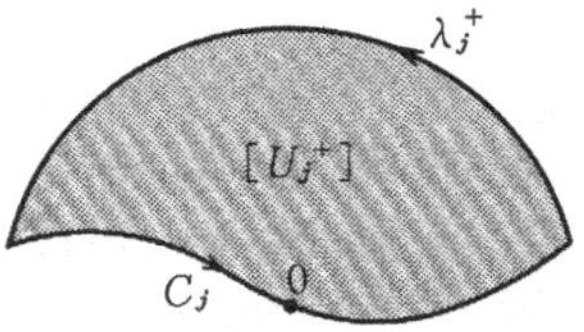

the piecewise smooth Jordan arc $C_j = C \cap [U_j]$ and the arc $\lambda_j^+ = \lambda_j \cap [D] : \partial[U_j^+] = C_j \cdot \lambda_j^+$. The integral of the continuous 2-form $\rho_j \omega$ over the closed region $[U_j^+]$ can be defined as

$$\int_{[U_j^+]} \rho_j \omega = \int_{U_j^+} \rho_j \omega. \tag{6.64}$$

Since, by (6.62) we have

$$[D] = [U_1^+] \cup \cdots \cup [U_h^+] \cup U_{h+1} \cup \cdots \cup U_m \tag{6.65}$$

we define the integral of ω over $[D]$ as:

$$\int_{[D]} \omega = \sum_{j=1}^{h} \int_{U_j^+} \rho_j \omega + \sum_{j=h+1}^{m} \int_{U_j} \rho_j \omega. \tag{6.66}$$

Then, by (6.63):

$$\int_{[D]} \omega = \int D\omega. \tag{6.67}$$

If ω is a continuous 2-form on $\mathcal{R}$ and $\int_{\mathcal{R}} |\omega| < +\infty$, then

$$\mathcal{R} - [D] = U_1^- \cup \cdots \cup U_h^- \cup U_{m+1} \cup U_{m+2} \cup \cdots,$$

hence

$$\int_{\mathcal{R}-[D]} \omega = \sum_{j=1}^{h} \int_{U_j^-} \rho_j \omega + \sum_{j \geqslant m+1} \rho_j \omega.$$

Since

$$\int_{[U_j^+]} \rho_j \omega + \int_{u_j^-} \rho_j \omega = \int_{U_j} \rho_j \omega$$

for each $j = 1, 2, \ldots, h$, we obtain the equality

$$\int_{\mathcal{R}} \omega = \int_{[D]} \omega + \int_{\mathcal{R}-[D]} \omega \tag{6.68}$$

from (6.66) and the definition of integral (6.51).

The above result can easily be extended to the case that the boundary of the region D consists of a finite number of mutually disjoint piecewise smooth Jordan curves.

Theorem 6.10 (Green's Theorem). Let $[D]$ be a compact, closed region on a Riemann surface $\mathcal{R}$ and let $C = \partial[D]$ consist of a finite number of mutually disjoint piecewise smooth Jordan curves. If φ is a continuously differentiable 1-form on D, then

$$\int_C \varphi = \int_{[D]} d\varphi, \qquad C = \partial[D].$$

Proof: We prove the theorem for the case that $C = \partial[D]$ is just one piecewise smooth Jordan curve. The proof of the general case is similar. Let $\mathcal{U} = \{U_j : j = 1, 2, 3, \ldots\}$ be the locally finite open covering given by (6.59) and let $\{\rho_j : j = 1, 2, 3, \ldots\}$ be a partition of unity subordinated to $\mathcal{U}$. Since, by (6.65)

$$[D] = [U_1^+] \cup \cdots \cup [U_h^+] \cup U_{h+1} \cup \cdots \cup U_m$$

we have $\varphi = \sum_{j=1}^m \rho_j \varphi$ identically on $[D]$, hence $d\varphi = \sum_{j=1}^m d(\rho_j \varphi)$. Therefore

$$\int_C \varphi = \sum_{j=1}^m \int_C \rho_j \varphi, \qquad \int_{[D]} d\varphi = \sum_{j=1}^m \int_{[D]} d(\rho_j \varphi).$$

Hence it suffices to show that the equality

$$\int_C \rho_j \varphi = \int_{[D]} d(\rho_j \varphi) \tag{6.69}$$

holds for all j with $1 \leqslant j \leqslant m$. Since $\operatorname{supp} \rho_j \subset U_j$, both $\rho_j \varphi$ and $d(\rho_j \varphi)$ vanish on $\mathcal{R} - U_j$. Hence we have, for $1 \leqslant j \leqslant h$

$$\int_{[D]} d(\rho_j \varphi) = \int_{[U_j^+]} d(\rho_j \varphi).$$

Considering $[U_j^+]$ as a closed domain in the z_j-plane and applying Green's Theorem (Theorem 6.2) we get

$$\int_{[U_j^+]} d(\rho_j \varphi) = \int_{\partial[U_j^+]} \rho_j \varphi = \int_{C_j} \rho_j \varphi + \int_{\lambda_j^+} \rho_j \varphi.$$

Since $C_j = C \cap [U_j]$, we have $\int_{C_j} \rho_j \varphi = \int_C \rho_j \varphi$ and since $\lambda_j^+ = \lambda_j \cap [D]$ and $\lambda_j = \partial[U_j]$, we have $\int_{\lambda_j^+} \rho_j \varphi = 0$. Therefore

$$\int_{[U_j^+]} d(\rho_j \varphi) = \int_C \rho_j \varphi. \tag{6.70}$$

Therefore (6.69) is valid for $1 \leqslant j \leqslant h$.

Since $[U_j] \subset D$ for $h + 1 \leqslant m$, we have

$$\int_{[D]} d(\rho_j \varphi) = \int_{[U_j]} d(\rho_j \varphi) = \int_{\partial[U_j]} \rho_j \varphi = 0.$$

Since obviously $\int_C \rho_j \varphi = 0$, we conclude that (6.69) also holds in this case.

For an arbitrary piecewise smooth Jordan curves, let $\mathcal{U} = \{U_j\}$, $U_j = U_{r(j)} q(j)$ be the locally finite open covering given by (6.59), let $q_j \in C$ for $j = 1, 2, \ldots, h$ and $[U_j] \cap C = \varnothing$ for $j \geqslant h + 1$. Let $\{\rho_j\}$ be a partition of unity subordinated to $\mathcal{U}$. Since $\mathcal{U}$ is locally finite, $V = \mathcal{R} - \bigcup_{j \geqslant h+1} [U_j]$ is an open set and

$$C \subset V \subset U(C) = \bigcup_{j=1}^{h} U_j.$$

If $p \in V$ then $\rho_j(p) = 0$ for $j \geqslant h+1$, hence

$$\rho_C(p) = \sum_{j=1}^{h} \rho_j(p)$$

is identically equal to 1 on the neighborhood V of C. Putting

$$\rho_C^+(p) = \begin{cases} \rho_C(p), & p \in U^+(C) \cup C, \\ 0, & p \notin U^+(C) \cup C, \end{cases}$$

ρ_C^+ is a function of class C^∞ on $\mathcal{R} - C$ and $\operatorname{supp} \rho_C^+ \subset U^+(C) \cup C$, while $\rho_C^+(p) = 1$ identically on $(U^+(C) \cup C) \cap V$. Therefore, $d\rho_C^+$ is a 1-form of class C^∞ on $\mathcal{R}$ and $\operatorname{supp} d\rho_C^+ \subset U^+(C) - V$. If ρ is a continuously differentiable 1-form on $\mathcal{R}$ with $d\varphi = 0$, then

$$\int_C \varphi = \int_{\mathcal{R}} d\rho_C^+ \wedge \varphi. \tag{6.71}$$

We derive this as follows. Since by assumption

$$d(\rho_j \varphi) = d\rho_j \wedge \varphi + \rho_j \, d\varphi = d\rho_j \wedge \varphi,$$

we have, by (6.70)

$$\int_{[U_j^+]} d\rho_j \wedge \varphi = \int_C \rho_j \varphi, \qquad 1 \leqslant j \leqslant h,$$

Hence:

$$\int_{[U_j^+(C)]} d\rho_C^+ \wedge \varphi = \int_C \varphi,$$

since $[U_j^+] \subset [U^+(C)]$ and $\sum_{j=1}^{h} \rho_j(p) = \rho_C^+(p)$ on $[U^+(C)]$, while $\sum_{j=1} \rho_j(p) = 1$ on C. Since $\sup d\rho_C^+ \subset U^+(C)$, this proves (6.71).

Theorem 6.11. Let φ be a continuously differentiable 1-form on the Riemann surface $\mathcal{R}$. If $\operatorname{supp} \varphi$ is compact, then

$$\int_{\mathcal{R}} d\varphi = 0.$$

Proof: Let $\mathfrak{U} = \{U_j\}$ be a locally finite open covering by coordinate neighborhoods U_j of $\mathcal{R}$, and let $\{p_j\}$ be a partition of unity subordinated to $\mathfrak{U}$. Since $\operatorname{supp} \varphi$ is compact, there exists an m such that $U_j \cap \operatorname{supp} \varphi = \emptyset$ for $j > m$, hence $\rho_j \varphi = 0$ for $j > m$. Therefore

$$\varphi = \sum \rho_j \varphi = \sum_{j=1}^{m} \rho_j \varphi,$$

so, by Green's Theorem (Theorem 6.1)

$$\int_{\mathcal{R}} d\varphi = \sum_{j=1}^{m} \int_{\mathcal{R}} d(\rho_j \varphi) = \sum_{j=1}^{m} \int_{\partial[U_j]} \rho_j \varphi = 0.$$

Let $\gamma : t \to \gamma(t)$, $a \leqslant t \leqslant b$, be a piecewise smooth curve on $\mathcal{R}$ and let φ be a closed continuously differentiable closed 1-form on $\mathcal{R}$. We want to consider $\int_\gamma \varphi$.

If $\gamma(t)$ is a point on γ such that there exists s with $s \neq t$ and $a \leqslant s \leqslant b$ such that $\gamma(s) = \gamma(t)$, then it is called a *multiple* point. If γ is a closed curve, its base point $\gamma(a) = \gamma(b)$ and $\gamma(t) = \gamma(a)$. If γ has no multiple points, then γ is a Jordan arc or a Jordan curve. If $p = \gamma(t)$ is a multiple point, then there exist only finitely many s with $\gamma(s) = p$. This follows because if there were infinitely many different s_n, $n = 1, 2, 3, \ldots,$ with $\gamma(s_n) = p$, there would be a convergent subsequence $\{s_n\}$, $n_1 < n_2 < \cdots < n_j < \cdots$, of the bounded sequence $\{s_n\}$. Putting $c = \lim_{j\to\infty} s_{n_j}$, we have $\gamma(s_{n_j}) = \gamma(c)$. Hence $\gamma'(c) = 0$, contradicting the assumption that γ is piecewise smooth. So, if γ has multiple points and if their number is finite there are only finitely many t such that $\gamma(t)$ is a multiple point, which we denote by $t_1, t_2, \ldots, t_l$, $a \leqslant t_1 < t_2 < \cdots < t_l \leqslant b$. Hence, for each t_i, there exists at least one t_k, $k \neq i$, with $\gamma(t_k) = \gamma(t_i)$. Among all pairs t_i, t_k, $t_i < t_k$, satisfying $\gamma(t_k) = \gamma(t_i)$, let t_j, t_h denote the pair such that $t_k - t_i$ is smallest and let us put $a_1 = t_j$ and $b_1 = t_h$. Since obviously $\gamma(a_1) = \gamma(b_1)$ and $\gamma(t) \neq \gamma(s)$ if $a_1 \leqslant t < s \leqslant b_1$, the curve $C_1 : t \to \gamma(t)$, $a_1 \leqslant t \leqslant b_1$, is a piecewise

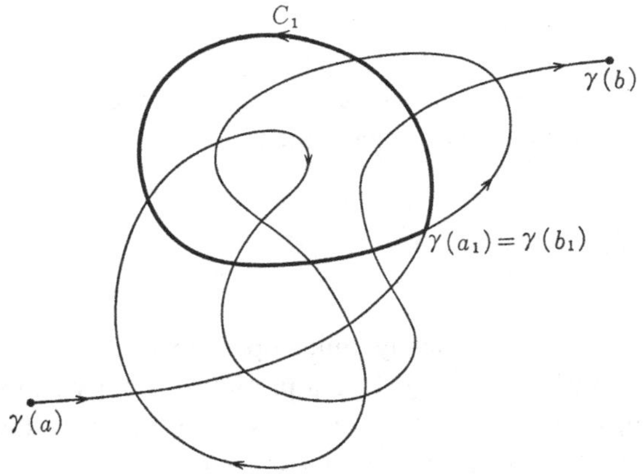

smooth Jordan curve. The curve γ_1, which is the result of removing C_1 from γ, is a piecewise smooth curve connecting $\gamma(a)$ and $\gamma(b)$, given by the parameter representation

$$\gamma_1 : t \to \gamma_1(t) = \begin{cases} \gamma(t), & a \leqslant t \leqslant a_1, \\ \gamma(t - a_1 + b_1), & a_1 \leqslant t \leqslant c_1, \end{cases}$$

where we have put $c_1 = b - b_1 + a_1$.

We obviously have

$$\int_\gamma \varphi = \int_{C_1} \varphi + \int_{\gamma_1} \varphi.$$

In a similar way, removing a piecewise smooth Jordan curve C_2 from γ_1 and denoting the remainder by γ_2, we have

$$\int_\gamma \varphi = \int_{C_1} \varphi + \int_{C_2} \varphi + \int_{\gamma_2} \varphi.$$

Repeating this procedure, if $\gamma(a) \neq \gamma(b)$, after a finite number of steps we arrive at a piecewise smooth Jordan arc γ_ν and we have

$$\int_\gamma \varphi = \int_{C_1} \varphi + \int_{C_2} \varphi + \cdots + \int_{C_\nu} \varphi + \int_{\gamma_\nu} \varphi.$$

If $\gamma(a) = \gamma(b)$, after a finite number of steps we are left with no remainder at all and we have

$$\int_\gamma \varphi = \int_{C_1} \varphi + \int_{C_2} \varphi + \cdots + \int_{C_\nu} \varphi.$$

In the notation of 1-chains (see Section 4.4) we have:

$$\gamma = C_1 + C_2 + \cdots + C_\nu + \gamma_\nu,$$

respectively:

$$\gamma = C_1 + C_2 + \cdots + C_\nu.$$

If γ has infinitely many multiple points, we first "transform" γ somewhat, so that it becomes a piecewise smooth curve $\hat{\gamma}$ with at most finitely many multiple points. To see how this can be done, let $\mathcal{U} = \{U_j\}$ be a locally finite open covering by coordinate disks U_j of $\mathcal{R}$. Since $|\gamma|$ is compact, it can be covered by a finite number of U_j. Therefore it is possible to find an m such that each curve $\gamma_n : t \to \gamma(t)$, $c_{n-1} \leqslant t \leqslant c_n$ with $c_n = a + n(b-a)/m$, $n = 1, 2, \ldots, m$, belongs to one $U_{j(n)}$, while obviously $\gamma = \gamma_1 \cdot \gamma_2 \cdot \cdots \cdot \gamma_n \cdot \cdots \cdot \gamma_m$. By Theorem 6.3, $\varphi = du_j$ for some twice continuously differentiable function u_j on each U_j. Hence

$$\int_{\gamma_n} \varphi = u_{j(n)}(\gamma(c_n)) - u_{j(n)}(\gamma(c_{n-1})).$$

In this way the value of $\int_{\gamma_n} \varphi$ is determined by the end points $\gamma(c_{n-1})$

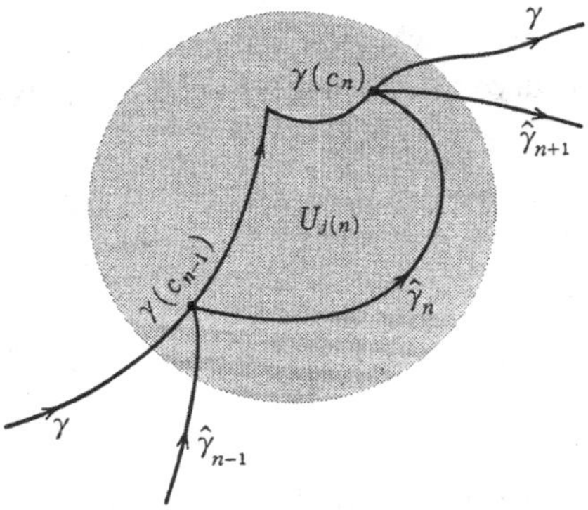

and $\gamma(c_n)$ of γ_n. Therefore the value of $\int_{\gamma_n}\varphi$ does not change if we replace γ_n by an arbitrary smooth Jordan arc $\hat{\gamma}_n$ which connects $\gamma(c_{n-1})$ and $\gamma(c_n)$ in $U_{j(n)}$: $\int_{\gamma_n}\varphi = \int_{\hat{\gamma}_n}\varphi$. Now select Jordan arcs $\hat{\gamma}_1, \hat{\gamma}_2, \ldots, \hat{\gamma}_n, \ldots, \hat{\gamma}_m$ in this order such that $\hat{\gamma}_n$ intersects $\hat{\gamma}_1 \cdot \hat{\gamma}_2 \cdot \cdots \cdot \hat{\gamma}_{n-1}$ in at most finitely many points and put $\hat{\gamma} = \hat{\gamma}_1 \cdot \hat{\gamma}_2 \cdot \cdots \cdot \hat{\gamma}_n \cdot \cdots \cdot \hat{\gamma}_m$. Now, $\hat{\gamma}$ is a piecewise smooth curve with at most finitely many multiple points connecting $\gamma(a)$ and $\gamma(b)$ and

$$\int_{\gamma}\varphi = \int_{\hat{\gamma}}\varphi.$$

Removing from $\hat{\gamma}$ piecewise smooth Jordan curves $C_1, C_2, C_3, \ldots$ in the way described above, we are left with a piecewise smooth Jordan arc γ_ν connecting $\gamma(a)$ and $\gamma(b)$, i.e.

$$\hat{\gamma} = C_1 + C_2 + \cdots + C_\nu + \gamma_\nu, \tag{6.72}$$

if $\gamma(a) \neq \gamma(b)$. If $\gamma(a) = \gamma(b)$, then $\hat{\gamma}$ is also a closed curve and we have

$$\hat{\gamma} = C_1 + C_2 + \cdots + C_\nu. \tag{6.73}$$

So we have proved:

Theorem 6.12. If γ is a piecewise smooth closed curve on $\mathcal{R}$, if φ is a continuously differentiable 1-form on $\mathcal{R}$ and if the piecewise smooth Jordan curves $C_1, C_2, \ldots, C_\nu$ are determined by (6.73), then

$$\int_{\gamma}\varphi = \int_{C_1}\varphi + \int_{C_2}\varphi + \cdots + \int_{C_\nu}\varphi.$$

Let C be a piecewise smooth Jordan curve on $\mathcal{R}$, then $\int_C\varphi = \int_{\mathcal{R}} d\rho_C^+ \wedge d\varphi$ by (6.71) and the $d\rho_C^+$ occurring in this equality is a closed 1-form of class C^∞ with compact support. Hence putting

$$\xi_\gamma = d\rho^+_{C_1} + d\rho^+_{C_2} + \cdots + \int_{d\rho^+_{C_\nu}}$$

we obtain the following theorem:

Theorem 6.13. If γ is a piecewise smooth Jordan curve on $\mathcal{R}$, then there exists a closed 1-form ξ_γ of class C^∞ with compact support such that

$$\int_\gamma \varphi = \int_{\mathcal{R}} \xi_\gamma \wedge \varphi \tag{6.74}$$

where φ is a continuously differentiable, closed 1-form defined on $\mathcal{R}$.

If the 1-form φ is exact, i.e. if $\varphi = du$ for some continuously differentiable function u on $\mathcal{R}$, (6.46), $\int_\gamma \varphi = 0$ where γ is an arbitrary piecewise smooth closed curve on $\mathcal{R}$. Conversely

Theorem 6.14. Let φ be a continuously differentiable closed 1-form on $\mathcal{R}$. If $\int_\gamma \varphi = 0$ for all piecewise smooth closed curves γ on $\mathcal{R}$, then φ is exact on $\mathcal{R}$.

Proof: Let p_0 be a fixed point of $\mathcal{R}$. For an arbitrary point $p \in \mathcal{R}$, let $\gamma : t \to \gamma(t)$, $a \leqslant t \leqslant b$, be a piecewise smooth curve connecting p_0 and p and put

$$u(p) = \int_\gamma \varphi.$$

The function $u(p)$ depends only on p, not on the choice of γ. This is so because if γ_1 is another piecewise smooth curve connecting p_0 and p, then $\gamma_1 \cdot \gamma^{-1}$ is a closed curve, hence by the assumption

$$\int_{\gamma_1}^{p} \varphi - \int_{\gamma}^{p} \varphi = \int_{\gamma_1 \cdot \gamma^{-1}} \varphi = 0.$$

Therefore we may write $u(p) = \int_{p_0}^{p} \varphi$. In order to prove that $du = \varphi$, we restrict our attention to the coordinate disk U_j, with center q_j.

By the Theorem 6.3 we can write $\varphi = du_j$, where u_j is a twice continuously differentiable function on U_j, hence, by (6.6):

$$u(p) = \int_{p_0}^{q_j} \varphi + \int_{q_j}^{p} du_j = \int_{p_0}^{q_j} \varphi + u_j(p) - u_j(q_j).$$

Therefore $du(p) = du_j(p) = \varphi$.

Corollary 1. If φ is a continuously differentiable closed 1-form such that

$\int_C \varphi = 0$ for all piecewise smooth Jordan curves C on $\mathcal{R}$, then φ is exact on $\mathcal{R}$.

Corollary 2. A continuously differentiable closed 1-form φ defined on $\mathcal{R}$ is exact if and only if $\int_{\mathcal{R}} \xi \wedge \varphi = 0$ for all continuously differentiable 1-forms ξ with a compact support.

Proof:

(1) If $\varphi = du$, then $\xi \wedge \varphi = -du \wedge \xi = -d(u\xi)$. Since $\operatorname{supp} u\xi$ is compact, we have by Theorem 6.11:

$$\int_{\mathcal{R}} \xi \wedge \varphi = -\int_{\mathcal{R}} d(u\xi) = 0.$$

(2) if conversely $\int_{\mathcal{R}} \xi \wedge \varphi = 0$ for all ξ, then $\int_\gamma \varphi = \int_{\mathcal{R}} \xi_\gamma \wedge \varphi = 0$ for all piecewise smooth closed curves γ on $\mathcal{R}$ by Theorem 6.13. Hence φ is exact.

If p and q are arbitrary points of a region $D \subset \mathcal{R}$, then it is possible to connect p and q by a piecewise analytic curve $\gamma : t \to \gamma(t)$, $a \leqslant t \leqslant b$, in D. Since D is a Riemann surface in its own right, we can apply (6.72) to γ considered as a curve in D to obtain

$$\hat{\gamma} = C_1 + C_1 + \cdots + C_\nu + \gamma_\nu$$

where γ_ν is a piecewise smooth Jordan arc connecting $p = \gamma(a)$ and $q = \gamma(b)$ in D. Furthermore, when constructing $\hat{\gamma} = \hat{\gamma}_1 \cdot \hat{\gamma}_2 \cdot \cdots \cdot \hat{\gamma}_n \cdot \cdots \cdot \hat{\gamma}_m$ it is possible to pick analytic Jordan arcs $\hat{\gamma}_n$ connecting $\gamma(c_{n-1})$ and $\gamma(c_n)$. Hence we have proved the following useful result:

Theorem 6.15. If p and q are two points of a region $D \subset \mathcal{R}$, then it is possible to connect p and q by a piecewise analytic Jordan arc in D.

6.4 Dirichlet's Principle

a. *Inner product and norm*

Let q_0 be a fixed point of a Riemann surface $\mathcal{R}$ and let $z : p \to z(p) = z_{q_0}(p)$ be a local coordinate with center q_0 and $U_0 = U_R(q_0) = \{p : |z(p)| < R\}$ a coordinate disk. By definition, $z(p)$ is a biholomorphic function, defined on a region containing the closed disk $[U_0]$. Since $z/R : p \to z(p)/R$ is also a local coordinate with center q_0, we may assume from the beginning that U_0 is the unit disk $\{p : |z(p)| < 1\}$. Let $S(z)$ be a holomorphic function defined on a region containing $[U]$ from which the point q_0 of a smooth Jordan arc $|\sigma| \subset U_0$ has been deleted. We write

$$S(z) = \Phi + i\Psi, \qquad \Phi = \Phi(x, y), \qquad z = x + iy.$$

For simplicity's sake, we will let $|\sigma|$ include the point q_0 even though q_0 has been deleted from the region of definition of $S(z)$. We call φ a *harmonic* 1-form on $\mathcal{R} - |\sigma|$ with the same singularity as $d\Phi$ if there exists a harmonic 1-form φ_0 on U_0 such that

$$\varphi = d\Phi + \varphi_0$$

on $U_0 - |\sigma|$. (In general, φ is not defined on $|\sigma|$.) In this section we will prove that there exists a harmonic 1-form φ on $\mathcal{R}$ with the same singularity as $d\Phi$ if

$$\Psi = \text{constant on } \partial[U_0]. \tag{6.75}$$

To give a typical example, consider

$$S(z) = \frac{1}{z} + z.$$

Since

$$\Psi(x, y) = \operatorname{Im} S(z) = y\left(1 - \frac{1}{x^2 + y^2}\right)$$

$\Psi(x, y) = 0$ if $x^2 + y^2 = 1$, i.e. Ψ satisfies condition (6.75). Let $\{z_0, z_2, z_2, \ldots\}$ be a system of local coordinates on $\mathcal{R}$, $\mathcal{U} = \{U_0, U_1, U_2, \ldots\}$ a locally finite open covering, where each U_j is a region, such that $[U_j]$ is compact and contained in the domain of z_j, while U_0 is a unit disk and $[U_j] \cap |\sigma| = \emptyset$ for $j \geqslant 1$ and let $\tau_{jk} : z_k \to \tau_{jk}(z_k)$ represent the coordinate transformations. A 1-form φ on $\mathcal{R}$ and a 2-form ω on $\mathcal{R}$ can be written as $\varphi(z_j) = \varphi_{j1}(z_j)dx_j + \varphi_{j2}(z_j)dy_j$ and $\omega_{12}(z)dx \wedge dy$ respectively. For a continuous 2-form ω on $\mathcal{R}$, $\int_{\mathcal{R}} \omega$ is defined by:

$$\int_{\mathcal{R}} \omega = \sum_j \int_{U_j} \rho_j \omega_{j12}(z_j) dx_j\, dy_j$$

where $\{\rho_j\}$ is a partition of unity subordinated to $\mathcal{U} = \{U_j\}$. On a region U_k we have

$$\omega_{k12}(z_k) = |\tau'_{jk}(z_k)|^2 \omega_{j12}(z_j) \frac{\partial(x_j, y_j)}{\partial(x_k, y_k)}$$

by (6.34) and (3.5), hence

$$\int_{U_k} \omega = \sum \int_{U_k} \rho_j \omega_{k12}\, dx_k\, dy_k = \int_{U_k} \omega_{k12}(z_k) dx_k\, dy_k. \tag{6.76}$$

Suppressing the index j we write

$$\int_{\mathcal{R}} \omega = \int_{\mathcal{R}} \omega_{12}(z) dx\, dy.$$

The *inner product* of two continuous, real 1-forms φ and ψ on $\mathcal{R}$ is defined by

$$(\varphi, \psi)_{\mathcal{R}} = \int_{\mathcal{R}} \varphi \wedge *\psi.$$

Since

$$\begin{aligned}\varphi \wedge *\psi &= (\varphi_1\, dx + \varphi_2\, dy) \wedge (-\psi_2\, dx + \psi_1\, dy) \\ &= (\varphi_1\psi_1 + \varphi_2\psi_2)dx \wedge dy\end{aligned}$$

then

$$(\varphi, \psi)_{\mathcal{R}} = \int_{\mathcal{R}} (\varphi_1\psi_1 + \varphi_2\psi_2)dx\, dy. \tag{6.77}$$

We call $\sqrt{(\varphi, \varphi)_{\mathcal{R}}}$ the *norm* of φ, denoted by $\|\varphi\|_{\mathcal{R}}$:

$$\|\varphi\|_{\mathcal{R}}^2 = \int_{\mathcal{R}} (\varphi_1^2 + \varphi_2^2)dx\, dy.$$

The integral on the right side of this equality either converges or diverges towards $+\infty$. If it converges, its value is $\geqslant 0$. Since

$$2|\varphi_1\psi_1 + \varphi_2\psi_2| \leqslant \varphi_1^2 + \varphi_2^2 + \psi_1^2 + \psi_2^2$$

$\|\varphi\|_{\mathcal{R}} < +\infty$ and $\|\psi\|_{\mathcal{R}} < +\infty$ implies that the integral occurring in (6.77) converges absolutely and that $(\varphi, \psi)_{\mathcal{R}}$ is determined. For a region $D \in \mathcal{R}$, the inner product $(\varphi, \psi)_D$ of two continuous 1-forms φ and ψ defined on D and the norm $\|\varphi\|_D$ of φ are defined by considering D as a Riemann surface. From now on we will write (φ, ψ) and $\|\varphi\|$ instead of $(\varphi, \psi)_{\mathcal{R}}$ and $\|\varphi\|_{\mathcal{R}}$ respectively.

Since

$$(\varphi_1 + \psi_1)^2 + (\varphi_2 + \psi_2)^2 \leqslant 2(\varphi_1^2 + \varphi_2^2 + \psi_1^2 + \psi_2^2)$$

$\|\varphi\| < +\infty$ and $\|\psi\| + \infty$ implies $\|\varphi + \psi\| < +\infty$. It is easy to verify that (φ, ψ) is linear in φ and ψ, that $\|\lambda\varphi\| = 0$ for $\lambda \in \mathbb{R}$ and that

$$(\varphi, \psi) = (\psi, \varphi).$$

Hence

$$\begin{aligned}\|\lambda\varphi + \mu\psi\|^2 &= (\lambda\varphi + \mu\psi, \lambda\varphi + \mu\psi) \\ &= \lambda^2\|\varphi\|^2 + 2\lambda\mu(\varphi, \psi) + \mu^2\|\psi^2\|^2\end{aligned}$$

for arbitrary real λ and μ. Since $\|\lambda\varphi + \mu\psi\|^2 \geqslant 0$ the right-hand side of this equality is a positive-definite quadratic form in the variables λ and μ. Hence its discriminant $(\varphi, \psi)^2 - |\varphi\|^2\|\psi\|^2$ is nonpositive, i.e.

$$|(\varphi, \psi)| \leqslant \|\varphi\|\, \|\psi\|.$$

Hence $\|\varphi+\psi\|^2 = \|\varphi\|^2 + 2(\varphi, \psi) + \|\psi\|^2 \leqslant (\|\varphi\| + \|\psi\|)^2$ and therefore

$$\|\varphi+\psi\| \leqslant \|\varphi\| + \|\psi\|.$$

Replacing φ by $\varphi - \psi$ we get $\|\varphi\| \leqslant \|\varphi-\psi\| + \|\psi\|$ i.e. $\|\varphi\| - \|\psi\| \leqslant \|\varphi - \psi\|$. Hence the inequality

$$|\,\|\varphi\| - \|\psi\|\,| \leqslant \|\varphi - \psi\| \tag{6.78}$$

holds.

Let $S(z) = \Phi + \iota\Psi$ be given. We want to see if it is possible to find a continuously differentiable 1-form φ on $\mathcal{R} - |\sigma|$ such that, on $U_0 - |\sigma|$,

$$\varphi = d\Phi + \varphi_0 \tag{6.79}$$

where φ_0 is some continuously differentiable 1-form on U_0. Put $K = [U_0]$. Since Φ is a harmonic function defined on a region $G \supset K$ from which $|\sigma|$ has been deleted, if such a φ exists, then $\varphi_0 = \varphi - d\Phi$ can be extended to a continuously differentiable 1-form on G. we denote this extension also by φ_0. We defined $\|\varphi\|_\Phi$ by:

$$\|\varphi\|_\Phi^2 = \|\varphi_0\|_K^2 + (\|\varphi\|_{\mathcal{R}-K})^2.$$

We denote the collection of all continuously differentiable 1-forms φ defined on $\mathcal{R} - |\sigma|$, satisfying condition (6.79) and

(i) $d\varphi = 0$
(ii) $\|\varphi\|_\Phi < +\infty$

by $\mathcal{L}_\Phi$. If $\varphi \in \mathcal{L}_\Phi$, then $d\varphi_0 = 0$ on U_0, because $d\varphi_0$ is continuous on U_0 and $d\varphi_0 = d(\varphi - d|\varphi_0) = 0$ on $U_0 - |\sigma|$. For $\varphi, \psi \in \mathcal{L}_\Phi$, $(\varphi, \psi)_\Phi$ is defined by

$$(\varphi, \psi)_\Phi = (\varphi_0, \psi_0)_K + (\varphi, \psi)_{\mathcal{R}-K}.$$

Obviously, $(\varphi, \psi)_\Phi = (\psi, \varphi)_\Phi$.

Lemma 6.1.

(1) If $\varphi, \psi \in \mathcal{L}_\Phi$ and λ and μ are real numbers satisfying $\lambda + \mu = 1$, then $\lambda\varphi + \mu\psi \in \mathcal{L}_\Phi$ and

$$\|\lambda\varphi + \mu\psi\|_\Phi^2 = \lambda^2\|\varphi\|_\Phi^2 + \mu^2\|\psi\|_\Phi^2 + 2\lambda\mu(\varphi, \psi)_\Phi. \tag{6.80}$$

(2) if $\varphi, \psi \in \mathcal{L}_\Phi$, then $\varphi - \psi$ can be extended to a continuously differentiable 1-form on $\mathcal{R}$. Denoting this 1-form by $\varphi - \psi$, we have

$$\|\varphi - \psi\|^2 = \|\varphi\|_\Phi^2 - 2(\varphi, \psi)_\Phi + \|\psi\|_\Phi^2. \tag{6.81}$$

Proof:

(1) By (6.79) we have $\varphi = d\Phi + \varphi_0$, $\psi = d\Phi + \psi_0$ on $U_0 - |\sigma|$, where φ_0 and ψ_0 are continuously differentiable on U_0. Since $\lambda + \mu = 1$, we have $\lambda\varphi + \mu\psi = d\Phi + \lambda\varphi_0 + \mu\psi_0$, i.e. $\lambda\varphi + \mu\psi$ satisfies condition (6.79). It is obvious that $d(\lambda\varphi + \mu\psi) = 0$ on $\mathcal{R} - |\sigma|$. Finally since

$$\|\lambda\varphi + \mu\psi\|_\phi^2 = \|\lambda\varphi_0 + \mu\varphi_0\|_K^2 + (\|\lambda\varphi + \mu\psi\|_{\mathcal{R}-K})^2 < +\infty$$

we conclude that $\lambda\varphi + \mu\psi \in \mathcal{L}_\Phi$.

Since further

$$\|\lambda\varphi_0 + \mu\psi_0\|_K^2 = \lambda^2\|\varphi_0\|_K^2 + 2\lambda\mu(\varphi_0, \psi_0)_K + \mu^2\|\psi_0\|_K^2,$$

$$(\|\lambda\varphi + \mu\psi\|_{\mathcal{R}-K})^2 = \lambda^2(\|\varphi\|_{\mathcal{R}-K})^2 + 2\lambda\mu(\varphi, \psi)_{\mathcal{R}-K} + \mu^2(\|\psi\|_{\mathcal{R}-K})^2$$

(6.80) is also true.

(2) Since $\varphi = d\Phi + \varphi_0$ and $\psi = d\Phi + \psi_0$ on $U_0 - |\sigma|$, we have $\varphi - \psi = \varphi_0 - \psi_0$, while $\varphi_0 - \psi_0$ is continuously differentiable on U_0. Hence $\varphi - \psi$ can be extended to a continuously differentiable 1-form on $\mathcal{R}$. This extention coincides with $\varphi_0 - \psi_0$ on K, hence

$$\|\varphi - \psi\|^2 = \|\varphi_0 - \psi_0\|_K^2 + (\|\varphi - \psi\|)_{\mathcal{R}-K}^2$$

by (6.68), proving (6.81).

Lemma 6.2. If η is a twice continuously differentiable function on $\mathcal{R}$ such that $\|\eta\| < +\infty$ and if $\varphi \in \mathcal{L}_\Phi$, then $\varphi + d\eta \in \mathcal{L}_\Phi$. If, moreover, $\operatorname{supp}\eta \cap |\sigma| = \emptyset$, then

$$\|\varphi + d\eta\|_\Phi^2 = \|\varphi\|_\Phi^2 + 2(\varphi, d\eta) + \|d\eta\|^2. \tag{6.82}$$

Proof: Since $\varphi + d\eta = d\Phi + \varphi_0 + d\eta$ and $\varphi_0 + d\eta$ is continuously differentiable on U_0, $\varphi + d\eta$ satisfies condition (6.79). Since obviously $d(\varphi + d\eta) = 0$ and

$$\|\varphi + d\eta\|_\Phi^2 = \|\varphi_0 + d\eta\|_K^2 + (\|\varphi + d\eta\|_{\mathcal{R}-K})^2 < +\infty$$

we conclude that $\varphi + d\eta \in \mathcal{L}_\Phi$.

If $\operatorname{supp}\eta \cap |\sigma| = \emptyset$, then

$$(\varphi_0, d\eta)_K = (\varphi - d\Phi, d\eta)_K = (\varphi, d\eta)_K - (d\Phi, d\eta)_K.$$

Since $S(z) = \Phi + \Psi$ is holomorphic on a region $G \supset K$ from which $|\sigma|$ has been deleted, the Cauchy–Riemann equations tell us that

$$d\Phi = \Psi_x\,dx + \Psi_y\,dy = -\Phi_y\,dx + \Phi_x\,dy = *d\Phi.$$

Hence

$$(d\Phi, d\eta)_K = \int_K d\eta \wedge *d\Phi = \int_K d\eta \wedge d\Psi = \int_K d(\eta\,d\Psi).$$

Let $\theta \to z = e^{i\theta}$, $0 \leqslant \theta \leqslant 2\pi$, be a parameter representation of the unit circle ∂K. Since by conditon (6.75) $\Phi = \text{constant}$ on ∂K we have $\partial\Psi/\partial\theta = 0$, hence by Green's Theorem (Theorem 6.1)

$$\int_K d(\eta\,d\Psi) = \int_{\partial K} \eta\,d\Psi = \int_0^{2\pi} \eta \frac{d\Psi}{d\theta}\,d\theta = 0,$$

and therefore

$$(d\Phi, d\eta)_K = 0. \tag{6.83}$$

Hence $(\varphi_0, d\eta)_K = (\varphi, d\eta)_K$ and so

$$\begin{aligned}
\|\varphi + d\eta\|_\Phi^2 &= \|\varphi_0 + d\eta\|_K^2 + (\|\varphi + d\eta\|_{\mathcal{R}-K})^2 \\
&= \|\varphi_0\|_K^2 + 2(\varphi, d\eta)_K + (\|\varphi\|_{\mathcal{R}-K})^2 + 2(\varphi, d\eta)_{\mathcal{R}-K} \\
&\qquad + (\|d\eta\|_{\mathcal{R}-K})^2 \\
&= \|\varphi\|_\Phi^2 + 2(\varphi, d\eta) + \|d\eta\|^2,
\end{aligned}$$

proving (6.82).

b. Dirichlet's Principle

Let η be a twice continuously differentiable function defined on $\mathcal{R}$ satisfying $\|d\eta\| < +\infty$. If $\psi \in \mathcal{L}_\Phi$, then $\psi + d\eta \in \mathcal{L}_\Phi$, hence $\|\psi + d\eta\|_\Phi$ is defined. For fixed ψ let κ be the infimum of all norms $\|\psi + d\eta\|_\Phi$, where η is a function as above:

$$\kappa = \inf_\eta \|\psi + d\eta\|_\Phi.$$

In fact, κ is the minimum value of the norms $\|\psi + d\eta\|_\Phi$, i.e. there exists a twice continuously differentiable function η_∞, defined on $\mathcal{R}$, such that $\|d\eta_\infty\| < +\infty$ and $\|\psi + d\eta\|_\Phi = \kappa$. $\varphi = \psi + d\eta_\infty$ is a harmonic 1-form on $\mathcal{R}$ with the same singularity as $d\varphi$.

We will prove this result in this section. Selecting a sequence $\{\eta_n\}$ of twice continuously differentiable functions η_n with $\|d\eta_n\| < +\infty$ such that $\lim_{n\to\infty} \|\psi + d\eta_n\| = \kappa$, we put

$$\varphi_n = \psi + d\eta_n, \qquad \delta_n = \sqrt{\|\varphi_n\|_\Phi^2 - \kappa^2} \tag{6.84}$$

and we obviously have $\lim_{n\to\infty} \delta_n = 0$. An important part of the proof of Dirichlet's Principle is contained in the proof of the following existence theorem.

Theorem 6.16. There exists a 1-form $\varphi \in \mathcal{L}_\Phi$ such that
$$\lim_{n\to\infty} \|\varphi_n - \varphi\| = 0.$$

Proof: Let us first prove the inequality
$$\|\varphi_m - \varphi_n\| \leqslant \delta_m - \delta_n. \tag{6.85}$$
For arbitrary λ, $\mu \in \mathbb{R}$ satisfying $\lambda + \mu = 1$, we have
$$\lambda\varphi_m + \mu\varphi_n = \psi + d(\lambda\eta_m + \mu\eta_n)$$
hence
$$\|\lambda\varphi_m + \mu\varphi_n\| \geqslant \kappa^2 = \lambda^2\kappa^2 + 2\lambda\mu\kappa^2 + \mu^2\kappa^2.$$
Therefore, by (6.80)
$$\lambda^2(\|\varphi_m\|_\Phi^2 - \kappa^2) + 2\lambda\mu((\varphi_m, \varphi_n)_\Phi - \kappa^2) + \mu^2(\|\varphi_n\|_\Phi^2 - \kappa^2) \geqslant 0.$$
Hence,
$$|(\varphi_m, \varphi_n)_\Phi - \kappa^2|^2 \leqslant (\|\varphi_m\|_\Phi^2 - \kappa^2),$$
i.e.
$$|(\varphi_m, \varphi_n)_\Phi - \kappa^2| \leqslant \delta_m \cdot \delta_n.$$
Using (6.81) we have
$$\begin{aligned}\|\varphi_m - \varphi_n\|^2 &= \|\varphi_m\|_\Phi^2 - 2(\varphi_m, \varphi_n)_\Phi + \|\varphi_n\|_\Phi^2 \\ &= \|\varphi_m\|_\Phi^2 - \kappa^2 - 2((\varphi_m, \varphi_n)_\Phi - \kappa^2) + \|\varphi_n\|_\Phi^2 - \kappa^2 \\ &\leqslant \|\varphi_m\|_\Phi^2 - \kappa^2 + 2|(\varphi_m, \varphi_n)_\Phi - \kappa^2| + \|\varphi_n\|_\Phi^2 - \kappa^2 \\ &\leqslant \delta_m^2 + 2\delta_m \cdot \delta_n^2 + \delta_n^2 = (\delta_m + \delta_n)^2,\end{aligned}$$
proving (6.85).

Next we prove that
$$|(\varphi_n, d\eta)| \leqslant \|d\eta\| \cdot \delta_n, \tag{6.86}$$
if $\operatorname{supp} \eta \cap |\sigma| = \varnothing$ and
$$|(\varphi_{n0}, d\eta)| \leqslant \|d\eta\| \cdot \delta_n \tag{6.87}$$
if $\operatorname{supp} \eta \subset U_0$.

For arbitrary $\lambda \in \mathbb{R}$ and assuming $\operatorname{supp} \eta \cap |\sigma| = \varnothing$ we have
$$\|\varphi_n\|_\Phi^2 + 2\lambda(\varphi_n, d\eta) + \lambda^2\|\partial\eta\|^2 = \|\varphi_n + \lambda\, d\eta\|_\Phi^2 \geqslant \kappa^2,$$
by (6.82), i.e.
$$\delta_n^2 + 2\lambda(\varphi_n, \partial\eta) + \lambda^2\|d\eta\|^2 \geqslant 0.$$
Therefore,
$$|(\varphi_n, d\eta)|^2 \leqslant \|d\eta\|^2 \cdot \delta_n^2,$$
proving (6.86).

If $\operatorname{supp}\eta \subset U_0 = K$, then

$$\|\varphi_{n0} + \lambda\, d\eta\|_K^2 + (\|\varphi_n\|_{\mathcal{R}-K})^2 = \|\varphi_n + \lambda\, d\eta\|_\Phi^2 \geqslant \kappa^2$$

hence

$$\|\varphi_n\|_\Phi^2 + 2\lambda(\varphi_{n0}, d\eta) + \lambda^2\|d\eta\|^2 \geqslant \kappa^2.$$

From this we obtain (6.87) in a similar way.

To prove the existence of a $\varphi \in \mathcal{L}_\Phi$ such that $\lim_{n\to\infty}\|\varphi_n - \varphi\| = 0$, it suffices to show that there exist for all $j = 1, 2, 3, \ldots$ continuously differentiable closed 1-forms $\varphi_{(j)}$ such that $\lim_{n\to\infty}\|\varphi_n - \varphi_{(j)}\|_{U_j} = 0$ on U_j and a continuously differentiable closed 1-form φ_0 such that $\lim_{n\to\infty}\|\varphi_{n0} - \varphi_0\|_{U_0} = 0$ on U_0. (Because it is easy to verify that if we put $\varphi = \varphi_{(j)}$ on all U_j, $f \geqslant 1$, and $\varphi = d\Phi + \varphi_0$ on $U_0 - |\sigma|$, we obtain the required $\varphi \in \mathcal{L}_\Phi$.)

First, if $U_j \cap U_k \neq \varnothing$, $j \geqslant 1$, $k \geqslant 1$, then

$$\begin{aligned}\|\varphi_{(j)} - \varphi_{(k)}\|_{U_j\cap U_k} &\leqslant \|\varphi_{(j)} - \varphi_n\|_{U_j\cap U_k} + \|\varphi_n - \varphi_{(k)}\|_{U_j\cap U_k}\\ &\leqslant \|\varphi_n - \varphi_{(j)}\|_{U_j} + \|\varphi_n - \varphi_{(k)}\|_{U_k} \to 0,\\ &\qquad \text{as } n \to \infty.\end{aligned}$$

Hence $\|\varphi_{(j)} - \varphi_{(k)}\|_{U_j\cap U_k} = 0$, i.e. $\varphi_{(j)}$ and $\varphi_{(k)}$ coincide on $U_j \cap U_k$. Similarly, if $U_j \cap U_0 \neq \varnothing$, $j \geqslant 1$, then $\varphi_n = d\Phi + \varphi_{n0}$ on $U_j \cap U_0$, hence

$$\|\varphi_{(j)} - d\Phi - \varphi_0\|_{U_j\cap U_0} \leqslant \|\varphi_n - \varphi_{(j)}\|_{U_j} + \|\varphi_{n0} - \varphi_0\|_{U_0} \to 0,$$
$$\text{as } n \to \infty.$$

Therefore, $\varphi_{(j)}$ and $d\Phi + \varphi_0$ coincide on $U_j \cap U_0$.

Hence, by putting $\varphi = \varphi_{(j)}$ on U_j, $j \geqslant 1$, and $\varphi = d\Phi + \varphi_0$ on $U_0 - |\sigma|$, we obtain a continuously differentiable 1-form φ defined on $\mathcal{R} - |\sigma|$. We can consider $\varphi_0 = \varphi - d\Phi$ as being extended to a continuously differentiable 1-form defined on a region containing $K = [U_0]$. Since

$$\|\varphi_n - \varphi\|^2 = \|\varphi_{n0} - \varphi_0\|_K^2 + (\|\varphi_n - \varphi\|_{\mathcal{R}-K})^2$$

and $K = [U_0]$, we have

$$\|\varphi_{n0} - \varphi_0\|_K = \|\varphi_{n0} - \varphi_0\|_{U_0} \to 0, \qquad \text{as } n \to \infty.$$

In order to prove that $\lim_{n\to\infty}\|\varphi_n - \varphi\|_{\mathcal{R}-k} = 0$, we write

$$\varphi \wedge *\varphi = |\varphi|^2 dx \wedge dy, \qquad |\varphi|^2 = \varphi_1^2 + \varphi_2^2,$$

where $\varphi - \varphi_1\, dx + \varphi_2\, dy$ is an arbitrary 1-form. Since $\rho_0 + \sum_{j\geqslant 1}\rho_j = 1$ and $\operatorname{supp}\rho_0 \subset U_0$, we have $\sum_{j\geqslant 1}\rho_j = 1$ on $\mathcal{R} - K$ and

$$(\|\varphi_n - \varphi\|_{\mathcal{R}-K})^2 = \sum_{j\geqslant 1}\int_{\mathcal{R}-K}\rho_j|\varphi_n - \varphi|^2 dx\, dy. \tag{6.88}$$

Since, for all j,

$$\|\varphi_n - \varphi\|_{U_j} = \|\varphi_n - \varphi_{(j)}\|_{U_j} \to 0 \text{ as } n \to \infty$$

we have

$$\int_{\mathcal{R}-K} \rho_j |\varphi_n - \varphi|^2 dx\, dy \leqslant \int_{U_j} |\varphi_n - \varphi|^2 dx\, dy = \|\varphi_n - \varphi\|^2_{U_j} \to 0,$$

as $n \to \infty$.

If $\mathcal{R}$ is compact, then $\mathcal{U} = \{U_j\}$ is an open covering and the sum in the right-hand side of (6.88) is a finite sum and we can conclude immediately that

$$\lim_{n\to\infty} \|\varphi_n - \varphi\|_{\mathcal{R}-K} = 0.$$

If $\mathcal{R}$ is not compact, then

$$\begin{aligned}\sum_{j=1}^{k} \int_{\mathcal{R}-K} \rho_j |\varphi_m - \varphi_n|^2 dx\, dy &\leqslant \int_{\mathcal{R}-K} |\varphi_m - \varphi_n|^2 dx\, dy \\ &= (\|\varphi_m - \varphi_n\|_{\mathcal{R}-K})^2 \\ &\leqslant \|\varphi_m - \varphi_n\|^2 \leqslant (\delta_m + \delta_n)^2\end{aligned}$$

by (6.85). Hence, multiplying both sides of the inequality

$$|\varphi_m - \varphi|^2 \leqslant 2|\varphi_m - \varphi_n|^2 + 2|\varphi_n - \varphi|^2$$

with $\sum_{j=1}^{k} \rho_j$ and integrating over $\mathcal{R} - K$, we obtain

$$\sum_{j=1}^{k} \int_{\mathcal{R}-K} \rho_j |\varphi_m - \varphi|^2 dx\, dy \leqslant 2(\delta_m + \delta_n)^2 + 2\sum_{j=1}^{k} \|\Phi_n - \varphi\|^2_{U_j}.$$

Since $\lim_{n\to\infty} \delta_n = 0$ and $\lim_{n\to\infty} \|\varphi_n - \varphi\|_{U_j} = 0$ we have

$$\sum_{j=1}^{k} \int_{\mathcal{R}-K} \rho_j |\varphi_m - \varphi|^2 dx\, dy \leqslant 2\delta_m^2.$$

Therefore

$$(\|\varphi_m - \varphi\|_{\mathcal{R}-K})^2 = \sum_{j=1}^{\infty} \int_{\mathcal{R}-K} \rho_j |\varphi_m - \varphi|^2 dx\, dy \leqslant 2\delta_m^2. \qquad (6.89)$$

Therefore

$$\lim_{m\to\infty} \|\varphi_m - \varphi\|_{\mathcal{R}-K} = 0,$$

hence

$$\lim_{m\to\infty} \|\varphi_m - \varphi\| = 0.$$

In order to prove that $\varphi \in \mathcal{L}_\Phi$ we have to prove that (i) $d\varphi = 0$ on $\mathcal{R} - |\sigma|$ and (ii) $\|\varphi\|_\Phi < \infty$. Since on all U_j, $j \geqslant 1$, $\varphi = \varphi_{(j)}$ and the $\varphi_{(j)}$

are closed 1-forms, we have $d\varphi = 0$. Since on $U_0 - |\sigma|$, $\varphi = d\Phi + \varphi_0$ and on U_0, $d\varphi_0 = 0$, we have $d\varphi = d(d\Phi + \varphi_0) = 0$. Hence by (6.89):

$$\|\varphi\|_{\mathcal{R}-K} \leqslant \|\varphi - \varphi_m\|_{\mathcal{R}-K} + \|\varphi_m\|_{\mathcal{R}-K}$$
$$\leqslant \sqrt{2}\delta_m + \|\varphi_m\|_\Phi < +\infty,$$

and hence, since obviously $\|\varphi_0\|_K < +\infty$, we have $\|\varphi\|_\Phi < +\infty$.

In this way, in order to prove our Theorem 6.16, it suffices to prove that there exists a continuously differentiable, closed 1-form φ_0 on U_0 such that $\lim_{n\to\infty}\|\varphi_{n0} - \varphi_0\|_{U_0} = 0$ and continuously differentiable closed 1-forms $\varphi_{(j)}$ on U_j such that $\lim_{n\to\infty}\|\varphi_n - \varphi_{(j)}\|_{U_j} = 0$ for $j \geqslant 1$.

By (6.85) we have:

$$\|\varphi_m - \varphi_n\|_{U_j} \leqslant \|\varphi_m - \varphi_n\| \leqslant \delta_m + \delta_n, \qquad j \geqslant 1,$$
$$\|\varphi_{m0} - \varphi_{n0}\|_{U_0} \leqslant \|\varphi_m - \varphi_n\|_{U_0} \leqslant \delta_m + \delta_n,$$

and by (6.86) and (6.87):

$$|(\varphi_n, d\eta)_{U_j}| \leqslant \|d\eta\|_{U_j} \cdot \delta_n, \qquad \operatorname{supp}\eta \subset U_j$$
$$|(\varphi_{n0}, d\eta)_{U_0}| \leqslant \|d\eta\|_{U_0} \cdot \delta_n, \qquad \operatorname{supp}\eta \subset U_0.$$

We may consider U_j as the unit disk with center at the origin in the z_j-plane. Hence Theorem 6.16 reduces to the following lemma concerning a sequence $\{\varphi_n\}$ of continuously differentiable 1-forms φ_n with $\|\varphi_n\|_U < +\infty$ defined on the unit disk U of the z-plane.

Lemma 6.3. If $\{\varphi_n\}$ satisfies the following two conditions with respect to a real sequence $\{\delta_n\}$ with $\delta_n > 0$ and $\lim_{n\to\infty}\delta_n = 0$:

(i) $\|\varphi_m - \varphi_n\|_U \leqslant \delta_m + \delta_n$,
(ii) $|(\varphi_n, d\eta)_U| \leqslant \|d\eta\|_U \cdot \delta_n$,

for an arbitrary twice continuously differentiable function η with $\operatorname{supp}\eta \subset U$, then there exists a continuously differentiable, closed 1-form φ on the unit disk U such that $\|\varphi\|_U < +\infty$ and $\lim_{n\to\infty}\|\varphi_n - \varphi\|_U = 0$.

Once the existence of such a φ has been established, it follows easily that φ is closed in the following way. For an arbitrary η with $\operatorname{supp}\eta \subset U$, determine $W = \{z : |z| < 1 - \alpha\}$, $\alpha < 0$, such that $\operatorname{supp}\eta \subset W$. By Green's Theorem (Theorem 6.1) we have

$$\int_U d(\eta\varphi) = \int_{[W]} d(\eta\varphi) = \int_{\partial[W]} \eta\varphi = 0$$

Therefore, since $d(\eta\varphi) = (d\eta \wedge \varphi) + (\eta \wedge d\varphi) = (\varphi \wedge **d\eta) + (\eta\, d\varphi)$, we have

$$\int_U \eta\, d\varphi = -\int_U \varphi \wedge **\, d\eta = -(\varphi, *d\eta)_U.$$

Since $d\varphi_n = 0$, we have $(\varphi_n, *d\eta)_U = -\int_U \eta\, d\varphi_n = 0$. Hence

$$|(\varphi, *d\eta)_U| = |(\varphi - \varphi_n, *d\eta)_U|$$
$$\leqslant \|\varphi - \varphi_n\|_U \cdot \|*d\eta\|_U \to 0, \qquad n \to \infty.$$

Therefore

$$\int_U \eta\, d\varphi = -(\varphi, *d\eta)_U = 0.$$

Since $\int_U \eta\, d\varphi = 0$ holds for arbitrary η with $\operatorname{supp}\eta \subset U$, we conclude that $d\varphi = 0$. Since

$$\|\varphi_n - \varphi\|_U^2 = \int_U (|\varphi_{n1} - \varphi_1|^2 + |\varphi_{n2} - \varphi_2|^2) dx\, dy,$$

in order to prove the existence of a continuously differentiable 1-form φ such that $\lim_{n\to\infty} \|\varphi_n - \varphi\|_U = 0$, it suffices to prove that there exist continuously differentiable functions φ_1 and φ_2 on U, such that for all function sequences $\{\varphi_{n1}\}$ and $\{\varphi_{n2}\}$ we have:

$$\int_U |\varphi_{n1} - \varphi_1|^2 dx\, dy \to 0, \qquad \int_U |\varphi_{n2} - \varphi_2|^2 dx\, dy \to 0,$$
$$n \to \infty.$$

To this end, we define

$$(f, g) = \int_U f(z) g(z) dx\, dy, \qquad \|f\|^2 = \int_U |f(z)|^2 dx\, dy$$

where $f = f(z) = f(x + iy)$ and $g = g(z) = g(x + iy)$ are real-valued continuous functions defined on $U = \{z : |z| < 1\}$. If η is a three times continuously differentiable function with $\operatorname{supp}\eta \subset U$, then

$$(\varphi_n, d\eta_x)_U = \int_U (\varphi_{n1}\eta_{xx} + \varphi_{n2}\eta_{xy}) dx\, dy,$$

$$(\varphi_n, *d\eta_y)_U = \int_U (-\varphi_{n1}\eta_{yy} + \varphi_{n2}\eta_{xy}) dx\, dy$$

hence

$$(\varphi_{n1}, \Delta\eta) = \int_U \varphi_{n1}(\eta_{xx} + \eta_{xy}) dx\, dy = (\varphi_n, d\eta_x)_U - (\varphi_n, *d\eta_y)_U.$$

Since $d\varphi_n = 0$, we have $(\varphi_n, *d\eta_y)_U = 0$. Therefore, by condition (ii),

$$|(\varphi_{n1}, \Delta\eta)| = |(\varphi_n, d\eta_x)_U| \leqslant \|d\eta_x\|_U \cdot \delta_n,$$

and similarly,

$$|(\varphi_{n2}, \Delta\eta)| = |(\varphi_n, d\eta_y)_U| \leqslant \|d\eta_y\|_U \cdot \delta_n.$$

Since $\|d\eta_x\|_U^2 = \|\eta_{xx}\|^2 + \|\eta_{xy}\|^2$ and $\|d\eta_y\|_U^2 = \|\eta_{yx}\|^2 + \|\eta_{yy}\|^2$, we have, putting

$$N(\eta) = \sqrt{\|\eta_{xx}\|^2 + \|\eta_{xy}\|^2 + \|\eta_{yy}\|^2}$$

that

$$|(\eta_{n1}, \Delta\eta)| \leqslant N(\eta) \cdot \delta_n, \qquad |(\varphi_{n2}, \Delta\eta)| \leqslant N(\eta) \cdot \delta_n.$$

Since $\|\varphi_{m1} - \varphi_{n1}\|^2 + \|\varphi_{m2} - \varphi_{n2}\|^2 = \|\varphi_m - \varphi_n\|_U^2$, we have by condition (i)

$$\|\varphi_{m1} - \varphi_{n1}\| \leqslant \delta_m + \delta_n, \qquad \|\varphi_{m2} - \varphi_{n2}\| \leqslant \delta_m + \delta_n.$$

Therefore, in order to prove Lemma 6.3 it suffices to prove the following lemma.

Lemma 6.4. If with respect to some sequence $\{\delta_n\}$ of positive real numbers satisfying $\lim_{n\to\infty}\delta_n = 0$, the sequence of continuous functions $\{f_n\}$ defined on U such that $\|f_n\| < +\infty$ satisfies the following conditions:

(i) $\|f_m - f_m\| \leqslant \delta_m + \delta_n$,
(ii) If η is an arbitrary three times continuously differentiable function with $\operatorname{supp}\eta \subset U$, then

$$|(f_n, \Delta\eta)| \leqslant N(\eta) \cdot \delta_n, \qquad N(\eta) = \sqrt{\|\eta_{xx}\|^2 + \|\eta_{xy}\|^2 + \|\eta_{yy}\|^2},$$

then there exists a continuously differentiable function f defined on U such that $\|f\| < +\infty$ and $\lim_{n\to\infty}\|f_n - f\| = 0$.

In order to prove this lemma, we put $P = |z|^2 = x^2 + y^2$. For sufficiently small positive real numbers α and β, $\beta < \alpha$, let $\rho(P)$ be a C^∞ function of P such that:

$$\begin{cases} \rho(P) = 1 & \text{if } P \leqslant \beta^2 \\ 0 \leqslant \rho(P) \leqslant 1 & \text{if } \beta^2 \leqslant P \leqslant \alpha^2 \\ \rho(P) = 0 & \text{if } \alpha^2 \leqslant P. \end{cases}$$

Next, we determine a three times continuously differentiable function $\eta_\varepsilon(P)$ for η satisfying $0 < \eta < \beta$ such that:

$$\begin{cases} \eta_\varepsilon(P) \text{ is a third degree polynomial in } P & \text{if } 0 \leqslant P \leqslant \varepsilon^2 \\ \eta_\varepsilon(P) = \dfrac{1}{4\pi}\rho(P)\log P & \text{if } \varepsilon^2 \leqslant P. \end{cases}$$

A simple calculation yields the following expression for the polynomial which defines $\eta_\varepsilon(P)$ when $0 \leqslant P \leqslant \varepsilon^2$:

$$\eta_\varepsilon(P) = \frac{1}{24\pi\varepsilon^6}(2P^3 - 9\varepsilon^2 P^2 + 18\varepsilon^4 P - 11\varepsilon^6) + \frac{1}{4\pi}\log\varepsilon^2.$$

Since $\partial^2\eta_\varepsilon(P)/\partial x^2 = 4x^2\eta_\varepsilon''(P) + 2\eta_\varepsilon'(P)$, we have

$$\Delta\eta_\varepsilon(P) = \left(\frac{\partial^2}{\partial x^2} + \frac{\partial^2}{\partial y^2}\right)\eta_\varepsilon(P) = 4(P\eta_\varepsilon''(P) + \eta_\varepsilon'(P)).$$

Hence, for $0 \leqslant P \leqslant \varepsilon^2$

$$\Delta\eta_\varepsilon(P) = \frac{3}{\pi\varepsilon^6}(P - \varepsilon^2)^2.$$

Therefore, putting

$$E_\varepsilon(P) = \begin{cases} \Delta\eta_\varepsilon(P), & 0 \leqslant P \leqslant \varepsilon^2, \\ 0, & P \geqslant \varepsilon^2, \end{cases}$$

we conclude that $E_\varepsilon(P)$ is a continuously differentiable function of P such that $E_\varepsilon(P) \geqslant 0$. Further,

$$\int E_\varepsilon(P)dx\,dy = 1 \tag{6.90}$$

because putting $z = re^{i\theta}$, we have $P = r^2$, hence $dP = 2r\,dr$ and

$$\int E_\varepsilon(P)dx\,dy = \int_0^\varepsilon\int_0^{2\pi} E_\varepsilon(P)r\,dr\,d\theta = \pi\int_0^{\varepsilon^2} E_\varepsilon(P)dP$$

$$= \frac{3}{\varepsilon^6}\int_0^{\varepsilon^2}(P - \varepsilon^2)^2 dp = \frac{1}{\varepsilon^6}(P - \varepsilon^2)^3\Big|_0^{\varepsilon^2} = 1.$$

We put

$$M(P) = \frac{1}{4\pi}\Delta(\rho(P)\log P) = \frac{1}{4\pi}\left(\Delta\rho(P)\cdot\log\frac{1}{P} - 8\rho'(P)\right).$$

Note that $M(P)$ is a C^∞ function of $P \geqslant 0$. Since $\rho(P) = 1$ for $0 \leqslant P \leqslant \beta^2$, we have $M(P) = 0$ for $P \geqslant \alpha^2$. Therefore, $\operatorname{supp} M(|z|^2)$ is contained in the annulus $\{z : \beta \leqslant |z| \leqslant \alpha\}$. Obviously

$$\Delta\eta_\varepsilon(P) = E_\varepsilon(P) - M(P). \tag{6.91}$$

Let $W \subset U$ be a disk with center 0 and radius $1 - \alpha$ and let us represent points of W by $w = u + iv$. We put

$$\eta_\varepsilon(z, w) = \eta_\varepsilon(|z - w|^2), \qquad E_\varepsilon(z, w) = E_\varepsilon(|z - w|^2),$$
$$M(z, w) = M(|z - w|^2).$$

Just as we sometimes write f for $f(z)$, supressing the variable z, for fixed w we will write the functions $\eta_\varepsilon(z, w)$, $E_\varepsilon(z, w)$ and $M_\varepsilon(z, w)$ as $\eta_\varepsilon(z, w)$, $E_\varepsilon(z, w)$ and $M(z, w)$ respectively.

If $w \in W$, then $\operatorname{supp}\eta_\varepsilon(z, w) \subset U$. Hence by condition (ii)

$$|(f_n, \Delta\eta_\varepsilon(z, w))| \leqslant N(\eta_\varepsilon(z, w)) \cdot \delta_n.$$

Since $\eta_\varepsilon(z, w) = \eta_\varepsilon(|z - w|^2)$, $N(\eta_\varepsilon(z, w))$ is not dependent on w. Hence, putting $N(\varepsilon) = N(\eta_\varepsilon(z, w))$, we have

$$|(f_n, \Delta\eta_\varepsilon(z, w))| \leqslant N(\varepsilon) \cdot \delta_n.$$

Therefore, by (6.91)

$$|(f_n, E_\varepsilon(z, w)) - (f_n, M(z, w))| \leqslant N(\varepsilon) \cdot \delta_n. \tag{6.92}$$

Putting

$$(E_\varepsilon g)(w) = (g, E_\varepsilon(z, w)) = \int_U E_\varepsilon(z, w) g(z) dx\, dy$$

where $g = g(z)$ is an arbitrary continuous function on U satisfying $\|g\| < +\infty$, we consider $(E_\varepsilon g)(w)$ as a function of w defined on W. We call $E_\varepsilon : g \to E_\varepsilon g$ an *integral operator.*

We will prove

$$\|E_\varepsilon g\|_W \leqslant \|g\|. \tag{6.93}$$

Writing $(E_\varepsilon g)(w)$ as

$$(E_\varepsilon g)(w) = \int_U \sqrt{E_\varepsilon(z, w)} \cdot \sqrt{E_\varepsilon(z, w)} g(z) dx\, dy$$

and applying Schwarz's inequality we obtain

$$|(E_\varepsilon g)(w)|^2 \leqslant \int_U E_\varepsilon(z, w) dx\, dy \int_U E_\varepsilon(z, w) |g(z)|^2 dx\, dy.$$

Since, by (6.90):

$$\int_U E_\varepsilon(z, w) dx\, dy = \int_U E_\varepsilon(|z - w|^2) dx\, dy = 1$$

we have:

$$|(E_\varepsilon g)(w)|^2 \leqslant \int_U E_\varepsilon(z, w) |g(z)|^2 dx\, dy.$$

Hence, since

$$\int_W E_\varepsilon(z, w) du\, dv \leqslant \int E_\varepsilon(|w - z|^2) dy\, dv \leqslant 1,$$

we arrive at

$$\int_W |(E_\varepsilon g)(w)|^2 du\, dv \leqslant \int_U |g(z)|^2 dx\, dy,$$

which proves (6.93).

We put

$$(Mg)(w) = (g, M(z, w)) = \int_U M(z, w) g(z) dx\, dy;$$

$(Mg)(w)$ is a continuous function of w on W. To prove this, let w and w_1 be two arbitrary points in W. Since

$$|(Mg)(w) - (Mg)(w_1)| \leqslant \|M(z, w) - M(z, w_1)\| \, \|g\|$$

and $M(P)$ is a C^∞ function of P satisfying $M(P) = 0$ if $P \geqslant \alpha^2$, $M(z, w) = M(|z - w|^2)$ is uniformly continuous in z and w. Therefore

$$\|M(z, w) - M(z, w_1)\|^2 - \int_U |M(z, w) - M(z, w_1)|^2 dx\, dy \to 0$$

is $w \to w_1$. Furthermore,

$$|(Mg)(w)| \leqslant \|M(z, w)\| \, \|g\|$$

and if $w \in W$ then supp $M(z, w) \subset U$ and $M(z, w) = M(|z - w|^2)$, hence $\|M(z, w)\|$ is not dependent on w. Putting $\|M\| = \|M(z, w)\|$ we have

$$|(Mg)(w)| \leqslant \|M\| \, \|g\|, \qquad w \in W. \tag{6.94}$$

By this inequality we have $|(Mf_m)(w) - (Mf_n)(w)| \leqslant \|M\| \, \|f_m - f_n\|$, hence

$$|(Mf_m)(w) - (Mf_n)(w)| \leqslant \|M\|(\delta_m + \delta_n)$$

by condition (i).

Since $\lim_{n\to\infty}\delta_n = 0$, the function sequence $(Mf_n)(w)$ converges uniformly on the disk W. Putting $f(W) = \lim_{n\to\infty}(Mf_n)(w)$, and observing that all $(Mf_n)(w)$ are continuous in w, we conclude that $f(w)$ too is continuous on W and

$$|(Mf_n)(w) - f(w)| \leqslant \|M\| \cdot \delta_m, \qquad w \in W. \tag{6.95}$$

On the other hand, by (6.92)

$$|(E_\varepsilon f_m)(w) - f(w)| \leqslant N(\varepsilon) \cdot \delta_m.$$

Hence

$$|(E_\varepsilon f_m)(w) - f(w)| \leqslant (\|M\| + N(\varepsilon)) \cdot \delta_m, \qquad w \in W.$$

If, in general, $|g(w)| \leqslant c$, then $\|g\|_W^2 = \int_W |g(w)|^2 du\, dr \leqslant \pi c^2$, hence

$$\|E_\varepsilon f_m - f\|_W \leqslant \sqrt{\pi}(\|M\| + N(\varepsilon)) \cdot \delta_m.$$

Since, by (6.93) and condition (i)

$$\|E_\varepsilon f_n - E_\varepsilon f_m\|_W \leqslant \|f_n - f_m\| \leqslant \delta_n + \delta_m$$

we obtain

$$\|E_\varepsilon f_n - f\|_W \leqslant \sqrt{\pi}(\|M\| + N(\varepsilon)) \cdot \delta_m + \delta_m + \delta_n.$$

Therefore, since $\lim_{m\to\infty}\delta_m = 0$, we have

$$\|E_\varepsilon f_n - f\|_W \leqslant \delta_n.$$

Because $E_\varepsilon(z, w) \geqslant 0$, $E_\varepsilon - 0$ if $|z - w| \geqslant \varepsilon$ and $\int E_\varepsilon(z, w) = 1$, while $f_n(z)$ is uniformly continuous on for example the closed disk $\{z : |z| \leqslant 1 - \alpha + \beta\}$, the function of w

$$(E_\varepsilon f_n)(w) = \int E_\varepsilon(z, w) f_n(z) dx\, dy$$

converges uniformly to $f_n(w)$ on W if $\varepsilon \to +0$. Therefore

$$\|f_n - f\|_W \leqslant \delta_n, \qquad n = 1, 2, 3, \ldots.$$

Designating the continuous function f by f_W, we see that for each disk $W = \{w : |w| < 1 - \alpha\}$, $0 < \alpha < 1$, there are continuous functions $f_W = f_W(w)$ on W satisfying the inequality

$$\|f_n - f_W\|_W \leqslant \delta_n.$$

If V is such that $W \subset V = \{w : |w| < 1 - \gamma\}$, $\gamma < \alpha < 1$, then:

$$\|f_V - f_W\|_W \leqslant \|f_V - f_n\|_V + \|f_n - f_W\|_W \leqslant 2\delta_n \to 0,$$
as $n \to \infty$

i.e. $f_V(w) = f_W(w)$ if $w \in W$. Hence there exists a continuous function f on U which coincides with f_W in each W, while

$$\|f_n - f\|_W \leqslant \delta_n.$$

Since δ_n is independent of W and $U = \bigcup_{0<\alpha<1} W$, we have

$$\|f_n - f\| \leqslant \delta_n, \qquad n = 1, 2, 3, \ldots. \tag{6.96}$$

Since $\|f_n\| < +\infty$, it follows immediately that $\|f\| < \infty$. In order to prove this function f is continuously differentiable on the unit disk U, it suffices to show that f is continuously differentiable on each $W = \{w : |w| < 1 - \alpha\}$, $0 < \alpha < 1$. Therefore we fix W again. by (6.96) and (6.94) we have

$$|(Mf_n)(w) - (Mf)(w)| \leqslant \|M\| \cdot \delta_n, \qquad w \in W.$$

Hence, $(Mf)(w) = \lim_{n\to\infty}(Mf_n)(w)$ for $w \in W$, while also $f(w) = \lim_{n\to\infty}(Mf_n)(w)$ by the definition of f. Therefore

$$f(w) = (Mf)(w) = (f, M(z, w)), \qquad w \in W. \tag{6.97}$$

Since $M(P)$ is a C^∞ function of P,

$$M(x + iy, u + iv) = M(z, w) = M(|z - w|^2), \qquad z = x + iy,$$
$$w = u + iv,$$

is a C^∞ function of X, Y, U and v. $m(p) = 0$ for $p \geqslant \alpha^2$, hence the partial derivative

$$\mu_U(Z, W) = \frac{\partial}{\partial U} M(x, +iy, u + iv) = M'(|z - w|^2) \cdot 2u$$

is a uniformly continuous function of z and w, for $w \in W$. Hence for each $\varepsilon > 0$ we can find a $\delta(\varepsilon) > 0$ such that

$$|M_u(z, w_1) - M_u(z, w)| \leqslant \varepsilon \text{ if } |w_1 - w| < \delta(\varepsilon). \tag{6.98}$$

Let $h \neq 0$ be a real number. By the Mean Value Theorem we have

$$\frac{M(z,\,u+h+iv)-M(z,\,u+iv)}{h}=M_u(z,\,u+\theta h+iv),$$
$$0<\theta<1,$$

hence, if $-\delta(\varepsilon)<h<\delta(\varepsilon)$ and $h\neq 0$ then

$$\left|\frac{M(z,\,w+h)-M(z,\,w)}{h}-M_u(z,\,w)\right|<\sqrt{\pi}\varepsilon,$$

and therefore

$$\left\|\frac{M(z,\,w+h)-M(z,\,w)}{h}-M_u(z,\,w)\right\|<\sqrt{\pi}\varepsilon.$$

Hence, by (6.97)

$$\left|\frac{f(w+h)-f(w)}{h}-(f,\,M_u(z,\,w))\right|\leqslant\|f\|\cdot\sqrt{\pi}\varepsilon.$$

Since this inequality holds for arbitrary $\varepsilon>0$ and h such that $-\delta(\varepsilon)<h<\delta(\varepsilon)$, $h\neq 0$, we conclude that

$$\lim_{h\to 0}\frac{f(w+h)-f(w)}{h}=(f,\,M_u(z,\,w)).$$

That is, $f(u+iv)$ is partially differentiable with respect to u on W and the partial derivative is given by

$$f_u(u+iv)=(f,\,M_u(z,\,u+iv)).$$

Since by (6.98)

$$|f_u(w_1)-f_u(w)|\leqslant\|f\|\,\|M_u(z,\,w_1)-M_u(z,\,w)\|\leqslant\|f\|\cdot\sqrt{\pi}\varepsilon,$$

if $|w_1-w|<\delta(\varepsilon)$, it is clear that $f_u(w)=(f,\,M_u(z,\,w))$ is continuous in w. Similarly, $f(w)=f(u+iv)$ is seen to be partially differentiable with respect to v and the partial derivative $f_v(w)=(f,\,M_v(z,\,w))$ a continuous function of w. Therefore, $f(w)=f(u+iv)$ is a continuously differentiable function of u and v on W.

Theorem 6.17. The 1-form $\varphi\in\mathcal{L}_\Phi$, occurring in Theorem 6.16, has the following properties:

(i) $\varphi=\psi+d\eta_\infty$, where η_∞ is a twice continuously differentiable function on $\mathcal{R}$ such that $\|d\eta_\infty\|<+\infty$,

(ii) $\|\varphi\|_\Phi=K$,

(iii) if η is twice differentiable on $\mathcal{R}$, such that $\|d\eta\|<+\infty$, $\operatorname{supp}\eta\cap|\sigma|=\varnothing$, then
$$(\varphi,\,d\eta)=0,$$

(iv) φ is a harmonic 1-form on $\mathcal{R}$ with the same singularities as $d\Phi$.

Proof:

(i) We can extend $\varphi - \psi$ to a twice continuously differentiable 1-form on the whole of $\mathcal{R}$. Denoting this 1-form also by $\varphi - \psi$, we have

$$\|\varphi - \psi - d\eta_n\| = \|\varphi - \varphi_n\| \to 0, \qquad n \to \infty.$$

If ξ is an arbitrary continuously differentiable closed 1-form with compact support on $\mathcal{R}$, we have by Corollary 2 to Theorem 6.14

$$(d\eta_n, *\xi) = \int_{\mathcal{R}} d\eta_n \wedge **\xi = \int_{\mathcal{R}} \wedge\, d\eta_n = 0.$$

Therefore

$$|(\varphi - \psi, *\xi)| = |(\varphi - \psi - d\eta_n, *\xi)|$$
$$\leqslant \|\varphi - \psi - d\eta_n\| \, \|*\xi\| \to 0,$$

and hence

$$\int_{\mathcal{R}} \xi \wedge (\varphi - \psi) = (\varphi - \psi, *\xi) = 0.$$

Therefore, by the same Corollary 2, $\varphi - \psi$ is an exact differential form on $\mathcal{R}$, i.e. we can represent $\varphi - \psi$ as $d\eta_\infty$ for some continuously differentiable function η_∞ on $\mathcal{R}$. Since $\varphi - \psi$ is continuously differentiable, η_∞ is twice continuously differentiable and $\|d\eta_\infty\| = \|\varphi - \psi\| < +\infty$.

(ii) Since by (6.78)

$$|\,\|\varphi_{n0}\|_K - \|\varphi_0\|_K| \leqslant \|\varphi_{n0} - \varphi_0\|_K \leqslant \|\varphi_n - \varphi\| \to 0,$$
$$|\,\|\varphi_n\|_{\mathcal{R}-K} - \|\varphi\|_{\mathcal{R}-K}| \leqslant \|\varphi_n - \varphi\|_{\mathcal{R}-K} \leqslant \|\varphi_n - \varphi\| \to 0$$

if $n \to \infty$, we have

$$\|\varphi\|_\Phi^2 = \|\varphi_0\|_K^2 + (\|\varphi\|_{\mathcal{R}-K})^2 = \lim_{n\to\infty} \|\varphi_n\|_\Phi^2 = \kappa^2.$$

(iii) For an arbitrary real λ we have by (6.82)

$$\|\varphi\|_\Phi^2 + 2\lambda(\varphi, d\eta) + \lambda^2\|d\eta\|^2 = \|\varphi + \lambda\, d\eta\|_\Phi^2.$$

Since $\varphi + \lambda\, d\eta = \psi + d(\eta_\infty + \lambda\eta)$, we have $\|\varphi + \lambda\, d\eta\|_\Phi^2 \geqslant \kappa^2$, while by (ii) $\|\varphi\|_\Phi^2 = \kappa^2$. Hence, $2\lambda(\varphi, d\eta) + \lambda^2\|d\eta\|^2 \geqslant 0$ and therefore, $(\varphi, d\eta) = 0$.

(iv) Selecting η such that $\operatorname{supp}\eta$ is compact and $\operatorname{supp}\eta \cap |\sigma| = \varnothing$, $\eta * \varphi$ is a continuously differentiable 1-form with compact support, hence $\int_{\mathcal{R}} d(\eta * \varphi) = 0$ by Theorem 6.11. Further, $\int_{\mathcal{R}} d\eta \wedge *\varphi = (\varphi, d\eta) = 0$ by (iii). Since $\eta d * \varphi = d(\eta * \varphi) - d\eta \wedge *\varphi$, we conclude that $\int_{\mathcal{R}} \eta d * \varphi = 0$. Since this is true for all η with compact support such that $\operatorname{supp}\eta \subset \mathcal{R} - |\sigma|$, we conclude that $d * \varphi = 0$ on

$\mathcal{R} - |\sigma|$. On the other hand, $d\varphi = 0$ on $\mathcal{R} - |\sigma|$ since $\varphi \in \mathcal{L}_\Phi$. Therefore, φ is a harmonic 1-form on $\mathcal{R} - |\sigma|$. So $\varphi = d\Phi - \varphi_0$ on $U_0 - |\sigma|$, where φ_0 is a continuously differentiable function on U_0 such that $d\varphi_0 = 0$. Since Φ is a harmonic function on $U_0 - |\sigma|$, $d * d\Phi = 0$, hence $d * \varphi_0 = d(*\varphi - *d\Phi) = 0$ on $U_0 - |\sigma|$, while $d * \varphi_0$ is continuous on U_0. Hence $d * \varphi_0 = 0$ on U_0, and therefore the 1-form φ_0 is harmonic on U_0. Hence φ is a harmonic 1-form on $\mathcal{R}$ with the same singularities as $d\Phi$.

This completes the proof of Dirichlet's Principle.

The 1-form $\varphi = \psi + d\eta_\infty$ satisfying $\|\psi + d\eta_\infty\| = \kappa$ is uniquely determined by ψ because we have

$$\|\varphi_1 - \varphi\| \leqslant \sqrt{\|\varphi_1\|_\Phi^2 - \kappa^2}$$

by (6.85) for an arbitrary $\varphi_1 = \varphi + d\eta$.

Remark 1: Let us denote by $\mathcal{F}_\Phi$ the collection of functions v that are twice continuously differentiable on $\mathcal{R} - |\sigma|$ and satisfy the following two conditions:

(i) $v = \Phi + v_0$ on $U_0 - |\sigma|$, where v_0 is a twice continuously differentiable function on U_0.

(ii)

$$\|dv_0\|_K^2 + (\|dv\|_{\mathcal{R}-K})^2 < +\infty$$

For $v \in \mathcal{F}_\Phi$, the expression

$$D(v) = \|dv_0\|_K^2 + (\|dv\|_{\mathcal{R}-K})^2$$

is called the *Dirichlet integral* of v. If $v \in \mathcal{L}_\Phi$, then $D(v) = \|dv\|_\Phi^2$, hence we obtain from the above Dirichlet's Principle for 1-forms the following Dirichlet's Principle for functions:

Dirichlet's Principle

There exists a function $u \in \mathcal{F}_\Phi$ such that $D(u) \leqslant D(v)$ for all $v \in \mathcal{F}_\Phi$. This function u is harmonic on $\mathcal{R} - |\sigma|$ and $u = \Phi + u_0$ on $U_0 - |\sigma|$, where u_0 is a harmonic function on U_0. (Weyl, 1955)

Remark 2: The above proof of Dirichlet's Principle can be simplified by invoking Weyl's method of orthogonal projection. (Weyl, 1940) A 1-form $\psi = \psi_1\, dx + \psi_2\, dy$ is called *Lebesgue measurable* if ψ_1 and ψ_2 are Lebesgue measurable functions of the local coordinates x and y. Let us denote by $\mathcal{H}_\Phi$ the collection of all Lebesgue measurable 1-forms ψ on $\mathcal{R}$, satisfying

$$\|\psi\|_\Phi^2 = |\psi_0\|_K^2 + (\|\psi\|_{\mathcal{R}-K})^2 < +\infty, \qquad \psi_0 = \psi - d\Phi.$$

By the method of orthogonal projection the collection of 1-forms under consideration is extended from $\mathcal{L}_\Phi$ to $\mathfrak{H}_\Phi$. Since $\|\varphi_m - \varphi_n\| \to 0$ is $m \to \infty$ and $n \to \infty$ by (6.85), the existence of an element $\varphi \in \mathfrak{H}_\Phi$ such that $\lim_{n\to\infty}\|\varphi_n - \varphi\| = 0$ follows immediately from the completeness of the Hilbert space $\mathfrak{H}_\Phi$. The difficulty is now to show that this φ is continuously differentiable on U_0. Let η be a twice continuously differentiable function on $\mathcal{R}$ with compact support, such that $\operatorname{supp}\eta \cap |\sigma| = \varnothing$; then $(\varphi, d\eta) = 0$ by (6.86) and $(\varphi, *d\eta) = \lim_{n\to\infty}(\varphi_n, *d\eta) = \lim_{n\to\infty}\int_{\mathcal{R}} d(\eta\varphi_n) = 0$ by Theorem 6.11. Similarly, $(\varphi_0, d\eta) = (\varphi_0, *d\eta) = 0$ if $\operatorname{supp}\eta \subset U_0$ by (6.87). Hence, in order to prove the following lemma, which now takes the place of our Lemma 6.4.

Weyl's Lemma

Let f be a Lebesgue measurable function defined on the unit disk U, satisfying $\|f\|^2 = \int_U |f|^2 dx\,dy < +\infty$. If for all three times continuously differentiable functions η with $\operatorname{supp}\eta \subset U$ we have $(f, \Delta\eta) = 0$, then f is continuously differentiable on U.

Proof: It suffices to prove that f is continuously differentiable on $W = \{z : |z| < 1 - \alpha\}$ for arbitrary α with $0 < \alpha < 1$. Let $\eta_\varepsilon(z, w)$, $E_\varepsilon(z, w)$ and $M(z, w)$ be defined as in the proof of Lemma 6.4. If $w \in W$, then $\operatorname{supp}\eta_\varepsilon(z, w) \subset U$, hence $(f, \Delta\eta_\varepsilon(z, w)) = 0$ by our assumption. Therefore, by (6.91)

$$(f, E_\psi(z, w)) - (f, M(z, w)) = 0.$$

i.e.

$$(E_\varepsilon f)(w) = (Mf)(w), \qquad w \in W.$$

Since $|(Mf)(w_1) - (Mf)(w)| \leqslant \|M(z, w_1) - M(z, w\| \cdot \|f\|$ for $w, w_1 \in W$, we have $(Mf)(w)$ is a continuous function of w on W. There exists a sequence $\{f_n\}$ of continuous functions on U such that $\lim_{n\to\infty}\delta_n = \lim_{n\to\infty}\|f_n - f\| = 0$. By (6.93) we have

$$\|E_\varepsilon f_n - Mf\|_W = \|E_\varepsilon f_n - E_\varepsilon f\|_W \leqslant \delta_n.$$

Since f_n is a continuous function, the function sequence $\{E_\varepsilon f_n\}$ converges uniformly to f_n if $\varepsilon \to 0$. Therefore $\|f_n - Mf\|_W \leqslant \delta_n$, hence

$$\|f - Mf\|_W \leqslant \|f - f_n\| + \|f_n - Mf\|_W \leqslant d\delta_n \to 0,$$

as $n \to \infty$,

i.e. equality (6.97) holds. Therefore, f is continuously differentiable on W.

In this way it is possible to give a relatively simple proof of Dirichlet's Principle. However, it seems that Weyl did not like the method of orthogonal projection. (See the foreword of Weyl, 1955). The author remembers that Weyl told him: "Maybe it is because I am old-fashioned, but I think that the method of orthogonal projection is not satisfactory." In the author's opinion this might have something to do with Weyl's attitude towards the "crisis" in mathematics, about which he wrote:

> *We have had it for nearly fifty years. Outwardly it does not seem to hamper our daily work, and yet I for one confess that it has had a considerable practical influence on my mathematical life: it directed my interests to fields I considered relatively "safe" . . .*
> *(Hermann Weyl: Gesammelte Abhandlungen, Band IV, p. 279)*

I first planned to prove Dirichlet's Principle using the method of orthogonal projection in this book. However, I did not like to have to use the concept of Lebesgue measurability only for the proof of Dirichlet's Principle and therefore I rewrote it in such a way that I did not have to.

c. *Analytic functions*

The concept of homotopy of curves in a region $\Omega \subset A$ is based only on continuity of mappings and can therefore be applied to a Riemann surface $\mathcal{R}$ without any change. Similarly, the idea of a 1-chain can be applied to Riemann surfaces, so that also the definition of homology can be used in the case of a Riemann surface without any change.

Let $p_0 \in \mathcal{R}$ be a fixed point, $p \in \mathcal{R}$ an arbitrary point, and $\gamma : t \to \gamma(t)$, $0 \leqslant t \leqslant b$, be a piecewise smooth curve in $\mathcal{R}$ connecting p_0 and p. Let φ be a continuously differentiable closed 1-form on $\mathcal{R}$ and consider $\int_\gamma \varphi$.

Let $\mathcal{U} = \{U_j\}$ be a locally finite open covering consisting of coordinate disks U_j of $\mathcal{R}$. By Theorem 6.3, $\varphi = du_j$ on each coordinate disk U_j, where u_j is a twice continuously differentiable function on U_j. Since $|\gamma|$ is covered by finitely many $U_j \in \mathcal{U}$ it is possible to partition γ into m curves $\gamma_n : t \to \gamma_n(t)$, $t_{n-1} \leqslant t \leqslant t_n$, $0 = t_0 < t_1 < t_2 < \cdots < t_m = b$, such that each γ_n is contained in one disk $U_{j(n)} : |\gamma_n| \subset U_{j(n)}$. Putting $p_n = \gamma(t_n)$, we have

$$\int_\gamma \varphi = \sum_{n=1}^{m} \int_{\gamma_n} du_{j(n)} = \sum_{n=1}^{m} (u_{j(n)}(p_n) - u_{j(n)}(p_{n-1}))$$

$$= -u_{j(1)}(p_0) + \sum_{n=1}^{m-1} (u_{j(n)}(p_n) - u_{j(n+1)}(p_n)) + u_{j(m)}(p).$$

Obviously, $p_n = \gamma(t_n) \in U_{j(n)} \cap U_{j(n+1)}$. Let V_n denote the connected component of $U_{j(n)} \cap U_{j(n+1)}$ containing p_n. Since $du_{j(n)} - du_{j(n+1)} = 0$, $c_n = u_{j(n)} - u_{j(n+1)}$ is constant on V_n. Therefore

$$\int_\gamma \varphi = u_{j(m)}(p) - u_{j(1)}(p_0) + \sum_{n=1}^{m-1} c_n. \tag{6.99}$$

The right-hand side of this equality is also meaningful if γ is not piecewise smooth. Therefore, (6.99) can be considered as the definition of the integral of φ along the arbitrary curve γ connecting p_0 and p. It is easy to verify that the value of the right-hand side of (6.99) is independent of the choice of open covering $\mathcal{U}$ and the partition of γ. Since the constants $c_n = u_{j(n)}(p_n) - u_{j(n+1)}(p_n)$, $1 \leqslant n \leqslant m-1$, are independent of p_n, $p_n \in V_n$, the value of the integral does not change if γ is transformed continuously in such a way that $p_0 = \gamma(0)$ and $p = \gamma(b)$ remain fixed and each γ_n remains within $U_{j(n)}$. Hence we have proved

Thorem 6.18 (Monodromy Theorem). Let φ be a continuously differentiable, closed 1-form on $\mathcal{R}$. If γ and γ_1 are two curves connecting the points p_0 and p of $\mathcal{R}$, such that γ and γ_1 are homotopic on $\mathcal{R}$, then

$$\int_{\gamma_1} \varphi = \int_\gamma \varphi.$$

If we want to consider $\int_\gamma \varphi$ as a function of the end point p of γ, we write $\int_\gamma^p \varphi$. Since $u(p) = \int_\gamma^p \varphi$ is determined by the homotopy class $[\gamma]$ of γ, in general $u(p)$ is a multi-valued function of p. If we limit the domain of p to one coordinate disk U_j, then $u(p) = u_j(p) +$ a constant, by (6.99). Hence, $u = u(p)$ is a twice continuously differentiable function satisfying $du = du_j = \varphi$. For the sake of simplicity, we will write

$$u = \int \varphi.$$

If $\mathcal{R}$ is simply connected, all curves connecting p_0 and p in $\mathcal{R}$ are homotopic and hence $u(p) = \int_\gamma^p \varphi$ is dependent only on p, not on γ. Writing

$$u(p) = \int_{p_0}^{p} \varphi$$

$u = u(p)$ is a twice continuously differentiable function on $\mathcal{R}$ satisfying $\varphi = du$. Considering a region $\Omega \subset \mathcal{R}$ as a Riemann surface in its own right, we get the following corollary by applying this result to Ω.

Corollary. If φ is a continuously differentiable closed 1-form on a simply connected region Ω, then φ can be represented as $\varphi = du$, where u is a twice continuously differentiable function.

If γ is a closed curve, such that $\gamma \simeq 0$, then $\int_\gamma \varphi = 0$ by Theorem 6.18. Hence, if the 1-cycle γ is homologous to 0, then

$$\int_\gamma \varphi = 0, \qquad \gamma \sim 0. \tag{6.100}$$

The above preparations are necessary in order to continue the preceding section. Let $\varphi = \psi + d\eta_\infty \in \mathcal{L}_\Phi$ be the harmonic 1-form of Theorem 6.16. Since $d\varphi = d * \varphi = 0$ on $\mathcal{R} - |\sigma|$, both

$$u = \int \varphi, \qquad v = \int * \varphi$$

are harmonic (not necessarily single-valued) functions on $\mathcal{R} - |\sigma|$. $\varphi + i * \varphi$ is a holomorphic 1-form on $\mathcal{R} - |\sigma|$, i.e. on a neighborhood of each point $q \in \mathcal{R} - |\sigma|$ we can write $\varphi + i * \varphi$ as

$$\varphi + i * \varphi = h_q(z_q) dz_q$$

where $h_q(z_q)$ is a holomorphic function of the local coordinate z_q around q. Hence

$$f(p) = u(p) + iv(p) = \int_0^{z_q(p)} h_q(z_q) dz + u(q) + iv(q)$$

is holomorphic, i.e. $f = u + iv$ is a holomorphic, but not necessarily single-valued analytic function on $\mathcal{R} - |\sigma|$.

Now $\varphi = d\Phi + \varphi_0$ on $U_0 - |\sigma|$, where φ_0 is a harmonic 1-form on U_0. Therefore, $\varphi_0 + i * \varphi_0 = h_0(z)dz$ is a holomorphic 1-form on U_0, hence

$$f_0(p) = \int^p (\varphi_0 + i * \varphi_0) = \int_0^{z(p)} h_0(z) dz$$

is a single-valued holomorphic function on U_0. Since

$$\varphi + i * \varphi = d\Phi + i\, d\Psi + \varphi_0 + i * \varphi_0$$

i.e. $df = dS(z) + df_0$, we have $f = S(z) + f_0 + c_0$ on $U_0 - |\sigma|$, where c_0

is a constant. Rewriting $f_0 + c_0$ as f_0, where f_0 is holomorphic on U_0, we have

$$f = S(z) + f_0 \tag{6.101}$$

on $U_0 - |\sigma|$, i.e. f is an analytic function on $\mathcal{R}$ with the same singularities as $S(z)$.

Theorem 6.19. Let $S(z) = \Phi + i\Psi$ and $\psi \in \mathcal{L}_\Phi$ be given, and let φ be the harmonic 1-form mentioned in Theorem 6.16. Let $p_0 \in \mathcal{R} - |\sigma|$ be a fixed point and let γ be a piecewise smooth curve on $\mathcal{R}$ connecting p_0 and an arbitrary point $p \in \mathcal{R} - |\sigma|$ without intersecting $|\sigma|$. Putting

$$f(p) = u(p) + iv(p) = \int_\gamma^p (\varphi + i * \varphi) \tag{6.102}$$

we see that $f = f(p)$ is an analytic, in general multi-valued, function on $\mathcal{R}$ with the same singularities as $S(z)$. If γ is closed, then

$$\mathrm{Re}\int_\gamma df = \int_\gamma \varphi = \int_\gamma \psi. \tag{6.103}$$

Proof: We have proved already that $f = f(p)$ is an analytic function on $\mathcal{R}$ with the same singularities as $S(z)$. Since $\varphi = \psi + d\eta_\infty$, we have $\int_\gamma \varphi = \int_\gamma \psi$.

Now φ is a harmonic function on $\mathcal{G} - |\sigma_0|$ for some region $\mathcal{G} \supset K = [U_0]$. Selecting a C^∞ function ρ with $\operatorname{supp}\rho \subset \mathcal{G}$ such that $\rho = 1$ identically on K and putting $\psi = d(\rho\Phi)$, we have $\psi_0 = \psi - d\Phi = 0$ on $U_0 - |\sigma|$. Hence $\psi \in \mathcal{L}_\Phi$. If $\varphi = d(\rho\Phi + \eta_\infty)$ is the harmonic form corresponding to this ψ, then

$$u = \int \varphi = \rho\Phi + \eta_\infty$$

is a single-valued harmonic function on $\mathcal{R} - |\sigma|$, thus:

Corollary. If we take $d(\rho\Phi)$ for ψ, the real part $u = \mathrm{Re}\, f$ of the analytic function $f = f(p)$ is a single-valued harmonic function on $\mathcal{R} - |\sigma|$. That is

$$d(f(\Gamma(t, s))d\Gamma(t, s)) = 0.$$

From

$$d(f(\Gamma(t, s))d\Gamma(t, s)) = f'(\Gamma(t, s))d\Gamma(t, s) \wedge d\Gamma(t, s) = 0$$

it is obvious that a holomorphic function $f(z)$ of z satisfies this equation. Letting K be the rectangle

$$K = \{t + is : a \leqslant t \leqslant b, 0 \leqslant s \leqslant 1\}$$

Theorem 6.2 yields

$$\int_{\partial K} f(\Gamma(t, s))d\Gamma(t, s) = \int_K d(f(\Gamma(t, s))d\Gamma(t, s)) = 0$$

and the left-hand member of this equality equals

$$\int_a^b f(\Gamma(t, 0))\Gamma_t(t, 0)dt + \int_0^1 f(\Gamma(b, s))\Gamma_s(b, s)ds$$

$$- \int_a^b f(\Gamma(t, 1))\Gamma_t(t, 1)dt - \int_0^1 f(\Gamma(a, s))\Gamma_s(a, s)ds$$

$$= \int_{\gamma 0} f(z)dz + \int_{b\gamma} f(z)dz - \int_{\gamma 1} f(z)dz - \int_{a\gamma} f(z)dz$$

This yields (6.44).

A 1-form ϕ that can be represented as $\phi = du$ with u a continuously differentiable function is called an *exact differential form*. A continuously differentiable 1-form ϕ satisfying $d\phi = 0$ is called a *closed differential form*. Since $ddu = 0$ by (6.3), every continuously differentiable exact form ϕ is closed. Locally, the converse of this statement is valid too.

Theorem 6.20. Let the 1-form ϕ be continuously differentiable on the region Ω such that $d\phi = 0$. Let c be an aritrary point of Ω and choose $r > 0$ such that $U_r(c) \subset \Omega$. Then there exists a twice continuously differentiable function u such that $\phi = du$ on the disk $U_r(c)$.

7

The structure of Riemann surfaces

7.1 Planar* Riemann surfaces

a. Planar Riemann surfaces

Let $\mathcal{R}$ be a Riemann surface: by definition, $\mathcal{R}$ is a connected Hausdorff space.

Theorem 7.1. Let C be a piecewise smooth Jordan curve on $\mathcal{R}$. Then $\mathcal{R} - C$ is either one region or the disjoint union of two regions.

Proof: Let $\mathfrak{U} = \{U_j\}$, $U_j = U_{r(j)}(q_j)$ be the locally finite open covering of (6.59). By (6.61), $U(C) = \bigcup_{q_j \in C} U_j$ is a region containing C and C divides $U(C)$ into two regions $U^+(C)$, to the left of C, and $U^-(C)$, to the right of C. If W is a connected component of the open set $\mathcal{R} - C$, then W has at least one boundary point, (because if $W = [W]$, then $\mathcal{R}$ could be written as the disjoint union of the open sets W and $\mathcal{R} - W$, contradicting the connectedness of $\mathcal{R}$). Since the boundary points of W are on C, $W \cap U(C) \neq \varnothing$, hence at least one of $W \cap U^+(C)$ and $W \cap U^-(C)$ is not empty. Suppose $W \cap U^+(C) \neq \varnothing$; then $U^+(C) \subset W$ since $U^+(C) \subset \mathcal{R} - C$ and U is connected. Similarly, if $W \cap U^-(C) \neq \varnothing$, then $U^-(C) \subset W$. Hence there are only three possibilities:

(1) $U^+(C) \subset W$, $U^-(C) \subset W$,
(2) $U^+(C) \subset W$, $W \cap U^-(C) = \varnothing$,
(3) $U^-(C) \subset W$, $W \cap U^+(C) = \varnothing$.

In case (1), $\mathcal{R} - C = W$ is one region. In case (2), we write W^+ for W and

* The author used *schlichtartig*, which we have termed planar, this terminology benefiting from the fact that such surfaces can only exist in the plane, as will be proved later.

denote by W^- another connected component of $\mathcal{R} - C$ which, in this case, exists. We have $U^-(C) \subset W^-$, $W^- \cap U^+(C) = \emptyset$, hence

$$\mathcal{R} - C = W^+ \cup W^-, \qquad W^+ \cap W^- = \emptyset.$$

Case (3) is treated similarly.

If $\mathcal{R} - C$ consists of two disjoint open sets, i.e. $\mathcal{R} - C = W^+ \cup W^-$, we say that the Jordan curve C *divides* $\mathcal{R}$ *into the two regions* W^+ and W^-.

Definition 7.1. If each piecewise smooth Jordan curve on $\mathcal{R}$ divides $\mathcal{R}$ into two regions, then $\mathcal{R}$ is called *planar.*

Theorem 7.2. A simply connected Riemann surface $\mathcal{R}$ is planar.

Proof: It suffices to prove that a piecewise smooth Jordan curve C on $\mathcal{R}$ divides $\mathcal{R}$ into two regions. Let $d\rho_C^+$ be the closed 1-form of class C^∞ used in (6.71) so ρ_C^+ is of class C^∞ on $U^+(C)$ and continuous on $U^+(C) \cup C$, while $\operatorname{supp} \rho_C^+ \subset U^+(C) \cup C$. There is a neighborhood V of C such that $\rho_C^+ = 1$ identically on $(U^+(C) \cup C) \cap V$. Now, let $\lambda : t \to \lambda(t)$, $0 \leqslant t \leqslant 1$, be a smooth Jordan arc crossing C at q from left to right, such that $\lambda(1/2) = q$, $\lambda(t) \in U^+(C)$ if $0 \leqslant t < 1/2$ and $\lambda(t) \in U^-(C)$ if $1/2 < t \leqslant 1$. If $\mathcal{R} - C$ consists of one region, then we connect $\lambda(1)$ and $\lambda(0)$

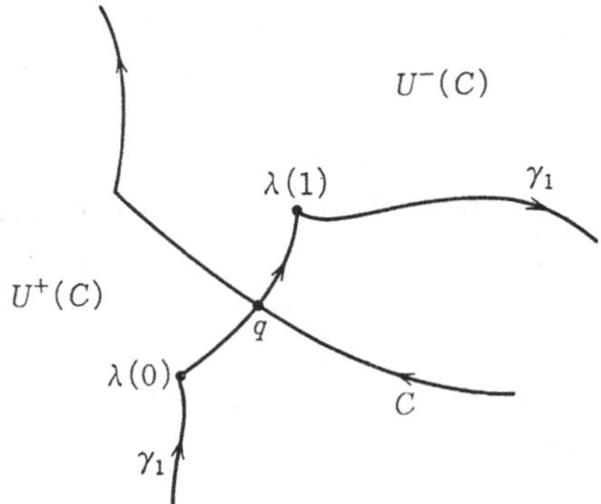

in $\mathcal{R} - C$ by a smooth curve γ, and $\gamma = \gamma_1 \cdot \lambda$ is then a piecewise smooth closed curve. Since $|\lambda| \subset V$ and $d\rho_C^+ = 0$ on V, we have:

$$\int_\gamma d\rho_C^+ = \int_{\gamma_1} d\rho_C^+ = \rho_C^+(\lambda(0)) - \rho_C^+(\lambda(1)).$$

Since $\lambda(0) \in U^+(C) \cap V$, we have $\rho_C^+(\lambda(0) = 1$ and since $\lambda(1) \in U^-(C)$, we have $\rho_C^+(\lambda(1)) = 0$, hence:

$$\int_\gamma d\rho_C^+ = 1. \tag{7.1}$$

On the other hand, since $\mathcal{R}$ is simply connected, we have $\gamma \simeq 0$. Hence $\int_\gamma d\rho_C^+ = 0$ (by Theorem 6.18) and so C does divide $\mathcal{R}$ into two regions.

Since the complex plane $\mathbb{C}$ is simply connected, it is planar. Hence a piecewise smooth Jordan curve C divides $\mathbb{C}$ into two regions. One of those regions is bounded, the other is not. The bounded region is called the *interior* of C, the unbounded region is the *exterior* of C.

Corollary 1. A piecewise smooth Jordan curve C in the complex plane $\mathbb{C}$ divides $\mathbb{C}$ into two regions: one called the interior, and the other the exterior of C.

Since the Riemann sphere **S** is also simply connected, it is planar. Obviously, regions on a planar Riemann surface are planar, hence:

Corollary 2. All regions on the Riemann sphere are planar.

We want now to prove the converse of Corollary 2, that is, a schlichtartig Riemenn surface is (biholomorphically equivalent to) a region on the Riemann sphere. This result is also valid for arbitrary, i.e. not necessarily piecewise smooth, Jordan curves. In that form the theorem is known as the *Jordan Curve Theorem*.

A closed set $F \in S$ is said to be *connected* if it is impossible to find two disjoint, nonempty closed sets F_1 and F_2 such that $F = F_1 \cup F_2$. Denoting by F_p the union of all connected closed subsets F containing the point $p \in F$, F_p is also a connected closed subset of F. We call F_p the connected component of F containing p and F is the union of finitely or infinitely many mutually disjoint connected components $F_p, F_q, F_r, \ldots$.

Let $\mathcal{R}$ be a planar Riemann surface, $S(z) = \Phi + i\Psi = 1/z + z$, and $\psi = d(\rho\Phi)$ and consider the analytic function

$$f = f(p) = u(p)iv(p) = \int_\gamma^p (\varphi + i * \varphi) \tag{7.2}$$

(compare Theorem 6.19). Here, $z : p \longrightarrow z(p)$ is a local coordinate around q_0, i.e. $z(q_0) = 0$ and f has the same singularities as $S(z)$. Considering $f(p)$ as a function of the local coordinate $z = z(p)$, we write $f(z)$, that is on the coordinate disk U_0 with center q_0, we have

$$f(z) = \frac{1}{z} + z + f_0(z), \qquad f_0(z) \text{ holomorphic on } U_0. \tag{7.3}$$

Thus $f(z)$ is holomorphic on $\mathcal{R} - \{q_0\}$ and has a pole of the first order at q_0. Hence (by the Corollary of Theorem 6.19) $u = \operatorname{Re} f$ is a harmonic single-valued function on $\mathcal{R} - \{q_0\}$. Next, we prove that $v = \operatorname{Im} f$ is also a harmonic single-valued function on $\mathcal{R} - \{q_0\}$. In order to prove that $v = \int *\varphi$ is single-valued it suffices, by Theorem 6.14, Corollary 1, to show that $\int_C * \varphi = 0$ for all piecewise smooth Jordan curves C that do not pass through q_0. Since $\mathcal{R}$ is planar, C divides $\mathcal{R}$ into two regions W^+ and W^-. By changing the orientation of C, if necessary, we may assume $q_0 \in W^+$. Choosing the locally finite open covering $\mathfrak{U} = \{U_j\}$, $U_j = U_{r(j)}(q_j)$ of (6.59) in such a way that $q_0 \notin U_j$ if $q_j \in C$, we achieve $q_0 \notin \operatorname{supp} \rho_C^+$, hence, by (6.71),

$$\int_C * \varphi = \int_{\mathcal{R}} d\rho_C^+ \wedge *\varphi = (d\rho_C^+, \varphi).$$

Since $[W^+] = W^+ \cup C$, $\rho_C^+ = 1$ identically on $[W^+] \cap V$ for some neighborhood V of C. Putting:

$$\eta(p) = \begin{cases} \rho_C^+(p), & p \in [W^+], \\ 1, & p \in W^-, \end{cases}$$

$\eta = \eta(p)$ is a C^∞ function on $\mathcal{R}$, $q_0 \notin \operatorname{supp} \eta$ and $d\eta = d\rho_C^+$; hence $\|d\eta\| > +\infty$. Since in this case, $|\sigma| = q_0$,

$$\int_C * \varphi = (\varphi, d\rho_C^+) = (\varphi, d\eta) = 0$$

by Theorem 6.17, (iii). Hence $v = \int *\varphi$ is a harmonic single-valued function on $\mathcal{R} - \{q_0\}$. We have proved that $f = u + iv$ is a holomorphic, single-valued function on $\mathcal{R} - \{q_0\}$ with a pole of the first order at q_0 and for this function, we have the following theorem.

Theorem 7.3. The function $f : p \to f(p)$ is a biholomorphic mapping of $\mathcal{R}$ onto a region $\Omega \subset \mathbf{S} = \mathbb{C} \cup \{\infty\}$ while $f(q_0) = \infty$. If $\mathcal{R}$ is compact, then $\Omega = \mathbf{S}$. If $\mathcal{R}$ is not compact, then $F = \mathbf{S} - \Omega$ is a bounded closed set in $\mathbb{C}$; F is either a point or consists of connected components, which are segments parallel to the real axis of $\mathbb{C}$.

Proof: (1) We prove that for arbitrary constants v_0 and $v_1 (v_0 < v_1)$ the open set $\mathrm{G} = \{p : v_0 < v(p) < v_1\}$ is connected in $\mathcal{R}$. We first investigate the shape of G near the pole q_0 of f. Since

$$\frac{1}{f(p)} = \frac{1}{f(z)} = \frac{z}{1 + zf_0(z) + z^2}$$

in a neighborhood of q_0 by (7.3), it is possible to use $1/f(p)$ as a local coordinate around q_0 instead of $z = z(p)$. So we define a new local coordinate by

$$z : p \to z(p) = \frac{1}{f(p)}.$$

Then on a neighborhood of q_0 we have

$$f = f(p) = u(p) + iv(p) = \frac{1}{z}, \qquad z = z(p).$$

Assuming that the range of the single-valued, holomorphic function $z(p)$ contains the closed disk $\{z : |z| \leqslant a\}$, $a > 0$, we redefine U_0 as $U_0 = \{p : |z(p)| < a\}$. In order to investigate the shape of G on U_0, we first consider $U_0 = \{z : |z| < a\}$ as a disk in the z-plane and we consider the function $w = u + iv = 1/z$ on the z-plane. The fractional linear transformation $w \to z = 1/w$ maps the line $v = v_0$ of the extended w-plane onto Γ_0, which is a circle or a line of the extended z-plane through 0. Hence

$$u \to z = \frac{1}{u + iv_0}, \qquad -\infty < u < +\infty,$$

is a parametric representation of $\Gamma_0 - \{0\}$. If $v_0 \neq 0$, then $u = \infty$ corresponds to $z = 0$, $u = 0$ to $z = -i/v_0$ and $u = v_0$ to $z = (1 - i)/2v_0$; hence Γ_0 is the circle through 0, $-i/v_0$, and $(1 - i)/2v_0$. Its center is the point $-i/2v_0$ on the imaginary axis. Since $0 \prec -i/v_0 \prec (1 - i)/2v_0$ with respect to the natural orientation of Γ_0, the point z, starting from 0, moves along Γ_0 in the positive direction if $v_0 > 0$ and in the negative direction if $v_0 < 0$ as the parameter u increases from $-\infty$ to $+\infty$. If $v_0 = 0$, then Γ_0 coincides with the real axis of the z-plane. The part of the curve on $\mathcal{R}$ defined by the equation $v(p) = v_0$ contained in U_0 is $\Gamma_0 \cap U_0$. Similarly, Γ_1 is defined in terms of v_1. If $v_1 \neq 0$, then Γ_1 is the circle through 0 with center $-i/2v_1$ in the z-plane and if $v_1 = 0$ then Γ_1 is the real axis. The part of the curve defined by the equation $v(p) = v_1$ contained in U_0 is $\Gamma_1 \cap U_0$. Therefore $G \cap U_0$ is the part of the disk U_0 contained between Γ_0 and Γ_1.

Assume that $R > 1/a$. Putting $z = x + iy$, we have $u = \mathrm{Re}(1/z) = x/(x^2 + y^2)$, hence the inequality $u \leqslant -R$ defines the closed disk K with center $-1/2R$ and radius $1/2R$ in the z-plane. Obviously, $K \subset U_0$ and q_0 is on the circumference of K. We put

$$G_R = G - K = \{p | u(p) > -R, \qquad v_0 < v(p) < v_1\}.$$

Now, let us assume that the open set G is not connected. Then G_R is not connected. There is only one connected component of G_R with q_0 as a

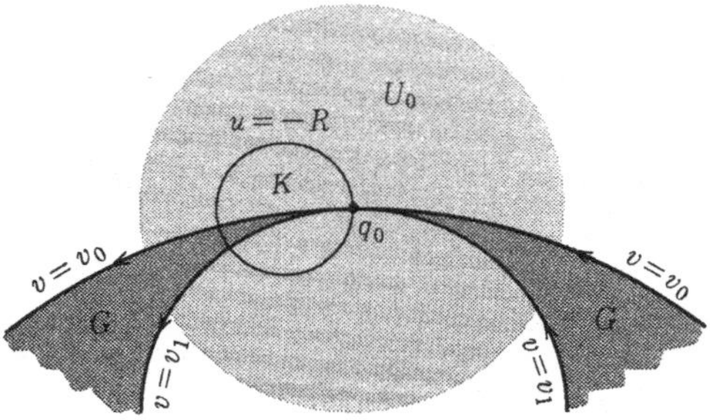

boundary point, so there exists a connected component W of G_R such that $q_0 \notin [W]$. Determine C^∞ functions $g(u)$ of u and $h(v)$ of v satisfying:

$$\begin{cases} g(-R) = g'(-R) = g''(-R) = 0, \\ |g(u)| < 1,\ 0 < g'(u) < 1, \text{ if } u \neq -R \end{cases}$$

and

$$\begin{cases} h(v) = 0, & \text{if } v \leqslant v_0 \\ 0 < h(v) \geqslant 1,\ |h'(v)| < 1, & \text{if } v_0 < v < v_1 \\ h(v) = 0, & \text{if } v \geqslant v_1 \end{cases}$$

respectively and put

$$\eta(p) = \begin{cases} g(u(p))h(v(p)), & p \in W, \\ 0, & p \notin W. \end{cases}$$

Now $\eta = \eta(p)$ is a twice continuously differentiable function on $\mathcal{R}$: this is easily seen as follows. Since $u = u(p)$ and $v = v(p)$ are harmonic on $\mathcal{R} - \{q_0\}$, $g(u)h(v)$ is a C^∞ function on $\mathcal{R} - \{q_0\}$. Since the boundary point p of the connected component W of G_R is a boundary point of $G_R : u(p) = -R$ or $v(p) = v_0$ or $v_1(p) = v$, hence $g(u)h(v) = 0$ on the boundary of W. Thus, since $q_0 \notin [W]$, η is continuous on $\mathcal{R}$. Considering $g(u(p))h(v(p))$ as a function of the local coordinates $x = x(p)$ and $y = y(p)$ of p, we have

$$\frac{\partial}{\partial x}(g(u)h(v)) = g'(u)h(v)u_x + g(u)h'(v)v_x,$$

$$\frac{\partial^2}{\partial x \partial y}(g(u)h(v)) = g'(u)h(v)u_{xy} + g(u)h'(v)v_{xy} + g''(u)h(v)u_y u_x$$
$$+ g'(u)h'(v)(u_x v_y + u_y v_x) + g(u)h''(v)v_x v_y$$

Further, since

$$g(-R) = g'(-R) = g''(-R) = 0,$$
$$h(v_0) = h'(v_0) = h''(v_0) = 0, \qquad h(v_1) = h'(v_1) = h''(v_1) = 0,$$

the first and second order partial derivatives of $g(u)h(v)$ vanish on the boundary of W. Hence η is twice continuously differentiable on $\mathcal{R}$.

Since $\operatorname{supp}\eta \subset [W]$, $q_0 \notin \operatorname{supp}\eta$. Further, $du = \varphi$ and $dv = *\varphi$, hence:

$$\begin{aligned} d\eta &= g'(u)h(v)\varphi + g(u)h'(v) * \varphi, \\ *d\eta &= g'(u)h(v) * \varphi - g(u)h'(v)\varphi. \end{aligned}$$

Therefore

$$\|d\eta\|^2 = \int_{\mathcal{R}} d\eta \wedge *d\eta = \int_W (g'(u)^2 h(v)^2 + g(u)^2 h'(v)^2)\varphi \wedge *\varphi.$$

Further, $\varphi \wedge *\varphi = (\varphi_1^2 + \varphi_2^2)dx \wedge dy$, $g'(u)^2 h(v)^2 + g(u)^2 h'(v)^2 < 2$ (from the definition of $g(u)$ and $h(v)$), $q_0 \notin [W]$ and $\|\varphi\|_\Phi < +\infty$. Therefore

$$\|d\eta\|^2 \leqslant 2\int_W \varphi \wedge *\varphi < +\infty.$$

Hence, by Thereom 6.17 (iii), $(\varphi, d\eta) = 0$ while also

$$(\varphi, d\eta) = \int_{\mathcal{R}} \varphi \wedge *d\eta = \int_W g'(u)h(v)\varphi \wedge *\varphi.$$

Now $g'(u) > 0$ if $u \neq -R$ and $h(v) > 0$ on W, hence if $(\varphi, d\eta) = 0$, then $g'(u) = 0$ identically on W, that is $u = -R =$ constant. If u is constant, then v is also constant by the Cauchy–Riemann equations, hence $f(p)$ is constant on W. Since this is a contradiction, we conclude that G is connected.

(2) The open set $G^+ = \{p : v(p) > v_0\}$ is connected, for, selecting v_1 such that $v_1 > 1/a$, we have $\Gamma_1 \subset U_0$, hence $G^+ = G \cup \{p \in U_0 : v(p) > v_0\}$. Similarly it is seen that $G_1^- = \{p : v(p) < v_1\}$ is also connected. By replacing v_1 by v_0, it is clear that $G^- = \{p : v(p) < v_0\}$ is connected as well.

(3) We prove that the function f is biholomorphic on a neighborhood of each point $q \neq q_0$ of $\mathcal{R}$. Because of Theorem 3.2, it suffices to prove that $f_q'(0) \neq 0$, where we consider $f(p)$ as a function of the local coordinate z_q around q and write: $f(p) = f_q(z_q)$ and $z_q = z_q(p)$. Assuming that $f_q'(0) = 0$, we have

$$f_q(z_q) = c_0 + z_q^m(c_m + c_{m+1}z_q + \cdots), \qquad c_m \neq 0, \qquad m \geqslant 2.$$

Since $z_q(c_m + c_{m+1}z_q + \cdots)^{1/m}$ is a single-valued holomorphic function of z_q, it is possible to use this expression as a local coordinate around q instead of z_q. Denoting this new local coordinate again by z_q, we have on a neighborhood of q:

$$f = f_q(z_q) = c_0 + z_q^m. \tag{7.4}$$

Putting $z_q = re^{i\theta}$ and $r = |z_q|$ and equating the imaginary parts of the left- and right-hand side of (7.4) we get:

$$v = v_0 + r^m \sin(m\theta), \qquad v_0 = \operatorname{Im} c_0, \qquad m \geqslant 2. \tag{7.5}$$

Now put $U_\delta(q) = \{p : |z_q(p)| < \delta\}$ for some sufficiently small δ. By (7.5), the curve determined by the equation $v(p) = v_0$, consists, on a neighborhood of q, of m segments passing through q, dividing the disk $U_\delta(p)$ into $2m$ sectors:

$$S_k = \{p | z_q(p) = re^{i\theta}, \quad 0 < r < \delta, \quad (k-1)\pi/m < \theta < k\pi/m\},$$
$$k = 1, 2, \ldots, 2m.$$

By (7.5), on $S_1, S_3, \ldots, S_{2n-1}, \ldots$ we have $v(p) > v_0$ and on $S_2, S_4, \ldots, S_{2n}, \ldots$ we have $v(p) < v_0$, i.e. $S_{2n-1} \subset G^+$ and $S_{2n} \subset G^-$. Next, in each sector S_k we pick a point p_k and we connect q with p_k by a line segment λ_k. Since G^+ is connected (see (2)) and p_1 and p_3 belong to G^+, by Theorem 6.15, there is a piecewise analytic Jordan arc γ_1 connecting p_1 and p_3. Selecting γ_1 in such a way that is intersects the polygonal line $\lambda_3^{-1} \cdot \lambda_1$ only in p_1 and p_3, $C = \gamma_1 \cdot \lambda_3^{-1} \cdot \lambda_1$ is a piecewise analytic Jordan curve. Since $\mathcal{R}$ is planar, C divides $\mathcal{R}$ into two regions W^+ and W^-. Since

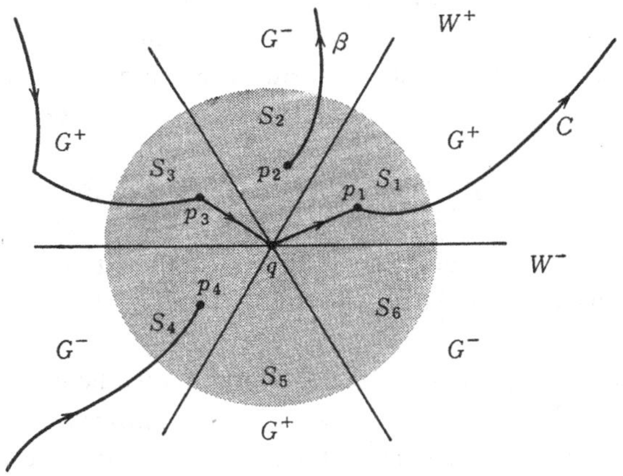

p_2 and p_4 belong to G^-, it is possible to connect p_2 and p_4 within G^- by a curve β. Since $|C| \subset G^+ \cup \{q\}$, β does not intersect C. This contradicts the fact that $p_2 \in W^+$ and $p_4 \in W^-$. Hence $f'_q(0) \neq 0$, proving that f is biholomorphic on a neighborhood of each point $q \neq q_0$ of $\mathcal{R}$.

(4) We now show that $f : p \to w = f(p)$, mapping $\mathcal{R}$ onto a region of S, is injective. Since q_0 is the only pole of q_0, it suffices to show that if p_1 and p_2 are two points differing from q_0 and if $p_1 \neq p_2$, then $f(p_1) \neq f(p_2)$. Let us assume the contrary; that is

$$f(p_1) = f(p_2) = w_2 = u_2 + iv_2, \qquad p_1 \neq p_2,$$

Since f is biholomorphic on neighborhoods of p_1 and p_2, we know, by (3), that there is a disk $U_\varepsilon = \{w : |w - w_2| < \varepsilon\}$, that determines disks U_1 and U_2 with centers p_1 and p_2 respectively which are mapped biholomorphically onto U_ε by f: that is $f(U_1) = f(U_2) = U_\varepsilon$. Of course, we may assume that f is biholomorphic on a region containing $[U_1]$ and on a region containing $[U_2]$. Let

$$l : t \to w = l(t) = w_2 + it, \qquad -\varepsilon \leqslant t \leqslant \varepsilon,$$

be the diameter of U_ε determined by the equation $u = u_2$ and let $\alpha = l(-\varepsilon) = u_2 + iv_0$ $(v_o = v_2 - \varepsilon)$ be its starting point and $\omega = l(\varepsilon) = u_2 + iv$, $(v_1 = v_2 + \varepsilon)$ it end point. The lines $v = v_0$ and $v = v_1$, parallel to the real axis in the ω-plane, pass through α and ω respectively. We denote the diameter of U_1 and U_2 corresponding to l under f^{-1} by l_1 and l_2 respectively, the starting points of l_1 and l_2 by α_1 and α_2 respectively, and

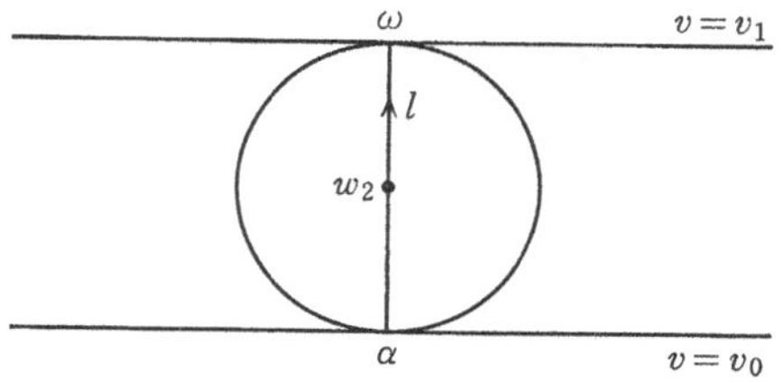

the end points of l_1 and l_2 by ω_1 and ω_2 respectively. Put

$$G = \{p | v_0 < v(p) < v_1\}, \qquad G_1^+ = \{p | v(p) > v_1\},$$

$$G_0^- = \{p | v(p) < v_0\}.$$

As $v(\omega_1) = v(\omega_2) = v_1$ and the open set G_1^+ is connected, (see (2)), it is possible to connect ω_1 and ω_2 by a piecewise analytic Jordan arc γ_1, which is in G_1^+ with the exception of its end points, (Theorem 6.15). Since $v(\alpha_1) = v(\alpha_2) = v_0$, it is also possible to connect α_2 and α_1 by a piecewise analytic Jordan arc γ_0 which is in G_0^- with the exception of its end points. Now, $\Gamma = \gamma_1 \cdot l_2^{-1} \cdot \gamma_0 \cdot l_1$ is a piecewise analytic Jordan curve on $\mathcal{R}$. Hence Γ divides the planar Riemann surface $\mathcal{R}$ into two regions W^+ and W^-. Since $q_0 \notin \Gamma$, q_0 belongs either to W^+ or to W^-. Let us denote the set from among W^+ and W^- not containing q_0 by W'', and put $G'' = G \cap W''$. Since $f(l_1) = f(l_2) = l$, if $p \in |l_1|$ or $p \in |l_2|$, then $u(p) = u_2$. Hence, for a boundary point p of G'', we have $u(p) = u_2$ or $v(p) = v_0$ or $v(p) = v_1$. Selecting C^∞ functions $g(n)$ and $h(v)$ such that:

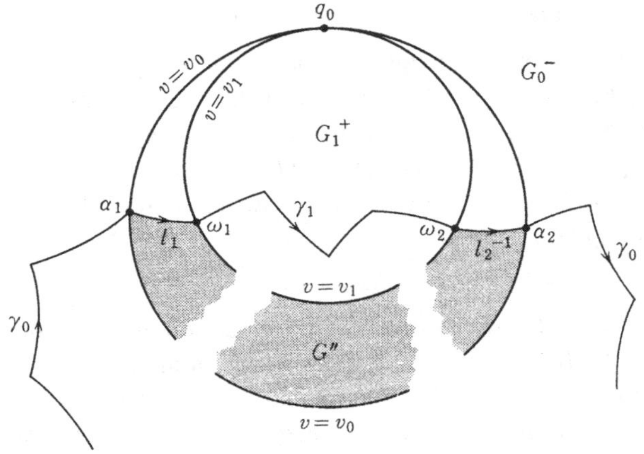

$$\begin{cases} g(u_1) = g'(u_2) = g''(u_2) = 0, & \\ |g(u)| < 1,\ 0 < g'(u) < 1, & \text{if } u \neq u_2 \end{cases}$$

$$\begin{cases} h(v) = 0, & \text{if } v \leqslant v_0 \\ 0 < h(v) < 1,\ |h'(v)| < 1, & \text{if } v_0 < v < v_1 \\ h(v) = 0, & \text{if } v \geqslant v_1, \end{cases}$$

we put:

$$\eta(p) = \begin{cases} g(u(p))h(v(p)), & p \in G'', \\ 0, & p \notin G''. \end{cases}$$

This function $\eta = \eta(p)$, just as the η of (1), is a twice differentiable function on $\mathcal{R}$, such that $q_0 \notin \operatorname{supp} \eta \subset [G'']$ and further $\|d\eta\| < +\infty$. Therefore, by Theorem 6.17 (iii), $(\varphi, d\eta) = 0$. Further,

$$(\varphi, d\eta) = \int_{G''} g'(u)h(v)\varphi \wedge *\varphi = \int_{G''} g'(u)h(v)(\varphi_1^2 + \varphi_2^2)dx\,dy$$

while $h(v) > 0$ on G''. Therefore $g'(u) = 0$ identically on G'', i.e. $u = u_2 = $ constant, contradicting the fact that $f = u + iv$ is not a constant. We conclude that $f : p \to f(p)$ is an injective map.

(5) Put $\Omega = f(\mathcal{R})$. Then Ω is a region on the Riemann sphere $\mathbf{S} = \mathbb{C} \cup \{\infty\}$ and $\infty = f(q_0) \in \Omega$, while f is a biholomorphic map mapping $\mathcal{R} - \{q_0\}$ onto $\Omega - \{\infty\}$. Considering $\mathbf{S}$ as a Riemann surface, a local coordinate around ∞ is given by $\hat{w} = 1/w$. Using this local coordinate, f can be represented as $f : z \to \hat{w} = z$ on a neighborhood of q_0. In this sense, f is also biholomorphic in a neighborhood of q_0, i.e. f is a biholomorphic map which maps $\mathcal{R}$ onto Ω. If $f = \mathbf{S} - \Omega$ is nonempty,

then F is a bounded closed set on $\mathbb{C}$. For arbitrary v_0 and v_1 with $v_0 < v_1$ the open set $G = \{p : v_0 < v(p) < v_1\}$ is connected, hence

$$\{w \in \Omega : v_0 < \operatorname{Im} w < v_1\} = f(G)$$

is also connected. In order to prove that all connected components of $F = \mathbb{C} - \Omega$ are either segments parallel to the real axis or single points, it suffices to prove that two points $w_0 = u_0 + iv_0$ and $w_1 = u_1 + iv_1$ satisfying $v_0 < v_1$, in F belong to different connected components of F. Putting $b = (v_0 + v_1/2)$ and denoting by C_R a circle with center ib and sufficiently large radius R, we have seen that F belongs to the interior of C_R. Denote the upper semicircle of C_R by C_R^+. Since $f(G) = \{w \in \Omega : v_0 < \operatorname{Im} w < v_1\}$ is connected, it is possible here, by Theorem 6.15, to connect the end points $-R + ib$ and $R + ib$ of C_R^+ in $f(G)$ by a piecewise analytic Jordan arc γ_1 intersecting C_R^+ only at its end points. Since $\gamma = \gamma_1 \cdot C_R^+$ is a

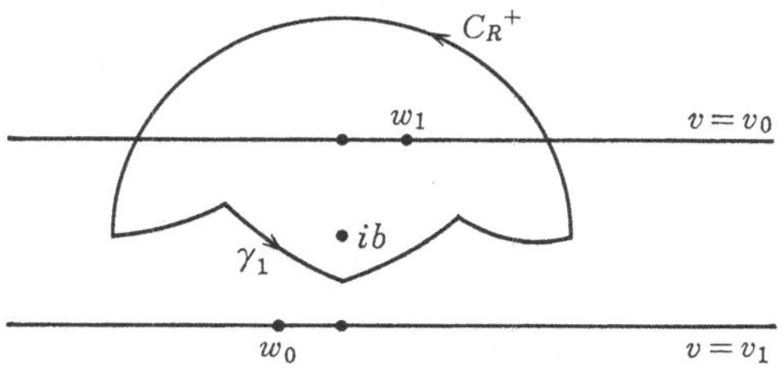

piecewise analytic Jordan curve, γ divides $\mathbb{C}$ into two regions, namely the interior V^+ and the exterior V^- of γ (Theorem 7.2, Corollary 1). Since obviously $|\gamma| \cap F = \varnothing$, we have $F \subset V^+ \cup V^-$, that if F is the union of two disjoint closed subsets $F^+ = F \cap V^+ = F - V^-$ and $F^- = F \cap V^- = F - V^+$. Since $w_1 \in F^+$ and $w_0 \in F^-$, we find that w_1 and w_0 belong to different connected components of F. Therefore each connected component of F is either a segment parallel to the real axis or a single point. In this case, $\mathcal{R} - f^{-1}(\Omega)$ is not compact.

If $F = \varnothing$, then $f(\mathcal{R}) = \Omega = \mathbf{S}$ and $\mathcal{R}$ is compact.

Since f is a biholomorphic map from $\mathcal{R}$ onto Ω, $\mathcal{R}$ and Ω can be identified as Riemann surfaces. If $F \neq \varnothing$ then each connected component of F is either a point or a horizontal segment, hence $\Omega = \mathbf{S} - F$ is called a *horizontal slit region*.

b. *Simply connected Riemann surfaces*

Let $\mathcal{R}$ be a simply connected Riemann surface. Since simply connected Riemann surfaces are planar (Theorem 7.2) compactness of $\mathcal{R}$ implies that $\mathcal{R}$ is the Riemann sphere S by the above result.

If $\mathcal{R}$ is not compact, then $= f(\mathcal{R}) = \mathbf{S} - F$ is a horizontal slit region, where f is the biholomorphic mapping of Theorem 7.3. The simply connected region $\Omega = \mathbf{S} - F$ is a region without holes (Definition 5.1), hence F is connected, as proved in Section 5.1e. Since in this case Ω is a horizontal slit region, that fact can also easily be verified directly as follows. Assume that F is not connected, then F has at least two different connected components F_1 and F_2. F_1 and F_2 are either single points or segments parallel to the real axis. Selecting points $\alpha_1 \in F_1$ and $\alpha_3 \in F_2$ we put:

$$\varphi + i * \varphi = \frac{1}{2\pi i}\left(\frac{dw}{w-\alpha_2} - \frac{dw}{w-\alpha_1}\right).$$

Using the local coordinate $\hat{w} = 1/w$ around $\infty \in \Omega$, φ can be represented as

$$\varphi + i * \varphi = \frac{(\alpha_1 - \alpha_2)d\hat{w}}{2\pi i(1-\alpha_2\hat{w})(1-\alpha_1\hat{w})};$$

hence φ is a harmonic 1-form on Ω. Since Ω is simply connected, all closed curves γ contained in Ω satisfy $\gamma \simeq 0$. Hence by the Monodromy Theorem, 6.18

$$\int_\gamma \varphi = 0. \tag{7.6}$$

If F is contained in one line parallel to the real axis, we may assume that F is on the real axis, $0 \notin F$, $F_1 \subset \mathbb{R}^-$ and $F_2 \subset \mathbb{R}^+$. Determine a disk U_R with center 0 and radius R such that $F \subset U_R$ and denote the right half of

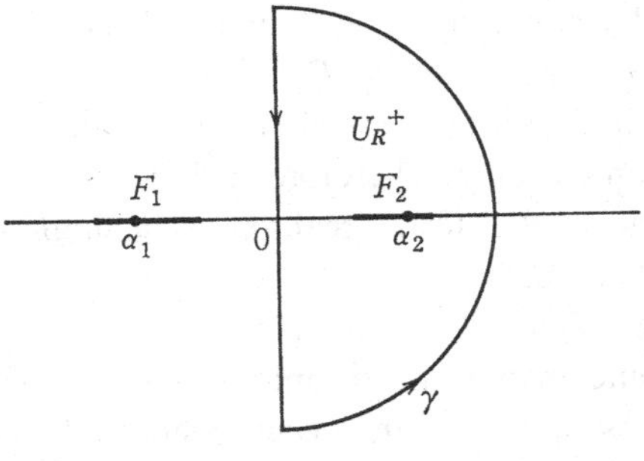

U_R by U_R^+. Putting $\gamma = \partial[U_R^+]$, F_1 is in the interior, and F_2 in the exterior, of γ. Hence, by Cauchy's Theorem and the Integral Formula we have:

$$\int_\gamma \varphi = \operatorname{Re}\frac{1}{2\pi i}\left(\int_\gamma \frac{dw}{w-\alpha_2} - \int_\gamma \frac{dw}{w-\alpha_1}\right) = 1.$$

Since $\gamma \subset \Omega$, this contradicts (7.6).

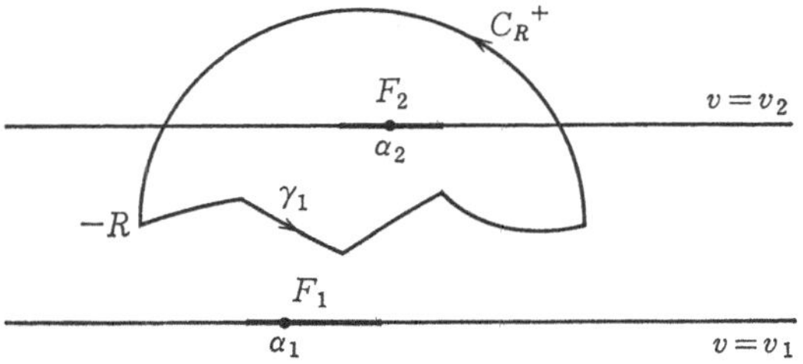

If F is not contained in one line parallel to the real axis, we may assume that F_1 and F_2 are on different lines. Let F_1 be on the line $v = v_1$ and F_2 on the line $v = v_2$ and assume $v_1 < 0 < v_2$. Determine a circle C_R with center 0 and radius R such that F is contained in the interior of C_R and denote the upper half of C_R by C_R^+. By the first part of the proof of Theorem 7.3, the open set $G = \{w \in \Omega : v_1 < \operatorname{Im} w < v_2\}$ is connected. Hence $G \cap U_R(0)$ is also connected. Hence it is possible to connect the points $-R$ and R with a piecewise analytic curve γ_1, which with the exception of its endpoints $-R$ and R, is contained in $G \cap U_R(0)$ · $\gamma = \gamma_1 \cdot C_R^+$ is a piecewise analytic Jordan curve in Ω, and F_1 is in the exterior and F_2 in the interior of γ. Hence $\int_\gamma \varphi = 1$ by Cauchy's Theorem and the Integral Formula. This contradicts (7.6). Hence F is a connected closed set, i.e. F is a segment parallel to the real axis or a single point.

If $F = \{\alpha\}$, the transformation $w \to z = 1/(w - \alpha)$ maps $\Omega = \mathbf{S} - F$ onto the z-plane. Hence, in this case $\mathcal{R}$ is biholomorphically equivalent to the z-plane. If F is the segment connecting α and $\beta = \alpha + \delta$, $\delta > 0$, the transformation $w \to 2/\delta(w - (\alpha + \beta)/z)$ carries F into the segment $[-1, 1]$. Hence we may assume from the begining that $F = [-1, 1]$. The transformation $w \to (w+1)/(w-1)$ maps -1 onto 0, $(-1, 1)$ on $\mathbb{R}^-$ and 1 onto ∞, hence $\mathbf{S} - F$ onto $\mathbb{C} - \mathbb{R}^- \cup \{0\}$. Let $\sqrt{\ }$ denote the principal value of the square root. The map $w \to \zeta = \sqrt{(w+1)/(w-1)}$ is biholomorphic map, from $\mathbf{S} - F$ onto the right half ζ-plane. Since $\zeta \to z = (\zeta - 1)/(\zeta + 1)$ is a conformal mapping the right half ζ-plane onto the unit disk $\mathbf{U}$ of the z-plane, $w \to z = (\zeta - 1)/(\zeta + 1)$ is a biholomorphic map from $\mathbf{S} - F$ onto $\mathbf{U}$. A simple calculation yields

$$z = w - \sqrt{w^2 - 1}. \tag{7.7}$$

(here too, $\sqrt{\ }$ denotes the principal value). Therefore, in this case $\mathcal{R}$ is biholomorphically equivalent to the unit disk $\mathbf{U}$. We have proved:

Theorem 7.4. A simply connected Riemann surface $\mathcal{R}$ is biholomorphically equivalent to the Riemann sphere **S**, the complex plane $\mathbb{C}$ or the unit disk **U**.

Corollary. (Riemann's Mapping Theorem). If the region is simply connected and $D \subset \mathbb{C}$, $D \neq \mathbb{C}$, then there exists a conformal mapping D onto the unit disk U.

Proof: Since D is simply connected Riemann surface, and obviously not compact, D is biholomorphically equivalent to $\mathbb{C}$, or **U**. Suppose that D is biholomorphically equivalent to $\mathbb{C}$, then there exists a biholomorphic map $f : z \to w = f(z)$ mapping $\mathbb{C}$ onto D. Now $f(z)$ is an entire function, but not transcendental. To see this, let $U_\varepsilon(0)$, $\varepsilon > 0$, be an ε-neighborhood of $0 \in \mathbb{C}$ and put $c = f(0)$. Then $W = f(U_\varepsilon(0))$ is a neighborhood of c by Theorem 3.2, and $f(z) \notin W$ if $|z| \geqslant \varepsilon$, contradicting Weierstrass' Theorem (Theorem 1.23). Hence $f(z)$ is a polynomial of z and so $D = f(\mathbb{C}) = \mathbb{C}$ by the Fundamental Theorem of Algebra, contrary to out assumption. We conclude that D is biholomorphically equivalent to **U**. Theorem 7.4 can be considered as an extension of Riemann's Mapping Theorem. It is also possible to derive Theorem 5.3 from Theorem 7.3:

Theorem 7.5. Let D be a region on the Riemann sphere **S** and let $A = S - D$ be a connected closed set. Let $z_0 \in D$. If A does not consist of a single point, there exists a biholomorphic map $f : z \to w = f(z)$ mapping D onto the unit disk **U** and satisfying $f(z_0) = 0$.

Proof: There exists a biholomorphic map $f : z \to w = f(z)$ mapping D onto the horizontal slit region $\Omega = \mathbb{C} \cup \{\infty\} - F$ satisfying $f(z_0) = \infty$ (using Theorem 7.3 with $\mathcal{R} = D$ and $q_0 = z_0$). In order to show that F consists of only one segment, we assume that F has at least two connected components F_1 and F_2. As above, it is proved that there exists a piecewise analytic Jordan curve $\gamma \subset \Omega$, such that F_1 is contained in the exterior and F_2 in the interior of γ. Since f is biholomorphic $f^{-1}(\gamma) \subset D = \mathbf{S} - A$ is a piecewise analytic Jordan curve. Since **S** is planar, by Theorem 7.2, $f^{-1}(\gamma)$ divides **S** into two regions W^+ and W^-. Since A is a connected closed set, A is contained in W^+ or W^-. Assuming that $A \subset W^+$, $f^{-1}(\gamma)$ divides the region D into two open sets $W^+ - A$ and W^-, hence $W^+ - A$, by Theorem 7.2, has to be a region also. Hence γ divides $\Omega = f(D)$ into two regions $f(W^+ - A)$ and $f(W^-)$. On the other hand, γ divides the Riemann sphere $\mathbb{C} \cup \{\infty\}$ into two regions V^+ and V^-. Hence $f(W^-) = V^+ - F$ or

$f(W^-) = V^- - F$. Since both cases are treated similarly we assume $f(W^-) = V^- - F$. Since $[W^-] = W^- \cup f^{-1}(|\gamma|)$ we have:

$$f([W^-]) = f(W^{-1}) \cup |\gamma| = (V^- \cup |\gamma|) - F$$

and, since $[W^-]$ is compact, by Theorem 6.7 $f([W^-])$ is also compact, that is $f([W^-])$ is closed. Hence:

$$(V^- \cup |\gamma|) - F = [(V^- \cup |\gamma|) - F] = V^- \cup |\gamma|,$$

and therefore $V^- \cap F = \varnothing$. This contradicts the fact that $F_1 \subset V^-$ or $F_2 \subset V^-$. Hence F consists of one connected component, that is F is a segment or a point.

If F is a single point, then $\Omega = \mathbb{C} \cup \{\infty\} - F$ and $\mathbb{C}$ are biholomorphically equivalent, while on the other hand, by moving A by a fractional linear transformation in such a way that $\infty \in A$, the regions $D = \mathbf{S} - A \subset \mathbb{C}$ and $\mathbb{C}$ are biholomorphically equivalent. Hence $A = \{\infty\}$ consists of only one point, contradicting our assumption. Hence F consists of one segment. Assuming $F = [-1, 1]$, the biholomorphic map (7.7)

$$h : w \to h(w) = w - \sqrt{w^2 - 1}$$

maps $\Omega = \mathbb{C} \cup \{\infty\} - F$ onto the unit disk and the point ∞ onto 0. Hence $h \circ f$ is a biholomorphic mapping which maps D onto the unit disk U and z_0 onto 0.

c. Multiply connected regions

Define f as in Theorem 7.3: $\Omega = f(\mathcal{R}) = \mathbf{S} - F$. If the planar Riemann surface $\mathcal{R}$ is not simply connected, F has at least two connected components. In this case, Ω, is called *multiply connected* and the number of connected components of F is called the *connectivity* of Ω. In this section we want to investigate the case of finite connectivity.

Let m be the connectivity of Ω and let $F_1, F_2, \ldots, F_k, \ldots, F_m$ denote the connected components of F. Each connected component F_k is either a segment parallel to the real axis or a single point.

If all connected components F_k are points β_k in the w-plane $\mathbb{C}$, the transformation

$$w \to z = \frac{1}{w - \beta_1}$$

maps Ω onto the region

$$D = \mathbb{C} - \{\alpha_2, \alpha_3, \ldots, \alpha_m\}$$

where $\alpha_k = 1/(\beta_k - \beta_1)$, $k = 2, 3, \ldots, m$. Hence $\mathcal{R}$ and D are biholomorphically equivalent. Putting

$$z(p) = \frac{1}{f(p) - \beta_1},$$

$z : p \to z(p)$ defines a global coordinate on $\mathcal{R}$, that is a coordinate that can be used on the whole of $\mathcal{R}$.

If at least one of the F_k, say F_1, is a segment, we assume $F_1 = [-1, 1]$ and apply the transformation (7.7):

$$w \to z = h(w) = w - \sqrt{w^2 - 1}$$

which maps $\mathbf{S} - F_1$ onto the unit disk of the z-plane and Ω onto the region

$$D = \mathbf{U} - A_2 \cup A_3 \cup \cdots \cup A_m$$

where $A_k = h(F_k)$, $k = 2, \ldots, m$. If F_k is a segment, $A_k = h(F_k)$ is an analytic Jordan arc, and if F_k is a point, obviously $h(F_k)$ is also a point. Further $\mathcal{R}$ and D are biholomorphically equivalent.

Solving the equation $z = h(w)$ for w, we put

$$w = h^{-1}(z) = g(z) = \frac{1}{2}\left(\frac{1}{z} + z\right). \tag{7.8}$$

In order to investigate the properties of the biholomorphic map h^{-1} which maps $\mathbf{U}$ onto $\mathbf{S} - F_1$, $F_1 = [-1, 1]$, we put $z = re^{i\theta}$, $0 < r < 1$. Then

$$w = u + iv = \frac{1}{2}\left(\frac{1}{r} + r\right)\cos\theta - \frac{1}{2}\left(\frac{1}{r} - r\right)i\sin\theta, \tag{7.9}$$

i.e.

$$u = \frac{1}{2}\left(\frac{1}{r} + r\right)\cos\theta, \qquad v = \frac{1}{2}\left(\frac{1}{r} - r\right)\sin(-\theta). \tag{7.10}$$

Elimination of θ yields

$$\frac{u^2}{a^2} + \frac{v^2}{b^2} = 1, \qquad a = \frac{1}{2}\left(\frac{1}{r} + r\right), \qquad b = \frac{1}{2}\left(\frac{1}{r} - r\right). \tag{7.11}$$

for fixed r, (7.11) is the equation of an ellipse in the w-plane and the biholomorphic map $g = h^{-1}$ maps the circle $|z| = r$ of the z-plane onto this ellipse. If the point $z = re^{i\theta}$ traverses the circle $|z| = r$ once, starting from r in the positive direction, the direction corresponding with increasing θ, the point $w = u + iv$ traverses the ellipse once in the negative direction starting from a. For brevity, we will write $r \uparrow 1$ if r increases monotonically to 1 and $a \downarrow 1$ is a decreases monotonically to 1. If $r \uparrow 1$, then $a \downarrow 1$ and $b \downarrow 0$ and the ellipse (7.11) "shrinks monotonically" and converges towards $F_1 = [-1, 1]$. Denoting the interior of the ellipse (7.11) by E_b, h^{-1} maps the region $\{z : r < |z| < 1\}$ onto $E_b - F_1$.

Elimination of r from the two formulas (7.10) yields

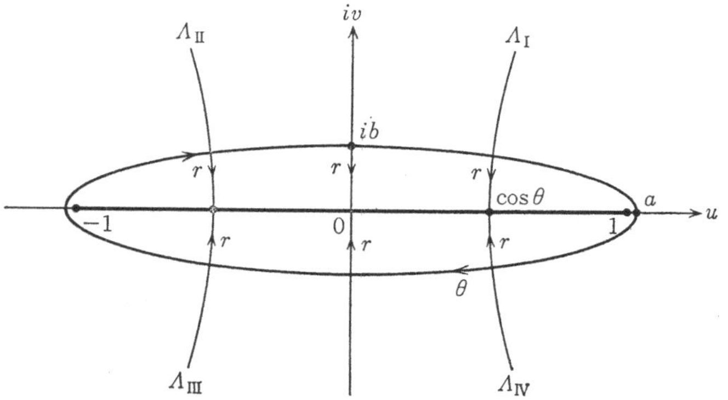

$$\frac{u^2}{\cos^2\theta} - \frac{v^2}{\sin^2\theta} = 1.$$

For fixed $\theta \neq n(\pi/2)$, where n is an integer this is the equation of a hyperbola in the w-plane. Denote this hyperbola by λ and denote by λ_I, λ_{II}, λ_{III} and λ_{IV} the parts of λ in the first, second, third and fourth quadrant respectively. For $0 < \theta < \pi/2$ we have $u \downarrow \cos\theta$ and $v \uparrow 0$ is $r \uparrow 1$ by (2.10), hence h^{-1} maps the radius $r \to z = re^{i\theta}$, $0 < r < 1$, of U into λ_{IV}. If r increases from 0 to 1, $w = u + iv$ traverses λ_{IV} from ∞ to $\cos\theta$. Similarly, if r increases from 0 to 1, $w = u + iv$ traverses λ_{III} if $\pi/2 < \theta < \pi$, λ_{II} if $\pi < \theta < 3\pi/2$ and λ_I if $3\pi/2 < \theta < 2\pi$, in each case from ∞ to $\cos\theta$. For $\theta = \pi/2$ we have $w = iv$ and $v \downarrow 0$ if $r \downarrow 1$, and for $\theta = 3\pi/2$ we have $w = iv$ and $v \downarrow 0$ if $r \uparrow 1$. For $\theta = 0$, we have $w = u \downarrow 1$ is $r \uparrow 1$, and for $\theta = \pi$, we have $w = u \uparrow -1$ if $r \uparrow 1$.

The biholomorphic map g which maps the unit disk **U** onto $\mathbf{S} - F_1$ can be extended in a natural way to a continuous map

$$\tilde{g} : z \to w = \tilde{g}(z) = \frac{1}{2}\left(\frac{1}{z} + z\right) \tag{7.12}$$

which maps [**U**] onto **S**. On $\partial[\mathbf{U}]$ we have $\tilde{g}(1) = 1$, $\tilde{g}(-1) = -1$ and $\tilde{g}(e^{i\theta}) = \tilde{g}(e^{-i\theta}) = \cos\theta$, hence $\tilde{g}$ maps $\partial[\mathbf{U}] - \{1, -1\}$ one-to-one onto the open interval $(-1, 1) = F_1 - \{-1, 1\}$.

Since $g(z) = \frac{1}{2}(1/z + z)$ is invariant under the transformation $z \to 1/z$, $g : z \to w = g(z)$ is a biholomorphic map which maps the exterior $\mathbb{C} \cup \{\infty\} - [\mathbf{U}]$ of the closed unit disk in the extended z-plane onto $S - F_1$ and $\tilde{g}$ is a continuous map which maps $\mathbb{C} \cup \{\infty\} - \mathbf{U}$ onto S. Now $g(\infty) = \infty$ and $g : \hat{z} = 1/z \to \hat{w} = 1/w = 2\hat{z}/(1 + \hat{z}^2)$ is biholomorphic, and therefore conformal, on a neighborhood of ∞. In this sense g is a conformal mapping which maps $\mathbb{C} \cup \{\infty\} - [\mathbf{U}]$ onto $\mathbf{S} - F_1$ and which

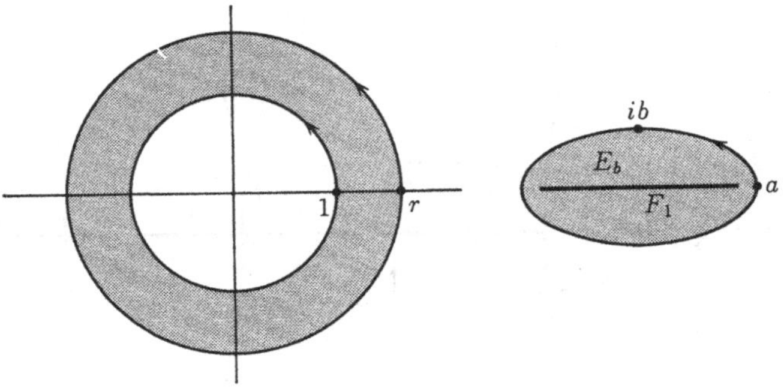

The z-plane The w-plane

can be extended to the continuous map $\tilde{g}$. If $r > 1$, then $z \to 1/z$ maps the circle $|z| = r$ onto the circle $|z| = 1/r$ reversing the orientation, hence, if z traverses the circles $|z| = r$ in the positive direction, $w = g(z)$ traverses the ellipse $\partial[E_b]$ (here $b = (r - 1/r)/2$) of (7.11) in the positive direction also. Further, g maps the region $\{z : 1 < |z| < r\}$ conformally onto the region $E_b - F_1$. The extension $\tilde{g}$ of g maps $\mathbb{C} \cup \{\infty\} - \mathbf{U}$ onto $\mathbf{S}$ and the closed region $\{z : 1 \leqslant |z| \leqslant r\}$ onto $[E_b]$. Solving $w = g(z) = (z + \frac{1}{z})/2$ for z we get: $z = w + \sqrt{w^2 - 1}$. Hence

$$h = g^{-1} : w \to z = h(w) = w + \sqrt{w^2 - 1} \tag{7.13}$$

is a conformal map mapping of $\mathbf{S} - F_1$ onto $\mathbb{C} \cup \{\infty\} - [\mathbf{U}]$ (here $\sqrt{\ }$ denotes the principal value of the square root). Obviously, $h(\infty) = \infty$.

Lemma 7.1. If W is a connected open set in $\mathbb{C}$ and A a piecewise smooth Jordan arc in W, then $W - A$ is also a connected open set.

Proof: The proof is similar to the proof of Theorem 7.1. Let a be the starting point and b the end point of the Jordan arc A. As explained in the beginning of Section 6.3e, for each $c \in A$, $c \neq a$, $c \neq b$, it is possible to find a (sufficiently small) $r(c) > 0$ such that A divides the disk $U_{r(c)}(c) \subset W$ into two regions: $U^+_{r(c)}(c)$, to the left of A and $U^-_{r(c)}(c)$ to the right of A. Further, it is possible to determine $r(a) > 0$ such that the circle $\lambda_r : \theta_r \lambda_r(\theta) = a + re^{i\theta}$, $0 \leqslant \theta \leqslant 2\pi$, with $0 < r \leqslant r(a)$ intersects A at only one point $\lambda_r(\alpha(r))$ (see Section 2.1b). Here, $\alpha(r)$, $0 < r \leqslant r(a)$, is a continuously differentiable function and we put $\alpha(0) = \lim_{r \to +0} \alpha(r)$ (see Section 5.2). Hence $U_{r(a)}(a) - A \subset W$ is one region. Similarly for the end point b, pick $r(b) > 0$ such that $U_{r(b)}(b) - A \subset W$ is one region. Since A

is covered by these disks $U_{r(c)}(c)$ $(c \in A)$, it is also covered by a finite number of them:

$$A \subset U(A), \qquad U(A) = \bigcup_{j=0}^{m} U_j \subset W,$$

where $U_j = U_{r(j)}(c_j)$ and $r(j) = r(c_j)$ for $j = 0, 1, 2, \ldots, m$. In order to verify that the open set $U(A) - A$ is connected, we assume that $U_0 = U_{r(a)}(a)$ and we consider $A_j = A \cap U_j$. By renumbering $U_1, \ldots, U_j, \ldots, U_m$, if necessary, we can obtain

$$(A_0 \cup A_1 \cup \cdots \cup A_j) \cap A_{j+1} \neq \varnothing, \qquad j = 0, 1, 2, \ldots, m-1. \tag{7.14}$$

Indeed, if for some j

$$(A_0 \cup A_1 \cup \cdots \cup A_j) \cap A_k = \varnothing, \qquad k = j+1, j+2, \ldots, m,$$

the set B given by

$$B = A_0 \cup A_1 \cup \cdots \cup A_j = A - \bigcup_{k=j+1}^{m} U_k$$

is a closed subset of A. Now $C = A_{j+1} \cup \cdots \cup A_m$ is also a closed subset of A and $A = B \cup C$ and $B \cap C \neq \varnothing$, contradicting the connectedness of A.

Since the open set $U_0 - A$ is connected and $A_0 \cap A_1 \neq \varnothing$, we see that $(U_0 - A) \cup U_1^+$ and $(U_0 - A) \cup U_1^-$ are connected, hence

$$(U_0 \cup U_1) - A = (U_0 - A) \cup U_1^+ \cup U_1^-$$

is a connected open set. Since, if $A_j \cap A_k \neq \varnothing$ both $U_j^+ \cup U_k^+$ and $U_j^- \cup U_k^-$ are connected, we observe that

$$\begin{aligned}(U_0 \cup U_1 \cup \cdots \cup U_j) - A \\ = (U_0 - A) \cup U_1^+ \cup U_1^- \cup \cdots \cup U_j^+ \cup U_j^-\end{aligned}$$

and deduce that if $(U_0 \cup \cdots \cup U_j) - A$ is connected, so is $(U_0 \cup \cdots \cup U_j \cup U_{j+1}) - A$. Therefore, $U(A) - A$ is see by induction to be a connected open set. Since W is connected and $W \supset U(A)$, $W - A$ is also a connected open set.

Lemma 7.2. Let A be a piecewise smooth Jordan arc on the Riemann sphere $\mathbf{S}$ such that $\infty \notin A$. There exists a conformal mapping $h : w \to z = h(w)$ which maps $\mathbf{S} - A$ onto $\mathbb{C} \cup \{\infty\} - [\mathbf{U}]$ and ∞ onto ∞.

Proof: Since $\mathbf{S} - A$ is a region by Lemma 7.1, there exists a biholomorphic map $f : w \to z = f(w)$ which maps $\mathbf{S} - A$ onto $\mathbf{U}$ and ∞ onto 0. Putting $h(w) = 1/f(w)$ we obtain the desired map.

In what follows we will denote by $\mathbf{S} - [U_1] - [U_2] - \cdots - [U_m]$ for example the open set obtained by deleting the closed domains $[U_1], [U_2], \ldots, [U_m]$ from the Riemann sphere $\mathbf{S}$.

Theorem 7.6. If $\Omega = \mathbf{S} - F$ is a horizontal slit region of connectivity m such that all connected components $F_1, F_2, \ldots, F_m$ of F are segments, then Ω is conformally equivalent to a bounded region $D \subset \mathbb{C}$ of the following form:

$$D = \mathbb{C} \cup \{\infty\} - [U_1] - [U_2] - \cdots - [U_m], \qquad \infty \in U_1. \tag{7.15}$$

Each U_k is a simply connected region on the Riemann sphere $\mathbb{C} \in \{\infty\}$ and its boundary $|\Gamma_k| = [U_k] - U_k$ is an analytic Jordan curve while the closed regions $[U_1], [U_2], \ldots, [U_m]$ are mutually disjoint. Orienting all Γ_k in such a way that $\Gamma_k = -\partial[U_k]$, we have

$$\partial[D] = \Gamma_1 + \Gamma_2 + \cdots + \Gamma_m. \tag{7.16}$$

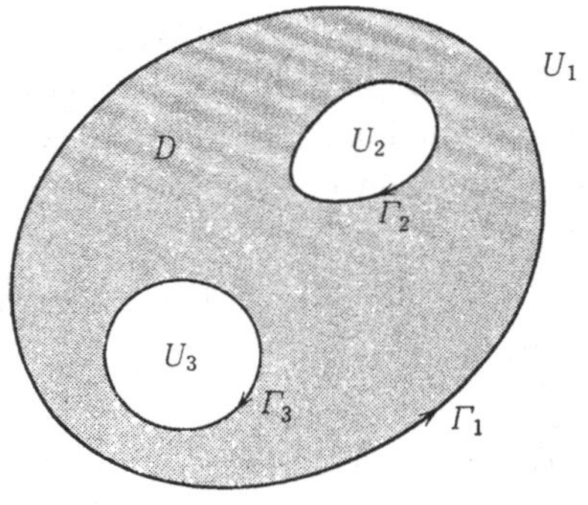

$m=3$

Proof: The proof is by induction on m.

(1) The case $m = 1$. Assuming $F = [-1, 1]$, the biholomorphic map (7.7)

$$h : w \to z = h(w) = w - \sqrt{w^2 - 1}$$

maps $\Omega = \mathbf{S} - F$ conformally onto the unit disk $\mathbf{U} \subset \mathbb{C}$. Putting $\hat{z} = 1/z$, $U_1 = \mathbb{C} \cup \{\infty\} - [\mathbf{U}]$ is the unit disk in the $\hat{z}$-plane. This unit disk is, of course, simply connected and its boundary $|\Gamma_1| = [U_1] - U_1$ is a circle. Hence, Ω is conformally equivalent to $D = \mathbf{U} = \mathbb{C} \cup \{\infty\} - [U_1]$.

(2) We assume the theorem is true for $m = n - 1$ and prove it for

$m = n$. By the induction hypothesis, there exists a conformal mapping g which maps $\mathbf{S} - F_1 - F_2 - \cdots - F_{n-1}$ onto

$$E = \mathbb{C} \cup \{\infty\} - [V_1] - [V_2] - \cdots - [V_{n-1}], \qquad \infty \in V_1.$$

Here, each V_k is a simply connected region on $\mathbb{C} \cup \{\infty\}$ and the boundaries $|B_k| = [V_k] - V_k$ are analytic Jordan curves, while $[V_j] \cap [V_k] = \emptyset$ if $j \neq k$. Now $A_n = g(F_n)$ is an analytic Jordan arc in E and g maps $\Omega = \mathbf{S} - F_1 - F_2 - \cdots - F_{n-1} - F_n$ onto $E - A_n$. By Lemma 7.2,

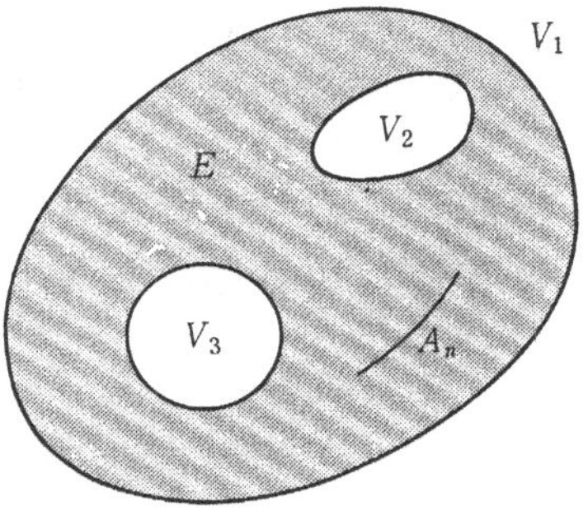

$n = 4$

there exists a conformal map $h : w \to z = h(w)$ mapping $\mathbb{C} \cup \{\infty\} - A_n$ onto $\mathbb{C} \cup \{\infty\} - [\mathbf{U}]$ and ∞ onto ∞. Since

$$E - A_n = \mathbb{C} \cup \{\infty\} - A_n - [V_1] - [V_2] - \cdots - [V_{n-1}],$$

h maps $E - A_n$ conformally onto

$$D = \mathbb{C} \cup \{\infty\} - [\mathbf{U}] - [U_1] - [U_2] - \cdots - [U_{n-1}],$$

where we have put $U_k = h(V_k)$. Obviously $\infty = h(\infty) \in h(V_1) = U_1$. Since V_k is simply connected, $U_k = h(V_k)$ is also simply connected. Since $|B_k| = [V_k] - V_k$ is an analytic Jordan curve, so is $|\Gamma_k| = [U_k] - U_k = h(|B_k|)$. Since $[V_j] \cap [V_k] = \emptyset$, $j \neq k$, $[U_j] \cap [U_k] = h([V_j]) \cap h([V_k]) = \emptyset$ and further, since $[V_j] \subset \mathbb{C} \cup \{\infty\} - A_n$, we have $[U_j] \subset \mathbb{C} \cup \{\infty\} - [\mathbf{U}]$; that is $[U_j] \cap [\mathbf{U}] = \emptyset$. Putting $U_n = \mathbf{U}$ we have $[U_j] \cap [U_n] = \emptyset$ for $j \neq n$, and

$$D = \mathbb{C} \cup \{\infty\} - [U_1] - [U_2] - \cdots - [U_n],$$

proving the theorem for $m = n$.

Corollary. Let $\Omega = \mathbf{S} - F$ be a horizontal slit region of connectivity $m + n$, let the connected components $F_1, F_2, \ldots, F_m$ of F be segments and connected components $F_{m+1}, F_{m+2}, \ldots, F_{m+n}$ points. Then Ω is conformally equivalent to a region D of the form

$$D = \mathbb{C} \cup \{\infty\} - [U_1] - \cdots - [U_m] - \{c_1, c_2, \ldots, c_n\},$$
$$\infty \in U_1,$$

where $c_1, c_2, \ldots, c_n$ are points.

7.2 Compact Riemann surfaces

a. *Cohomology groups*

Let $\mathcal{R}$ be an arbitrary Riemann surface. Let $\mathfrak{W} = \{W_m\}$ be a locally finite open covering consisting of coordinate disks W_m, let w_m be a local coordinate on W_m and let $\{\rho_m\}$ be a partition of unity subordinated to $\mathfrak{W}$. For $p, q \in \mathcal{R}$ we define the distance $d(p, q)$ between p and q by

$$d(p, q) = \sum_m (|\rho_m(p) - \rho_m(q)| + |\rho_m(p)w_m(p) - \rho_m(q)w_m(q)|).$$

Here we assume that $\rho_m(p)w_m(p)$ has been extended to a C^∞ function on $\mathcal{R}$ which vanishes whenever $p \notin W_m$. The distance $d(p, q)$ enjoys the following properties:

(i) $d(p, q)$ is a continuous function of p and q and $d(p, q) = d(q, p)$.
(ii) $d(p, p) = 0$ and $d(p, q) > 0$ if $p \neq q$.
(iii) $d(p, q) \leqslant d(p, r) + d(r, q)$ for three points p, q and r.

Proof: It is obvious that $d(p, q)$ satisfies (i) and (iii) and that $d(p, p) = 0$. If $d(p, q) = 0$, then $\rho_m(p) = \rho_m(q)$ for all m. Since $\sum_m \rho_m(p) = 1$, we have $\rho_m(p) = \rho_m(q) > 0$ for at least one m, i.e. p and q belong to the same W_m. Since further $\rho_m(p)w_m(p) = \rho_m(q)w_m(q)$, we have $w_m(p) = w_m(q)$. Hence $p = q$. In general, a Hausdorff space on which is defined a function satisfying (i), (ii) and (iii), is called a *metric space*, and $d(p, q)$ is called the *distance* between p and q. In this sense, a Riemann surface is a metric space. For a metric space Σ the ε-neighborhood of $q \in \Sigma$ is defined as $U_\varepsilon(q) = \{p : d(p, q) < \varepsilon\}$. A subset U of Σ is open if and only if for each $q \in U$ there exists an $\varepsilon > 0$ such that $U_\varepsilon(q) \subset U$. Therefore the distance function $d(p, q)$ determines a topology on Σ.

Lemma 7.3. It is possible to choose a locally finite open covering $\mathcal{U} = \{U_j\}$ of $\mathcal{R}$ consisting of coordinate disks U_j such that:

$$\text{if } U_j \cap U_k \neq \emptyset, \text{ then there exists a } W_m \in \mathfrak{W} \text{ such that } U_j \cup U_k \subset W_m. \tag{7.17}$$

Proof: For $q \in \mathcal{R}$ there are only finitely many $W_m \in \mathfrak{W}$ such that $q \in W_m$. Therefore

$$\text{if } q \in W_m, \quad d(p, q) < \delta(q) \text{ then } p \in W_m \tag{7.18}$$

for a sufficiently small $\delta(q) > 0$. Let z_q be a local coordinate around q and select a coordinate disk $U_{R(q)}(q) = \{p : |z_q(p)| < R(q)\}$ such that:

$$\text{if } p \in U_{R(q)}(q) \text{ then } d(p, q) < \frac{1}{2}\delta(q). \tag{7.19}$$

Selecting such a coordinate disk $U_{R(q)}(q)$ for all points $q \in \mathcal{R}$, it is possible, by Theorem 6.8, to choose a finite or infinite number of coordinate disks:

$$U_j = U_{r(j)}(q_j), \qquad 0 < r(j) \leqslant R(q_j), \qquad j = 1, 2, 3, \ldots,$$

such that $\mathfrak{U} = \{U_j : j = 1, 2, 3, \ldots\}$ is a locally finite open covering of $\mathcal{R}$. If $\mathcal{R}$ is compact, this is obvious. In order to verify that $\mathfrak{U}$ satisfies condition (7.17), we assume that $U_j \cap U_k \neq \varnothing$ and we pick an arbitrary point $p_1 \in U_j \cap U_k$. Assuming further that $\delta(q_j) \geqslant \delta(q_k)$, we have $q_j \in W_m$. Since $p_1 \in U_j = U_{r(j)}(q_j)$, $r(j) \leqslant R(q_j)$, we have by (7.19)

$$d(p_1, q_j) < \frac{1}{2}\delta(q_j),$$

and similarly

$$d(p_1, q_k) < \frac{1}{2}\delta(q_k).$$

Therefore

$$d(q_k, q_j) \leqslant d(q_k, p_1) + d(p_1, q_j) < \frac{1}{2}\delta(q_k) + \frac{1}{2}\delta(q_j) \leqslant \delta(q_j).$$

Hence by (7.18), $q_k \in W_m$. Assume $p \in U_j \cup U_k$. If $p \in U_j$, we have $d(p, q_j) < \delta(q_j)/2$ hence $p \in W_m$. If $p \in U_k$, as $q \in W_m$ and $d(p, q_k) < \delta(q_k)/2$, we have $p \in W_m$. Therefore $U_j \cup U_k \subset W_m$.

Let $\mathcal{L}$ be the collection of all closed 1-forms φ continuously differentiable on $\mathcal{R}$ and let $\mathcal{E}$ be the subset of $\mathcal{L}$ consisting of all exact differential forms du, where u is a twice continuously differentiable function on $\mathcal{R}$. Clearly, $\mathcal{L}$ is a vector space over $\mathbb{R}$ and $\mathcal{E}$ is a subspace of $\mathcal{L}$.

Definition 7.2. The quotient space $\mathcal{L}/\mathcal{E}$ is called the *first d-cohomology group* of the Riemann surface $\mathcal{R}$ and denoted by $H_d^{1\times 2}(\mathcal{R})$. The element in $H_d^1(\mathcal{R})$ corresponding to $\psi \in \mathcal{L}$ is called the *d-cohomology class* of ψ. If $\mathcal{R}$ is compact, $\mathcal{L}$ coincides with the space $\mathcal{L}_0$ obtained by taking for Φ in (6.79) the function that is identically 0. We choose a locally finite open covering $\mathfrak{U} = \{U_j\}$ satisfying:

$$\begin{array}{l}\text{each } U_j \text{ is a simply connected region and if } U_j \cap U_k \neq \varnothing, \\ \text{then there exists a simply connected region } W, \\ \text{depending on } U_j \text{ and } U_k \text{ such that } U_j \cup U_k \subset W.\end{array} \tag{7.20}$$

(Such an open covering exists by Lemma 7.3). If $\psi \in \mathcal{L}$, then by Theorem 6.18 we can write $\psi = du_j$, where u_j is a C^2 function on the simply connected region U_j. If $U_j \cap U_k \neq \varnothing$, then $c_{jk} = u_k - u_j$ is constant on $U_j \cap U_k$. To see this, consider a simply connected region W satisfying $W \supset U_j \cup U_k$. Then $\psi = du$ on W, where u is a C^2 function on W. Hence $d(u_j - u) = 0$ on U_j and therefore $u_j - u$ is constant. Similarly, we see that $u_k - u$ is constant on U_k.

$$c_{jk} = u_k - u_j = (u_k - u) - (u_j - u)$$

is constant. Writing $\psi = du_j$ on all U_j, $\psi \in \mathcal{L}$, and putting $c_{jk} = u_k - u_j$ on $U_j \cap U_k \neq \varnothing$, we have set up a correspondence between the real numbers c_{jk}, and all pairs (j, k), such that $U_j \cap U_k \neq \varnothing$, that is ψ determines a real-valued function c_{jk} defined on $\{(j, k) : U_j \cap U_k \neq \varnothing\}$. It is easily verified that two functions c_{jk} satisfies the following conditions:

(i) $c_{kj} = -c_{jk}$,
(ii) if $U_i \cap U_j \cap U_k \neq \varnothing$ then $c_{ij} + c_{jk} + c_{ki} = 0$.

A real-valued function c_{jk} defined on $\{(j, k) : U_j \cap U_k \neq \varnothing\}$ satisfying conditions (i) and (ii) is called a 1-*cocycle* (with real coefficients) on $\mathcal{U}$. If the 1-cocycle c_{jk} can be represented as

$$c_{jk} = c_k - c_j$$

where c_j is a function of j, then c_{jk} is said to be *cohomologous* to 0. The collection of all 1-cocycles on $\mathcal{U}$ is a vector space over $\mathbb{R}$ and denoted by $L(\mathcal{U})$. The collection of all 1-cocycles cohomologous to 0 is a subspace of $L(\mathcal{U})$ and denoted by $E(\mathcal{U})$.

Definition 7.3. The quotient space $L(\mathcal{U})/E(\mathcal{U})$ is called the *first cohomology group* of the open covering $\mathcal{U}$ and is denoted by $H^1(\mathcal{U})$. The element of $H^1(\mathcal{U})$ corresponding with the 1-cocycle $c_{jk} \in L(\mathcal{U})$ is called the *cohomology class* of c_{jk}.

Theorem 7.7. (de Rham's Theorem). Let $\mathcal{U} = \{U_j\}$ be a locally finite open covering of $\mathcal{R}$ satisfying condition (7.20). The d-cohomology group $H_d^1(\mathcal{R})$ of $\mathcal{R}$ and the cohomology group $H^1(\mathcal{U})$ of $\mathcal{U}$ are isomorphic as real vector spaces:

$$H_d^1(\mathcal{R}) \cong H^1(\mathcal{U}) \tag{7.21}$$

Proof: Assign to each closed 1-form $\psi \in \mathcal{L}$ a 1-cocycle $c_{jk} = u_k - u_j$ as described above. The C^2 function u_j on U_j satisfying $du_j = \psi$ is determined uniquely up to a constant, that is if $du'_j = du_j = \psi$, then $u'_j = u_j + c_j$ where c_j is a constant. Therefore

$$c'_{jk} = u'_k - u'_j = u_k - u_j + c_k - c_j = c_{jk} + c_k - c_j;$$

that is the cohomology class of the 1-cocycle c_{jk} is uniquely determined by ψ. Hence the correspondence $\psi \to c_{jk}$ yields a linear map from $\mathcal{L}$ into $H^1(\mathfrak{U}) = L(\mathfrak{U})/E(\mathfrak{U})$.

If $\psi \in \mathcal{E}$, then $\psi = du$ for some C^2 function u on $\mathcal{R}$. Hence $d(u_j - u) = 0$ and therefore $u_j = u + c_j$, c_j a constant, on U_j. Hence $c_{jk} = u_k - u_j = c_k - c_j$. Conversely, if $c_{jk} = c_k - c_j$, then $u_j - c_j = u_k - c_k$ on $U_j \cap U_k \neq \emptyset$, hence the function u defined by putting $u = u_j - c_j$ on all U_j is a C^2 function on $\mathcal{R}$ and $du = du_j = \psi$, i.e. $\psi \in \mathcal{E}$. Therefore, $\psi \to c_{jk}$ is a one-to-one linear mapping from $H^1_d(\mathcal{R}) = \mathcal{L}/\mathcal{E}$ into $H^1(\mathfrak{U})$. To prove that $H^1_d(\mathcal{R}) \cong H^1(\mathfrak{U})$ it suffices to show that for an arbitrary 1-cocycle $c_{jk} \in L(\mathfrak{U})$ there exists a closed 1-form ψ corresponding with c_{jk}. Let $\{\rho_i\}$ be a partition of unity subordinated to $\mathfrak{U} = \{u_i\}$ and define a C^∞ function $u_j = u_j(p)$ on each U_j by

$$u_j(p) = \sum_i \rho_i(p) c_{ij}, \qquad p \in U_j.$$

If $U_i \cap U_j \cap U_k \neq \emptyset$, then $c_{jk} = c_{ik} - c_{ij}$, by conditions (i) and (ii), hence

$$u_k(p) - u_j(p) = \sum_i \rho_i(p)(c_{ik} - c_{ij}) = \sum_i \rho_i(p) c_{jk} = c_{jk}$$

if $p \in U_j \cap U_k$. Hence $du_k = du_j$ on $U_j \cap U_k \neq \emptyset$. Defining $\psi = du_j$ on each U_j, we have found the desired $\psi \in \mathcal{L}$ corresponding with c_{jk}.

The isomorphism $H^1_d(\mathcal{R}) \cong H^1(\mathfrak{U})$ shows that $H^1(\mathfrak{U})$ is independent of the choice of $\mathfrak{U}$ provided the locally finite open covering $\mathfrak{U}$ satisfies condition (7.20). Since only the concept of continuity is needed for the formulation of condition (7.20), $H^1(\mathfrak{U})$ is uniquely determined by the topology of $\mathcal{R}$. Hence $H^1_d(\mathcal{R})$ is also uniquely determined by the topology of $\mathcal{R}$.

The dimension of $H^1_\alpha(\mathcal{R})$ as a vector space over $\mathbb{R}$ is called the *first Betti number* of $\mathcal{R}$ and denoted by $b_1 = b_1(\mathcal{R})$:

$$b_1 = b_1(\mathcal{R}) = \dim_{\mathbb{R}} H^1_d(\mathcal{R}). \tag{7.22}$$

Of course, $b_1 = b_1(\mathcal{R})$ is uniquely determined by the topology of $\mathcal{R}$, that is the Betti number $b_1 = b_1(\mathcal{R})$ is a topological invariant.

If $\mathcal{R}$ is compact, then $\mathfrak{U}$ is a finite covering and $L(\mathfrak{U})$ a finite dimensional vector space. Hence $H^1(\mathfrak{U}) = L(\mathfrak{U})/E(\mathfrak{U})$ is also finite dimensional. Hence:

Corollary. The d-cohomology group $H^1_d(\mathcal{R})$ of the compact Riemann surface $\mathcal{R}$ is a finite dimensional vector space over $\mathbb{R}$ and the Betti number $b_1(\mathcal{R})$ is finite.

b. Structure of compact Riemann surfaces

Let $\mathcal{R}$ be a compact Riemann surface. If $\mathcal{R}$ is schlichtartig, then by Theorem 7.3, $\mathcal{R}$ is (biholomorphically equivalent to) the Riemann sphere **S**. Hence we assume the $\mathcal{R}$ is not planar for the remainder of this section.

Since $\mathcal{R}$ is not planar, there is a piecewise smooth Jordan curve C_1 on $\mathcal{R}$, which does not divide $\mathcal{R}$ into two regions. Putting $\mathcal{R}' = \mathcal{R} - |C_1|$, $\mathcal{R}'$ is again a Riemann surface. If $\mathcal{R}'$ is not planar, there exists a piecewise smooth Jordan curve C_2 on $\mathcal{R}'$ which does not divide $\mathcal{R}'$ into two regions and $\mathcal{R}''' = \mathcal{R}' - |C_2|$ is again a Riemann surface. If $\mathcal{R}''$ is not planar, there exists a piecewise smooth Jordan curve C_3 on $\mathcal{R}''$ which does not divide $\mathcal{R}''$ into two regions and $\mathcal{R}''' = \mathcal{R}'' - |C_3|$ is again a Riemann surface. We now show that by *repeating this procedure, the process terminates after a finite number of steps at a planar* Riemann surface $\mathcal{R}^{(m)} = \mathcal{R}^{(m-1)} - |C_m|$, i.e.

$$\mathcal{R}^{(m)} = \mathcal{R} - |C_1| - |C_2| - \cdots - |C_m|. \tag{7.23}$$

Proof: By the corollary of de Rham's Theorem (Theorem 7.7), the Betti number $b_1 = b_1(\mathcal{R})$ of $\mathcal{R}$ is finite. Hence it suffices to show that if the open set $\mathcal{R}^{(m)} = \mathcal{R} - |C_1| - \cdots - |C_m|$, where $C_1, \ldots, C_m$ are piecewise smooth Jordan curves in $\mathcal{R}$, is connected, then $m \leqslant b_1$, Let $d(p, q)$ be the distance function introduced in Section 7.2a and for each $q \in \mathcal{R}$, select an $e(q) > 0$ satisfying:

$$\text{if } q \notin C_n, \text{ and } d(p, q) \leqslant e(q) \text{ then } p \notin C_n$$

(if $m = 1$, then we put $e(q) = 1$ for $q \in C_1$). Next we fix a local coordinate $z_q : p \to z_q(p)$ around q for each point q and we select (a sufficiently small) coordinate disk $U_{r(q)}(q) = \{p \in \mathcal{R}: |z_q(p)| < r(q)\}$, $r(q) > 0$, such that:

$$\text{if } p \in U_{r(q)}(q) \text{ then } d(p, q) < \tfrac{1}{2}e(q).$$

Then $[U_{r(q)}(q)] \cap C_n = \varnothing$ if $q \notin C_n$; also $[U_{r(q)}(q)] \cap [U_{r(s)}(s)] = \varnothing$ if $q \in C_l$ and $s \in C_n$, $n \neq l$. These follow because if $p \in [U_{r(q)}(q)] \cap [U_{r(w)}(s)]$, then $d(p, q) \leqslant e(q)/2$ are $d(p, s) \leqslant e(s)/2$. hence, assuming that $e(q) \geqslant e(s)$, we have $d(s, q) \leqslant e(q)$. Therefore, $s \notin C_n$ since $q \notin C_n$ contradicting the assumption.

Since $\mathcal{R}$ is compact, the open covering $\{U_{r(q)}(q): q \in \mathcal{R}\}$ has a finite subcovering

$$\mathfrak{U} = \{U_j | j = 1, 2, 3, \cdots\}, \qquad U_j = U_{r(j)}(q_j), \qquad r(j) = r(q_j).$$

If $q_j \notin C_n$ then $[U_j] \cap C_n = \emptyset$, hence, if $U_j \cap C_n \neq \emptyset$, then $q_j \in C_n$. Further, if $q_j \in C_l$ and $q_k \in C_n$, $n \neq l$, then $[U_j] \cap [U_k] = \emptyset$. Now (by (6.61)) C_n divides U_j, $q_j \in C_n$, into two regions: U_j^+ to the left and U_j^- to the right of C_n, so

$$U_j - C_n = U_j^+ \cup U_j^-, \qquad U_j^+ \cap U_j^- = \emptyset.$$

We put

$$U(C_n) = \bigcup_{q_j \in C_n} U_j, \quad U^+(C_n) = \bigcup_{q_j \in C_n} U_j^+, \quad U^-(C_n) = \bigcup_{q_j \in C_n} U_j^-;$$

$U(C_n)$ is a region containing C_n, and C_n divides $U(C_n)$ into two regions $U^+(C_n)$ and $U^-(C_n)$. If $n \neq h$, then $U(C_n) \cap U(C_h) = \emptyset$.

Now let $\{\rho_j\}$ be a partition of unity subordinated to $\mathfrak{U} = \{U_j\}$. For $q_j \in C_n$ we define

$$\rho_j^+(p) = \begin{cases} \rho_j(p), & p \in U_j^+ \cup C_n, \\ 0, & p \notin U_j^+ \cup C_n. \end{cases}$$

Further, we put

$$\rho_{C_n}^+(p) = \sum_{q_j \in C_n} \rho_j^+(p)$$

so $\rho_{C_n}^+$ is a C^∞ function on $\mathcal{R} - |C_n|$, and $\operatorname{supp} \rho_{C_n}^+ \subset U^+(C_n) \cup C_n$ and $\rho_{C_n}^+(p) = 1$ at all points sufficiently close to C_n. Hence $d\rho_{C_n}^+$ is a closed 1-form of class C^∞ on $\mathcal{R}$ and $\operatorname{supp} d\rho_{C_n}^+ \subset U^+(C_n)$. Choose $p_n \in C_n$ such that C_n is smooth at p_n and select an analytic Jordan arc $\lambda_n : t \to \lambda_n(t)$, $0 \leqslant t \leqslant 1$, intersecting C_n from left to right such that $\lambda_n(1/2) = p_n$, $\lambda_n(t) \in U^+(C_n)$ if $0 \leqslant t < 1/2$ and $\lambda_n(t) \in U^-(C_n)$ if $1/2 < t \leqslant 1$ (compare the proof of Theorem 7.2). By Theorem 6.15, it is possible to connect $\lambda_n(1)$ and $\lambda_n(0)$ by a piecewise analytic Jordan arc γ_{n1} in $\mathcal{R}^{(m)}$ such that $\gamma_n = \gamma_{n1} \cdot \lambda_n$ becomes a piecewise smooth Jordan curve. Now γ_n intersects C_n only at p_n and does not intersect C_h, $h \neq n$. Also, by (7.1) $\int_{\gamma_n} d\rho_{C_n}^+ = 1$. Since $\rho_{C_n}^+$ is a C^∞ function on $\mathcal{R} - |C_n|$ for $h \neq n$, (by (6.46)) $\int_{\gamma_n} d\rho_{C_n}^+ = 0$. Putting $\psi_h = d\rho_{C_n}^+$, we have

$$\int_{\gamma_n} \psi_h = \delta_{nh}. \tag{7.24}$$

From this, it follows that the d-cohomology classes of $\psi_1, \psi_2, \ldots, \psi_m$ are

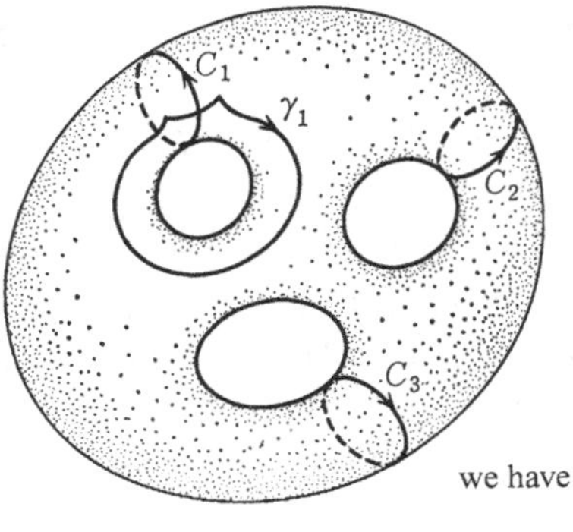

linearly independent over $\mathbb{R}$ (because, if $c_1\psi_1 + c_2\psi_2 + \cdots + c_m\psi_m = du$, where $c_1, \ldots, c_m$ are constants and u is a C^2 function on $\mathcal{R}$, then $c_n = \int_{\gamma_n} du = 0$). Hence $m \leqslant b_1 = \dim_{\mathbb{R}} H_d^1(\mathcal{R})$.

From the collection of all sets $\{C_1, C_2, \ldots, C_m\}$ of mutually disjoint piecewise smooth Jordan curves such that $\mathcal{R} - |C_1| - |C_2| - \cdots - |C_m|$ is a connected open set, we select one with the greatest number m of elements and put

$$\mathcal{R}^{(m)} = \mathcal{R} - |C_1| - |C_2| - \cdots - |C_m|.$$

Of course, $\mathcal{R}^{(m)}$ is planar. Since each $p \in C_n$ is a boundary point of $\mathcal{R}^{(m)}$, p can be represented as $p = \lim_{k\to\infty} p_k$, $p_k \in \mathcal{R}^{(m)}$. If the p_k tend to p from the left of C_n, we write $p^+ = \lim_k p_k$, $p_k \in U^+(C_n)$, and if the p_k tend to p from the right of C_n we write $p^- = \lim_k p_k$, $p_k \in U^-(C_n)$, and we consider p^+ and p^- as different boundary points of $\mathcal{R}^{(m)}$. By adding the boundary points p^+ and p^- to $\mathcal{R}^{(m)}$ we obtain a surface $\tilde{\mathcal{R}}$ that is different from $\mathcal{R}$. In order to define this surface $\tilde{\mathcal{R}}$ accurately we consider the function $\rho^+_{C_n}$. Now $\rho^+_{C_n}$ is a C^∞ function on $\mathcal{R} - |C_n|$ and $\operatorname{supp} \rho^+_{C_n} \subset U^+(C_n) \cup C_n$. Moreover, it is continuous on $U^+(C_n) \cup C_n$ and $\rho^+_{C_n}(p) = 1$ if $p \in C_n$. Observing that $U(C_h) \cap U(C_n) = \varnothing$ if $h \neq n$, we put

$$\rho^+ = \sum_{n=1}^{m} \rho^+_{C_n}$$

so ρ^+ is a C^∞ function on $\mathcal{R}^{(m)}$. Suppose $\lim p_k = p \in C_n$; then $\lim \rho^+(p_k) = 1$ if $p_k \in U^+(C_n)$ while $\lim \rho^+(p_k) = 0$ if $p_k \in U^-(C_n)$. We define a C^∞ map μ from $\mathcal{R}^{(m)}$ into $\mathcal{R} \times \mathbb{R}$ by

$$\mu : p \to (p, \rho^+(p)).$$

Let $\tilde{\omega}$ represent the projection from $\mathcal{R} \times \mathbb{R}$ onto $\mathcal{R}$, then $\tilde{\omega}\mu(p) = p$ hence

$\mathcal{R}^{(m)}$ and $\mu(\mathcal{R}^{(m)})$ are diffeomorphic. $\tilde{\mathcal{R}}$ is defined as the closure of $\mu(\mathcal{R}^{(m)})$ in $\mathcal{R} \times \mathbb{R}$

$$\tilde{\mathcal{R}} = [\mu(\mathcal{R}^{(m)})].$$

As $\tilde{\mathcal{R}}$ is a closed subset of the compact Hausdorff space $\mathcal{R} \times [0, 1]$, we see that $\tilde{\mathcal{R}}$ is compact. If $P \in \tilde{\mathcal{R}} - \mu(\mathcal{R}^{(m)})$, then $P = \lim_{k\to\infty}(p_k, \rho^+(p_k))$ where $p_k \in \mathcal{R}^{(m)}$ and $\lim_{k\to\infty} p_k = p = \tilde{\omega}(P)$. Suppose $p \in \mathcal{R}^{(m)}$; then $\lim \rho^+(p_k) = \rho^+(p)\rho^+(p)$, hence $P = (p, \rho^+(p)) \in \mu(\mathcal{R}^{(m)})$, contradicting our assumption. Hence $p \in C_n$ for some C_n. Since $\lim_{k\to\infty} \rho^+(p_k)$ exists, either $p_k \in U^+(C_n)$ for sufficiently large k and $\lim \rho^+(p_k) = 1$ or $p_k \in U^-(C_n)$ for sufficiently large k and $\lim \rho^+(p_k) = 0$. Therefore, $P = (p, 1)$ or $P = (p, 0)$, $p \in C_n$. Hence, putting $C_n^+ = C_n \times 1$ and

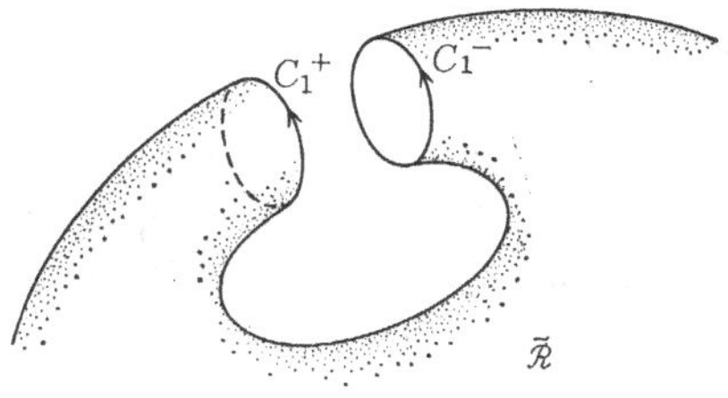

$C_n^- = C_n \times 0$, we have: Writing $P = (p, \rho^+(P))$ and $\tilde{\omega}(P)$ for $P \in \tilde{\mathcal{R}} = \mathcal{R} \times \mathbb{R}$, $\rho^+(P)$ is a continuous function on $\tilde{\mathcal{R}}$ and the distance between P and $Q \in \tilde{\mathcal{R}}$ is given by

$$\tilde{d}(P, Q) = \sqrt{d(p, q)^2 + |\rho^+(P) - \rho^+(Q)|^2},$$
$$p = \tilde{\omega}(P), \quad q = \tilde{\omega}(Q). \tag{7.25}$$

If $P \in \mu(\mathcal{R}^{(m)})$, then $\rho^+(P) = \rho^+(p)$; if $P = (p, 1) \in C_n^+$, then $\rho^+(P) = 1$ and if $P = (p, 0) \in C_n^-$ then $\rho^+(P) = 0$.

Since there is a one-to-one correspondence between points $P = \mu(p) \in \mu(\mathcal{R}^{(m)})$ and points $p = \tilde{\omega}(P) \in \mathcal{R}^{(m)}$, we can identify $P \in \mu(\mathcal{R}^{(m)})$ and $p = \tilde{\omega}(P) \in \mathcal{R}^{(m)}$. Thus, $\mu(\mathcal{R}^{(m)})$ and $\mathcal{R}^{(m)}$ become the same Riemann surface, only the distance function $d(p, q)$ has to be replaced by

$$\tilde{d}(p, q) = \sqrt{d(p, q)^2 + |\rho^+(p) - \rho^+(q)|^2}. \tag{7.26}$$

Hence

$$\tilde{\mathcal{R}} = \mathcal{R}^{(m)} \cup C_1^+ \cup C_1^- \cup \cdots \cup C_m^+ \cup C_m^-.$$

For $p \in C_n$ we put $p^+ = (p, 1)$ and $p^- = (p, 0)$. Then

$$C_n^+ = \{p^+ | p \in C_n\}, \qquad C_n^- = \{p^- | p \in C_n\},$$

and $\rho^+(p^+) = 1$ and $\rho^+(p^-) = 0$. Let $q \in C_n$. For sufficiently small $\varepsilon > 0$, if $d(p, q) < \varepsilon$ then $\rho^+(p) = 1$ for $p \in U^+ + (C_n)$ and $\rho^+(p) = 0$ for $p \in U^-(C_n)$. Hence, if $\mathcal{U}_\varepsilon(q) = \{p : d(p, q) < \varepsilon\}$ is an ε-neighborhood of q on $\mathcal{R}$, the ε-neighborhood $\tilde{\mathcal{U}}_\varepsilon(q^+) = \{p | \tilde{d}(p, q) < \varepsilon\}$ on $\tilde{\mathcal{R}}$ is given by

$$\tilde{\mathcal{U}}_\varepsilon(q^+) = (U^+(C_n) \cap \mathcal{U}_\varepsilon(q)) \cup \{p^+ \in C_n^+ | p \in \mathcal{U}_\varepsilon(q)\}. \tag{7.27}$$

Similarly

$$\tilde{\mathcal{U}}_\varepsilon(q^-) = (U^-(C_n) \cap \mathcal{U}_\varepsilon(q)) \cup \{p^- \in C_n^- | p \in \mathcal{U}_\varepsilon(q)\}. \tag{7.28}$$

Suppose that the sequence $\{p_k\}$, $p_k \in \mathcal{R}^{(m)}$, converges to q on $\mathcal{R}$. For sufficiently large k, if $p_k \in U^+(C_n)$ then $p_k \in \tilde{\mathcal{U}}_\varepsilon(q^+)$ and $p_k \notin \tilde{\mathcal{U}}_\varepsilon(q^-)$, while if $p_k \in U^-(C_n)$ then $p_k \in \tilde{\mathcal{U}}_\varepsilon(q^-)$ and $p_k \notin \tilde{\mathcal{U}}_\varepsilon(q^+)$. Hence q^+ is the limiting position if p_k tends to q from the left of C_n and q^- is the limiting position if p_k tends to q from the right of C_n. Since $\tilde{d}(q^+, q^-) = 1$, we have $q^+ \neq q^-$.

Let $q_0 \in \mathcal{R}^{(m)}$ be a fixed point. Since $\mathcal{R}^{(m)}$ is planar, by Theorem 2.3 there exists a biholomorphic map $f : p \to w = f(p)$, mapping $\mathcal{R}^{(m)}$ onto the horizontal slit region $\Omega = \mathbf{S} - F$ (where $\mathbf{S}$ is the Riemann sphere) such that $f(q_0) = \infty$.

Lemma 7.4. We can extend f to a continuous map $\tilde{f} : p \to w = \tilde{f}(p)$ mapping $\tilde{\mathcal{R}} = [\mathcal{R}^{(m)}]$ onto $\mathbf{S}$.

Proof: For sufficiently large f, put $U_{1/R}(q_0) = \{p \in \tilde{\mathcal{R}} : |f(p)| > R\}$ and $W_R = \{w \in \mathbb{C} : |w| < R\}$. Thus f maps $\mathcal{R}^{(m)} - U_{1/R}(q_0)$ onto $[W_R] - F$ conformally. It is sufficient to show that the conformal map

$$f : \mathcal{R}^{(m)} - U_{1/R}(q_0) \to [W_R] - F \tag{7.29}$$

can be extended to a continuous map

$$\tilde{f} : \tilde{\mathcal{R}} - U_{1/R}(q_0) \to [W_R]. \tag{7.30}$$

The proof of this is similar to the proof of Theorem 5.5. It suffices to prove that f is uniformly continuous on $\mathcal{R}^{(m)} - U_{1/R}(q_0)$ i.e. that if $\varepsilon > 0$ is arbitrarily given, then there exists a $\delta > 0$ such that

$$\text{if } \tilde{d}(p, q) < \delta(\varepsilon) \text{ then } |f(p) - f(q)| < \varepsilon.$$

Let us suppose that f is not uniformly continuous. Since $\tilde{\mathcal{R}} - U_{1/R}(q_0)$ is compact, there exists (by (2) of the proof of Theorem 5.5) some $\varepsilon_0 > 0$ and some sequences $\{p_k\}$ and $\{q_k\}$ in $\mathcal{R}^{(m)} - U_{1/R}(q_0)$ such that

$$\lim_{k\to\infty} p_k = \lim_{k\to\infty} q_k = \tilde{q}, \qquad \tilde{q} \in \tilde{\mathcal{R}},$$

$$\lim_{k\to\infty} f(p_k) = P, \qquad \lim_{k\to\infty} f(q_k) = Q, \qquad |P - Q| \geqslant \varepsilon_0.$$

Suppose $\tilde{q} \in \mathcal{R}^{(m)}$, then $f(\tilde{q}) = P$ and $f(\tilde{q}) = Q$, which is impossible since $|P - Q| \geqslant \varepsilon_0$. Hence $\tilde{q} \in \tilde{\mathcal{R}} - \mathcal{R}^{(m)}$, i.e. $\tilde{q} \in C_n^+$ or $\tilde{q} \in C_n^-$ for some n. Since both cases are treated similarly, we may assume $\tilde{q} = q^+ \in C_n^+$ and $q \in C_n$. Let $U(q)$ be a sufficiently small coordinate disk around q on $\mathcal{R}$ and let $U^+(q)$ be the part of $U(q)$ to the left of C_n. Since $\lim_k p_k = q^+$ on $\tilde{\mathcal{R}}$, we have by (7.27) $p_k \in U^+(q)$ with the exception of a finite number. Similarly, $q_k \in U^+(q)$ with the exception of a finite number. Hence the conformal mapping f maps $U^+(q)$ onto the bounded region $W_R - F$ and there are sequences $\{p_k\}$ and $\{q_k\}$ in $U^+(q)$ such that $\lim_k p_k = \lim_k q_k = q$, $\lim_k f(p_k) = P$, $\lim_k f(q_k) = Q$ and $|P - Q| \geqslant \varepsilon_0$. This contradicts the fact that the area of $f(U^+(q)) \subset W_R - F$ is bounded (compare (2) of the proof of Theorem 5.5). Hence f is uniformly continuous on $\mathcal{R}^{(m)} - U_{1/R}(q_0)$.

Since $\tilde{\mathcal{R}}$ is compact, $\tilde{f}(\tilde{\mathcal{R}})$ is, by Theorem 6.7, also compact and therefore closed. Hence $\tilde{f}(\tilde{\mathcal{R}}) \supset [\Omega] = \mathbf{S}$, that is $\tilde{f}(\tilde{\mathcal{R}}) = \mathbf{S}$.

Observe that $\tilde{f}(C_n^+) \subset F$. Indeed, if $q \in C_n^+$, we can write $q = \lim_{k\to\infty} p_k$ with $p_k \in \mathcal{R}^{(m)}$, hence $\tilde{f}(q) = \lim_k f(p_k)$ where $f(p_k) \in \Omega$. Now suppose that $\tilde{f}(q) \in \Omega$. Then as f^{-1} is holomorphic on Ω we have

$$q = \lim_k p_k = \lim_k f^{-1}(f(p_k)) = f^{-1}(\tilde{f}(q)) \in f^{-1}(\Omega) = \mathcal{R}^{(m)}$$

which contradicts the assumption that $q \in C_n^+$.

Hence, since $\tilde{f}(\tilde{\mathcal{R}}) = S$, we have

$$F = \tilde{f}(C_1^+) \cup \tilde{f}(C_1^-) \cup \cdots \cup \tilde{f}(C_m^+) \cup \tilde{f}(C_m^-).$$

In order to prove that the connected closed sets $\tilde{f}(C_1^+)$, $\tilde{f}(C_1^-)$, $\tilde{f}(C_2^+), \ldots, \tilde{f}(C_m^-)$ are all different connected components of F, we assume that for connected component F_1 we have, for example

$$\tilde{f}^{-1}(F_1) = C_1^+ \cup C_2^-.$$

Assuming that $F_1 = [-1, 1]$ and denoting the interior of the ellipse (7.11) by E_b, we have $E_b \supset F_1$ and $\bigcap_{0<b<1} [E_b] = F_1$. Hence

$$\bigcap_{0<b<1} \tilde{f}^{-1}([E_b]) = C_1^+ \cup C_2^-.$$

So $\tilde{f}^{-1}([E_b])$ is compact and $\tilde{f}^{-1}([E_c]) \supset \tilde{f}^{-1}([E_b])$ if $c > b > 0$. As

$$\tilde{U} = U^+(C_1) \cup C_1^+ \cup U^-(C_2) \cup C_2^- \supset C_1^+ \cup C_2^-$$

is an open set in $\tilde{\mathcal{R}}$, $\tilde{f}^{-1}([E_b]) \subset \tilde{U}$ for sufficiently small b. Therefore

$$f^{-1}(E_b - F_1) \subset \tilde{U} - C_1^+ \cup C_2^- = U^+(C_1) \cup U^-(C_2).$$

Since f^{-1} is holomorphic and the open set $E_b - F_1$ is connected, this contradicts the fact that $U^+(C_1) \cap U^-(C_2) = \varnothing$. We have assumed F_1 to be a segment; if F_1 is a point, we take for E_b a disk with center F_1 and radius b and arrive at a contradiction in a similar way.

Hence F consists of $2m$ connected components $F_n^+ = \tilde{f}(C_n^+)$ and $F_n^- = \tilde{f}(C_n^-)$, $n = 1, 2, \ldots, m$. Let F_n^+ be a segment of length $2l_n$ parallel to the real axis, let B_n be the ellipse with the midpoint of F_n^+ as center, major axis $2al_n$, and minor axis $2bl_n$, where $a = (1/r + r)/2$ and $b = (1/r - r)/2$, $r < 1$, and let $E_b(F_n^+)$ denote the interior of B_n. Denoting the midpoint of F_n^+ by c_n, B_n is given by

$$\theta \to w = c_n + \frac{l_n}{2}\left(\frac{1}{r}e^{-i\theta} + re^{i\theta}\right), \qquad 0 \leqslant \theta \leqslant 2\pi. \tag{7.31}$$

If F_n^+ is a point, then we let B_n be the circle with center F_n^+ and radius b and $E_b(F_n^+)$ the interior of the circle B_n. For sufficiently small b, we have $B_n \subset \Omega$, hence $\Gamma_n = f^{-1}(B_n)$ is an analytic Jordan curve on $\mathcal{R}^{(m)} \subset \mathcal{R}$ and $\Gamma_n \cap \Gamma_h = \varnothing$ for $n \neq h$.

Lemma 7.5. $\mathcal{R} - |\Gamma_1| - |\Gamma_2| - \cdots - |\Gamma_m|$ is a planar Riemann surface.

Proof: It suffices to show that the open set $\mathcal{R} - |\Gamma_1| - \cdots - |\Gamma_m|$ is connected. Let us suppose

$$\mathcal{R} - |\Gamma_1| - \cdots - |\Gamma_m| = U \cup V, \qquad U \cap V = \varnothing,$$
U, V open sets.

Since we assume that b is sufficiently small, the ellipses $B_1, B_2, \ldots, B_m$ divide the sphere $\mathbf{S}$ into $m+1$ regions:

$$\mathbf{S} - |B_1| - \cdots - |B_m| = E_b(F_1^+) \cup \cdots \cup E_b(F_m^+) \cup W.$$

Since $W'' = W - F_1^- - \cdots - F_m^- \subset \Omega$ is connected and f^{-1} is biholomorphic on Ω, $f^{-1}(W'')$ is a connected open set in $\mathcal{R}$. Hence $f^{-1}(W'')$ is contained in either U or V, say $f^{-1}(W'') \subset U$. Since all points of C_n^{-1} are boundary points of $f^{-1}(W'')$ on $\tilde{\mathcal{R}}$, all points $p \in C_n$ are boundary points of $f^{-1}(W'')$ on $\mathcal{R}$. If $p \notin U$, then p belongs to the boundary $|\Gamma_1| \cup \cdots \cup |\Gamma_m|$ of U, contradicting $p \in C_n$. Hence $C_n \subset U$. Since $E_b(F_n^+) - F_n^+ \subset \Omega$ is a connected open set, $f^{-1}(E_b(F_n^+) - F_n^+)$ is also a connected open set in $\mathcal{R}$. Since $C_n \subset U$, we conclude $f^{-1}(E_b(F_n^+) - F_n^+) \subset U$.

As

$$\mathcal{R} - |\Gamma_1| - \cdots - |\Gamma_m| = \bigcup_{n=1}^{m} (f^{-1}(E_b(F_n^+) - F_n^+) \cup C_n) \cup f^{-1}(W'')$$

we have $\mathcal{R} - |\Gamma_1| - \cdots - |\Gamma_m| = U$. Therefore, $\mathcal{R} - |\Gamma_1| - \cdots - |\Gamma_m|$ is connected.

Each Jordan curve Γ_n has a neighborhood of simple form in $\mathcal{R}$. To see this, put

$$\beta_n(z) = c_n + \frac{l_n}{2}\left(\frac{1}{rz} + rz\right).$$

The parametric representation (7.31) of the ellipse B_n now takes the form: $\theta \to w = \beta_n(e^{i\mathrm{H}})$. For sufficiently small $\varepsilon > 0$ we put

$$Z_\varepsilon = \left\{z \in \mathbb{C} \,\Big|\, \frac{1}{1+\varepsilon} < |z| < 1 + \varepsilon\right\}.$$

Now $\beta_n : z \to w = \beta_n(z)$ maps Z_ε conformally onto the region $\beta_n(Z_\varepsilon) \supset B_n$ of the w-plane. Hence $U(\Gamma_n) = f^{-1}(\beta_n(Z_\varepsilon))$ is a neighborhood of Γ_n on $\mathcal{R}$, while

$$z_n : p \to z_n(p) = \beta_n^{-1}(f(p))$$

gives a local complex coordinate defined on $U(\Gamma_n)$. The range of z_n, of course, is the region Z_ε and $|\Gamma_n| = \{p : |z_n(p)| = 1\}$. By Lemma 7.5 it is possible to take analytic Jordan curves $\Gamma_1, \Gamma_2, \ldots \Gamma_m$ for the curves C_1, $C_2, \ldots, C_m$ occurring in (7.23). Doing so, each Jordan curve C_n satisfies the following condition:

> there exists a local complex coordinate $z_n : p \to z_n(p)$ of $\mathcal{R}$ defined on the neighborhood $U(C_n)$ such that:
> $\{z_n(p) | p \in U(C_n)\} = Z_\varepsilon, \qquad |C_n| = \{p|\, |z_n(p)| = 1\}.$

We want to consider the Riemann surface

$$\mathcal{R}^{(m)} = \mathcal{R} - |C_1| - |C_2| - \cdots - |C_m|$$

again, in the light of the above condition. Let $z_n = z_n(p)$ denote the local coordinate of the point $p \in U(C_n)$, that is C_n is the unit circle in the z_n-plane. Let the orientation of C_n be the positive orientation of the unit circle, then

$$U^+(C_n) = \{z_n \in Z_\varepsilon \| z_n| < 1\}, \quad U^-(C_n) = \{z_n \in Z_\varepsilon \| z_n| > 1\}.$$

On

$$\tilde{\mathcal{R}} = [\mu(\mathcal{R}^{(m)})] = \mathcal{R}^{(m)} \cup C_1^+ \cup C_1^- \cup \cdots \cup C_m^+ \cup C_m^-$$

the open sets

$$\tilde{U}(C_n^+) = U^+(C_n) \cup C_n^+ \text{ and } \tilde{U}(C_n^-) = U^-(C_n) \cup C_n^-$$

are neighborhoods on C_n^+ and C_n^- respectively.

Putting $z_n(p^+) = z_n(p)$ where $p = \tilde{\omega}(p^+)$, for each $p^+ \in C_n^+$, and extending the local coordinate z_n on $U^+(C_n)$ to $\tilde{U}(C_n^+)$, we have

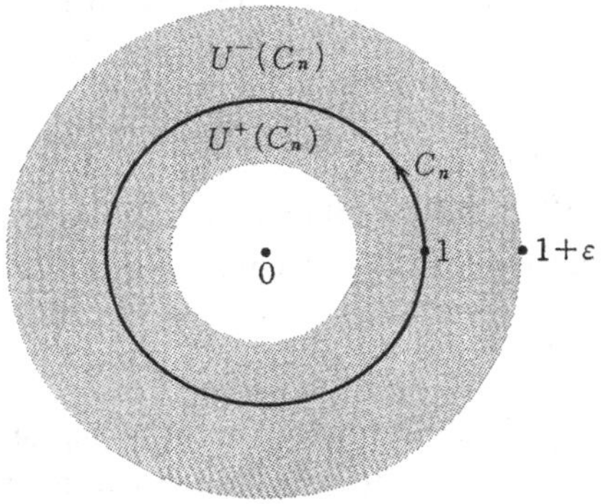

The z-plane

$$\tilde{U}(C_n^+) = \left\{ z_n \in \mathbb{C} \middle| \frac{1}{1+\varepsilon} < |z_n| \leqslant 1 \right\}. \tag{7.32}$$

Hence, each $z_n \in C_n^+$ is an interior point of $\tilde{\mathcal{R}}$ and a boundary point of $\tilde{U}(C_n^+)$ in the z_n-plane; also C_n^+ is the unit circle $|z_n| = 1$. Similarly for C_n^-

$$\tilde{U}(C_n^-) = \{ z_n \in \mathbb{C} | 1 < |z_n| \leqslant 1 + \varepsilon \}. \tag{7.33}$$

We call C_n^+ and C_n^- the *borders* of $\tilde{\mathcal{R}}$ and $\tilde{\mathcal{R}}$ is called a *bordered Riemann surface* (see Ahlfors and Sario, 1960, p. 117).

Fixing a point $q_0 \in \mathcal{R}^{(m)}$, let $f : p \to w = f(p)$ be a biholomorphic map which maps $\mathcal{R}^{(m)}$ onto the horizontal slit region $\Omega = \mathbf{S} - F$ and q_0 onto ∞. Now extend f to a continuous map $\tilde{f} : p \to w = \tilde{f}(p)$ mapping $\tilde{\mathcal{R}}$ onto $\mathbf{S}$. We claim that all connected components $F_n^+ = \tilde{f}(C_n^+)$ and $F_n^- = \tilde{f}(C_n^-)$, $n = 1, 2, \ldots, m$, are segments parallel to the real axis. To prove this let us suppose that $F_n^+ = c$ is a point. Now $\tilde{f}(z_n)$ is a continuous function of z_n on $\tilde{U}(C_n^+) = \{z_n : (1+\varepsilon)^{-1} < |z_n| \leqslant 1\}$, holomorphic for $|z_n| < 1$ and $\tilde{f}(z_n) = c$ if $|z_n| = 1$. Writing z for z_n, and denoting the circle with center 0 and radius r by C_r, we have by Cauchy's integral formula

$$f(z) = \frac{1}{2\pi i} \int_{C_1} \frac{\tilde{f}(\zeta)}{\zeta - z} \, d\zeta - \frac{1}{2\pi i} \int_{C_r} \frac{f(\zeta)}{\zeta - z} \, d\zeta$$

if $(1+\varepsilon)^{-1} < r < |z| < 1$. Since $\tilde{f}(\zeta) = c$ on C_1, the first integral on the right equals c. The second integral

$$f_2(z) = \frac{1}{2\pi i} \int_{C_r} \frac{f(\zeta)}{\zeta - z} \, d\zeta$$

is a holomorphic function of z for $|z| > r$. For $|z| > 1$ we have

$$f_2(z) = \frac{1}{2\pi i}\int_{C_1} \frac{\tilde{f}(\zeta)}{\zeta - z}\, d\zeta = \frac{1}{2\pi i}\int_{C_1} \frac{c}{\zeta - z}\, d\zeta = 0,$$

Hence, by Theorem 3.1, $f_2(z) = 0$ for all z satisfying $|z| > r$. Hence $f(z) = c$ identically if $r < |z| < 1$, contradicting the fact that f is biholomorphic.

Since, in this way, all connected components $F_1^+, F_1^-, \ldots, F_m^+, F_m^-$ of F are segments parallel to the real axis, there exists by Theorem 7.6 a conformal map $h : w \to z = h(w)$, which maps Ω onto a bounded domain of the form (7.15):

$$D = \mathbb{C} \cup \{\infty\} - [U_1] - [U_2] - \cdots - [U_{2m}], \qquad \infty \in U_1.$$

Putting $g(p) = h(f(p))$, g is a biholomorphic map which maps $\mathcal{R}^{(m)}$ onto D. We proved in lemma 7.4 that the biholomorphic map which maps $\mathcal{R}^{(m)}$ onto Ω can be extended to a continuous map mapping $\tilde{\mathcal{R}} = [\mu(\mathcal{R}^{(m)})]$ onto $\mathbf{S} = [\Omega]$ and it can be proved in exactly the same way that g can be extended to a continuous map $\tilde{g}$ mapping $\tilde{\mathcal{R}}$ onto $[D]$. Now $(\tilde{g}((\tilde{\mathcal{R}} - \mathcal{R}^{(m)}) = [D] - D$ and for each C_n^+ and C_n^-, $(\tilde{g}(|C_n^+|)$ and $(\tilde{g}(|C_n^-|)$ are contained in one of the connected components $|\Gamma_k|$, $\Gamma_k = -\partial[U_k]$ of $[D] - D$; hence we may assume that $(\tilde{g}(|C_n^+|) = |\Gamma_{2n-1}|$ and $(\tilde{g}(|C_n^-|) = |\Gamma_{2n}|$.

As is clear from the proof of Theorem 7.6, for each U_k there exists a conformal mapping $h_k : \zeta \to z = h_k(\zeta)$ which for some sufficiently small positive δ maps $\Delta = \{\zeta \in \mathbb{C} : |\zeta| < 1 + \delta\}$, onto a region N_k such that $[U_k] \subset N_k$ and $N_k - [U_k] \subset D$, where $U_k = \{h_k(\zeta) : |\zeta| < 1\}$. We put $\zeta_k(z) = h_k^{-1}(z)$; that is ζ_k is a local coordinate defined on the region N_k of the z-plane and U_k is the unit coordinate disk. Hence, representing the local coordinate of $z \in N_k$ by $\zeta_k = \zeta_k(z)$, we have

$$[D] \cap N_k = \{\zeta_k | 1 + \delta > |\zeta_k| \geqslant 1\}. \tag{7.34}$$

On $\tilde{\mathcal{R}}$, a neighborhood of C_n^+ is given by $\tilde{U}(C_n^+) = \{z_n \in Z_\varepsilon : |z_n| \leqslant 1\}$ by (7.32). For sufficiently small $\varepsilon > 0$, $\tilde{g}$ maps $\tilde{U}(C_n^+)$ onto the interior of $[D] \cap N_k$, $k = 2n - 1$. Representing $\tilde{g}$ as

$$\tilde{g} : z_n \to \zeta_k = \tilde{g}(z_n),$$

the continuous function $\tilde{g}(z_n)$ of z_n, defined on $\tilde{U}(C_n^+)$, is biholomorphic for $|z_n| < 1$ and $\tilde{g}(z_n) = 1$ if $|z_n| = 1$. We next prove that it is possible to extend the continuous map $\tilde{g} : z_n \to \tilde{g}(x_n)$ to a conformal mapping G, that maps $Z_\varepsilon \supset \tilde{U}(C_n^+)$ onto the interior of N_k. Let us denote the reflection of z with respect to the unit circle by z^* (compare Section 5.3a). If $|z_n| = 1$ then $\tilde{U}(z_n^*)^* = \tilde{U}(z_n)$, hence, putting

$$G(z_n) = \begin{cases} \tilde{U}(z_n), & (1+\varepsilon)^{-1} < |z_n| \leqslant 1, \\ \tilde{U}(z_n^*)^*, & 1 \leqslant |z_n| < 1+\varepsilon, \end{cases} \tag{7.35}$$

we obtain a function G continuous on Z_ε and biholomorphic on $Z_\varepsilon - |C_n^+|$. It can be proved that $G(z_n)$ is holomorphic on the unit circle $|C_n^+|$ also (compare the proof of Theorem 5.8). Let r be fixed such that $1 < r < 1+\varepsilon$ and let C_r and C_s be circles in the z_n-plane with center 0 and radius r and $s = 1/r$ respectively and put

$$H(z_n) = \frac{1}{2\pi i}\int_{C_r} \frac{G(z)}{z - z_n}\,dz - \frac{1}{2\pi i}\int_{C_s} \frac{G(z)}{z - z_n}\,dz.$$

Then $H(z_n)$ is holomorphic for $s < |z_n| < r$. Writing

$$H(z_n) = \frac{1}{2\pi i}\int_{C_r - C_1} \frac{G(z)}{z - z_n}\,dz + \frac{1}{2\pi i}\int_{C_1 - C_s} \frac{G(z)}{z - z_n}\,dz,$$

it is easily seen that $H(z_n) = G(z_n)$ if $|z_n| \neq 1$. Hence, by the continuity of $G(z_n)$, $G(z_n) = H(z_n)$ identically for $s < |z_n| < r$ that is $G(z_n)$ is a holomorphic function on Z_ε. Since $G(z_n)$ is single-valued on $Z_\varepsilon - |C_n^+|$, it is also single-valued on Z_ε. To see this, suppose that $G(c_1) = G(c_2) = \alpha$ for two different points $c_1, c_2 \in C_n^+$. For sufficiently small neighborhoods $U(c_1)$, $U(c_2)$ and $W(\alpha)$ of x_1, c_2 and α we have (by Theorems 3.2 and 3.3) $G(U(c_1)) = G(U(c_2)) = G(W(\alpha))$, contradicting the single-valuedness of $G(z_n)$ on $Z_\varepsilon - |C_n^+|$. Now $G : z_n \to \zeta_{2n-1} = G(z_n)$ is a conformal mapping which maps Z_ε into the region N_{2n-1}. We denote this conformal mapping by G_n^+. Next we write z_n^+ for z_n and

$$Z_n^+ = \{z_n^+ | (1+\varepsilon)^{-1} < |z_n^+| < 1+\varepsilon\} \tag{7.36}$$

for Z_ε. By extending the neighborhood $\tilde{U}(C_n^+) = \{z_n^+ \in Z_n^+ : |z_n^+| \leqslant 1\}$ of the border C_n^+ of $\tilde{\mathcal{R}}$ to Z_n^+, it is possible to extend $\tilde{\mathcal{R}}$ to $\tilde{\mathcal{R}} \cup Z_n^+$. The part added by this extension is of course

$$\tilde{\mathcal{R}} \cup Z_n^+ - \tilde{\mathcal{R}} = \{z_n^+ || z_n^+| > 1\}.$$

Hence we have extended the continuous mapping $\tilde{g} : z_n^+ \to \zeta_{2n-1} = \tilde{g}(z_n^+)$ defined on $\tilde{U}(C_n^+) = \{z_n^+ \in Z_n^+ : |z_n^+| \leqslant 1\}$ to a conformal mapping $G_n^+ : z_n^+ \to \zeta_{2n-1} = G_n^+(z_n^+)$ defined on Z_n^+. In the same way one introduces the variable z_n^- and the region

$$Z_n^- = \{z_n^- | (1+\varepsilon)^{-1} < |z_n^-| < 1+\varepsilon\}$$

and one proves that the continuous mapping $\tilde{g} : z_n^- \to \zeta_{2n} = \tilde{g}(z_n^-)$ defined on $\tilde{U}(C_n)$ $[z_n^- \omega \in Z_n^- : |z_n^-| \leqslant 1\}$ can be extended to a conformal mapping $G_n^- : z_n^- \to \zeta_{2n} = G_n^-(z_n^-)$ defined on Z_n^-. By extending each neighborhood $\tilde{U}(C_n^+)$ and $\tilde{U}(C_n^-)$ to Z_n^+. and Z_n^- respectively, $\tilde{\mathcal{R}}$ is extended to a Riemann surface

$$\mathcal{W} = \tilde{\mathcal{R}} \cup Z_n^+ \cup Z_n^- \cup \cdots \cup Z_m^+ \cup Z_m^-;$$

$\tilde{\mathcal{R}}$ is a closed domain in $\mathcal{W}$ and its boundary is given by

$$\partial\tilde{\mathcal{R}} = C_1^+ - C_1^- + C_2^+ - C_2^- + \cdots + C_m^+ - C_m^-.$$

If the continuous map $\tilde{g}$ on $\tilde{\mathcal{R}}$ is extended to conformal mappings G_n^+ and G_n^- on regions Z_n^+ and Z_n^- respectively, a conformal map $\mathcal{G}$ on $\mathcal{W}$ is obtained. Thus $\mathcal{G} : p \to z = \mathcal{G}(p)$ is a conformal map which maps $\mathcal{W}$ onto the region $W = \mathcal{G}(\mathcal{W}) \supset [D]$ in the z-plane and $\tilde{g} = \mathcal{G}|_{\tilde{\mathcal{R}}}$. Hence $\tilde{g}$ is a homeomorphism between $\tilde{\mathcal{R}}$ and $[D]$. Since $\tilde{g}$ is the restriction of the conformal map $\mathcal{G}$, we will call $\tilde{g}$ a conformal map mapping $\tilde{\mathcal{R}}$ onto $[D]$. Since $[D] = \mathcal{G}(\tilde{\mathcal{R}})$, we will say that $\tilde{\mathcal{R}}$ and $[D]$ are *conformally equivalent*. We can identify $\tilde{\mathcal{R}}$, as a bordered Riemann surface, with the bounded, closed region $[D]$. Since $\mathcal{G}$ is conformal, $\tilde{g}$ maps the oriented boundary $\partial\tilde{\mathcal{R}}$ of $\tilde{\mathcal{R}}$ onto the oriented boundary $\partial[D]$ of $[D]$. Hence, observing that $\tilde{g}(|C_n^+|) = |\Gamma_{2n-1}|$ and $\tilde{g}(|C_n^-|) = |\Gamma_{2n}|$, and comparing (7.37) and (7.16), we conclude that

$$\tilde{g}(C_n^+) = \Gamma_{2n-1}, \qquad \tilde{g}(C_n^-) = -\Gamma_{2n}. \tag{7.38}$$

Putting

$$V(\Gamma_{2n-1}) = \mathcal{G}(Z_n^+), \qquad V(\Gamma_{2n}) = \mathcal{G}(Z_n^-),$$

$V(\Gamma_k)$ is a region in the z-plane containing $|\Gamma_k|$ for $k = 1, 2, \ldots, 2m$. Hence

$$W = [D] \cup V(\Gamma_1) \cup V(\Gamma_2) \cup \cdots \cup V(\Gamma_{2m}). \tag{7.39}$$

The analytic Jordan curve $\Gamma_k = -\partial[U_k]$ divides $V(\Gamma_k)$ into two regions, namely $V^+(\Gamma_k) = V(\Gamma_k) \cap D$ to the left of Γ_k and $V^-(\Gamma_k) = V(\Gamma_k) \cap U_k$ to the right of Γ_k. The projection $\tilde{\omega} : \tilde{\mathcal{R}} \to \mathcal{R}$ is the identity map on $\mathcal{R}^{(m)} \subset \tilde{\mathcal{R}}$ mapping $\mathcal{R}^{(m)}$ onto $\mathcal{R}^{(m)} \subset \mathcal{R}$, while on the border of $\tilde{\mathcal{R}}$ we have $\tilde{\omega}(p^+) = \tilde{\omega}(p^-) = p$ where $p^+ = (p, 1) \in C_n^+$ and $p^- = (p, 0) \in C_n^-$. Hence $\mathcal{R}$ can be obtained from $\tilde{\mathcal{R}}$ by piecing together C_n^+ and C_n^-, $n = 1, 2, \ldots, m$, such that $p^+ \in C_n^+$ and $p^- \in C_n^-$ coincide. We express this by saying that $\mathcal{R}$ is obtained from $\tilde{\mathcal{R}}$ by identifying corresponding points $p^+ \in C_n^+$ and $p^- \in C_n^-$ or, for short, by identifying the borders C_n^+ and C_n^-. It is a natural consequence of the definition of $\tilde{\mathcal{R}}$ that $\mathcal{R}$ can be obtained from $\tilde{\mathcal{R}}$ by identifying C_n^+ and C_n^-.

Since $\tilde{\mathcal{R}}$ can be considered as the same Riemann surface as

$$[D] = \tilde{g}(\tilde{\mathcal{R}}) = \mathbb{C} \cup \{\infty\} - U_1 - U_2 - \cdots - U_{2m}, \qquad \infty \in U_1,$$

$\mathcal{R}$ can be obtained by identifying the points $\tilde{g}(p^+) \in \tilde{g}(C_n^+) = \Gamma_{2n-1}$ and $\tilde{g}(p^-) \in \tilde{g}(C_n^-) = -\Gamma_{2n}$ of the boundary of the bounded closed domain; that is $\mathcal{R}$ can be obtained from $[D]$ by identifying Γ_{2n-1} and $-\Gamma_{2n}$,

$n = 1, 2, \ldots, m$. By this identification a point $z \in [D]$ becomes a point $p = \tilde{\omega}\tilde{g}^{-1}(z) \in \mathcal{R}$. Identification of Γ_{2n-1} and $-\Gamma_{2n}$ means that the Jordan curves Γ_{2n-1} and Γ_{2n} are pasted together with opposite orientations.

We show next that the map $\tilde{\omega}\tilde{g}^{-1} : z \to p = \tilde{\omega}\tilde{g}^{-1}(z)$, which maps $[D]$ onto $\mathcal{R}$ can be extended to a holomorphic map which maps the region $W \supset [D]$ onto $\mathcal{R}$. To see this, note that the conformal map $\tilde{g}^{-1}$ mapping $[D]$ onto $\tilde{\mathcal{R}}$ can be extended to a conformal map $\mathcal{G}^{-1}$, mapping W onto the Riemann surface $\mathcal{W} \subset \tilde{\mathcal{R}}$. On the other hand, the projection $\tilde{\omega} : \tilde{\mathcal{R}} \to \mathcal{R}$ can be extended in a natural way to a holomorphic map, mapping $\mathcal{W}$ onto $\mathcal{R}$. Actually, let $z_n = z_n(p)$ represent the local coordinate of a point $p \in U(C_n)$, where $U(C_n)$ is a neighborhood of C_n on $\mathcal{R}$. On the neighborhood $\tilde{U}(C_n^+) = \{z_n^+ \in Z_n^+ : |z_n^+| \leqslant 1\}$ of C_n^+ on $\tilde{\mathcal{R}}$ we have $\tilde{\omega} : z_n^+ \to z_n = z_n^+$. Hence $\tilde{\omega}$ can be extended to a holomorphic mapping $z_n^+ \to z_n = z_n^+$ which maps Z_n^+ onto $U(C)$. Similarly, the mapping $\tilde{\omega} : z_n^- \to z_n = z_n^-$ on $\tilde{U}(C_n^-) = \{z_n^- \in Z_n^- : |z_n^-| \geqslant 1\}$ can be extended to a holomorphic mapping $z_n^- \to z_n = z_n^-$ which maps Z_n^- onto $U(C_n)$. In this way, we extend $\tilde{\omega} : \tilde{\mathcal{R}} \to \mathcal{R}$ to a holomorphic mapping which maps $\mathcal{W}$ onto $\mathcal{R}$. We denote this new map also by $\tilde{\omega}$:

$$\tilde{\omega} : \mathcal{W} \to \mathcal{R}.$$

Thus we have extended the map $\tilde{\omega}\tilde{g}^{-1}$, which maps $[D]$ onto $\mathcal{R}$, to a holomorphic map $\tilde{\omega}\mathcal{G}^{-1}$ mapping W onto $\mathcal{R}$.

This extended holomorphic map will be denoted by

$$\Pi = \tilde{\omega}\mathcal{G}^{-1} : W \to \mathcal{R}. \tag{7.40}$$

Since $\tilde{\omega}^{-1}(U(C_n)) = Z_n^+ \cup Z_n^-$, $\mathcal{G}(Z_n^+) = V(\Gamma_{2n-1})$ and $\mathcal{G}(Z_n^-) = V(\Gamma_{2n})$, we have

$$\Pi^{-1}(U(C_n)) = V(\Gamma_{2n-1}) \cup V(\Gamma_{2n}) \cup V(\Gamma_{2n}), \qquad n = 1, 2, \ldots, m.$$

Since $\mathcal{G}$ is conformal and $\tilde{\omega}$ maps Z_n^+ and Z_n^- conformally onto $U(C_n)$, Π maps $V(\Gamma_{2n-1})$ and $V(\Gamma_{2n})$ conformally onto $U(C_n)$. Hence, for each $p \in U(C_n)$ there exist one point $z \in V(\Gamma_{2n-1})$ such that $\Pi(z) = p$ and one point $w \in v(\Gamma_{2n})$ such that $\Pi(w) = p$. For $z \in V(\Gamma_{2n-1})$ we denote by $\iota_n(z)$ those $w \in V(\Gamma_{2n})$ satisfying $\Pi(w) = \Pi(z)$. The map $\iota_n : z \to \iota_n(z)$ is a conformal map which maps $V(\Gamma_{2n-1})$ onto $V(\Gamma_{2n})$, while

$$\Pi(\iota_n(z)) = \Pi(z), \qquad z \in V(\Gamma_{2n-1}), \qquad \iota_n(z) \in V(\Gamma_{2n}).$$

If $p \in C_n$, then $\tilde{g}(p^+) \in |\Gamma_{2n-1}|$, $\tilde{g}(p^-) \in |\Gamma_{2n}|$, $\Pi(\tilde{g}(p^+)) = \tilde{\omega}(p^+) = p$ and $\Pi(\tilde{g}(p^-)) = \tilde{\omega}(p^-) = p$. Hence, putting $z = \tilde{g}(p^+)$, we have $\tilde{g}(p^-) = \iota_n(z)$: thus ι_n maps $|\Gamma_{2n-1}|$ onto $|\Gamma_{2n}|$. Taking orientations into consideration, we have

$$\iota_n(\Gamma_{2n-1}) = -\Gamma_{2n}$$

by (7.38), that is if the point z traverses Γ_{2n-1} once in the positive direction, then $\iota_n(z)$ traverses Γ_{2n} once in the negative direction. Hence ι_n maps $V^+(\Gamma_{2n-1})$ [which is to the left of Γ_{2n-1}] onto $V^-(\Gamma_{2n})$ [which is to the right of Γ_{2n}], and $V^-(\Gamma_{2n-1})$ onto $V^+(\Gamma_{2n})$ [for if $z \in V^+(\Gamma_{2n-1}) = V(\Gamma_{2n-1}) \cap D$, then $\Pi(z) = \tilde{\omega}\tilde{g}^{-1}(z) \in U^+(C_n)$ and if $w \in V^+(\Gamma_{2n}) = V(\Gamma_{2n}) \cap D$, then $\Pi(\omega \in U^-(C_n)$.]

We have seen that the Riemann surface $\mathcal{R}$ can be obtained from the bounded closed domain $[D]$ by identifying its boundary points $\tilde{g}(p^+) \in |\Gamma_{2n-1}|$ and $\tilde{g}(p^-) \in |\Gamma_{2n}|$. Putting $z = \tilde{g}(p^+)$ we have $\tilde{g}(p^-) = \iota_n(z)$; $\mathcal{R}$ can be obtained from $[D]$ by identifying the points $z \in |\Gamma_{2n-1}|$ and $\iota_n(z) \in |\Gamma_{2n}|$ for $n = 1, 2, \ldots, m$. We have now proved the following theorem:

Theorem 7.8. Let $\mathcal{R}$ be a compact Riemann surface that is not simply connected. There exists a closed region in the complex plane

$$[D] = \mathbb{C} \cup \{\infty\} - U_1 - U_2 - \cdots - U_{2m}, \qquad \infty \in U_1,$$

where the U_k, $k = 1, 2, \ldots, 2m$ are disjoint simply connected regions, where boundaries $|\Gamma_k| = [U_k] - U_k$ are analytic Jordan curves such that $\mathcal{R}$ can be obtained from $[D]$ by identifying $z \in |\Gamma_{2n-1}|$ with $\iota_n(z) \in |\Gamma_{2n}|$, $n = 1, \ldots, m$. Here, $\iota_n : z \longrightarrow \iota_n(z)$ is a conformal mapping which maps a region $V(\Gamma_{2n-1}) \supset |\Gamma_{2n-1}|$ onto a region $B(\Gamma_{2n}) \supset |\Gamma_{2n}|$. Determining orientations by $\Gamma_k = -\partial[U_k]$ we have

$$\iota_n(\Gamma_{2n-1}) = -\Gamma_{2n}. \tag{7.41}$$

Since $\iota_n(\Gamma_{2n-1}) = -\Gamma_{2n}$, as explained above, ι_n maps $V^+(\Gamma_{2n-1}) = V(\Gamma_{2n-1}) \cap D$ onto $V^-(\Gamma_{2n}) = V(\Gamma_{2n}) \cap U_{2n}$ and $V^-(\Gamma_{2n-1}) = V(\Gamma_{2n-1})$

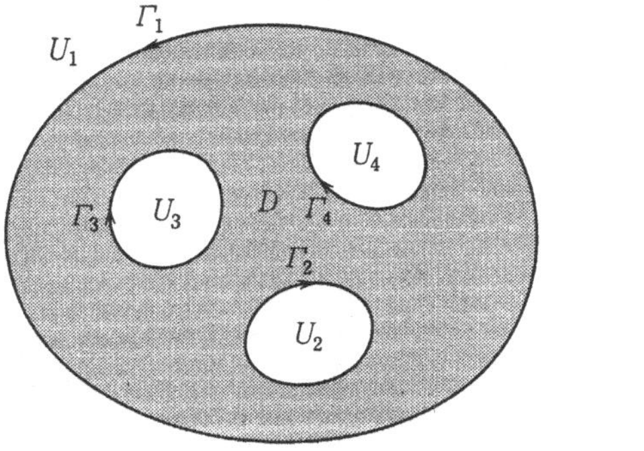

$\cap U_{2n-1}$ onto $V^+(\Gamma_{2n}) = V(\Gamma_{2n}) \cap D$. Hence, enlarging the scope of points to be identified from $[D]$ to the region

$$W = [D] \cup V(\Gamma_1) \cup V(\Gamma_2) \cup \cdots \cup V(\Gamma_{2m}),$$

it becomes clear that $\mathcal{R}$ can be obtained from W by identifying $z \in V(\Gamma_{2n-1})$ and $\iota_n(z) \in V(\Gamma_{2n})$, $n = 1, 2, \ldots, m$. If $z \in W$ corresponds with $p \in \mathcal{R}$, we put $p = \Pi(z)$. Since ι_n is conformal, $\Pi : z \to \Pi(z)$ is a holomorphic map mapping W onto $\mathcal{R}$. The holomorphic map is the same as the one (7.40): $\Pi = \tilde{\omega}\tilde{g}^{-1}$. Π is biholomorphic on $V(\Gamma_{2n-1})$ and $V(\Gamma_{2n})$ and

$$\Pi(\iota_n(z)) = \Pi(z), \qquad z \in V(\Gamma_{2n-1}), \qquad \iota_n(z) \in V(\Gamma_{2n}). \tag{7.42}$$

Hence, by (7.41), $C_n = \Pi(\Gamma_{2n-1}) = -\Pi(\Gamma_{2n})$ are analytic Jordan curves on $\mathcal{R}$ and Π maps $V(\Gamma_{2n-1})$ and $V(\Gamma_{2n})$ conformally onto the neighborhood $U(C_n)$ of C_n. Therefore

$$\begin{aligned} \Pi(V^+(\Gamma_{2n-1})) &= (V^-(\Gamma_{2n})) = U^+(C_n), \\ \Pi(V^+(\Gamma_{2n})) &= \Pi(V^-(\Gamma_{2n-1})) = U^-(C_n). \end{aligned} \tag{7.43}$$

From the above, it is clear that $\mathcal{R}$ is homeomorphic to a sphere with m handles attached to it. Actually, using the local coordinate $z_n = z_n(p)$, we have

$$U(C_n) = \{z_n \in \mathbb{C} | a < |z_n| < b\}, \qquad a = 1/b, \qquad b = 1 + \varepsilon,$$

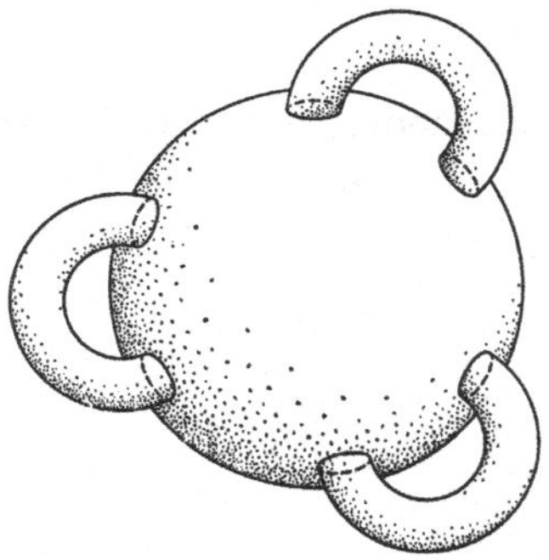

$m = 3$

where C_n is given by: $\theta \to z_n = e^{i\theta}$, $0 \leqslant \theta \leqslant 2\pi$. Since we may assume that $z_n(p)$ is a single-valued holomorphic function on a region containing $[U(C_n)]$, we have $[U(C_n)] = \{z_n \in \mathbb{C} : a \leqslant |z_n| \leqslant b\}$. Hence

$$[U(C_n)] = \{re^{i\theta} | a \leqslant r \leqslant b, 0 \leqslant \theta \leqslant 2\pi\} = [a, b] \times C_n$$

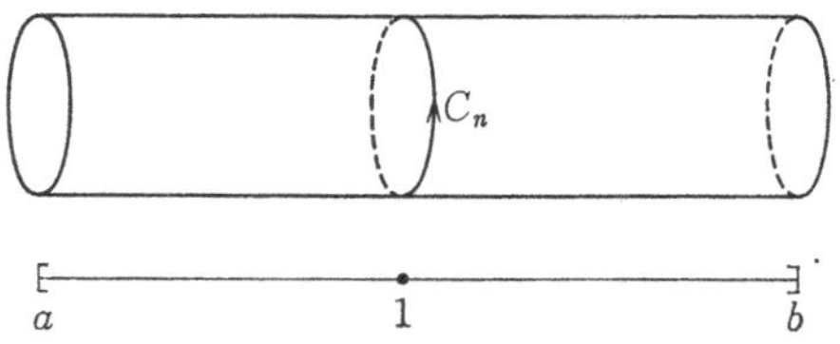

is a cylinder and its boundary is given by

$$\partial[U(C_n)] = (b \times C_n) - (a \times C_n).$$

Looking at this cylinder as a handle, we consider

$$\mathcal{S} = \mathcal{R} - U(C_1) - U(C_2) - \cdots - U(C_m);$$

$\mathcal{S}$ is a closed subregion of $\mathcal{R}^{(m)} = \mathcal{R} - |C_1| - \cdots - |C_m|$ and its boundary is given by

$$\partial\mathcal{S} = (a \times C_1) - (b \times C_1) + \cdots + (a \times C_m) - (b \times C_m).$$

Since Π maps D conformally onto $\mathcal{R}^{(m)}$, $\Pi^{-1}(\mathcal{S})$ is a closed region in the interior of D and its boundary consists of $2m$ analytic Jordan curves $B_{2n-1} = \Pi^{-1}(a \times C_n)$ and $B_{2n} = \Pi^{-1}(b \times C_n)$, $n = 1, 2, \ldots, m$. By (7.43)

$$\Pi^{-1}(\mathcal{S}) = D - V^+(\Gamma_1) - V^+(\Gamma_2) - \cdots - V^+(\Gamma_{2m-1}) - V^+(\Gamma_{2m}).$$

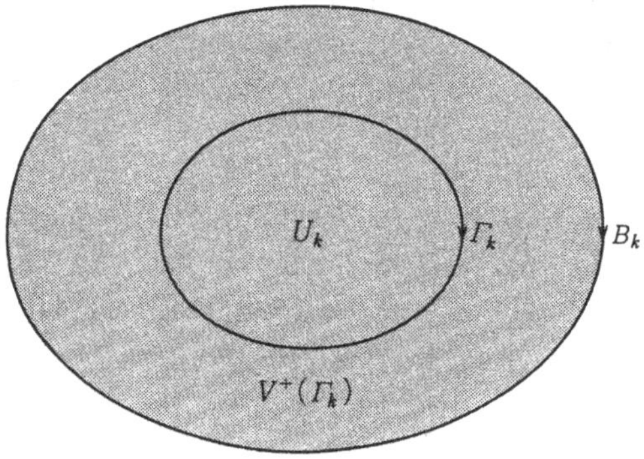

Hence, putting

$$V_k = U_k \cup V(\Gamma_k) = [U_k] \cup V^+(\Gamma_k)$$

we have $D = \mathbb{C} \cup \{\infty\} - [U_1] - [U_2] - \cdots - [U_{2m}]$ and therefore

$$\Pi^{-1}(\mathcal{S}) = \mathbb{C} \cup \{\infty\} - V_1 - V_2 - \cdots - V_{2m-1} - V_{2m}, \; \infty \in V_1.$$

Since each V_k is a simply connected region and $\partial[V_k] = -B_k$ is an analytic Jordan curve, we conclude from the Riemann Mapping Theorem

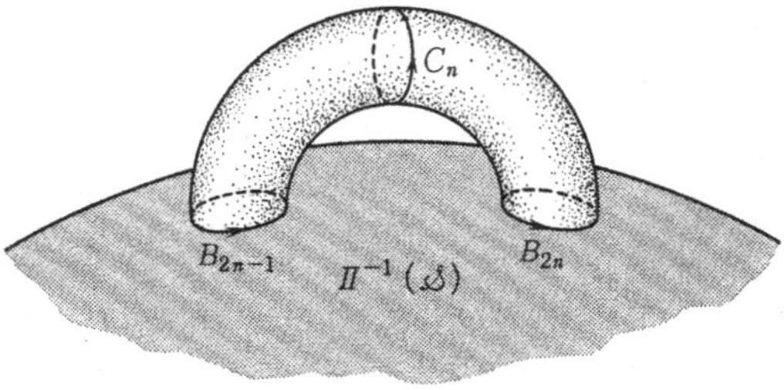

and Theorem 5.5, that V_k is conformally equivalent to the unit disk and $[V_k]$ is homeomorphic to the closed unit disk. Since $\mathcal{S}$ and $\Pi^{-1}(\mathcal{S})$ are conformally equivalent, we can identify the bordered Riemann surface $\mathcal{S}$ with the closed region $\Pi^{-1}(\mathcal{S})$. Hence

$$\mathcal{R} = \Pi^{-1}(\mathcal{S}) \cup [U(C_1)] \cup \{U(C_2)] \cup \cdots \cup [U(C_m)]$$

is the surface obtained by making m circular holes V_{2n-1} and V_{2n}, $n = 1, \ldots, m$, in the Riemann sphere $\mathbb{C} \cup \{\infty\}$ and attaching m handles $[U(C_n)]$ by identifying the borders B_{2n-1} and B_{2n} with the borders $a \times C_n$ and $-b \times C_n$ respectively of $[U(C_n)] = [a, b] \times C_n$.

c. *Homology groups*

As explained at the beginning of Section 6.4c, the definitions and theorems concerning homotopy of curves in a region $\Omega \subset \mathbb{C}$ and concerning 1-chains and 1-cycles in Ω can be applied without any difficulty to a Riemann surface $\mathcal{R}$.

For a given Riemann surface $\mathcal{R}$, the collection of all 1-cycles constitutes an additive group, which we denote by $Z_1(\mathcal{R})$; it is a subgroup of the group $C_1(\mathcal{R})$ of 1-chains (see Section 4.4). By the Definition 4.5 of homology, the collection $\{\gamma \in Z_1(\mathcal{R}) : \gamma \sim 0\}$ is a subgroup of $Z_1(\mathcal{R})$. The quotient group $Z_1(\mathcal{R})/\{\gamma \in Z_1(\mathcal{R}) : \gamma \sim 0\}$ is called the *first homology group* of $\mathcal{R}$ and denoted by $H_1(\mathcal{R})$:

$$H_1(\mathcal{R}) = Z_1(\mathcal{R})/\{\gamma \in Z_1(\mathcal{R}) | \gamma \sim 0\}. \tag{7.44}$$

The element of $H_1(\mathcal{R})$ corresponding with a 1-cycle $\gamma \in Z_1(\mathcal{R})$ is called the *homology class* of γ.

Let us investigate $H_1(\mathcal{R})$ when $\mathcal{R}$ is compact. By Theorem 4.13, an arbitrary 1-cycle γ on $\mathcal{R}$ can be represented as a finite sum $\gamma = n_1\gamma_1 + n_2\gamma_2 + n_3\gamma_3 + \cdots$, where $\gamma_1, \gamma_2, \gamma_3, \ldots$ are closed curves and $n_1, n_2, n_3, \ldots$ are integers. If $\mathcal{R}$ is simply connected, then $H_1(\mathcal{R}) = 0$: let us assume, therefore, that $\mathcal{R}$ is not simply connected. We consider the closed region

$$[D] = \mathbb{C} \cup \{\infty\} - U_1 - U_2 - \cdots - U_{2m-1} - U_{2m}, \qquad \infty \in U_1,$$

and its boundary

$$\partial[D] = \Gamma_1 + \Gamma_2 + \cdots + \Gamma_k + \cdots + \Gamma_{2m}, \qquad \Gamma_k = -\partial[U_k],$$

and define regions $V(\Gamma_k)$ containing $|\Gamma_k|$, and conformal maps $\iota_n : z \to \iota_n(z)$ mapping the regions $V(\Gamma_{2n-1})$ onto $V(\Gamma_{2n})$, where $\iota_n(\Gamma_{2n-1}) = -\Gamma_{2n}$. By Theorem 7.8, $\mathcal{R}$ can be obtained from $[D]$ by identifying its boundary points $z \in |\Gamma_{2n-1}|$ and $\iota_n(z) \in |\Gamma_{2n}|$. Further, let Π be the holomorphic mapping which maps the region

$$W = [D] \cup V(\Gamma_1) \cup V(\Gamma_2) \cup \cdots \cup V(\Gamma_{2m-1}) \cup V(\Gamma_{2m})$$

onto $\mathcal{R}$ (see (7.40)). The $C_n = \Pi(\Gamma_{2n-1}) = -\Pi(\Gamma_{2n})$ are analytic Jordan curves on $\mathcal{R}$ and Π maps $V(\Gamma_{2n-1})$ and $V(\Gamma_{2n})$ onto $U(C_n)$. By the identification of $z \in |\Gamma_{2n-1}|$ with $\iota_n(z) \in |\Gamma_{2n}|$, $z \in [D]$ can be considered as a point $p = \Pi(z) \in \mathcal{R}$. In the following, we will consider $H_1(\mathcal{R})$ by representing $\mathcal{R}$ as $\mathcal{R} = \Pi([D])$. The restriction $\Pi|_{[D]}$ will also be denoted by Π. Hence, by (7.43),

$$\Pi^{-1}(U^+(C_n)) = V^+(\Gamma_{2n-1}), \qquad \Pi^{-1}(U^-(C_n)) = V^+(\Gamma_{2n}).$$

First consider the segment $\ell_n : r \to z_n = re^{i\omega}$, $a \leqslant r \leqslant b$, ω constant, in $[U(C_n)] = \{z_n \in \mathbb{C};\ a \leqslant |z_n| \leqslant b\}$, $a = 1/b$ and $b = 1 + \varepsilon$ which intersects the unit circle orthogonally. Now $\Pi^{-1}(\ell_n)$ consists of two analytic Jordan arcs, namely $\mu_{2n-1} : r \to \mu_{2n-1}(r)$, $a \leqslant r \leqslant 1$, contained in the interior of $[B^+(\Gamma_{2n-1})]$ and $\lambda_{2n} : r \to z = \lambda_{2n}(r)$, $1 \leqslant r \leqslant b$, contained in the interior of $[V^+(\Gamma_{2n})]$. Putting $\lambda_{2n-1}(r) = \mu_{2n-1}(1/r)$, $(\mu_{2n-1})^{-1}$ is respresented as $\lambda_{2n-1} : r \to \lambda_{2n-1}(r)$, $1 \leqslant r \leqslant b$. For each λ_k we have $\lambda_k(1) \in |\Gamma_k|$ and λ_k intersects Γ_k orthogonally at $\lambda_k(1)$. Each λ_k is contained in D apart from the point $\lambda_k(1)$. For the sake of simplicity, we assume that $0 \in D$ and $0 \notin [V^+(\Gamma_k)]$, $k = 1, 2, \ldots, 2m$. Let U_0 be a disk

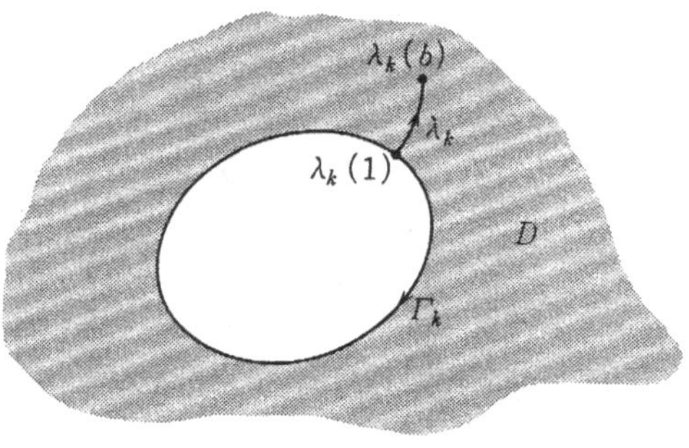

with center 0 and with sufficiently small radius ε_0 and put $e_k = \varepsilon_0 e^{k\pi i/m}$, $k = 1, 2, \ldots, 2m$. Put

$$G = D - [U_0] - |\lambda_1| - |\lambda_2| - \cdots - |\lambda_{2m-1}| - |\lambda_{2m}|;$$

G is a connected open set. Hence, by Theorem 6.15, it is possible to connect $\lambda_k(b)$ and e_k by a piecewise smooth Jordan arc $\beta_k : t \to \beta_k(t)$, $0 \leqslant t \leqslant 1$, $\beta_k(0) = \lambda_k(b)$ and $\beta_k(1) = e_k$, which, with the exception of its end points, is in G for $k = 1, 2, \ldots, 2m$. The curve $\gamma_k = \lambda_k \cdot \beta_k$ is a piecewise smooth Jordan arc, which, with the exception of is end points is contained in $D - [U_0]$. A parametric representation of γ_k is given by $t \to z = \gamma_k(t)$, where $\gamma_k(t) = \lambda_k(1 + 2(b-1)t)$ if $0 \leqslant t \leqslant 1/2$ and $\gamma_k(t) = \beta_k(2t-1)$ if $1/2 \leqslant t \leqslant 1$.

Lemma 7.6. It is possible to choose $\beta_1, \beta_2, \ldots, \beta_{2m}$ in such a way that none of the $\gamma_1, \gamma_2, \ldots, \gamma_{2m}$ intersect.

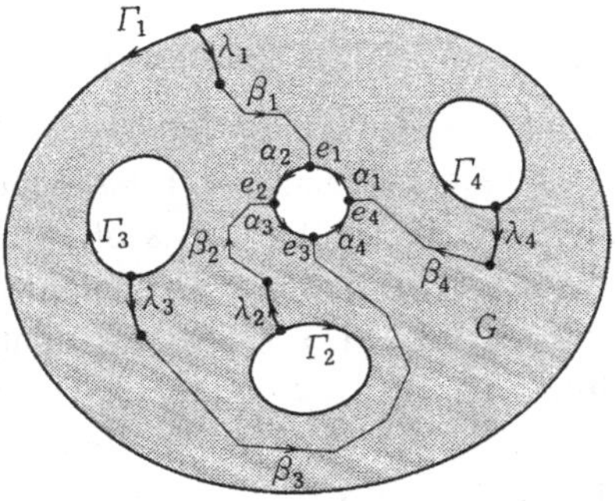

$m=2$

Proof: (1) We prove that, if one chooses β_1 in such a way that it intersects the circle $\partial[U_0]$ orthogonally at e_1, then the open set $G - |\beta_1|$ is connected. For $q \in |\gamma_1|$, take a disk $U(q)$ with center q and sufficiently small radius. Now $|\gamma_1|$ divides $U(q) \cap G$ into two regions. Covering $|\gamma_1|$ with finitely many of these disks $U(q_1)$, $U(q_2)$, $U(q_3)$, $\ldots$, we put

$$U(\gamma_1) = (U(q_1) \cup U(q_2) \cup U(q_3) \cup \cdots) \cap G;$$

$U(\gamma_1)$ is a region containing $|\beta_1 \cap G = \{\beta_1(t) : 0 < t < 1\}$ and γ_1 divides $U(\gamma_1)$ into two regions, namely, $U^+(\gamma_1)$ to the left and $U^-(\gamma_1)$ to the right of γ_1. Since the open set $U^+(C_1) - |\ell_1|$ is connected, so is $V^+(\Gamma_1) - |\lambda_1| = \Pi^{-1}(U^+(C_1) - |\ell_1|)$ and this set has a nonempty intersection with $U^+(\gamma_1)$ or $U^-(\gamma_1)$. Hence

$$W_1 = U(\gamma_1) \cup V^+(\Gamma_1) - |\gamma_1|$$
$$= U^+(\gamma_1) \cup U^-(\gamma_1) \cup (V^+(\Gamma_1) - |\lambda_1|)$$

is a connected open subset of $G - |\beta_1|$. From this fact, it follows almost at once that $G - |\beta_1|$ is connected. Indeed, suppose that $G - |\beta_1| = G_0 \cup G_1$ where G_0 and G_1 are open disjoint sets. W_1 is contained in either G_0 or G_1, say $W_1 \subset G_0$. Hence $U(\gamma_1) - |\gamma_1| \subset G_0$, and therefore $U(\gamma_1) \subset [G_0]$. Since $[G_0] \cap G_1 = \emptyset$, we have $U(\gamma_1) \cap G_1 = \emptyset$. Hence, since $|\beta_1| \cap G \subset U(\gamma_1)$, we have

$$G = G_0 \cup U(\gamma_1) \cup G_1, \qquad (G_0 \cup U(\gamma_1)) \cap G_1 = \emptyset.$$

Since G is connected, $G_1 = \emptyset$. Hence $G - |\beta_1|$ is a connected open set.

(2) As $G - |\beta_1|$ is connected, it is possible to connect $\lambda_2(b)$ and e_2 by a piecewise analytic Jordan curve β_2 which apart from its end points is contained in $G - |\beta_1|$. Choosing β_2 in such a way that it intersects $\partial[U_0]$ orthogonally at e_2, it is proved, as above, that $G - |\beta_1| - |\beta_2|$ is connected. Hence there is a piecewise analytic Jordan arc β_3, which connects $\lambda_3(b)$ and e_3 and which, apart from its end points, is contained in $G - |\beta_1| - |\beta_2|$. It is possible to choose β_3 in such a way that it intersects $\partial[U_0]$ orthogonally at e_3. Repeating this procedure for $k = 4, 5, 6, \ldots, 2m$, we arrive at the desired curves $\beta_1, \ldots, \beta_{2m}$.

Lemma 7.7. With the piecewise analytic Jordan curves $\gamma_k = \lambda_k \cdot \beta_k$ as above,

$$\mathcal{D} = D - [U_0] - |\gamma_1| - |\gamma_2| - \cdots - |\gamma_{2m-1}| - |\gamma_{2m}|$$

is a simply connected region.

Proof: It is clear from the proof of Lemma 7.6 that $\mathcal{D}$ is a region. The complement

$$[U_0] \cup [U_1] \cup \cdots \cup [U_{2m}] \cup |\gamma_1| \cup |\gamma_2| \cup \cdots \cup |\gamma_{2m}|$$

of $\mathcal{D}$ with respect to the Riemann sphere $\mathbb{C} \cup \{\infty\}$ is a connected closed set. Hence $\mathcal{D}$ is a region without holes (Definition 5.1) and so is conformally equivalent to the unit disk U (by Theorem 7.5 or 5.3) thus proving the assertion.

Let

$$\alpha_k : \theta \to z = \varepsilon_0 e^{i\theta}, \qquad \frac{(k-1)\pi}{m} \leqslant \theta \leqslant \frac{k\pi}{m},$$

be the arc of $\partial[U_0]$ connecting e_{k-1} and e_k and put

$$C_{m+n} = \Pi(\tilde{C}_{m+n}), \qquad \tilde{C}_{m+n} = \gamma_{2n-1} \cdot \alpha_{2n} \cdot (\gamma_{2n})^{-1}. \tag{7.45}$$

Now $\tilde{C}_{m+n} = \lambda_{2n-1} \cdot \beta_{2n-1} \cdot \alpha_{2n} \cdot (\beta_{2n})^{-1} \cdot (\lambda_{2n})^{-1}$ is a piecewise analytic arc connecting $\lambda_{2n-1}(1) \in |\Gamma_{2n-1}|$ and $\lambda_{2n}(1) \in |\Gamma_{2n}|$, which, apart from its end points, is contained in D. Since $\Pi(\lambda_{2n-1}(1)) = \Pi(\lambda_{2n}(1))$, C_{m+n} is a piecewise analytic Jordan curve on $\mathcal{R}$, which intersects C_n at only one point $p_n = \Pi(\lambda_{2n-1}(1)) = \Pi(\lambda_{2n}(1))$. By the definition of λ_{2n-1} and λ_{2n}, since

$$\Pi(\lambda_{2n-1})^{-1} \cdot \Pi(\lambda_{2n}) = \ell_n,$$

C_{m+n} coincides with ℓ_n^{-1} on a neighborhood of p_n. Hence, C_{m+n} crosses C_n at p_n from right to left and C_n crosses C_{m+n} from left to right. Since $\tilde{C}_{m+n}$, apart from its end points, belongs to D, C_{m+n} does not

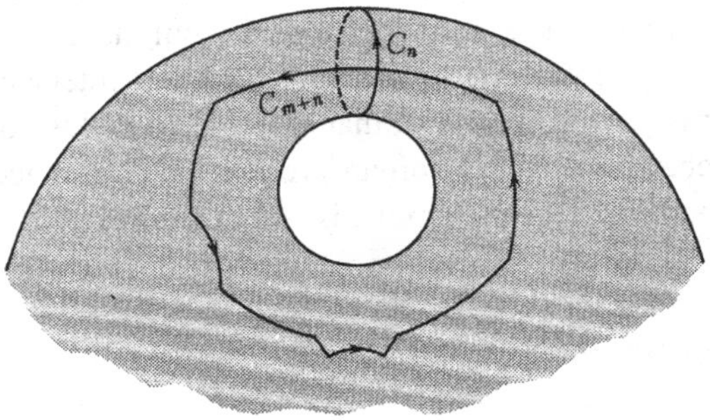

intersect C_h if $h \neq n$. Obviously, C_{m+n} does not intersect C_{m+h} if $h \neq n$, and C_n does not intersect C_h if $h \neq n$. For each analytic Jordan curve C_n, $n = 1, 2, \ldots, m$, there is defined an annular region $U(C_n)$. For each piecewise smooth Jordan curve C_h, $h = m+1, \ldots, 2m$, we put $U(C_h) = \bigcup_{q_j \in |C_h|} U_j$, where $\mathcal{U} = \{U_j\}$ is a finite open covering consisting of sufficiently small coordinate disks $U_j = U_{rj}(q_j)$ of $\mathcal{R}$. For each C_k, $k = 1, 2, \ldots, m, m+1, \ldots, 2m$, let $d\rho_{C_k}^+$ be the C^∞ 1-form used in (6.71). Consider the integral $\int_{C_h} d\rho_{C_k}^+$. We have $\operatorname{supp} d\rho_{C_k}^+ \subset U^+(C_k)$, hence $d\rho_{C_k}^+ = 0$ on C_k, and so $\int_{C_k} d\rho_{C_k}^+ = 0$. Since $\rho_{C_k}^+$ is a C^∞ function on $\mathcal{R} - |C_k|$, from (6.46) we have $\int_\gamma d\rho_{C_k}^+ = 0$, where γ is a piecewise smooth closed curve not intersecting C_k. The curve C_h intersects C_k only if $h = n$ and $k = m+n$ or if $h = m+n$ and $k = n$. Hence

$$\int_{C_h} d\rho_{C_k}^+ = 0, \qquad |h-k| \neq m. \tag{7.46}$$

So C_n and C_{m+n} intersect at only one point, and C_n crosses C_{m+n} from left to right and C_{m+n} crosses C_n from right to left at this point. Hence we obtain

$$\int_{C_n} d\rho^+_{C_{m+n}} = 1, \qquad \int_{C_{m+n}} d\rho^+_{C_n} = -1 \tag{7.47}$$

by the same method as used for (7.1).

Theorem 7.9. If γ is a 1-cycle on $\mathcal{R}$ then

$$\gamma \sim n_1C_1 + n_2C_2 + \cdots + n_kC_k + \cdots + n_{2m}C_{2m}. \tag{7.48}$$

where $n_1, n_2, \ldots, n_k, \ldots, n_{2m}$ are uniquely determined integers.

Proof: (1) We first prove that the coefficients $n_1, \ldots, n_{2m}$ are uniquely determined by γ. If φ is a continuously differentiable closed 1-form on $\mathcal{R}$ then $\int_\gamma \varphi$, where γ is an arbitrary curve on $\mathcal{R}$, is defined by (6.99), hence $\int_\gamma \varphi$ is also defined for an arbitrary 1-chain γ. If γ is a 1-cycle, then by (6.100) $\int_\gamma \varphi$ is uniquely determined by the homology class of γ. Hence by (7.48):

$$\int_\gamma \varphi = n_1\int_{C_1} \varphi + n_2\int_{C_2} \varphi + \cdots + n_{2m}\int_{C_{2m}} \varphi.$$

Substituting $\varphi = d\rho^+_{C_k}$, we get from (7.46) and (7.47),

$$n_h = \int_\gamma d\rho^+_{C_{m+h}}, \qquad n_{m+h} = -\int_\gamma d\rho^+_{C_h}, \qquad h = 1, \ldots, m.$$

(2) A 1-cycle on $\mathcal{R}$ is a finite linear combination with integer coefficients of closed curves. Hence, in order to prove that an arbitrary 1-cycle γ can be represented as (7.48), it suffices to prove this when γ is a closed curve. In this case it is possible to find a piecewise smooth closed curve $\hat{\gamma} \simeq \gamma$, which intersects each Jordan curve C_h transversally at most finitely many points. Hence we may assume that γ is a piecewise smooth closed curve intersecting each Jordan curve C_h transversally at most finitely many points. We want to prove that there exists a closed curve β, which does not intersect $C_1, C_2, \ldots, C_m$, such that

$$\beta \sim \gamma - \sum_{h=1}^{m} n_{m+h}C_{m+h}, \qquad n_{m+h} \in \mathbb{Z}. \tag{7.49}$$

Consider $[U(C_1)] = \{z_1 : a \leqslant |z_1| \leqslant b\}$ $a = 1/b$, $b = 1 + \varepsilon$, and assume for the sake of simplicity that the part of C_{m+1} contained in $[U(C_1)]$ is given by $(\ell_{m+1})^{-1} : r \to z_1 = r$, $a \leqslant r \leqslant b$. Assume that γ intersects C_1 and call the point of intersection q_1 since γ intersects C_1 transversally, on a neighborhood of $q_1 \in |\gamma|$, with γ given by $t \to z_1 = \gamma(t) = r(t)e^{i\theta(t)}$, $-\delta \leqslant t \leqslant \delta$, and $\gamma(0) = e^{i\theta(0)} = q_1$. If γ crosses C_1 from left to right at q_1, then $r(t)$ is a monotonic increasing function, with $r(-\delta) = a$ and $r(\delta) = b$ if γ crosses C_1 from right to left at q_1, then $r(t)$ is a monotonic decreasing

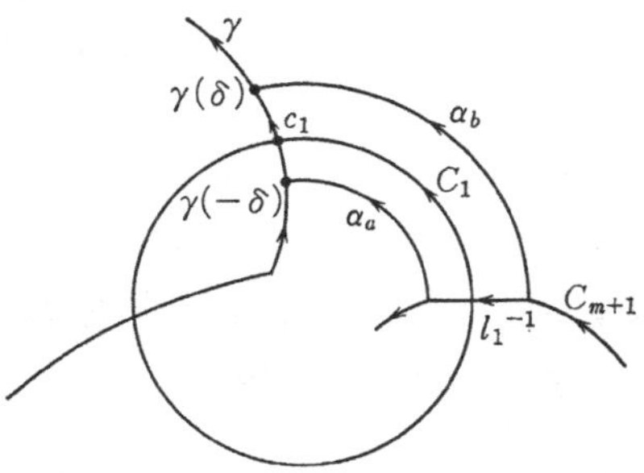

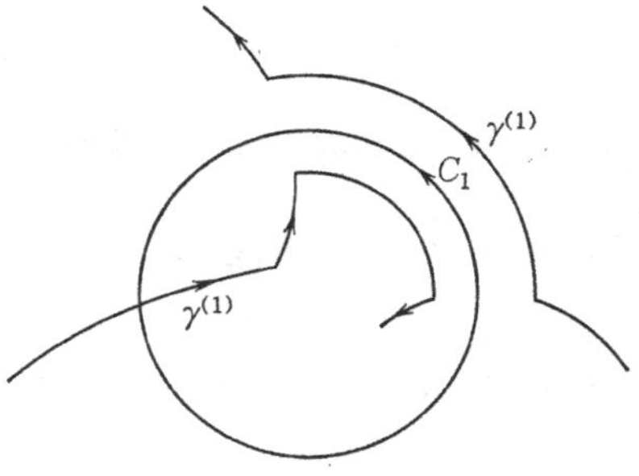

function with $r(-\delta) = b$ and $r(\delta) = a$. Since both cases are similar, we assume that γ crosses C_1 from left to right. Let v be the number of points of intersection of γ and C_1. We will construct a piecewise smooth closed curve $\gamma^{(1)}$ intersecting C_1 at $\nu - 1$ points and satisfying

$$\gamma^{(1)} \sim \gamma + C_{m+1}.$$

Let γ_* be the curve obtained by deleting $\{\gamma(t) : -\delta < t < \delta\}$ from γ. Then γ_* is a curve with starting point $\gamma(\delta)$ and end point $\gamma(-\delta)$. If $\gamma_\delta : t \to \gamma(t)$, $-\delta \leqslant t \leqslant \delta$, then $\gamma = \gamma_* \cdot \gamma_\delta$. Similarly, the curve $C_* = \Pi(\beta_1 \cdot \alpha_2 \cdot \beta_2^{-1})$, obtained from C_{m+1} by deleting $\{r : a < r < b\}$ is a Jordan arc with starting point a and end point b and $C_{m+1} = C_* \cdot \ell_1^{-1}$. Let $\alpha_a : \theta \to ae^{i\theta}$, $0 \leqslant \theta \leqslant \theta(-\delta)$, be the arc connecting a and $\gamma(-\delta)$ and let $\alpha_b : \theta \to be^{i\theta}$, $0 \leqslant \theta \leqslant \theta(\delta)$, be the arc connecting b and $\gamma(\delta)$. Then $\zeta = \ell_1 \cdot \alpha_b \cdot \gamma_\delta^{-1} \cdot \alpha_a^{-1}$ is a closed curve and obviously $\zeta \simeq 0$ on $[U(C_1)] \subset \mathcal{R}$. The curve $\gamma^{(1)} = \gamma_* \cdot C_* \cdot \alpha_b$ is a piecewise smooth closed curve intersection C_1 at $\nu - 1$ points, satisfying

$$\gamma^{(1)} = \gamma + C_{m+1} + \zeta, \qquad \zeta \sim 0,$$

(see Definitions 4.34 and 4.35), hence $\gamma^{(1)} \sim \gamma + C_{m+1}$. If γ crosses C_1 from right to left, then there is a piecewise smooth closed curve $\gamma^{(1)}$ intersecting C_1 at $\nu - 1$ points and satisfying $\gamma^{(1)} \sim \gamma - C_{m+1}$.

From $\gamma^{(1)}$ one obtains a curve $\gamma^{(2)} \sim \gamma^{(1)} \pm C_{m+1}$, intersecting C_1 at $\nu - 2$ points in the same way. Repeating this procedure one finally arrives at a closed curve $\gamma^{(\nu)}$ satisfying $\gamma^{(\nu)} \sim \gamma^{(\nu-1)} \pm C_{m+1}$ and which does not intersect C_1. Obviously, $\gamma^{(\nu)} \sim \gamma - n_{m+1} C_{m+1}$, for some integer n_{m+1}. Outside $[U(C_1)]\gamma^{(\nu)}$ coincides with the original γ. Hence by performing the same procedure for $C_2, \ldots, C_m$ we arrive at the desired curve β. It is clear that β intersects each C_{m+h} transversally at most finitely many points.

(3) To prove (7.48) it suffices to prove that the closed curve β can be represented as

$$\beta \sim n_1 C_1 + n_2 C_2 + \cdots + n_m C_m, \qquad n_1, n_2, \ldots, n_m \in \mathbb{Z}. \tag{7.50}$$

As explained above, $\mathcal{R} = \Pi([D])$. Since the domain of Π has been restricted to $[D]$, Π^{-1} maps $\mathcal{R}^{(m)} = \mathcal{R} - |C_1| - \cdots - |C_m|$ onto the region D. Since β is in $\mathcal{R}^{(m)}$, $\Pi^{-1}(\beta)$ is a piecewise smooth closed curve in $\Pi^{-1}(\beta)$. Choosing γ in such a way that it does not pass through the closed disk $\Pi([U_0\})$ on $\mathcal{R}$, we find that $\Pi^{-1}(\beta)$ is in the region $D - [U_0]$. Put $\sigma = \Pi^{-1}(\beta)$ and let $s \to z = \sigma(s)$, $0 \leqslant s \leqslant 1$, where $\sigma(0) = \sigma(1)$, be a parameter representation of σ. By Lemma 7.7,

$$\mathcal{D} = D - [U_0] - |\gamma_1| - |\gamma_2| - \cdots - |\gamma_{2m-1}| - |\gamma_{2m}|$$

is a simply connected domain. Hence, if σ does not intersect any of the Jordan arcs $\gamma_1, \gamma_2, \ldots, \gamma_{2m}$, we have $\sigma \simeq 0$ in $\mathcal{D}$, hence $\beta = \Pi(\sigma) \simeq 0$ in $\mathcal{R}$ and therefore $\beta \sim 0$. If σ intersects at least one of the γ_k, then $\sigma = \Pi^{-1}(\beta)$ intersects each γ_k transversally at most finitely many points, since β intersects each $C_{m+h} = \Pi(\gamma_{2h-1} \cdot \alpha_{2h} \cdot \gamma_{2h}^{-1})$ transversally at most finitely many points. Denote the points of intersection of σ and $\gamma_1, \gamma_2, \ldots, \gamma_{2m}$ by $\sigma(s_1), \sigma(s_2), \ldots, \sigma(s_{\upsilon-1})$, $0 < s_1 < s_2 < \cdots < s_{\nu-1} < 1$, and put $s_0 = 0$ and $s_\nu = 1$. The curve σ is divided into ν curves

$$\sigma_j : s \to z = \sigma(s), \qquad s_{j-1} \leqslant s \leqslant s_j, \qquad j = 1, 2, \ldots, \nu;$$

that is

$$\sigma = \sigma_1 \cdot \sigma_2 \cdots \sigma_j \cdots \sigma_\nu.$$

Let $t \to z = \gamma_k(t)$, where $\gamma_k(0) \in |\Gamma_k|$ and $\gamma_k(1) = e_k$, be a parametric representation of γ_k. Each $\sigma(s_j)$ determines a $k = k(j)$ and t_j, $0 < t_j < 1$, such that $\sigma(s_j) = \gamma_k(t_j)$. Let

$$\gamma_{kj} : t \to z = \gamma_k(t), \qquad 0 \leqslant t \leqslant t_j, \qquad k = k(j),$$

be the part of γ_k between $\gamma_k(0)$ and $\gamma_k(t_j)$. Consider Γ_k as a closed curve with base point $\gamma_k(0)$. Define $\varepsilon(j) = +1$ if σ crosses γ_k from right to left at $\sigma(s_j)$ and $\varepsilon(j) = -1$ if σ crosses γ_k from left to right at $\sigma(s_j)$. The curve

$$\xi_j = \gamma_{kj}^{-1} \cdot \Gamma_k^{s(j)} \cdot \gamma_{kj}, \qquad k = k(j), \tag{7.51}$$

is a piecewise smooth closed curve with base point $\sigma(s_j)$. Since $\sigma(s_j)$ is the end point of σ_j and the starting point of σ_{j+1}, the curve

$$\tau = \sigma_1 \cdot \xi_1 \cdot \sigma_2 \cdot \xi_2 \cdot \sigma_3 \cdots \sigma_{\nu-1} \cdot \xi_{\nu-1} \cdot \sigma_\nu \tag{7.52}$$

is a closed curve in $[D]$ not passing through the closed disk $[U_0]$ and

$$\Pi(\tau) = \Pi(\sigma_1) \cdot \Pi(\xi_1) \cdot \Pi(\sigma_2) \cdot \Pi(\xi_2) \cdots \Pi(\xi_{\nu-1}) \cdot \Pi(\sigma_\nu).$$

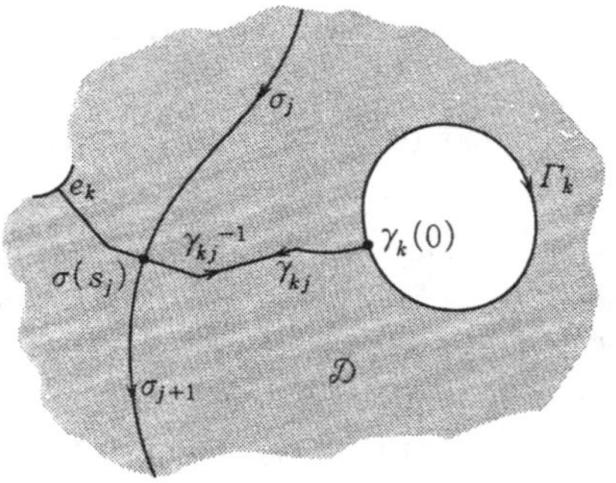

$\xi_j = \gamma_{kj}^{-1} \cdot \Gamma_k \cdot \gamma_{kj}$

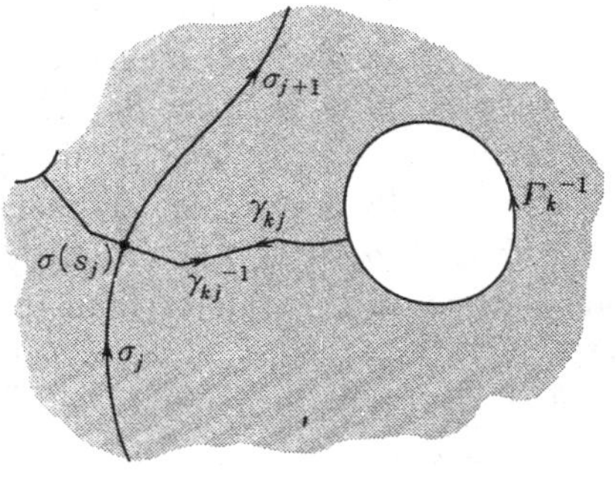

$\xi_j = \gamma_{kj}^{-1} \cdot \Gamma_k^{-1} \cdot \gamma_{kj}$

We have (compare Definitions 4.34 and 4.35)

$$\Pi(\tau) = \sum_{j=1}^{\nu} \Pi(\sigma_j) + \sum_{j=1}^{\nu-1} \Pi(\xi_j), \qquad \Pi(\xi_j) = \varepsilon(j)\Pi(\Gamma_{k(j)})$$

as 1-cycles on $\mathcal{R}$. Hence

$$\sum_{j=1}^{\nu} \Pi(\sigma_j) = \Pi(\sigma_1 \cdot \sigma_2 \cdots \sigma_\nu) = \Pi(\sigma) = \beta, \qquad \Pi(\Gamma_{2h-1}) = C_h$$

and $\Pi(\Gamma_{2h}) = -C_h$

and therefore

$$\beta = \sum_{h=1}^{m} n_h C_h + \Pi(\tau), \qquad n_h = \sum_{k(j)=2h} \varepsilon(j) - \sum_{k(j)=(2h-1)} \varepsilon(j).$$

So, in order to prove (7.50), it suffices to show that $\Pi(\tau) \simeq 0$ on $\mathcal{R}$. In order to prove this, it suffices to show that $\tau \simeq 0$ on $[\mathcal{D}] = [D] - U_0$. All points $p \in |\gamma_k|$ are boundary points of $\mathcal{D} = D - [U_0] - |\gamma_1| - \cdots - |\gamma_{2m}|$.

If we consider p as a boundary point of $U^+(\gamma_k)$ (to the left of γ_k) we denote p by p^+; if we consider p as a boundary point of $U^-(\gamma_k)$ we denote it by p^-. Considering p^+ and p^- as different points, we denote the closure

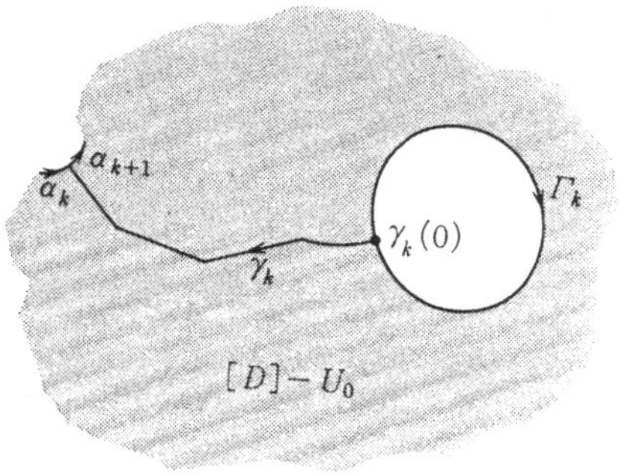

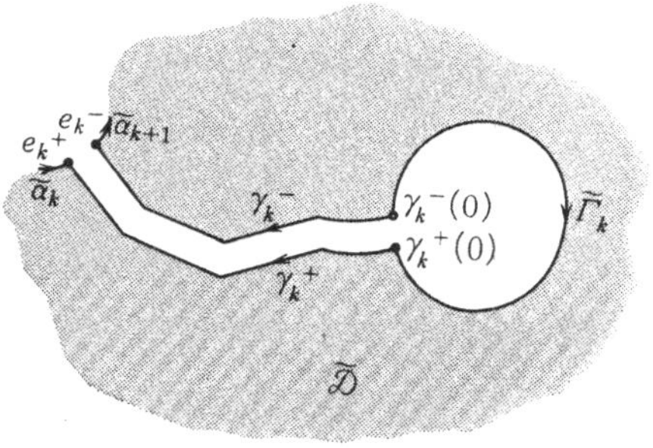

of $\mathcal{D}$ by $\tilde{\mathcal{D}}$. The figure shows a part of $[D] - U_0$ and the corresponding part of $\tilde{\mathcal{D}}$. To give a rigorous definition of $\tilde{\mathcal{D}}$, we follow the same procedure as used to define $\tilde{\mathcal{R}}$. For each γ_k, determine a C^∞ function $\rho_k = \rho_k(z)$ on $D - [U_0]$ satisfying $\operatorname{supp} \rho_k \subset U(\gamma_k)$, $\rho_k(z) \geqslant 0$ for all z and $\rho_k(\gamma_k(t)) = 1, 0 < t < 1$, and define

$$\rho^+(z) = \sum_{k=1}^{2m} \rho_k^+(z), \qquad \rho_k^+(z) = \begin{cases} \rho_k(z), & z \notin U^-(\gamma_k), \\ 0, & z \in U^-(\gamma_k). \end{cases}$$

Since $\rho_k^+(z)$ is a C^∞ function on $\mathcal{D}$, so is $\rho^+(z)$. Hence

$$\mu : z \to \mu(z) = (z, \rho^+(z))$$

is a diffeomorphism mapping $\mathcal{D}$ into $\mathbb{C} \times \mathbb{R}$ and we define

$$\tilde{\mathcal{D}} = [\mu(\mathcal{D})]$$

where $[\mu(\mathcal{D})]$ is the closure of $\mu(\mathcal{D})$ in $\mathbb{C} \times \mathbb{R}$.

Let $\tilde{\omega}$ denote the restriction of the projection of $\mathbb{C}\times\mathbb{R}$ onto $\mathbb{C}$ to $\tilde{\mathcal{D}}$. Of course, $\tilde{\omega}\mu(z) = z$ for $z \in \mathcal{D}$. Now $\tilde{\mathcal{D}}$ is a compact Hausdorff space and $\tilde{\omega} : p \to z = \tilde{\omega}(p)$ is a continuous map which maps $\tilde{\mathcal{D}}$ onto $[\mathcal{D}] = [D] - U_0$. The projection $\tilde{\omega}$ is one-to-one on a sufficiently small neighborhood of each $q \in \tilde{\mathcal{D}}$. The inverse image $\tilde{\omega}^{-1}(\gamma_k(t))$ of a point $\gamma_k(t)$ on γ_k consists of two different points $\gamma_k^+(t) = (\gamma_k(t),\ 1)$ and $\gamma_k^-(t) = (\gamma_k(t),\ 0)$. The inverse image of a point $z \in [D] - U_0 - |\gamma_1| - \cdots - |\gamma_{2m}|$ consists of one point $p = \tilde{\omega}^{-1}(z)$. The inverse image $\tilde{\omega}(\gamma_k)$ of γ_k consists of two Jordan arcs

$$\gamma_k^+ : t \to \gamma_k^+(t), \qquad \gamma_k^- : t \to \gamma_k^-(t), \qquad 0 \leqslant t \leqslant 1.$$

The end point of γ_k^+ is $\gamma_k^-(1) = e_k^+$, the end point of γ_k^- is $\gamma_k^-(1) = e_k^-$. The inverse image $\tilde{\Gamma}_k = \tilde{\omega}^{-1}(\Gamma_k)$, of Γ_k is a Jordan arc with starting point $\gamma_k^-(0)$ and end point $\gamma_k^+(0)$. Hence

$$\tilde{\eta}_k = (\gamma_k^-)^{-1} \cdot \tilde{\Gamma}_k \cdot \gamma_k^+ \tag{7.53}$$

is a Jordan arc with starting point e_{k-1}^- and end point e_k^+. The inverse image $\tilde{\omega}^{-1}(\alpha_k)$ of the arc α_k consists of two points and a Jordan arc, denoted by $\tilde{\alpha}_k$. The starting point of $\tilde{\alpha}_k$ is e_k^-, its end point is e_k^+. Hence

$$\tilde{\Xi} = \tilde{\eta}_{2m} \cdot \tilde{\alpha}_{2m}^{-1} \cdot \tilde{\eta}_{2m-1} \cdots \tilde{\alpha}_3^{-1} \cdot \tilde{\eta}_2 \cdot \tilde{\alpha}_2^{-1} \cdot \tilde{\eta}_1 \cdot \tilde{\alpha}_1^{-1} \tag{7.54}$$

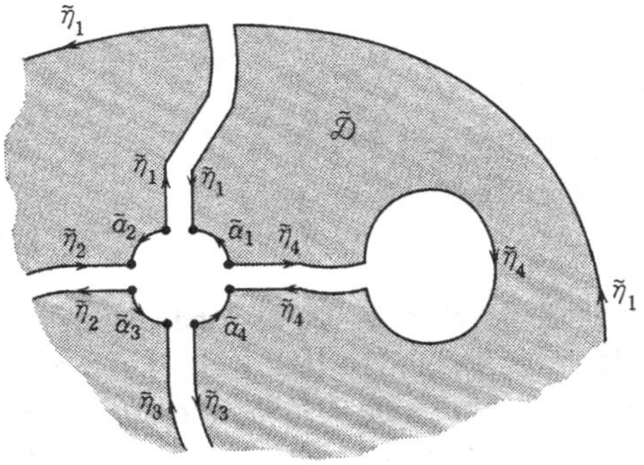

is a Jordan curve. Since $|\tilde{\Xi}| = \tilde{\omega}^{-1}([\mathcal{D}] - \mathcal{D})$ and $\mu(\mathcal{D}) = \tilde{\omega}^{-1}(\mathcal{D})$, we have

$$\tilde{\mathcal{D}} = \mu(\mathcal{D}) \cup |\tilde{\Xi}|, \qquad \mu(\mathcal{D}) \cap |\tilde{\Xi}| = \varnothing.$$

Identifying $z \in \mathcal{D}$ and $\mu(z) = \tilde{\omega}^{-1}(z)$, $\mu(\mathcal{D})$ and $\mathcal{D}$ can be regarded as the same Riemann surface, hence

$$\tilde{\mathcal{D}} = \mathcal{D} \cup |\tilde{\Xi}|, \qquad \mathcal{D} \cap |\tilde{\Xi}| = \varnothing. \tag{7.55}$$

Hence the Jordan arc $|\tilde{\Xi}|$ is the boundary of $\mathcal{D}$ in $\tilde{\mathcal{D}}$ and $\tilde{\mathcal{D}}$ is the surface obtained from $[\mathcal{D}] = [D] - U_0$ by replacing its boundary $[\mathcal{D}] - \mathcal{D}$ by $|\tilde{\Xi}|$. The C^∞ function $\rho^+(z)$ on $\mathcal{D}$ can be extended in a natural way to a continuous function $\tilde{\rho}^+(p)$ on $\tilde{\mathcal{D}}$. Denoting the restriction of the projection of $\mathbb{C} \times \mathbb{R}$ onto $\mathbb{R}$ to $\tilde{\mathcal{D}}$ by $\tilde{\rho}^+ : p \to \tilde{\rho}^+(p)$, then $\tilde{\rho}^+(p)$ is a continuous map on $\tilde{\mathcal{D}}$ and $\tilde{\rho}^+(p) = \rho^+(z)$ for $p = \mu(z) = z \in \mathcal{D}$. The distance $\tilde{d}(p, q)$ between two points $p, q \in \tilde{\mathcal{D}}$ is given by

$$\tilde{d}(p, q) = \sqrt{|\tilde{\omega}(p) - \tilde{\omega}(q)|^2 + |\tilde{\rho}^+(p) - \tilde{\rho}^+(q)|^2}.$$

As shown in the proof of Lemma 7.7, $\mathcal{D}$ is a region without holes and there exists a conformal map $f : z \to w = f(z)$ which maps $\mathcal{D}$ onto the unit disk $\mathbb{U}$. Exactly as in the proof of Lemma 7.4, it is possible to extend f to a continuous map $\tilde{f}$, which maps $\tilde{\mathcal{D}}$ onto $[\mathbb{U}]$. Similarly, it is possible to extend the conformal map $g = f^{-1}$ (which maps $\mathbb{U}$ onto $\mathcal{D}$) to a continuous map $\tilde{g}$ mapping $[\mathbb{U}]$ onto $[\mathcal{D}]$. The map $\tilde{g} \circ \tilde{f}$ maps $\tilde{\mathcal{D}}$ onto $[\mathcal{D}]$ and $\tilde{g} \circ \tilde{f}$ coincides with $\tilde{\omega}$ on D sine $(\tilde{g} \circ \tilde{f})(z) = g(f(z)) = z$ for $z \in D \subset \tilde{\mathcal{D}}$. Hence $\tilde{g} \circ \tilde{f}$ and $\tilde{\omega}$ coincide on $\tilde{\mathcal{D}}$:

$$\tilde{\omega} = \tilde{g} \circ \tilde{f}. \tag{7.56}$$

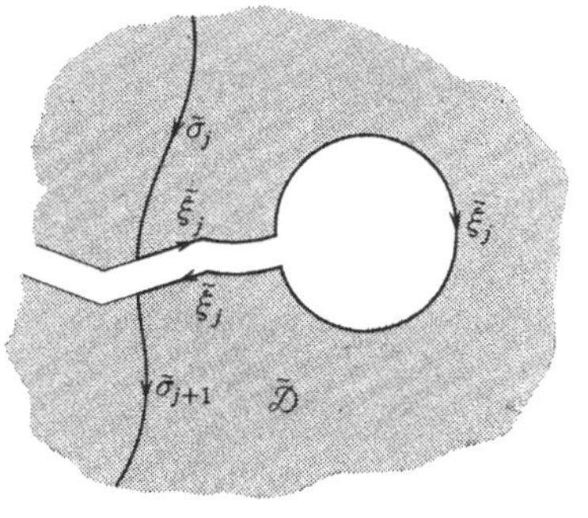

$\varepsilon(j) = +1$

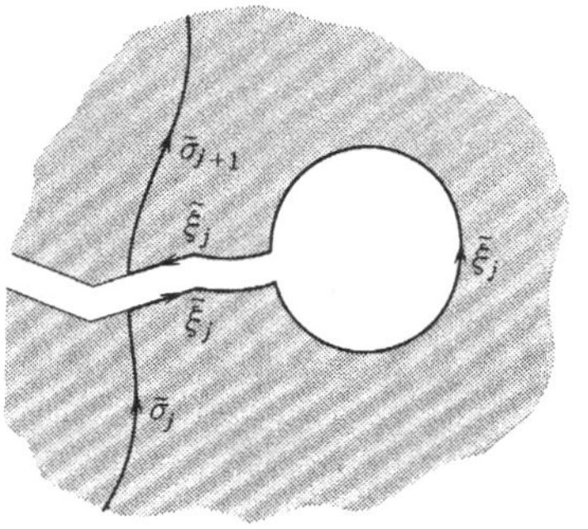

$\varepsilon(j) = -1$

We want to consider the closed curve $\tau = \sigma_1 \cdot \xi_1 \cdot \sigma_2 \cdots \xi_{\nu-1} \cdot \sigma_\nu$ on $[\mathcal{D}]$ (see (7.52)). The curve σ_j has $\sigma(s_{j-1})$ as its starting point and $\sigma(s_j)$ as its end point and $\tilde{\omega}^{-1}(\sigma_j)$ consists of two points and a curve $\tilde{\sigma}_j$ (just as $\tilde{\omega}^{-1}(\alpha_k)$). If σ crosses γ_k at $\sigma(s_j) = \gamma_k(t_j)$, $k = k(j)$, from right to left, then the end point of $\tilde{\sigma}_j$ is $\gamma_k^-(t_j)$; if σ crosses from left to right then the end point of $\tilde{\sigma}_j$ is $\gamma_k^+(t_j)$. Hence the end point of $\tilde{\sigma}_j$ is given by $\gamma_k^{-\varepsilon(j)}(t_j)$, where $\varepsilon(j)$ is defined in (7.51). Similarly, the starting point of $\tilde{\sigma}_{j+1}$ is $\gamma_k^{\varepsilon(j)}(t_j)$. The inverse image $\tilde{\omega}^{-1}(\gamma_{kj})$ of the Jordan arc γ_{kj} consists of two Jordan arcs

$$\gamma_{kj}^+ : t \to \gamma_k^+(t) \text{ and } \gamma_{kj}^- : t \to \gamma_k^-(t), \qquad 0 \leqslant t \leqslant t_j, \qquad k = k(j).$$

For ξ_j, as defined in (7.51), we put

$$\tilde{\xi}_j = (\gamma_{kj}^{-\varepsilon(j)})^{-1} \cdot \tilde{\Gamma}_{kj}^{\varepsilon(j)} \cdot \gamma_{kj}^{\varepsilon(j)}, \qquad k = k(j),$$

and $\tilde{\xi}_j$ is a Jordan arc on $\tilde{\mathcal{D}}$; its starting point is $\gamma_k^{-\varepsilon(j)}(t_j)$ and its end point is $\gamma_k^{\varepsilon(j)}(t_j)$. Hence, by joining together the curves $\tilde{\sigma}_1$, $\tilde{\xi}_1$, $\tilde{\sigma}_2$, $\tilde{\xi}_2$, $\tilde{\sigma}_3, \ldots$ in this order we obtain the closed curve

$$\tilde{\tau} = \tilde{\sigma}_1 \cdot \tilde{\xi}_1 \cdot \tilde{\sigma}_2 \cdot \tilde{\xi}_2 \cdot \tilde{\sigma}_3 \cdots \tilde{\sigma}_{\nu-1} \cdot \tilde{\xi}_{\nu-1} \cdot \tilde{\sigma}_\nu$$

on $\tilde{\mathcal{D}}$. Since $\tilde{\omega}(\tilde{\sigma}_j) = \sigma_j$ and $\tilde{\omega}(\tilde{\xi}_j) = \xi_j$, we have $\tilde{\omega}(\tilde{\tau}) = \tau$. Since $\tilde{\omega} = \tilde{g} \circ \tilde{f}$ by (7.56), we have

$$\tau = \tilde{g}(\tilde{f}(\tilde{\tau})).$$

Since $\tilde{f}(\tilde{\tau})$ is a closed curve on the closed unit disk $[\mathbf{U}]$, we have $\tilde{f}(\tilde{\tau}) \simeq 0$ in $[\mathbf{U}]$. Therefore, since $\tilde{g}$ maps $[\mathbf{U}]$ onto $[\mathcal{D}]$ continuously, we conclude $\tau = \tilde{g}(\tilde{f}(\tilde{\tau})) \simeq 0$ on $[\mathcal{D}]$.

We want to study the homology of a Riemann surface $\mathcal{R}$ from which the points $q_1, \ldots, q_r$ have been deleted. For simplicity, we assume that $q_k \notin \Pi([U_0]$ and $q_h \notin |C_k|$, $h = 1, \ldots, r$ and $k = 1, \ldots, 2m$. Since Π^{-1} maps $\mathcal{R}^{(m)}$ conformally onto D, it maps $\mathcal{R} - \Pi([U_0]) - |C_1| - \cdots - |C_{2m}|$ conformally onto $\mathcal{D}$. Hence

$$f \circ \Pi^{-1} : p \to w = f(\Pi^{-1}(p))$$

maps $\mathcal{R} - \Pi([U_0]) - |C_1| - \cdots - |C_{2m}|$ conformally onto the unit disk U. Put $w_h = f(\Pi^{-1}(q_h))$ and, 0, for sufficiently small $\varepsilon > 0$, let $U_\varepsilon(w_h)$ be an ε-neighborhood of w_h in the w-plane; that is $[U_\varepsilon(w_h)] \subset \mathbf{U}$. Regarding $w = f(\Pi^{-1}(p))$ as a local coordinate of p, then $\Pi(f^{-1}(U_\varepsilon(w_h)))$ is a coordinate disk with center q_h on $\mathcal{R}$, the circumference of which we denote by q_h.

Theorem 7.10. An arbitrary 1-cycle γ on $\mathcal{R} - \{q_1, \ldots, q_r\}$ is homologous to a linear combination with integer coefficients of $C_1, C_2, \ldots, C_{2m}$, $Q_1, Q_2, \ldots, Q_r$; that is

$$\gamma \sim n_1 C_1 + \cdots + n_{2m} C_{2m} + n_{2n+1} Q_1 + \cdots + n_{2m=r} Q_r. \tag{7.57}$$

Proof: From the proof of Theorem 7.9, we have

$$\gamma \sim \beta + n_{m+1} C_{m+1} + n_{m+2} C_{m+2} + \cdots + n_{2m} C_{2m},$$

on $\mathcal{R} - \{q_1, \ldots, q_r\}$, where β is a closed curve in $\mathcal{R}^{(m)} - \{q_1, \ldots, q_r\}$. Further

$$\beta = \sum_{h=1}^{m} n_h C_h + \Pi(\tau), \qquad \tau = \tilde{\omega}(\tilde{\tau}),$$

and $\tilde{\tau}$ is a closed curve on $\tilde{\mathcal{D}} = \mathcal{D} \cup |\tilde{\Xi}|$ not passing through $\Pi^{-1}(q_h) \in \mathcal{D}$. Hence $\tilde{f}(\tilde{\tau})$ is closed curve in the closed unit disk [**U**] not passing through $w_h = f(\Pi^{-1}(q_h))$, $h = 1, 2, \ldots, r$.

Now $B_h = \partial[U_\varepsilon(w_h)]$ is a circle with center w_h and radius ε, $Q_h = \Pi(g(B_h)) = (\Pi \circ \tilde{g})(B_h)$, $g = f^{-1}$, and $\Pi \circ \tilde{g}$ is a continuous map mapping [**U**] onto $\mathcal{R} - U_0$, while $(\Pi \circ \tilde{g})(w_h) = q_h$. Hence $\Pi \circ \tilde{g}$ maps $[\mathbf{U}] - \{w_1, w_2, \ldots, w_r\}$ onto $\mathcal{R} - U_0 - \cdots \{q_1, q_2, \ldots, q_r\}$ continuously. To prove (7.57) it suffices to prove that

$$\Pi(\tau) \sim n_{2m+1} Q_1 + n_{2m+2} Q_2 + \cdots + n_{2m+r} Q_r$$

on $\mathcal{R} - \{q_1, q_2, \ldots, q_r\}$. Since $\tau = \tilde{g}(\tilde{f}(\tilde{\tau}))$ and hence $\Pi(\tau) = (\Pi \circ \tilde{g})(\tilde{f}(\tilde{\tau}))$, it suffices to prove that

$$\tilde{f}(\tilde{\tau}) \sim n_{2m+1} B_1 + n_{2m+2} B_2 + \cdots + n_{2m+r} B_r \tag{7.58}$$

on $[\mathbf{U}] - \{w_1, w_2, \ldots, w_r\}$. Now (7.58) is easily verified by the method used in part (2) of the proof of Theorem 7.9. Fixing $w_0 \in \partial[\mathbf{U}]$, let L_h be

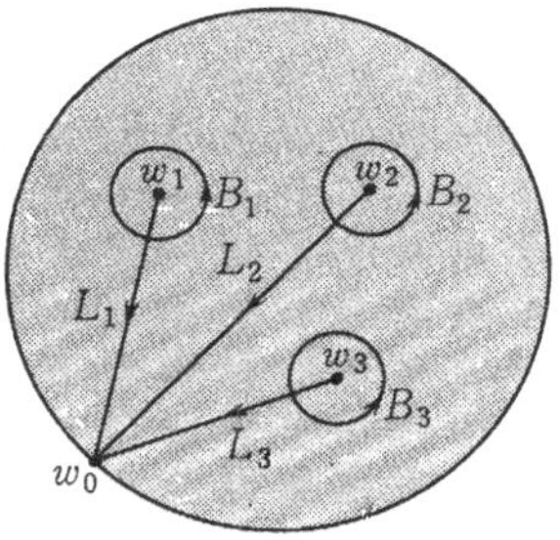

the segment connecting w_h and w_0. By choosing w_0 suitably, we may assume that any two of $L_1, L_2, \ldots, L_r$ intersect only at w_0. We choose a piecewise smooth closed curve μ in $\mathbf{U} - \{w_1, \ldots, w_r\}$ which intersects each L_h transversally at most finitely many points, such that $\mu \sim \tilde{f}(\tilde{\tau})$ on $[\mathbf{U}] - \{w_1, \ldots, w_r\}$: μ can be obtained by a slight continuous deformation of $\tilde{f}(\tilde{\tau})$.

Let $t \to w = \mu(t)$, $0 \leqslant t \leqslant 1$, $\mu(1) = \mu(0)$, be a parametric representation of μ. Let $\mu(t_1)$, $0 < t_1 < 1$, be a point of intersection of μ and L_h. For sufficiently small δ, divide μ into the Jordan arc $\mu_\delta : t \to w = \mu(t)$, $t_1 - \delta \leqslant t \leqslant t_1 + \delta$, and the remaining curve $\mu_* : \mu = \mu_* \cdot \mu_\delta$. Replacing μ_δ by a Jordan arc λ_1, connecting $\mu(t_1 - \delta)$ and $\mu(t_1 + \delta)$ in $\mathcal{U} - |L_1| - |L_2| - \cdots - |L_r|$, we obtain a closed curve

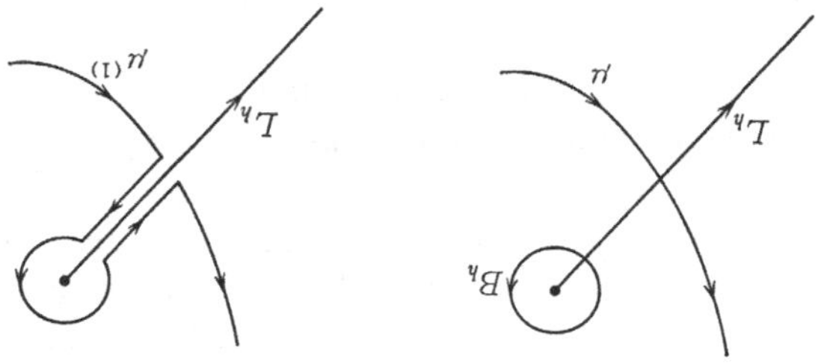

$$\mu^{(1)} = \mu_* \cdot \lambda_1 \sim \mu \pm B_h$$

(compare (2) in the proof of Theorem 7.9). The total number of points of intersection of $\mu^{(1)}$ and $L_1, L_2, \ldots, L_r$ is one fewer then the total number of points of intersection of μ and $L_1, L_2, \ldots, L_r$. By repeating this procedure a finite number of times, we arrive at a closed curve

$$\mu^{(\nu)} \sim \mu - n_{2m+1}B_1 - n_{2m+2}B_2 - \cdots - n_{2m+r}B_r$$

in $\mathbf{U}$, which does not intersect any of the segments $L_1, L_2, \ldots, L_r$. Thus $\mathbf{U} - |L_1| - \cdots - |L_r|$ is a simply connected region and so $\mu^{(\nu)} \simeq 0$ on that region. Hence, since $\tilde{f}(\tilde{\tau}) \sim \mu$, (7.58) is true.

According to Theorem 7.9, each element of $H_1(\mathcal{R})$ can be represented in only one way as the homology class of a linear combination with integer coefficients of Jordan curves $C_1, C_2, \ldots, C_{2m}$. We say $\{C_1, \ldots, C_{2m}\}$ constitutes a base of $H_1(\mathcal{R})$. Denoting, as usual the additive group $\{(n_1, n_2, \ldots, n_{2m}) : n_i \in \mathbb{Z}, i = 1, \ldots, 2m\}$ by $\mathbb{Z}^{2m}$, we can summarize the above as: $H_1(\mathcal{R})$ is isomorphic to $\mathbb{Z}^{2m}$. Since obviously $H_1(\mathcal{R})$ is determined uniquely by the topology of $\mathcal{R}$, m is a topological invariant of $\mathcal{R}$. We shall see later that $2m = b_1(\mathcal{R})$. For an arbitrary continuously differentiable closed 1-form ψ, we call $\int_{C_k} \psi$ the *period of* ψ on the 1-cycle C_k. If $\int_\gamma \psi = 0$ for all piecewise smooth closed curves γ on $\mathcal{R}$, then by Theorem 6.14 $\psi = d\eta$ for some twice continuously differentiable function η on $\mathcal{R}$. Since $\gamma \sim \sum_{k=1}^{2m} n_k C_k$ (Theorem 7.9), we have $\int_\gamma \psi = \sum_{k=1}^{2m} n_k \int_{C_k} \psi$ by (6.100). Hence, if all periods $\int_\gamma \psi$, $k = 1, 2, \ldots, 2m$, vanish, then ψ is exact. Let us put $\rho_k^+ = \rho_{C_k}^+$ then $d\rho_k^+$ is a closed 1-form of class C^∞ on $\mathcal{R}$.

Theorem 7.11. A continuously differentiable closed 1-form ψ on $\mathcal{R}$ can be represented as:

$$\psi = a_1\, d\rho_1^+ + a_2\, d\rho_2^+ + \cdots + a_{2m}\, d\rho_{2m}^+ + d\eta, \tag{7.59}$$

where η is a twice continuously differentiable function and where the coefficients $a_1, a_2, \ldots, a_{2m}$ are uniquely determined by ψ.

Proof: If ψ can be written in the form (7.59), then

$$a_h = -\int_{C_{m+h}} \psi, \qquad a_{m+h} = \int_{C_h} \psi \tag{7.60}$$

$h = 1, 2, \ldots, 2m$, by (7.46) and (7.47). Conversely, for a given ψ, define $a_1, a_2, \ldots, a_{2m}$ by (7.60). Putting $\varphi = \psi - \sum_{k=1}^{2m} a_k\, d\rho_k^+$, we have $\int_{C_k} \varphi = 0$, $k = 1, 2, \ldots, 2m$. Hence $\varphi = d\eta$ for some function η, thus proving (7.59).

It follows from this theorem that the dimension of the d-cohomology group $H_d^1(\mathcal{R})$ of $\mathcal{R}$ is $2m$ where $2m = b_1(\mathcal{R})$. We call $m = b_1(\mathcal{R})/2$ *the genus* of $\mathcal{R}$ and denote it by $g = g(\mathcal{R})$:

$$g = g(\mathcal{R}) = m = \frac{1}{2} b_1(\mathcal{R}). \tag{7.61}$$

Corollary. This first homology group $H_1(\mathcal{R})$ of $\mathcal{R}$ is isomorphic to $\mathbb{Z}^{b_1}$.

8

Analytic functions on a closed Riemann surface

In this chapter we will investigate analytic functions on a closed Riemann surface $\mathcal{R}$ which we will always suppose to be compact.

8.1 Abelian differentials of the first kind

a. Harmonic 1-forms of the first kind

Let $\mathcal{R}$ be a compact Riemann surface. The only holomorphic single-valued functions on $\mathcal{R}$ are constants. For suppose $f(p)$ is holomorphic on $\mathcal{R}$. Since $\mathcal{R}$ is compact, the continuous function $|f(p)|$ assumes a maximum at some point $p_0 \in \mathcal{R}$, hence $f(p)$ has to be a constant by the maximum principle (Theorem 1.21). However, for multi-valued functions the situation is different. For example, if φ is a harmonic 1-form on $\mathcal{R}$ (not identically equal to 0), then the integral of (6.102)

$$f(p) = \int_{\gamma}^{p} (\varphi + i * \varphi) \tag{8.1}$$

defines a multi-valued analytic function, holomorphic on $\mathcal{R}$. A 1-form which is harmonic on the whole of $\mathcal{R}$ is called *a harmonic* 1-*form of the first kind.* If φ is a harmonic 1-form of the first kind, then $\varphi + i * \varphi$ is a 1-form which is holomorphic on the whole of $\mathcal{R}$. A 1-form which is holomorphic on the whole of $\mathcal{R}$ is called *an Abelian differential of the first kind.* Writing an Abelian differential ω of the first kind as $\omega = \varphi + i * \varphi$, then φ and $*\varphi$ are harmonic 1-forms of the first kind.

Let $g = g(\mathcal{R})$ be the genus of $\mathcal{R}$. By (7.61), $g = m$, where m is defined in Section 7.2b. Choosing analytic Jordan curves $C_1, C_2, \ldots, C_g$, as described in Section 7.2b, piecewise analytic Jordan curves $C_{g+1}, C_{g+2}, \ldots, C_{2g}$ are determined by (7.45). Let λ, ν be two natural numbers less than or equal to g and let j, k be two natural numbers less than or equal to $2g$. Put $\rho_j^+ = \rho_{C_j}^+$. We have, by (7.46) and (7.47)

$$\int_{C_\nu} d\rho^+_{g+\nu} = -\int_{C_{g+\nu}} d\rho^+_\nu = 1,$$
$$\text{if } |j-k| \neq g \text{ then } \int_{C_k} d\rho^+_j = 0. \tag{8.2}$$

The linear space $\mathcal{L}$ of all continuously differentiable closed 1-forms on $\mathcal{R}$ coincides with the space $\mathcal{L}_\phi$, where ϕ is the function which vanishes identically, i.e. $\mathcal{L} = \mathcal{L}_0$ (compare (6.79)). Hence, by Dirichlet's Principle for each $\psi \in \mathcal{L}$ there exists a harmonic 1-form of the first kind φ such that $\varphi = \psi + d\eta$ where η is a twice continuously differentiable function on $\mathcal{R}$. (In general, $\varphi = \psi + d\eta_\infty$ is a harmonic 1-form with the same singularities as $d\phi$. Since in this case $d\phi = 0$, we have φ is harmonic on $\mathcal{R}$.)

Lemma 8.1. If the harmonic 1-form of the first kind φ is exact, then $\varphi = 0$ throughout $\mathcal{R}$.

Proof: By assumption, $\varphi = d\eta$ for some twice continuously differentiable function η on $\mathcal{R}$. Since $d * \varphi = 0$, we have, (from Theorem 6.11),

$$\|\varphi\|^2 = \int_{\mathcal{R}} \varphi \wedge *\varphi = \int_{\mathcal{R}} d\eta \wedge *\varphi = \int_{\mathcal{R}} d(\eta * \varphi) = 0.$$

Hence $\varphi = 0$.

For each $\psi \in \mathcal{L}$, the harmonic 1-form of the first kind φ satisfying $\varphi = \psi + d\eta$ is unique. To see this, let $\varphi_1 = \psi + d\eta_1$ and $\varphi_2 = \psi + d\eta_2$ be harmonic 1-forms of the first kind. Then $\varphi_2 - \varphi_1 = d(\eta_2 - \eta_1) = 0$ identically, by the above Lemma. The harmonic 1-form of the first kind $\varphi = \psi + d\eta$ is called the *harmonic part* of ψ.

Since $d\rho^+_j$ is a closed 1-form of class C^∞, we have $d\rho^+_j \in \mathcal{L}$. Letting e_j denote the harmonic part of $d\rho^+_j$, we have

$$e_j = d\rho^+_j = d\eta_j, \qquad j = 1, 2, \ldots, 2g. \tag{8.3}$$

The function η_j is of class C^∞ on $\mathcal{R}$. Since $\int_{C_k} d\eta_j = 0$ by (6.46), we have

$$\int_{C_\nu} e_{g+\nu} = 1, \qquad \int_{C_{g+\nu}} e_\nu = -1, \qquad \text{if } |j-k| \neq g \text{ then } \int_{C_k} e_j = 0. \tag{8.4}$$

Theorem 8.1. A harmonic 1-form of the first kind φ on $\mathcal{R}$ can be written in exactly one way as a linear combination of $e_1, e_2, \ldots, e_{2g}$:

$$\varphi = a_1 e_1 + a_2 e_2 + \cdots + a_g e_g + a_{g+1} e_{g+1} + \cdots + a_{2g} e_{2g}. \tag{8.5}$$

The coefficients are given by

$$a_\nu = -\int_{C_{g+\nu}} \varphi, \qquad a_{g+\nu} = \int_{C_\nu} \varphi, \qquad \nu = 1, 2, \ldots, g. \tag{8.6}$$

Proof: Assuming (8.5), (8.6) follows from (8.4). Conversely, we define $a_1, a_2, \ldots, a_{2g}$ by (8.6) and put

$$\psi = \varphi - a_1 e_1 - a_2 e_2 - \cdots - a_{2g} e_{2g}.$$

Then $\int_{C_k} \psi = 0$, $k = 1, 2, \ldots, 2g$, by (8.4). Now ψ is harmonic, hence closed, and the arbitrary closed curve γ on $\mathcal{R}$ is homologous to a linear combination of $C_1, C_2, \ldots, C_{2g}$ with integer coefficients. Hence $\int_\gamma \psi = 0$ by (6.100), therefore ψ is exact by Theorem 6.14. Hence $\psi = 0$ by Lemma 8.1, and we are done.

For arbitrary $\psi \in \mathcal{L}$, we have

$$\int_{C_j} \psi = \int_{\mathcal{R}} d\rho_j^+ \wedge \psi \tag{8.7}$$

by (6.71). Substituting e_k for ψ and observing that $\int_{\mathcal{R}} d\eta_j \wedge e_k = \int_{\mathcal{R}} d(\eta_j e_k) = 0$, we have

$$\int_{C_j} e_k = \int_{\mathcal{R}} d\rho_j^+ \wedge e_k = \int_{\mathcal{R}} e_j \wedge e_k.$$

Hence, by (8.4),

$$\int_{\mathcal{R}} e_\nu \wedge e_{g+\nu} = 1, \qquad \text{if } |j-k| \neq g \text{ then} \int_{\mathcal{R}} e_j \wedge e_k = 0. \tag{8.8}$$

The subspace of $\mathcal{L}$ consisting of all harmonic 1-forms of the first kind will be denoted by $\mathcal{H}$. By Theorem 8.1, $\dim_{\mathbb{R}} \mathcal{H} = 2g$ and $\{e_1, e_2, \ldots, e_{2g}\}$ is a basis of $\mathcal{H}$. The conjugate form $*e_k$ is also a harmonic 1-form of the first kind. Hence $*e_\lambda$ can be represented as:

$$*e_\lambda = \sum_{\nu=1}^{g} \alpha_{\lambda\nu} e_{g+\nu} + \sum_{\nu=1}^{g} \beta_{\lambda\nu} e_\nu, \qquad \lambda = 1, 2, \ldots, g, \tag{8.9}$$

where, by (8.8),

$$\alpha_{\lambda\nu} = \int_{\mathcal{R}} e_\nu \wedge *e_\lambda = (e_\nu, e_\lambda). \tag{8.10}$$

Hence $a_{\nu\lambda} = a_{\lambda\nu}$ and $\sum_{\lambda\nu=1}^{g} a_{\lambda\nu} x_\lambda x_\nu = \|\sum_{\lambda=1}^{g} x_\lambda e_\lambda\|^2 \geqslant 0$ with $\|\sum x_\lambda e_\lambda\| > 0$ whenever $x_1, x_2 \ldots x_{\mathrm{j}}$ are not all 0. Hence $\det(\alpha_{\lambda\nu}) \neq 0$, and each $e_{g+\nu}$, $\nu = 1, 2, \ldots, g$ can be represented as a linear combination of $e_1, e_2, \ldots, e_g, *e_1, *e_2, \ldots, *e_g$ by (8.9). It follows that $\{e_1, e_2, \ldots, e_g, *e_1, *e_2, \ldots, *e_g\}$ is another basis of $\mathcal{H}$.

b. Abelian differential of the first kind

An Abelian differential of the first kind on $\mathcal{R}$ can be written as $\omega = \varphi + i * \varphi$ with $\varphi \in \mathcal{H}$. Since $\varphi \in \mathcal{H}$ can be represented uniquely as

$$\varphi = \sum_{\lambda=1}^{g} a_\lambda e_\lambda + \sum_{\lambda=1}^{g} b_\lambda * e_\lambda,$$

ω can be written in a unique way as

$$\omega = \sum_{\lambda=1}^{g} c_\lambda (e_\lambda + i * e_\lambda), \qquad c_\lambda = a_\lambda - i b_\lambda.$$

We have, by (8.9),

$$e_\lambda + i * e_\lambda = \sum_{\nu=1}^{g} (\delta_{\lambda\nu} + i\beta_{\lambda\nu}) e_\nu + \sum_{\nu=1}^{g} i\alpha_{\lambda\nu} e_{g+\nu}$$

and $\det(\alpha_{\lambda\nu}) \neq 0$. Putting $(A_{\lambda\nu}) = (\alpha_{\lambda\nu})^{-1}$ and

$$\omega_\nu = \sum_{\lambda=1}^{g} A_{\nu\lambda}(e_\lambda + i * e_\lambda) = \sum_{\lambda=1}^{g} a_{\nu\lambda} e_\lambda + i e_{g+\nu}, \tag{8.11}$$

we obtain the following theorem.

Theorem 8.2. An Abelian differential of the first kind ω can be represented uniquely as

$$\omega = c_1 \omega_1 + c_2 \omega_2 + \cdots + c_g \omega_g$$

where $c_1, c_2, \ldots, c_g$ are complex numbers.

We have, by (8.4) and (8.11)

$$\int_{C_\lambda} \omega_\nu = i\delta_{\nu\lambda}, \qquad \int_{C_{g+\lambda}} \omega_\nu = -a_{\nu\lambda}. \tag{8.12}$$

8.2 Abelian differentials of the second and third kind

a. Meromorphic functions

A function $f = f(p)$ which is holomorphic and single-valued on $\mathcal{R}$ with the exception of a finite number of poles, is called *meromorphic*. Let $\mathfrak{q}_1, \ldots, \mathfrak{q}_t$ be the poles of f with orders $m_1, \ldots, m_t$ and let $\mathfrak{p}_1, \ldots, \mathfrak{p}_s$ be the zeros of f with orders $n_1, \ldots, n_s$. The 0-chain:

$$n_1\mathfrak{p}_1 + n_2\mathfrak{p}_2 + \cdots + n_s\mathfrak{p}_s - m_1\mathfrak{q}_1 - m_2\mathfrak{q}_2 - \cdots - m_t\mathfrak{q}_t$$

is called the *divisor* of f and denoted by (f). Generally, a 0-chain $\mathfrak{d} = \sum_k m_k \mathfrak{p}_k$ with integer coefficients is called a *divisor* of $\mathcal{R}$; $\sum_k m_k$ is called the *degree* of $\mathfrak{d}$ and denoted by $\deg \mathfrak{d}$. If all $m_k \geqslant 0$, then we write $\mathfrak{d} \geqslant 0$. If $m_k \geqslant 0$ and at least on $m_k > 0$, then $\mathfrak{d}$ is called *positive* and we write $\mathfrak{d} > 0$.

b. *Abelian differentials of the second and third kind*

Next we want to consider a 1-form ω which is holomorphic on $\mathcal{R}$ with the exception of a finite number of points $\mathfrak{q}_1, \ldots, \mathfrak{q}_r$. Fix $p_0 \in \mathcal{R}$ and for arbitrary $p \in \mathcal{R}$ let γ be a piecewise smooth Jordan arc connecting p_0 and p. The function

$$f(p) = \int_\gamma^p \omega$$

is a holomorphic analytic function on $\mathcal{R} - \{\mathfrak{q}_1, \ldots, \mathfrak{q}_r\}$. We assume that $p \neq \mathfrak{q}_k$ and $p_0 \neq \mathfrak{q}_k$ and that γ does not pass through any $\mathfrak{q}_k$, $k = 1, \ldots, r$. Of course, f is in general a multi-valued function and $\omega = df$, that is ω is locally the differential of the analytic function f. Let z_k: $p \to z_k(p)$ be a local coordinate around $\mathfrak{q}_k$ and let us write $f(z_k)$ for $f(p)$ as a function of $z_k = z_k(p)$. Then

$$\omega = f'(z_k)dz_k$$

and $0 = z_k(\mathfrak{q}_k)$ is an isolated singularity of $f'(z_k)$. If $z_k = 0$ is a pole of $f'(z_k)$ for $k = 1, \ldots, r$, then ω is called an *Abelian differential* on $\mathcal{R}$. If $z_k = 0$ is a pole of the m_kth order, $\mathfrak{q}_k$ is called a pole of the m_kth order of ω. On a neighborhood of $\mathfrak{q}_k$, ω can be represented as

$$\omega = \left(\frac{a_{k,-m}}{z_k^m} + \cdots + \frac{a_{k,-1}}{z_k} + a_{k,0} + \cdots\right) dz_k,$$

$$a_{k,-m} \neq 0, \qquad m = m_k,$$

The coefficient $a_{k,-1}$ is called the *residue* of ω at $\mathfrak{q}_k$ and is denoted by $\mathrm{Res}_{\mathfrak{q}_k}[\omega]$.

$$\mathrm{Res}_{\mathfrak{q}_k}[\omega] = \frac{1}{2\pi i}\oint_{\mathfrak{q}_k} \omega.$$

If $\mathrm{Res}_{\mathfrak{q}_k}[\omega] = 0$ for all poles $\mathfrak{q}_k$, then ω is called *an Abelian differential of the second kind*. If $\mathrm{Res}_{\mathfrak{q}_k}[\omega] \neq 0$ for some $\mathfrak{q}_k$ then ω is called an *an Abelian differential of the third kind*. The integral $\int_\gamma^p \omega$ is called an *Abelian integral* of the first (second, third) kind if ω is an Abelian differential of the first (second, third) kind. Considering the Abelian integral $f = f(p)$ as an indefinite integral of ω, we write $f = \int \omega$. The Abelian differential ω can be represented as $\omega = df$. Using a local coordinate, the Abelian differential $\omega = df$ can be written as $\omega = f'(z)\,dz$. If $z(\mathfrak{p})$ is a zero of the nth order of $f'(z)$, then $\mathfrak{p}$ is called a *zero* of the nth order of ω. Let ω have poles $\mathfrak{p}_1, \ldots, \mathfrak{q}_t$ with orders $m_1, \ldots, m_t$ and zeros $\mathfrak{p}_1, \ldots, \mathfrak{p}_s$ with orders $n_1, \ldots, n_s$. Exactly as for meromorphic functions, we define the divisor (ω) of ω by

$$(\omega) = n_1\mathfrak{p}_1 + n_2\mathfrak{p}_2 + \cdots + n_s\mathfrak{p}_s - m_1\mathfrak{q}_1 - m_2\mathfrak{q}_2 - \cdots - m_t\mathfrak{q}_t.$$

Theorem 8.3 (Residue Theorem). If ω is an Abelian differential of the third kind on $\mathcal{R}$ with poles $\mathfrak{q}_1, \ldots, \mathfrak{q}_r$, then $\sum_{k=1}^{r} \mathrm{Res}_{\mathfrak{q}_k}[\omega] = 0$.

Proof: Let $[D] = \mathcal{R} - U_1 - U_2 - \cdots - U_r$ be the closed region obtained by deleting from $\mathcal{R}$ sufficiently small disks $U_k = \{p | z_k(p)| < \varepsilon\}$ with center $\mathfrak{q}_k$. Since $d\omega = 0$, we have

$$2\pi i \sum_{k=1}^{r} \mathrm{Res}_{\mathfrak{q}_k}[\omega] = \sum_{k=1}^{r} \oint_{\mathfrak{q}_k} \omega = -\int_{\partial[D]} \omega = -\int_{[D]} d\omega = 0$$

by Green's Theorem (Theorem 6.10).

It is obvious that if f is a nonconstant meromorphic function on $\mathcal{R}$, then df is an Abelian differential of the second kind, and df/f is an Abelian differential of the third kind. Indeed if $(f) = \sum_{k=1}^{r} m_k\mathfrak{q}_k$, then f can be written as $f = z_k^{m_k} h_k(z_k)$, where z_k is a local coordinate around $\mathfrak{q}_k$ and $h_k(z_k)$ is a holomorphic function of z_k, $h_k(0) \neq 0$. Hence,

$$\frac{df}{f} = \left(\frac{m_k}{z_k} + \frac{h'_k(z_k)}{h_k(z_k)}\right) dz_k, \tag{8.13}$$

i.e. $\mathfrak{q}_k$ is a pole of the first order of df/f throughout. Since it is obvious that df/f is holomorphic on $\mathcal{R} - \{\mathfrak{q}_k, \ldots, \mathfrak{q}_r\}$, df/f is an Abelian differential of the third kind. Since $\mathrm{Res}_{\mathfrak{q}_k}[df/f] = m_k$ by (8.13), we deduce, from the above Theorem, that $\sum_{k=1}^{r} m_k = 0$.

We have proved:

Theorem 8.4. If f is a meromorphic function on $\mathcal{R}$, then $\deg(f) = 0$.

Since $\mathfrak{q}_k$ is a zero of the m_kth order if $m_k > 0$, and a pole of the $|m_k|$th order if $m_k < 0$, this theorem shows that the number of zeros of a meromorphic function equals of its poles.

8.3 The Riemann–Roch Theorem

a. Existence theorem

Fix $\mathfrak{q} \in \mathcal{R}$ and let $z_{\mathfrak{q}} : p \to z_{\mathfrak{q}}(p)$ be a local coordinate around $\mathfrak{q}$ such that its range in the $z_{\mathfrak{q}}$-plane contains the closed unit disk and put $U_0 = \{p : |z_{\mathfrak{q}}(p)| < 1\}$. We also put

$$S_m(z_{\mathfrak{q}}) = \Phi_m + i\Psi_m = \frac{1}{z_{\mathfrak{q}}^m} + z_{\mathfrak{q}}^m$$

where m is a natural number. Since $S_m(e^{i\theta}) = 2\cos m\theta$, we have $\Psi_m = 0$ on $\partial[U_0]$. Hence $S_m(z_{\mathfrak{q}})$ satisfies condition (6.75). Let $\psi = d(\rho\Phi_m)$ (compare with the Corollary of Theorem 6.19) and let $\varphi = \psi + d\eta_\infty$ be the harmonic 1-form of Theorem 6.16. The 1-form

$$d\tau_{\mathfrak{q},m} = \varphi + i * \varphi$$

is holomorphic on $\mathcal{R} - \{\mathfrak{q}\}$ and has the same singularity at $\mathfrak{q}$ as $dS_m(z_{\mathfrak{q}})$. Since $\oint_{\mathfrak{q}} dS_m(z_{\mathfrak{q}}) = 0$, $d\tau_{\mathfrak{q},m}$ is an Abelian differential of the second kind and $\tau_{\mathfrak{q},m} = \int(\phi + i * \phi)$ is an Abelian integral of the second kind. This proves:

Theorem 8.5. Let $\mathfrak{q} \in \mathcal{R}$. The there exists an Abelian differential of the second kind $d\tau_{\mathfrak{q},m}$ which is holomorphic on $\mathcal{R} - \{\mathfrak{q}\}$ and which, on a neighborhood of $\mathfrak{q}$, can be written as

$$d\tau_{\mathfrak{q},m} = d\left(\frac{1}{z_{\mathfrak{q}}^m}\right) + \text{a holomorphic 1-form.} \tag{8.14}$$

b. *The Riemann–Roch Theorem*

Let $\mathfrak{q}_1, \ldots, \mathfrak{q}_t$ be points of $\mathcal{R}$, let $z_{\mathfrak{q}_k} = z_k$ be a local coordinate around $\mathfrak{q}_k$, $k = 1, \ldots, t$, and let us write $d\tau_{k,m}$ instead of $d\tau_{\mathfrak{q}_k,m}$. Further let $\omega_\nu = e_\nu + i * e_\nu$ be the Abelian differential of the first kind introduced in Section 8.1b. The linear combination

$$df = \sum_{k=1}^{t} \sum_{m=1}^{m_k} c_{k,m}\, d\tau_{k,m} + \sum_{\nu=1}^{g} c_\nu \omega_\nu \tag{8.15}$$

with complex coefficients is an Abelian differential of the second kind and $f = \int df$ an Abelian integral of the second kind. The Abelian integral f is, of course, in general a multi-valued function on $\mathcal{R}$. If f is single-valued, then f is holomorphic on $\mathcal{R} - \{\mathfrak{q}_1, \ldots, \mathfrak{q}_t\}$ and, by (8.14),

$$f = \sum_{k=1}^{t} \sum_{m=1}^{m_k} \frac{c_{k,m}}{z_k^m} + \text{a holomorphic function} \tag{8.16}$$

on a neighborhood of each $\mathfrak{q}_k$. Hence f is meromorphic on $\mathcal{R}$, holomorphic on $\mathcal{R} - \{\mathfrak{q}_1, \ldots, \mathfrak{q}_t\}$ and has a pole of at most m_kth order at each $\mathfrak{q}_k$.

Conversely, if the meromorphic function f is holomorphic on $\mathcal{R} - \{\mathfrak{q}_1, \ldots, \mathfrak{q}_t\}$ and has a pole of at most m_kth order at $\mathfrak{q}_k$, then f is of the form (8.16) on a neighborhood of each $\mathfrak{q}_k$, hence df can be written in the form (8.15). What conditions must the coefficients $c_{k,m}$ and c_ν satisfy for the Abelian integral $f = \int df$ to be single-valued? It is single-valued if and only if $\int_\gamma df = 0$ for all piecewise smooth closed curves γ not passing through $\mathfrak{q}_1, \mathfrak{q}_2, \ldots, \mathfrak{q}_t$. For given $\mathfrak{q}_1, \ldots, \mathfrak{q}_t$ it is possible to choose the Jordan curves $C_1, C_2, \ldots, C_{2g}$ of Theorem 7.9 in such a way that none of them passes through $\mathfrak{q}_1, \ldots, \mathfrak{q}_t$. Actually, since C_n is the unit circle, which is contained in the region $\{z_n | a < |z_n| < b\}$, $b = 1/a > 1$, of $\mathcal{R}$, $n = 1, \ldots, g$, if suffices to replace C_n by a circle $|z_n| = r$ for some suitably chosen r, $a < r < b$, if $\mathfrak{q}_k \in |C_n|$. The choice of the curves $C_{g+1}, \ldots, C_{2g}$ is thus restricted further. Hence

$$\gamma \sim n_1 C_1 + n_2 C_2 + \cdots + n_{2g} C_{2g} + n_{2g+1} Q_1 + \cdots + n_{2g+t} Q_t$$

where Q_k is a sufficiently small circle $|z_k| = \varepsilon > 0$ with center $\mathfrak{q}_k$, in $\mathcal{R} - \{\mathfrak{q}_1, \ldots, \mathfrak{q}_t\}$ (see Theorem 7.10). Since df is an Abelian differential of the second kind, we have $\int_{Q_k} df = \oint_{\mathfrak{q}_k} df = 2\pi i \operatorname{Res}_{\mathfrak{q}_k}[df] = 0$. Hence

$$\int_\gamma df = n_1 \int_{C_1} df + n_2 \int_{C_2} df + \cdots + n_{2g} \int_{C_{2g}} df.$$

Therefore, f is single-valued if and only if

$$\int_{C_j} df = 0, \qquad j = 1, 2, \ldots, 2g. \tag{8.17}$$

We regard (8.17) as a system of linear equations in the unknowns $c_{k,m}$ and c_ν.

Putting

$$df = d\tau + \sum_{\nu=1}^{g} c_\nu \omega_\nu, \qquad d\tau = \sum_{k=1}^{t} \sum_{m=1}^{m_k} c_{k,m}\, d\tau_{k,m},$$

(8.17) can be written as

$$\int_{C_\lambda} d\tau + i c_\lambda = 0, \quad \int_{C_{g+\lambda}} d\tau - \sum_{\nu=1}^{g} a_{\nu\lambda} c_\nu = 0, \quad \lambda = 1, 2, \ldots, g, \tag{8.18}$$

by (8.12). An arbitrary Abelian differential ω can be represented as $\omega = h(z_\mathfrak{q})\, dz_\mathfrak{q}$, wherre $z_\mathfrak{q}$ is a local coordinate on a neighborhood of an arbitrary $\mathfrak{q} \in \mathcal{R}$. Since $dz_\mathfrak{q} \wedge dz_\mathfrak{q} = 0$, the exterior product of two Abelian differentials vanishes identically. Therefore,

$$i(a_{\lambda\nu} - a_{\nu\lambda}) = \int_{\mathcal{R}} \omega_\lambda \wedge \omega_\nu = 0,$$

by (8.11) and (8.8), so that $(a_{\lambda\nu})$ is a symmetric matrix. Hence, by eliminating c_λ from (8.18), we obtain

$$i\int_{C_{g+\lambda}} d\tau + \sum_{\nu=1}^{g} a_{\lambda\nu}\int_{C_\nu} d\tau = 0, \qquad \lambda = 1, 2, \ldots, g. \tag{8.19}$$

Since $\mathfrak{q}_k \notin |C_j|$, it is possible to choose $\rho_j^+ = \rho_{C_j}^+$ in such a way that $\operatorname{supp}\rho_j^+ \cap [U_\varepsilon(\mathfrak{q}_k)] = \varnothing$, When $U_\varepsilon(\mathfrak{q}_k) = \{z_k : |z_k| < \varepsilon\}$ for some sufficiently small ε. Since e_j is uniquely determined by condition (8.4), it is independent of the choice of ρ_j^+. Put

$$G_\varepsilon = \mathcal{R} - [U_\varepsilon(\mathfrak{q}_1)] - [U_\varepsilon(\mathfrak{q}_2)] - \cdots - [U_\varepsilon(\mathfrak{q}_t)]$$

Since $\operatorname{supp}\rho_j^+ \subset G_\varepsilon$ and $d\tau$ is holomorphic on G_ε, $\int_{C_j} d\tau = \int_{G_\varepsilon} d\rho_j^+ \wedge d\tau$. Hence, putting

$$\zeta_\lambda = i\eta_{g+\lambda} + \sum_{\nu=1}^{g} a_{\lambda\nu}\eta_\nu$$

we have

$$id\rho_{g+\lambda}^+ + \sum_{\nu=1}^{g} a_{\lambda\nu} d\rho_\nu^+ = \omega_\lambda - d\zeta_\lambda$$

by (8.3) and (8.11). Therefore (8.19) becomes

$$\int_{G_\varepsilon} \omega_\lambda \wedge d\tau - \int_{G_\varepsilon} d\zeta_\lambda \wedge d\tau = 0.$$

The exterior product $\omega_\lambda \wedge d\tau$ of the Abelian differentials ω_λ and $d\tau$ vanishes identically. As $\zeta_\lambda b_\lambda$ is a C^∞ function $\mathcal{R}$ and since $\partial[U_\varepsilon(\mathfrak{q}_k)] = Q_k$, we have

$$\partial[G_\varepsilon] = -Q_1 - Q_2 - \cdots - Q_t.$$

Hence, using Green's Theorem,

$$\int_{G_\varepsilon} d\zeta_\lambda \wedge d\tau = \int_{G_\varepsilon} d(\zeta_\lambda\, d\tau) = -\sum_{k=1}^{t}\int_{Q_k} \zeta_\lambda\, d\tau.$$

In general, $\tau = \int d\tau$ is a multi-valued function, but all branches of τ over $[U_\varepsilon(\mathfrak{q}_k)]$ are single-valued functions:

$$\tau = \sum_{m=1}^{m_k} \frac{c_{k,m}}{z_k^m} + h(z_k), \qquad h(z_k) \text{ 1 holomorphic function.}$$

Since further $\operatorname{supp}\rho_j^+ \cap [U_\varepsilon(\mathfrak{q}_k)] = \varnothing$, $d\zeta_\lambda = \omega_\lambda$ on $[U_\varepsilon(\mathfrak{q}_k)]$. Hence, since $\int_{Q_k} d(\zeta_\lambda\tau) = 0$, we have

$$-\int_{Q_k} \zeta_\lambda\, d\tau = \int_{Q_k} \tau\, d\zeta_\lambda = \sum_{m=1}^{m_k} c_{k,m}\int_{Q_k} \frac{1}{z_k^m}\omega_\lambda + \int_{Q_k} h(z_k)\omega_\lambda$$

and $\int_{Q_k} h(z_k)\omega_\lambda = 0$ by Cauchy's Theorem. Hence, writing $\oint_{\mathfrak{q}_k}$ for $\int_{Q_k}$, (8.19) takes the form

$$\sum_{k=1}^{t}\sum_{m=1}^{m_k} c_{k,m}\oint_{\mathfrak{q}_k}\frac{1}{z_k^m}\omega_\lambda = 0, \qquad \lambda = 1, 2, \ldots, g. \tag{8.20}$$

Thus if the integral $f = \int df$ of the Abelian differential of (8.15) is single-valued on $\mathcal{R}$, then the coefficients $c_{k,m}$, $m = 1, \ldots, m_k$; $k = 1, \ldots, t$, are solutions of the system of linear equations (8.20). Conversely, if the $c_{k,m}$ are solutions of (8.20), then $d\tau = \sum_k \sum_m c_{k,m}\, d\tau_{k,m}$ satisfies (8.19). So, putting $c_\lambda = i\int_{c_\lambda} d\tau$ we get (8.18) and therefore $f = \int df$ is single-valued.

Let $\mathfrak{q}_1, \ldots, \mathfrak{q}_t \in \mathcal{R}$ and $m_1, \ldots, m_t$ be natural numbers. The collection of all meromorphic functions on $\mathcal{R}$, which are holomorphic on $\mathcal{R}-\{\mathfrak{q}_1, \ldots, \mathfrak{q}_t\}$ and have poles of order at most m_k at $\mathfrak{q}_k$, constitute a linear space. Putting $\mathfrak{d} = m_1\mathfrak{q}_1 + \cdots + m_t\mathfrak{q}_t$, this space is denoted by $\mathcal{F}(\mathfrak{d})$. Assuming that f does not vanish identically, a meromorphic function f belongs to $\mathcal{F}(\mathfrak{d})$ if and only if $(f) + \mathfrak{d} \geqslant 0$. Since the divisor of the function which is identically equal to 0 is not defined, we will formally agree that $(0) + \mathfrak{d} \geqslant 0$. Hence

$$\mathcal{F}(\mathfrak{d}) = \{f|(f) + \mathfrak{d} \geqslant 0), f \text{ a meromorphic function}\}. \tag{8.21}$$

The function $f \in \mathcal{F}(\mathfrak{d})$ is uniquely determined by df apart from a constant and df is uniquely determined by the solution $c_{k,m}$ of the system of linear equations (8.20). Hence $\dim_{\mathbb{C}} \mathcal{F}(\mathfrak{d}) - 1$ equals the number of linear independent solutions of (8.20).

The linear system with unknowns $c_1, c_2, \ldots, c_g$

$$\sum_{\lambda=1}^{g} c_\lambda \oint_{\mathfrak{q}_k}\frac{1}{z_k^m}\omega_\lambda = 0,\ m = 1, 2, \ldots, m_k,\ k = 1, 2, \ldots, t, \tag{8.22}$$

is the transposed equation of (8.20). Representing $\omega = c_1\omega_1 + c_2\omega_2 + \cdots + c_g\omega_g$ on a neighborhood of q_k by

$$\omega = (a_{k,0} + a_{k,1}z_k + a_{k,2}z_k^2 + \cdots)\, dz_k,$$

Equation (8.22) means that for $k = 1, 2, \ldots, t$, we have $k = 1, 2, \ldots, t$, we have $a_{k,m} = 0$ when $m = 0, 1, 2, \ldots,\ m_{k-1}$ i.e. q_k is a zero of ω of at least m_kth order; in another way $(\omega) - \mathfrak{d} \geqslant 0$. The number of linearly independent Abelian differentials satisfying $(\omega) - \mathfrak{d} \geqslant 0$ is called the *index of speciality* of $\mathfrak{d}$ and denoted by $i(\mathfrak{d})$:

$$i(\mathfrak{d}) = \dim\{\omega|(\omega) - d \geqslant 0\}. \tag{8.23}$$

(Since in this case $\mathfrak{d} > 0$, any Abelian differential satisfying $(\omega) - \mathfrak{d} \geqslant 0$, is of the first kind.) Observe that $i(\mathfrak{d})$ equals the number of linearly independent solutions of the system of linear equations (8.22). Hence

$$i(\mathfrak{d}) = g - r \tag{8.24}$$

where r is the rank of the matrix of coefficients $\oint_{\mathfrak{q}_k}(1/z_k^m)\omega_\lambda$ of (8.22). Since the number of unknowns $c_{k,m}$ of (8.20) equals $m_1 + m_2 + \cdots + m_k = \deg \mathfrak{d}$ and the rank of its coefficient matrix also equals r, we have

$$\dim \mathcal{F}(\mathfrak{d}) - 1 = \deg \mathfrak{d} - r. \tag{8.25}$$

Elimination of r from (8.24) and (8.25) yields the following theorem:

Theorem 8.6 (The Riemann–Roch Theorem).

$$\dim \mathcal{F}(\mathfrak{d}) = \deg \mathfrak{d} - g + i(\mathfrak{d}) + 1. \tag{8.26}$$

The Riemann–Roch Theorem is one of the basic theorems concerning meromorphic functions on a Riemann surface. Since $i(\mathfrak{d}) \geqslant 0$, we have

$$\dim \mathcal{F}(\mathfrak{d}) \geqslant \deg \mathfrak{d} - g + 1.$$

We conclude from this equality that there exists at least one nonconstant meromorphic function f on $\mathcal{R}$ that is holomorphic on $\mathcal{R} - \{\mathfrak{q}_1, \ldots, \mathfrak{q}_{g+1}\}$ and has poles of at most first order at $\mathfrak{q}_1, \ldots, \mathfrak{q}_{g+1}$ for arbitrarily given points $\mathfrak{q}_1, \ldots, \mathfrak{q}_{g+1}$ on $\mathcal{R}$. This is because $\dim \mathcal{F}(\mathfrak{q}_1 + \mathfrak{q}_2 + \cdots + \mathfrak{q}_{g+1}) \geqslant 2$. If $\mathfrak{q}_1, \ldots, \mathfrak{q}_g$ are g points of $\mathcal{R}$ "in general position" and if $\mathfrak{d} = \mathfrak{q}_1 + \cdots + \mathfrak{q}_g$, then it can be shown that $i(\mathfrak{d}) = 0$. Hence by the Riemann–Roch Theorem $\dim \mathcal{F}(\mathfrak{d}) = 1$. Thus the only meromorphic functions which are holomorphic on $\mathcal{R} - \{\mathfrak{q}_1, \ldots, \mathfrak{q}_g\}$ and which have poles of at most first order at $\mathfrak{q}_1, \ldots, \mathfrak{q}_g$, are constants.

We have proved the Riemann–Roch Theorem under the assumption that $\mathfrak{d}$ is positive. If $\mathfrak{d}$ is not positive, we still define $\mathcal{F}(\mathfrak{d})$ and $i(\mathfrak{d})$ by (8.21), and (8.23) and the Theorem still holds. For the general case, let $\mathfrak{d} = \sum_{k=1}^{t} m_k \mathfrak{q}_k - \sum_{j=1}^{s} n_j \mathfrak{p}_j$, ($\mathfrak{p}_j \neq \mathfrak{q}_k$; m_k, $n_j \in \mathbb{N}$). Now $\mathcal{F}(\mathfrak{d})$ is the linear space consisting of all meromorphic functions f which are holomorphic on $\mathcal{R} - \{\mathfrak{q}_1, \ldots, \mathfrak{q}_t\}$, have poles of at most m_kth order at $\mathfrak{q}_k$ and zeros of order of at least n_j at $\mathfrak{p}_j$: $i(\mathfrak{d})$ is the number of linearly independent Abelian differentials which are holomorphic on $\mathcal{R} - \{\mathfrak{p}_1, \ldots, \mathfrak{p}_s\}$ and have poles of at most n_jth order at $\mathfrak{p}_j$ and zero of at least m_kth order at $\mathfrak{q}_k$. The general case of the Riemann–Roch Theorem follows easily from the special case $\mathfrak{d} > 0$.

(1) The case $\mathfrak{d} = 0$. Since $\mathcal{F}(0) = \mathbb{C}$, $\dim \mathcal{F}(0) = 1$ and $\{\omega \colon (\omega) \geqslant 0\}$ is the linear space consisting of all Abelian differentials of the first kind on $\mathcal{R}$, where $i(0) = g$. Hence (8.26) is valid in this case.

(2) The divisor of an arbitrary Abelian differential ω on $\mathcal{R}$ is called a *canonical divisor* and denoted by $\mathfrak{k} : \mathfrak{k} = (\omega)$ (we assume that ω is not identically equal to 0). Fixing an Abelian differential ω_0, we put $\mathfrak{k} = (\omega_0)$.

For an arbitrary meromorphic function f, the divisor of the Abelian differential $\omega = f \cdot \omega_0$ is given by

$$(\omega) = (f) + \mathfrak{f},$$

hence $(\omega) - \mathfrak{d} \geqslant 0$ is equivalent to $(f) + \mathfrak{f} - \mathfrak{d} \geqslant 0$. Thus the linear map $f \to \omega = f \cdot \omega_0$ is an isomorphism mapping $\mathcal{F}(\mathfrak{f} - \mathfrak{d})$ onto $\{\omega : (\omega) - \mathfrak{d} \geqslant 0\}$, hence

$$i(\mathfrak{d}) = \dim \mathcal{F}(\mathfrak{f} - \mathfrak{d}).$$

Therefore, (8.26) can be written as

$$\dim \mathcal{F}(\mathfrak{d}) - \dim \mathcal{F}(\mathfrak{f} - \mathfrak{d}) = \deg \mathfrak{d} - g + 1. \qquad (8.27)$$

By Theorem 8.6 and (1) above, (8.27) is true when $\mathfrak{d} \geqslant 0$.

(3) Two divisors $\mathfrak{d}_1$ and $\mathfrak{d}_2$ on $\mathcal{R}$ are called *linearly equivalent* if there exists a meromorphic function f such that $\mathfrak{d}_1 - \mathfrak{d}_2 = (f)$. We write this as $\mathfrak{d}_1 \approx \mathfrak{d}_2$. It is obvious that $\approx$ is an equivalence relation. Indeed, $\mathfrak{d} \approx \mathfrak{d}$ since $\mathfrak{d} - \mathfrak{d} = (1) = 0$. If $\mathfrak{d}_1 \approx \mathfrak{d}_2$, then $\mathfrak{d}_2 \approx \mathfrak{d}_1$ because if $\mathfrak{d}_1 - \mathfrak{d}_2 = (f)$, then $\mathfrak{d}_2 - \mathfrak{d}_1 = (1/f)$. If $\mathfrak{d}_1 - \mathfrak{d}_2 = (f_1)$ and $\mathfrak{d}_2 - \mathfrak{d}_3 + (f_2)$, then $\mathfrak{d}_1 - \mathfrak{d}_3 = (f_1 \cdot f_2)$, and so $\mathfrak{d}_1 \approx \mathfrak{d}_3$.

If $\mathfrak{d}_1 - \mathfrak{d}_2 = (f)$, then $\deg \mathfrak{d}_1 - \deg \mathfrak{d}_2 = \deg(f) = 0$ by Theorem 8.4, hence

$$\text{if } \mathfrak{d}_1 \approx \mathfrak{d}_2 \text{ then } \deg \mathfrak{d}_1 = \deg \mathfrak{d}_2. \qquad (8.28)$$

If $\mathfrak{d}_1 - \mathfrak{d}_2 = (f_0)$ and f is an arbitrary meromorphic function, then

$$(f \cdot f_0) + \mathfrak{d}_2 = (f) + (f_0) + \mathfrak{d}_2 = (f) + \mathfrak{d}_1$$

and the linear map $f \to f \cdot f_0$ maps $\mathcal{F}(\mathfrak{d}_1)$ isomorphically onto $\mathcal{F}(\mathfrak{d}_2)$. Thus

$$\text{if } \mathfrak{d}_1 \approx \mathfrak{d}_2 \text{ then } \dim \mathcal{F}(\mathfrak{d}_1) = \dim \mathcal{F}(\mathfrak{d}_2). \qquad (8.29)$$

If $\mathfrak{f}_1$ and $\mathfrak{f}_2$ are two canonical divisors on $\mathcal{R}$, then

$$\mathfrak{f}_1 \approx \mathfrak{f}_2 \qquad (8.30)$$

for if $\mathfrak{f}_1 = (\omega_1)$ and $\mathfrak{f}_2 = (\omega_2)$, then $\omega_1 = f \cdot \omega_2$, where f is a meromorphic function; hence $\mathfrak{f}_1 - \mathfrak{f}_2 = (f)$. Thus by (8.28)

$$\deg \mathfrak{f}_1 = \deg \mathfrak{f}_2. \qquad (8.31)$$

(4) The case $\dim \mathcal{F}(\mathfrak{d}) \geqslant 1$. Pick a (nonzero) $f_0 \in \mathcal{F}(\mathfrak{d})$ and put $\mathfrak{d}_0 = (f_0) + \mathfrak{d}$. Since $\mathfrak{d}_0 \geqslant 0$, (8.27) is valid so

$$\dim \mathcal{F}(\mathfrak{d}_0) - \dim \mathcal{F}(\mathfrak{f} - \mathfrak{d}_0) = \deg \mathfrak{d}_0 - g + 1.$$

If $g \geqslant 1$, at least one (nonzero) Abelian differential ω of the first kind exists. Since $\mathcal{F} = (\omega) \geqslant 0$, we have

$$\dim \mathcal{F}(\mathfrak{f}) - \dim \mathcal{F}(0) = \deg \mathfrak{f} - g + 1.$$

Since $\dim \mathcal{F}(\mathfrak{f}) = i(0) = g$ and $\dim \mathcal{F}(0) = 1$, we have

$$\deg \mathfrak{f} = 2g - 2. \tag{8.32}$$

If $g = 0$, then $\mathcal{R}$ is the Riemann sphere $S = \mathbb{C} \cup \{\infty\}$. Regarding $\mathbb{C}$ as the z-plane, z is a meromorphic function on S that is holomorphic on $\mathbb{C}$ and has a pole of first order at ∞. Let $z_\infty = 1/z$ be the local coordinate around ∞. The Abelian differential $\omega = dz$ is holomorphic on $\mathbb{C}$ and has n zeros on $\mathbb{C}$. Since $\omega = d(1/z_\infty) = -dz_\infty/dz_\infty^2$, ω has a pole of the second order at ∞. Hence $\mathfrak{f} = (\omega) = -2 \cdot \infty$, hence $\deg \mathfrak{f} = -2 = 2g - 2$. Hence, by (8.31), the degree of the canonical divisor $\mathfrak{f}$ is always given by (8.32).

(5) The case $\dim \mathcal{F}(\mathfrak{f} - \mathfrak{d}) \geqslant 1$. By (4), the formula

$$\dim \mathcal{F}(\mathfrak{f} - \mathfrak{d}) - \dim \mathcal{F}(\mathfrak{d}) = \deg(\mathfrak{f} - \mathfrak{d}) - g + 1$$

holds for the divisor $\mathfrak{f} - \mathfrak{d}$. Hence by (8.32),

$$\dim \mathcal{F}(\mathfrak{f} - \mathfrak{d}) - \dim \mathcal{F}(\mathfrak{d}) = -\deg \mathfrak{d} + g - 1$$

and this is nothing but (8.27).

(6) The case $\dim \mathcal{F} - (\mathfrak{d}) = \dim \mathcal{F}(\mathfrak{f} - \mathfrak{d})$. We first verify that $\dim \mathcal{F}(\mathfrak{d}) \geqslant 1$ if $\deg \mathfrak{d} \geqslant g$. If $\mathfrak{d} \geqslant 0$, then $\mathcal{F}(\mathfrak{d}) \supset \mathbb{C}$; that is $\dim \mathcal{F}(\mathfrak{d}) \geqslant 1$. If $\mathfrak{d} \geqslant 0$ is not true, we put $\mathfrak{d} = \sum_{k=1}^{t} m_k \mathfrak{q}_k - \sum_{j=1}^{s} n_j \mathfrak{p}_j$, $\mathfrak{p}_j \neq \mathfrak{q}_k$; $m_k, n_j \in \mathbb{N}$, $\mathfrak{d}^+ = \sum_{k=1}^{t} m_k \mathfrak{q}_k$ and $\mathfrak{d}_s^- = \sum_{j=1}^{s} n_j \mathfrak{p}_j$. We put $\mathfrak{p} = \mathfrak{p}_s$ and let $z_\mathfrak{p}$ denote a local coordinate around $\mathfrak{p}$. In order for $f \in \mathcal{F}(\mathfrak{d}^+ - \mathfrak{d}_{s-1}^-)$ to belong to $\mathcal{F}(\mathfrak{d}^+ - \mathfrak{d}_s^-) = \mathcal{F}(\mathfrak{d}^+ - \mathfrak{d}_{s-1}^- - n_s\mathfrak{p})$, f has to satisfy the n_s linear equations,

$$\oint_\mathfrak{p} f(z_\mathfrak{p}) \frac{dz_\mathfrak{p}}{z_\mathfrak{p}^n} = 0, \qquad n = 1, 2, \ldots, n_s$$

Hence, $\dim \mathcal{F}(\mathfrak{d}^+ - \mathfrak{d}_s^-) \geqslant \dim \mathcal{F}(\mathfrak{d}^+ - \mathfrak{d}_{s-1}^-) - n_s$, therefore by induction with respect to s we have

$$\dim \mathcal{F}(\mathfrak{d}^+ - \mathfrak{d}_s^-) \geqslant \dim \mathcal{F}(\mathfrak{d}^+) - \sum_{j=1}^{s} n_j.$$

Since $\sum_{k=1}^{t} m_k - \sum_{j=1}^{s} n_j = \deg \mathfrak{d} \geqslant g$, we have $\mathfrak{d}^+ > 0$. Hence,

$$\dim \mathcal{F}(\mathfrak{d}^+) \geqslant \sum_{k=1}^{t} m_k - g + 1,$$

by Theorem 8.6, and therefore,

$$\dim \mathcal{F}(\mathfrak{d}) \geqslant \deg \mathfrak{d} - g + 1 \geqslant 1.$$

Thus, since in this case $\dim \mathcal{F}(\mathfrak{d}) = 0$, $\deg \mathfrak{d} \leqslant g - 1$. Further, since $\dim \mathcal{F}(\mathfrak{f} - \mathfrak{d}) = 0$, we have $\deg(\mathfrak{f} - \mathfrak{d}) \leqslant g - 1$, i.e. $\deg \mathfrak{d} \geqslant \deg \mathfrak{f} - g + 1 = g - 1$. Thus $\deg \mathfrak{d} = g - 1$, proving (8.27).

This finishes the proof of the Riemann–Roch Theorem for an arbitrary divisor $\mathfrak{d}$ on $\mathcal{R}$.

8.4 Abel's Theorem

a Existence theorem

We want to prove that for arbitrary $\mathfrak{p}, \mathfrak{q} \in \mathcal{R}$, $\mathfrak{p} \neq \mathfrak{q}$, there exists an Abelian differential of the third kind which is holomorphic on $\mathcal{R} - \{\mathfrak{p}, \mathfrak{q}\}$ and has poles of the first order at $\mathfrak{p}$ and $\mathfrak{q}$. We first consider the case when $\mathfrak{p}$ and $\mathfrak{q}$ are contained in the same coordinate disk $U_0 = \{\mathfrak{p} : |z(\mathfrak{p})| < 1\}$. We introduce the function

$$S(z) = \Phi + i\Psi = i \log\left(\frac{z-\beta}{1-\overline{\beta}z} \Big/ \frac{z-\alpha}{1-\overline{\alpha}z}\right), \quad \alpha = z(\mathfrak{p}),\ \beta = z(\mathfrak{q}),$$

As $S(z)$ is a holomorphic single-valued function on $U_0 - |\sigma|$, when σ is the segment in U_0 connecting α and β, and on $\partial[U_0]$, we have

$$\Psi = \log\left|\frac{z-\beta}{1-\overline{\beta}z}\right| - \log\left|\frac{z-\alpha}{1-\overline{\alpha}z}\right| = 0;$$

that is $S(z)$ satisfies condition (6,75). For arbitrary $\varphi \in \mathcal{L}_\Phi$, let $\varphi = \psi + d\eta_\infty$ be the harmonic 1-form of Theorem 6.16. Then $\varphi + i * \varphi$ is an Abelian differential on $\mathcal{R}$ with the same singularities as $dS(z)$, by Theorem 6.19. Put

$$\omega_{\mathfrak{p}\mathfrak{q}} = -i(\varphi + i * \varphi).$$

Since we have

$$-idS(z) = -\frac{dz}{z - z(\mathfrak{p})} + \frac{dz}{z - z(\mathfrak{q})} + \text{holomorphic 1-form}$$

on U_0, it follows that $\omega_{\mathfrak{p}\mathfrak{q}}$ is an Abelian differential of the third kind with poles of the first order at $\mathfrak{p}$ and $\mathfrak{q}$ and holomorphic on $\mathcal{R} - \{\mathfrak{p}, \mathfrak{q}\}$. We have

$$\frac{dz}{z - z(\mathfrak{p})} = \frac{dz_\mathfrak{p}}{z_\mathfrak{p}} + \text{holomorphic 1-form},$$

where z is a local coordinate around $\mathfrak{p}$, and a similar expression with regard to a local coordinate around $\mathfrak{q}$; hence $\omega_{\mathfrak{p}\mathfrak{q}}$ can be represented as

$$\begin{aligned} \omega_{\mathfrak{p}\mathfrak{q}} &= \frac{dz_\mathfrak{p}}{z_\mathfrak{p}} + \text{holomorphic 1-form}, \\ \omega_{\mathfrak{p}\mathfrak{q}} &= \frac{dz_\mathfrak{q}}{z_\mathfrak{q}} + \text{holomorphic 1-form} \end{aligned} \tag{8.33}$$

on a neighborhood of $\mathfrak{p}$ and $\mathfrak{q}$ respectively. If $\mathfrak{p}$ and $\mathfrak{q}$ do not belong to the same coordinate disk, we choose a sequence of points: $\mathfrak{p}_0 = \mathfrak{p}$, $\mathfrak{p}_1$, $\mathfrak{p}_2$, $\ldots$, $\mathfrak{p}_n$, $\mathfrak{p}_{n+1} = \mathfrak{q}$ such that $\mathfrak{p}_k$ and $\mathfrak{p}_{k+1}$ belong to the same coordinate disk for all k. Defining $\omega_{\mathfrak{p}_k\mathfrak{p}_{k+1}}$ as above, we put

$$\omega_{\mathfrak{p}\mathfrak{q}} = \omega_{\mathfrak{p}\mathfrak{p}_1} + \omega_{\mathfrak{p}_1\mathfrak{p}_2} + \cdots + \omega_{\mathfrak{p}_n\mathfrak{q}}.$$

Then $\omega_{\mathfrak{pq}}$ is an Abelian differential of the third kind which is holomorphic on $\mathcal{R} - \{\mathfrak{p}, \mathfrak{q}\}$ and can be written as

$$\omega_{\mathfrak{pq}} = -\frac{dz_{\mathfrak{p}}}{z_{\mathfrak{p}}} + \text{holomorphic 1-form}, \ \omega_{\mathfrak{pq}} = \frac{dz_{\mathfrak{q}}}{z_{\mathfrak{q}}} + \text{holomorphic 1-form}$$

on a neighborhood of $\mathfrak{p}$ and $\mathfrak{q}$ respectively. Let $C_1, C_2, \ldots, C_{2g}$ be the Jordan curves of Theorem 7.9, chosen in such a way that they do not pass through $\mathfrak{p}$ and $\mathfrak{q}$. Put: $U_\varepsilon(\mathfrak{p})\{z_{\mathfrak{p}} : |z_{\mathfrak{p}}| < \varepsilon\}$, $U_\varepsilon(\mathfrak{q}) = \{z_{\mathfrak{q}} : |z_{\mathfrak{q}}| < \varepsilon\}$, $P_\varepsilon = \partial[U_\varepsilon(\mathfrak{p})]$ and $Q_\varepsilon = \partial[U_\varepsilon(\mathfrak{q})]$ for some sufficiently small $\varepsilon > 0$. If γ is a piecewise smooth closed curve in $\mathcal{R} \cdot \{\mathfrak{p}, \mathfrak{q}\}$, then, by Theorem 7.10,

$$\gamma \sim \sum_{j=1}^{2g} n_j C_j + nP_\varepsilon + mQ_\varepsilon.$$

Hence, by (8.34), we have

$$\int_\gamma \omega_{\mathfrak{pq}} = \sum_{j=1}^{2g} n_j \int_{c_j} \omega_{\mathfrak{pq}} + 2\pi i(m - n). \tag{8.35}$$

The 1-form:

$$\omega = \sum_{\nu=1}^{g} r_\nu(e_{g+\nu} + i * e_{g+\nu}) - \sum_{\nu=1}^{g} r_{g+\nu}(e_\nu + i * c_\nu)$$

where $r_j = \mathrm{Re} \int_{c_j} \omega_{\mathfrak{pq}}$, is an Abelian differential of the first kind and $\mathrm{Re} \int_{c_j} \omega = r_j$. Hence $\mathrm{Re} \int_\gamma (\omega_{\mathfrak{pq}} - \omega) = 0$ for arbitrary γ by (8.35). Thus the real part of the Abelian integral $\int(\omega_{\mathfrak{pq}} - \omega)$ is single-valued on $\mathcal{R}$. Writing $\omega_{\mathfrak{pq}}$ for $\omega_{\mathfrak{pq}} - \omega$ we put $u = \mathrm{Re} \int \omega_{\mathfrak{pq}}$. We have proved the following theorem:

Theorem 8.7. For arbitrary $\mathfrak{p}, \mathfrak{q} \in \mathcal{R}$, $\mathfrak{p} \neq \mathfrak{q}$ there exists an Abelian differential of the third kind

$$\omega_{\mathfrak{pq}} = du + i * du$$

which is holomorphic on $\mathcal{R} - \{\mathfrak{p}, \mathfrak{q}\}$ and has poles of the first order at $\mathfrak{p}$ and $\mathfrak{q}$. Then $\omega_{\mathfrak{pq}}$ can be written as

$$\omega_{\mathfrak{pq}} = -\frac{dz_{\mathfrak{p}}}{z_{\mathfrak{p}}} + \text{holomorphic 1-form}, \ \omega_{\mathfrak{pq}} = \frac{dz_{\mathfrak{q}}}{z_{\mathfrak{q}}} + \text{holomorphic 1-form}$$

on a neighborhood of $\mathfrak{p}$ and $\mathfrak{q}$ and $u = \mathrm{Re} \int \omega_{\mathfrak{pq}}$ is harmonic and single-valued on $\mathcal{R} - \{\mathfrak{pq}\}$.

We write $\exp \omega$ for e^ω. Now $\int_\gamma e_j$, where γ is an arbitrary piecewise smooth closed curve, is an integer by Theorem 7.9 and (8.4), hence $\exp(2\pi i \int_{\mathfrak{p}}^{\mathfrak{q}} e_j)$ is independent of the choice of the path of integration connection $\mathfrak{p}$ and $\mathfrak{q}$.

Theorem 8.8.

$$\exp\left(\int_{C_j}\omega_{\mathfrak{pq}}\right) = \exp\left(2\pi i\int_{\mathfrak{p}}^{\mathfrak{q}} e_j\right). \tag{8.36}$$

Proof: We put $G_\varepsilon = \mathcal{R} - [U_\varepsilon(\mathfrak{p})] - [U_\varepsilon(\mathfrak{q})]$. By (6.71) and (8.3), we have

$$\int_{C_j}\omega_{\mathfrak{pq}} = \int_{G_\varepsilon} d\rho_j^+ \wedge \omega_{\mathfrak{pq}} = \int_{G_\varepsilon} e_j \wedge \omega_{\mathfrak{pq}} - \int_{G_\varepsilon} d\eta_j \wedge \omega_{\mathfrak{pq}}.$$

Since $\omega_{\mathfrak{pq}} = du + i * du$ and $e_j \wedge *du = du \wedge *e_j$, we have $e_j \wedge \omega_{\mathfrak{pq}} = -du \wedge \overline{\omega}_j$ where $\overline{\omega}_j = e_j - i * e_j$, hence $d\overline{\omega}_j = 0$. Therefore

$$\int_{G_\varepsilon} e_j \wedge \omega_{\mathfrak{pq}} = -\int_{G_\varepsilon} du \wedge \overline{\omega}_j = -\int_{G_\varepsilon} d(u\overline{\omega}_j) = \int_{P_\varepsilon} u\overline{\omega}_j + \int_{Q_\varepsilon} u\overline{\omega}_j,$$

by Green's Theorem.

Since $u = \log|Z_{\mathfrak{p}}| +$ a harmonic function on a neighborhood of $\mathfrak{p}$ by (8.34), and $\omega_j = e_j + i * e_j$ is a holomorphic 1-form, we have $\int_{P_\varepsilon}\omega_j = 0$. Therefore

$$\lim_{\varepsilon\to 0}\int_{P_\varepsilon} u\overline{\omega}_j = -\lim_{\varepsilon\to 0}\int_{P_\varepsilon} \log|zh_{\mathfrak{p}}|\overline{\omega}_j = -\lim_{\varepsilon\to 0}\log\varepsilon\int_{P_\varepsilon}\overline{\omega}_j = 0,$$

and similarly $\lim_{\varepsilon\to 0}\int_{Q_\varepsilon} u\overline{\omega}_j = 0$. Again, By Green's Theorem, we have

$$-\int_{G_\varepsilon} d\eta_j \wedge \omega_{\mathfrak{pq}} = -\int_{G_\varepsilon} d(\eta_j\omega_{\mathfrak{pq}}) = \int_{P_\varepsilon}\eta_j\omega_{\mathfrak{pq}} + \int_{Q_\varepsilon}\eta_j\omega_{\mathfrak{pq}}.$$

Hence

$$\int_{C_j}\omega_{\mathfrak{pq}} = \lim_{\varepsilon\to 0}\left(-\int_{P_\varepsilon}\eta_j\frac{dz_{\mathfrak{p}}}{z_{\mathfrak{p}}} + \int_{Q_\varepsilon}\eta_j\frac{dz_{\mathfrak{q}}}{z_{\mathfrak{q}}}\right) = -2\pi i[\eta_j(\mathfrak{p}) - \eta_j(\mathfrak{q})].$$

Since $d\eta_j = e_j$ on $\mathcal{R} - \operatorname{supp}\rho_j^+$, we have

$$\eta_j(\mathfrak{q}) - \eta_j(\mathfrak{p}) = \int_{\mathfrak{p}}^{\mathfrak{q}} e_j$$

where $\int_{\mathfrak{p}}^{\mathfrak{q}} e_j$ is taken along a path of integration not passing through $\operatorname{supp}\rho_j^+$. Hence (8.36) is proved.

b Abel's Theorem

Let $\mathfrak{d} = (f)$ be the divisor of a nonconstant meromorphic function f on $\mathcal{R}$. By Theorem 8.4, we have $\deg\mathfrak{d} = 0$. Conversely, if a divisor $\mathfrak{d}$ on $\mathcal{R}$ is given such that $\deg\mathfrak{d} = 0$, then it is not necessarily true that there exists a meromorphic function f such that $\mathfrak{d} = (f)$. Abel's Theorem gives a necessary and sufficient condition for the existence of such a meromorphic function.

Let $\mathfrak{d} = \sum_{k=1}^{t} m_k \mathfrak{q}_k$ be a divisor on $\mathcal{R}$ such that $\deg \mathfrak{d} = \sum_{k=1}^{t} m_k = 0$. Let $z_k = z_{\mathfrak{q}_k}$ be a local coordinate around $z_{\mathfrak{q}}$. Fixing $\mathfrak{p}_0 \notin \mathfrak{q}_k$ we put

$$\omega_{\mathfrak{d}} = \sum_{k=1}^{t} m_k \omega_{\mathfrak{p}_0 \mathfrak{q}_k}. \tag{8.37}$$

Of course, $\omega_{\mathfrak{p}_0 \mathfrak{q}_k}$ represents the Abelian differential of the third kind of Theorem 8.7 corresponding to $\mathfrak{p}_0$ and $\mathfrak{q}_k$. Now $\omega_{\mathfrak{d}}$ is an Abelian differential of the third kind which is holomorphic on $\mathcal{R} - \{\mathfrak{q}_1, \dots, \mathfrak{q}_t\}$ and can be represented as

$$\omega_{\mathfrak{d}} = \frac{m_k dz_k}{z_k} + \text{a holomorphic 1-form} \tag{8.38}$$

on a neighborhood of each $\mathfrak{q}_k$.

The Abelian integral $w(p) = \int_{\mathfrak{p}_0}^{\mathfrak{p}} \omega_{\mathfrak{d}}$, where $p_0 \in \mathcal{R}$, is a holomorphic, multi-valued analytic function on $\mathcal{R} - \{\mathfrak{q}_1, \dots, \mathfrak{q}_t\}$ and all branches of $w(p)$ can be represented as

$$w(p) = m_k \log z_k(p) + \text{a holomorphic function}$$

on a neighborhood of $\mathfrak{q}_k$, by (8.38). Hence, the function

$$f(p) = \exp\left(\int_{p_0}^{p} \omega_{\mathfrak{d}}\right) \tag{8.39}$$

is also a holomorphic, nonzero, multi-valued analytic function on $\mathcal{R}-\{\mathfrak{q}_1, \dots, \mathfrak{q}_t\}$ and

$$f(p) = z_k(p)^{m_k} \cdot h_k(p), \text{where } h_k(p) \text{ is holomorphic and } h_k(\mathfrak{q}_k) \neq 0, \tag{8.40}$$

on a neighborhood of $\mathfrak{q}_k$.

By analytic continuation along a closed curve γ not passing through $\mathfrak{p}_0$, $\mathfrak{q}_1, \dots, \mathfrak{q}_t$, $f(p)$ becomes:

$$\chi_{\mathfrak{d}}(\gamma) f(p), \qquad \chi_{\mathfrak{d}}(\gamma) = \exp\left(\int_{\gamma} \omega_{\mathfrak{d}}\right).$$

If $\gamma \sim \sum_{j=1}^{2g} n_j C_j$ then, by (8.35),

$$\int_{\gamma} \omega_{\mathfrak{d}} = \sum_{j=1}^{2g} n_j \int_{C_j} \omega_{\mathfrak{d}} + 2\pi i \cdot m, \qquad m \in \mathbb{Z}.$$

Hence

$$\chi_{\mathfrak{d}}(\gamma) = \prod_{j=1}^{2g} \chi_{\mathfrak{d}}(C_j)^{n_j}.$$

Therefore, $f(p)$ is single-valued on $\mathcal{R}$ if and only if $\chi_{\mathfrak{d}}(C_j) = 1$, $j = 1, 2, \ldots, 2g$. Since $\omega_{\mathfrak{d}} = \sum_{k=1}^{t} m_k \omega_{\mathfrak{p}_0 \mathfrak{q}_t}$, we have

$$\chi_{\mathfrak{d}}(C_j) = \exp\left(2\pi i \sum_{k=1}^{t} m_k \int_{\mathfrak{p}_0}^{\mathfrak{q}_k} e_j \right).$$

by (8.36) and therefore $|\chi_{\mathfrak{d}}(\gamma)| = 1$. If $f(p)$ is single-valued then, by (8.40), $f = f(p)$ is meromorphic and $(f) = \mathfrak{d}$. Conversely, if f_0 is a meromorphic function satisfying $(f_0) = \mathfrak{d}$, then $h = f/f_0$ is a holomorphic analytic function on the whole of $\mathcal{R}$ and $|h(p)|$ is single-valued since $|\chi_{\mathfrak{d}}(\gamma)| = 1$. Hence $h(p) =$ constant by the maximum principle. Therefore $f(p)$ is single-valued too and $\chi_{\mathfrak{d}}(C_j) = 1$. We have proved:

Theorem 8.9. Let $\mathfrak{d} = \sum_{k=1}^{t} m_k \mathfrak{q}_k$ be a divisor on $\mathcal{R}$ such that $\deg \mathfrak{d} = 0$. There exists a meromorphic function f such that $(f) = \mathfrak{d}$ if and only if

$$\chi_{\mathfrak{d}}(C_j) = \exp\left(2\pi i \sum_{k=1}^{t} m_k \int_{\mathfrak{p}_0}^{\mathfrak{q}_k} e_j \right) = 1, \qquad j = 1, 2, \ldots, 2g.$$

Since $\sum_{k=1}^{t} m_k = 0$, we have $\chi_{\mathfrak{d}}(C_j)$ is uniquely determined by $\mathfrak{d}$ and does not depend on $\mathfrak{p}_0$. The analytic function $f = f(p)$ defined above is in general a multivalued function on $\mathcal{R}$, but its many-valuedness is not arbitrary: by analytic continuation along a closed curve $\gamma_{\mathfrak{d}}$, the value $f(p)$ becomes $\chi_{\mathfrak{d}}(\gamma) \cdot f(p)$. This is expressed by saying that f is a multiplicative function. Since $|\chi_{\mathfrak{d}}(\gamma)| = 1$, the poles and zeros of f and their orders do not change under analytic continuation, hence the divisor (f) of f is well defined and $(f) = \mathfrak{d}$ by (8.40). In this way, there always exists a multiplicative function f satisfying $(f) = \mathfrak{d}$ for an arbitrary divisor $\mathfrak{d}$ with $\deg \mathfrak{d} = 0$. Abel's Theorem gives necessary and sufficient conditions for this f to be single-valued.

Problems

The problems presented here comprise simple exercises and results that could not be mentioned in the text. The latter are marked with an asterisk.

Chapter 1

1. Prove that the equality $z_1^2 + z_2^2 + z_3^2 = z_2 z_3 + z_3 z_1 + z_1 z_2$ is valid if and only if the points z_1, z_2, and z_3 are the vertices of a regular triangle. (Ahlfors, 1966, p. 15.)

2*. If $2x = (z + \bar{z})$, $2iy = (z - \bar{z})$ and $z = x + iy$, z and $\bar{z}$ are not independent variables. A formal application of the chain rule for partial derivatives yields

$$\frac{\partial}{\partial z} = \frac{\partial x}{\partial z}\frac{\partial}{\partial x} + \frac{\partial y}{\partial z}\frac{\partial}{\partial y} = \frac{1}{2}\left(\frac{\partial}{\partial x} - i\frac{\partial}{\partial y}\right),$$

$$\frac{\partial}{\partial \bar{z}} = \frac{\partial x}{\partial \bar{z}}\frac{\partial}{\partial x} + \frac{\partial y}{\partial \bar{z}}\frac{\partial}{\partial y} = \frac{1}{2}\left(\frac{\partial}{\partial x} + i\frac{\partial}{\partial y}\right),$$

which we use to define the partial derivatives $\partial/\partial z$ and $\partial/\partial \bar{z}$ with respect to z and $\bar{z}$ respectively. If $u(x, y)$ and $v(x, y)$ are continuously differentiable, real-valued functions on some region, then $f(z) = u(x, y) + iv(x, y)$ is a holomorphic function of $z = x + iy$ if and only if $(\partial f(z)/\partial \bar{z}) = 0$. Prove this.

3. Find the power series expansion in z of $f(z) = 1/(1-z)^m$ (m is a natural number).

4. Find the radius of convergence of the power series:

$$\arcsin z = z + \frac{1}{2}\cdot\frac{z^3}{3} + \frac{1}{2}\cdot\frac{3}{4}\cdot\frac{z^5}{5} + \frac{1}{2}\cdot\frac{3}{4}\cdot\frac{5}{6}\cdot\frac{z^7}{7} + \cdots$$

5*. If the radius of convergence r of the power series $\sum_{n=0}^{\infty} a_n z^n$ satisfies $0 < r < +\infty$, and if ζ is a point on the circle of convergence such that $\sum_{n=0}^{\infty} a_n \zeta^n$ converges, then the function $f(t) = \sum_{n=0}^{\infty} a_n t^n \zeta^n$, $0 \leqq t \leqq 1$, is a continuous function of t. Prove this.

6. Let $\phi(t)$ and $\psi(t)$ be continuous functions defined on the closed interval

$[a, b]$ and let $|\phi(t)| < 1$ for all $t \in [a, b]$. Prove that

$$\int_{\gamma} dz \int_a^b \frac{\psi(t)}{z - \phi(t)}\, dt = 2\pi i \int_a^b \psi(t)\, dt,$$

where γ is the unit circle in the z-plane.

7. Prove that $\lim_{n\to\infty} (1+z)(1+z^2)(1+z^4)\cdots(1+z^{2^n}) = 1 - z$ for $|z| < 1$.

8. For which z does the series $\sum_{n=1}^{\infty} (z/(1+z))^n$ converge? (Ahlfors, 1966, p. 41.)

9. A series of the form $\sum_{n=1}^{\infty} a_n[z^n/(1-z^n)]$ is called a *Lambert series*. If $\sum_{n=1}^{\infty} a_n$ converges, then the Lambert series $\sum_{n=1}^{\infty} a_n[z^n/(1-z^n)]$ converges for all z with $|z| \neq 1$ and its sum $S(z)$ is holomorphic on the domain of convergence. Prove this.

10. Let $f(z)$ be an entire function of z. Prove that if $|f(z)|/|z|^n$ is bounded for $|z| \geqq 1$ and some natural number n then $f(z)$ is a polynomial of degree at most n.

11. Let $f(z)$ be a nonconstant entire function, and let $M(r) = \max_\theta |f(re^{i\theta})|$ be the maximum value assumed by $f(z)$ on a circle with center 0 and radius r. Prove that $M(r)$ is a monotone increasing function of r.

12. Let $f_1(z), f_2(z), \ldots, f_n(z)$ be holomorphic functions defined on a region D and put

$$\sigma(z) = |f_1(z)| + |f_2(z)| + \cdots + |f_n(z)|.$$

If not all functions $f_1(z), f_2(z), \ldots, f_n(z)$ are constants, then the function $\sigma(z)$ does not assume a maximum value on D. Prove this. (Pólya and Szegö, 1978, Problem 300.)

Chapter 2

We have seen that the proof of Cauchy's Theorem for a sufficiently general bounded region D is quite complicated, but for a region D of simple shape a simple proof of Cauchy's Theorem is possible, as is shown by the proof of Cauchy's integral formula for a circle given in Section 1.3c.

As another example, let us assume that the boundary $C = \partial[D]$ of D is a smooth Jordan curve, that the projection of C on the real axis is the closed interval $[a, b]$, and that for all $x \in (a, b)$, the line through x parallel to the imaginary axis intersects C transversely in exactly two points $x + i\phi(x)$ and $x + i\psi(x)$ with $\phi(x) < \psi(x)$. Putting $D_\xi = \{z \in D : \mathrm{Re}\, z < \xi\}$ for $a < \xi \leqq b$, the boundary $C_\xi = \partial[D_\xi]$ of D_ξ is a Jordan curve consisting of two smooth Jordan arcs and at most two segments parallel to the imaginary axis. Let $f(z)$ be a function holomorphic on $[D]$.

13. Prove that

$$\int_{C_\xi} f(z)dz = \int_{\phi(\xi)}^{\psi(\xi)} f(\xi+iy)i\,dy - \int_{\phi(a)}^{\psi(a)} f(a+iy)i\,dy$$
$$+ \int_a^{\xi} f(x+i\phi(x))(1+i\phi'(x))\,dx$$
$$- \int_a^{\xi} f(x+i\psi(x))(1+i\psi'(x))\,dx.$$

(The third and fourth integrals on the right are improper integrals.)

14. Prove that $(d/d\xi)\int_{C_\xi} f(z)\,dz = 0$.

15. Prove that $\int_C f(z)\,dz = 0$.

In Cauchy's Theorem (Theorem 2.2), we assumed that the boundary $\partial[D]$ of the bounded region D consists of a finite number of mutually disjoint piecewise smooth Jordan curves. For application it often suffices to consider regions D where boundary $C = \partial[D]$ satisfies the following condition:

> Let $[a, b]$ be the projection of $[D]$ onto the real axis. Then there exist a finite number of points $a_0 = a < a_1 < a_2 < \cdots < a_n = b$ such that for each $x \in (a_{k-1}, a_k)$ a line through x parallel to the imaginary axis intersects C transversely in $2m_k$ different points.

Under this assumption, the region $[D]$ is divided into a finite number of closed regions $[D_\lambda]$, $\lambda = 1, 2, \ldots, \nu$, by the lines $l_1, l_2, \ldots, l_{n-1}$ passing through $a_1, a_2, \ldots, a_{n-1}$, respectively, and parallel to the imaginary axis.

16. Prove that $\int_C f(z)\,dz = \sum_{\lambda=1}^{\nu} \int_{C_\lambda} f(z)\,dz = 0$, where $C_\lambda = \partial[D_\lambda]$.

17. Evaluate

(i) $$\int_{-\infty}^{+\infty} \frac{\cos(ax) - \cos(bx)}{x^2}\,dx, \qquad a > 0, b > 0.$$

(ii) $$\int_0^{+\infty} \left(\frac{\sin(ax)}{x}\right)^2 dx, \qquad a > 0.$$

(iii) $$\int_{-\infty}^{+\infty} \frac{x^2\,dx}{(x^2+1)^3}.$$

(iv) $$\int_0^{+\infty} \frac{dx}{x^4+1}.$$

(v) $$\int_0^{2\pi} \frac{d\theta}{\cos\theta + a}, \qquad a > 1.$$

(In order to evaluate integrals of the type $\int_0^{2\pi} R(\sin\theta, \cos\theta)\,d\theta$, where $R(\sin\theta, \cos\theta)$ is a rational expression of $\sin\theta$ and $\cos\theta$, one puts $z = e^{i\theta}$ and considers a suitable integral along the unit circle. Ahlfors, 1966, p. 154.)

(vi) $\displaystyle\int_0^{\pi} \frac{d\theta}{\sin^2\theta + a}, \qquad a > 0.$

(vii) $\displaystyle\int_0^{+\infty} \frac{x^{m-1}}{1+x^n}\,dx,$ where m, n are natural numbers $m < n$.

Chapter 3

18. Let $f(z)$ be holomorphic for $|z| < 1$ and assume $|f(z)| < 1$ if $|z| < 1$. Prove

$$|f(z) - f(0)| \leqq |z| \cdot \frac{1 - |f(0)|^2}{1 - |f(0)| \cdot |z|}, \qquad |z| < 1$$

(Pólya and Szegö, 1978, Problem 282.)

19. Let $f(z)$ be holomorphic for $|z| < 1$ and assume $|f(z)| < 1$ if $|z| < 1$. If α and β are such that $|\alpha| < 1$, $|\beta| < 1$, $\alpha \neq \beta$, and $f(\alpha) = f(\beta) = 0$, then

$$|f(z)| \leqq \left| \frac{1-\alpha}{1-\bar{\alpha}z} \cdot \frac{1-\beta}{1-\bar{\beta}z} \right|, \qquad |z| < 1.$$

Prove this. (Pólya and Szegö, 1978, Problem 294.)

20*. Let $f(z)$ be holomorphic on the closed disk $\{z: |z| \leqq R\}$. Deduce Poisson's formula:

$$f(re^{i\phi}) = \frac{1}{2\pi} \int_0^{2\pi} f(Re^{i\theta}) \cdot \frac{R^2 - r^2}{R^2 - 2Rr\cos(\theta - \phi) + r^2}\,d\theta$$

from the Mean Value Theorem with the help of the linear fractional transformation $g: z \to \zeta = g(z) = ((z - c)/(R - \bar{c}z/R))$, $c = re^{i\phi}$, $0 \leqq r < R$. (We have

$$f(g^{-1}(0)) = \frac{1}{2\pi i} \int_{|\zeta|=1} \frac{f(g^{-1}(\zeta))}{\zeta}\,d\zeta$$

by the Mean Value Theorem. Now substitute $g(z)$ for ζ in the expression.)

21. Prove that the linear fractional transformations

$$z \to z^1 = \frac{\cos\theta \cdot z - \sin\theta}{\sin\theta \cdot z + \cos\theta} \quad \text{and } z \to z^1 = e^{2i\theta} \cdot z$$

represent rotations of the Riemann sphere S through an angle of 2θ around the ξ_2-axis and the ξ_3-axis respectively.

22. Prove that a linear fractional transformation of the form $z \to z^1 = (\alpha z + \beta)/(-\bar{\beta}z + \bar{\alpha})$, $|\alpha|^2 + |\beta|^2 = 1$, represents a rotation of the Riemann sphere and that, conversely, all rotations of the Riemann sphere are represented by such linear fractional transformations. (First prove that all unitary matrices $\begin{pmatrix} \alpha & \beta \\ \beta & \alpha \end{pmatrix}$ can be written as:

$$\begin{pmatrix} \alpha & \beta \\ -\bar{\beta} & \bar{\alpha} \end{pmatrix} = \begin{pmatrix} e^{i\phi} & 0 \\ 0 & e^{-i\phi} \end{pmatrix} \begin{pmatrix} \cos\theta & -\sin\theta \\ \sin\theta & \cos\theta \end{pmatrix} \begin{pmatrix} e^{i\psi} & 0 \\ 0 & e^{-i\psi} \end{pmatrix}.$$

23*. Prove that the stereographic projection $\xi \to z$ maps the sphere $S - \{N\}$, from which the north pole has been deleted, "conformally" onto the z-plane.

24. Find a conformal mapping that maps $\{z: |z| < 1 \text{ and } \operatorname{Im} z > 0\}$ onto the unit disk.

25. Find a conformal mapping which maps the region $\{z: |z-i| < \sqrt{2}, |z+i| < \sqrt{2} \text{ and } \operatorname{Re} z < 0\}$ onto the unit disk.

26. Find a conformal mapping which maps the region $\{x+iy: x^2 - y^2 > a, \text{ and } x > 0\}$, $a > 0$, onto the unit disk.

Chapter 4

27. If the function $f(z)$ is holomorphic and without zeros on the right half-plane, $\{z: \operatorname{Im} z > 0\}$, and satisfies $f(z+1) = zf(z)$, $\operatorname{Im} z > 0$, then $f(z)$ can be analytically extended to a function holomorphic on $\mathbb{C} - \{0, -1, -2, \ldots, -n, \ldots\}$ and $0, -1, -2, \ldots, -n, \ldots$ are first-order poles of $F(z)$. Prove this.

28. Let $g(z)$ be an entire function and let C be a smooth Jordan arc connecting 0 and 1. Put $f(z) = \int_C g(\zeta)d\zeta/(\zeta - z)$. The function $f(z)$ is holomorphic on $\mathbb{C} - |C|$. Replacing C by another curve C_1, such that $C_1 \simeq C$ in $\mathbb{C} - \{z\}$, does not affect $f(z)$. Prove that $f(z)$ is freely analytically continuable on $\mathbb{C} - \{0, 1\}$. Prove also that the analytic continuation of $f(z)$ along a smooth closed curve γ which does not pass through 0 and 1 is $f(z) + 2\pi i m g(z)$, where

$$m = \frac{1}{2\pi i} \int_\gamma d\zeta/\zeta(\zeta - 1).$$

29. Evaluate $\int_0^\infty \log x\, dx/(x^2+1)^2$. (Ahlfors, 1966, p. 160.)

30. If $f(z)$ is holomorphic on the closed unit disk, then

$$\int_C f(z) \log z dz = 2\pi i \int_0^1 f(x)\, dx$$

where C is the unit circle. Prove this. (Pólya and Szegö, 1925, Problem 167.)

Chapter 5

31*. Let $\lambda: t \to \lambda(t)$, $a \leqq t \leqq b$ be an analytic Jordan arc. Prove that the reflection $z = \lambda(w) \to z^* = \lambda(\bar{w})$ is invariant under a parameter transformation: $t = \phi(\tau)$, $\alpha \leqq \tau \leqq \beta$, with $\phi(\tau)$ a real analytic function satisfying $\phi'(\tau) > 0$.

32*. Let $\phi: w \to z = \phi(w)$ be a conformal mapping from the upper half-plane H^+ onto the interior D of an n-gon such that $\tilde{\phi}(\infty)$ is a vertex of $[D]$. Prove that $\phi(w)$ is given by

$$\phi(w) = C_0 \int_{w_0}^{w} \prod_{\lambda=1}^{n-1} (w - a_\lambda)^{-\theta_\lambda/\pi} \, dw + C_1.$$

(Transform formula (5.38) using a linear fractional transformation from H^+ onto H^+ and a_n onto ∞.)

33*. Prove that a conformal mapping from the unit disk onto the interior D as an n-gon can be represented as

$$\psi: w \to z = \psi(w) = C_0 \int_0^w \prod_{\lambda=1}^{n} (w - b_\lambda)^{-\theta_\lambda/\pi} \, dw + C_1, \qquad |b_\lambda| = 1.$$

Considering $\psi(w)$ as a function defined on the closed unit disk, prove that $\psi: w \to z = \psi(w)$ is a homeomorphism from the closed unit disk onto $[D]$ and that $\psi(b_\lambda)$, $\lambda = 1, 2, \ldots, n$, is a vertex of $[D]$.

34. Prove that $w \to z = \int_0^w (1 - w^n)^{-2/n}$ is a conformal mapping from the unit disk onto the interior of a regular n-gon.

Chapter 6

35. Deduce Cauchy's Theorem (Theorem 2.2) directly from Green's Theorem (Thereom 6.1).

36*. Let $u(z)$ be a function which is continuous for $|z| \leqslant r$ and harmonic for $|z| < R$. Prove Poisson's formula:

$$u(re^{i\varphi}) = \frac{1}{2\pi}\int_0^{2\pi} u(\mathrm{R}e^{i\theta})\frac{R^2 - r^2}{R^2 - 2Rr\cos(\theta - \varphi) + r^2}\,d\theta,\ 0 \leqslant r < R;$$

compare Exercise 20*.

37. Consider $\omega \in \mathbb{C}$ such that $\operatorname{Im}\omega > 0$ then $G = \{m + n\omega : m, n \in \mathbb{Z}\}$ is an additive subgroup of $\mathbb{C}$. Show that "in a natural way" the quotient group $\mathcal{R} = \mathbb{C}/G$ is a Riemann surface.

38. Prove the equalities $\int_{[U_j^+]} d(\rho_j\varphi) = \int_C \rho_j\varphi$, (6.70) and $\int_{U_j} d(\rho_j\varphi) = 0$, if $[U_j] \subset D$, directly by using polar coordinates.

N.B. If one proves equality (6.69) directly in this way, Green's Theorem (Theorem 6.10) is proved by using partitions of unity reather than cell decompositions. Theorem 6.1 is a special case of Theorem 6.10. Hence, it is also possible to base the proof of Cauchy's Theorem on partitions of unity.

39*. Prove that the value of the right-hand side of (6.99) does not depend on the choice of the open cover $\mathcal{U} = \{U_j\}$ or on the decomposition of γ.

40. Define $\int_\gamma f(z)\,dz$, where $f(z)$ is a function holomorphic on a region $\Omega \subset \mathbb{C}$ and γ is an arbitrary curve in Ω in the same way as (6.99) and use this to prove Theorem 4.14 Cauchy's Theorem).

Chapter 7

41. Let φ be a continuously differentiable closed 1-form defined on the region $D = \{z : a < |z| < b\}$, $0 \leqslant a < b$ and let $\gamma : \theta \to z = re^{i\theta}$, $0 \leqslant \theta \leqslant 2\pi$, be a circle with center 0 and radius r, $a < r < b$. Prove that if $\int_\gamma \varphi = 0$, then φ is exact on D.

42. Let $\Omega = S - F$ be a horizontal slit region with connectivity m and let all connected components F_k, $k = 1, \ldots, m$, of F be segments connecting the points α_k and β_k ($\operatorname{Im}\alpha_k = \operatorname{Im}\beta_k$, $\operatorname{Re}\alpha_k < \operatorname{Re}\beta_k$) and put

$$g_k(z) = \frac{1}{4}(\beta_k - \alpha_k)\left(z + \frac{1}{z}\right) + \frac{1}{2}(\alpha_k + \beta_k).$$

Write $E_k = \{g_k(z) : 1 \leqslant |z| < 1 + \varepsilon\}$ as an elliptic neighborhood of F_k and $E_k - F_k \subset \Omega$, while $g_k : z \to w = g_k(z)$ maps $D = \{z : 1 < |z| < 1 + \varepsilon\}$ conformally onto $E_k - F_k$ for some sufficiently small $\varepsilon > 0$. The parametric representation $\gamma_k : \theta \to w = g_k(re^{i\theta})$, $0 \leqslant \theta \leqslant 2\pi$, where $r = 1 + \varepsilon/2$, defines an ellipse γ_k, where $|\gamma_k| \subset E_k - F_k$. Let φ be a continuously differentiable closed 1-form defined on Ω. Prove that φ is exact on Ω if $\int_{\gamma_k} \varphi = 0$ for $k = 2, 3, \ldots, m$.

43. Let $\Omega = S - F$ be a horizontal slit region of connectivity m. Prove that the first Betti number $b_1(\Omega) = m - 1$.

44*. (See Section 7.2c). The composition of the continuous map $\tilde{g}$, mapping $[\mathcal{U}]$ onto $[\mathcal{D}]$ and the holomorphic map $\mathfrak{q}$, mapping $[\mathcal{D}]$ onto $\mathcal{R}$ is a continuous map which maps $[\mathcal{U}]$ onto $\mathcal{R} - \mathfrak{q}(U_0)$ and maps $\mathcal{U}$ conformally onto $\mathcal{R} - \mathfrak{q}([\mathcal{U}_0]) - |C_1| - \cdots - |C_{2m}|$. Hence $\mathfrak{q} \circ \tilde{g}$ is one-to-one on $\mathcal{U}$ and $\mathcal{R} - \mathfrak{q}(U_0)$ can be obtained from $[\mathcal{U}]$ by identifying the points w_1 and $w_2 \in [\mathcal{U}] - \mathcal{U}$ if $(\mathfrak{q} \circ \tilde{g})(w_1) = (\mathfrak{q} \circ \tilde{g})(w_2)$. Using this fact, prove that $\mathcal{R}$, as a Hausdorff space, is uniquely determined by its genus $g(\mathcal{R}) = m$.

References

Ahlfors, L. V. 1966. *Complex Analysis*, 2nd ed. New York: McGraw-Hill.

Ahlfors, L. V. and L. Sario. 1960. *Riemann Surfaces*, Princeton University Press.

Hurwitz, A. 1929. *Vorlesungen über Allgemeine Funktionentheorie und Elliptische Funktionen*, 3rd ed. Berlin: Springer-Verlag.

Pólya, G. and G. Szegő. 1978. *Problems and Theorems in Analysis*. Berlin: Springer-Verlag.

Weyl, H. 1955. *The Concept of a Riemann Surface*, 3rd ed. Reading MA: Addison Wesley. Translated from the German, *Die Idee der Riemannschen Fläche*, Dritte Auflage. Leipzig: Teubner (1955).

Weyl, H. 1940. Method of orthogonal projection in potential theory. *Duke Math J.*, **7**, 411–444.

Index

Printed in the United States
by Baker & Taylor Publisher Services